Catherine Goldstein, Daniel Huybrechts, Joachim Schwermer (Eds.)
Crelle Bicentennial Edition

Crelle Bicentennial Edition

Edited by
Catherine Goldstein, Daniel Huybrechts and Joachim Schwermer

DE GRUYTER

Mathematics Subject Classification 2020
Primary: 00B55, 00B60, 01A06, 01A75
Secondary: 01A55, 01A60, 10A61, 11-03, 11R37, 14-03, 14H60, 17-03, 34-03, 37F25

Editors
Prof. Dr. Daniel Huybrechts
Mathematisches Institut
Universität Bonn
Endenicher Allee 60
53115 Bonn
Germany
huybrech@math.uni-bonn.de

Prof. Dr. Catherine Goldstein
Histoire des Sciences Mathématiques
Institut de Mathématiques de Jussieu-Paris Rive Gauche
4 Place Jussieu
75252 Paris
France
catherine.goldstein@imj-prg.fr

Prof. Dr. Joachim Schwermer
Faculty of Mathematics
University of Vienna
Oskar-Morgenstern Platz 1
1090 Vienna
Austria
Joachim.Schwermer@univie.ac.at

This book is available as an open access publication via https://www.degruyterbrill.com.

ISBN 978-3-11-125965-9
ISBN 978-3-11-126044-0 (PDF)
ISBN 978-3-11-126113-3 (EPUB)
DOI https://doi.org/10.1515/9783111260440

Library of Congress Control Number: 2026939988

Bibliographic information published by the Deutsche Nationalbibliothek
The Deutsche Nationalbibliothek lists this publication in the Deutsche Nationalbibliografie;
detailed bibliographic data are available on the Internet at http://dnb.dnb.de.

De Gruyter and Walter de Gruyter GmbH are part of De Gruyter Brill.
www.degruyterbrill.com

Questions about General Product Safety Regulation:
productsafety@degruyterbrill.com

Printing: CPI books GmbH, Leck

MIX
Papier | Fördert gute Waldnutzung
FSC® C083411

CONTENTS

Crelle Bicentennial Edition: Preface ... i
Acknowledgements ... ii
C. Goldstein, Crelle's *Journal* and its peers ... 1

1826–1850

G. Eisenstein, Geometric proof of the fundamental theorem of quadratic residues ... 21
J. Schwermer, Eisenstein's lens – a geometric view of the fundamental theorem for quadratic residues 25

1851–1875

H. A. Schwarz, On the cases when the Gaussian hypergeometric series represents an algebraic function of its fourth element ... 43
C. Goldstein, At the crossroads: H. A. Schwarz on linear differential equations ... 77

1876–1900

L. Kronecker, On the theory of abelian equations ... 103
N. Schappacher, On "Kronecker's journal", handwritten lecture notes, and Jacobi sums ... 128

1901–1925

J. Schur, On the representation of finite groups by fractional linear substitutions ... 141
G. Williamson, Schur and the birth of projective representations ... 164

1926–1950

R. Brauer, H. Hasse, E. Noether, Proof of a main theorem in the theory of algebras ... 183
J. Schwermer, Class field theory – Cyclic algebras, Brauer groups, and a local-global principle ... 189

1951–1975

B. J. Birch, H. P. F. Swinnerton-Dyer, Notes on elliptic curves I & II ... 211
Y. Kezuka, D. Zagier, The conjecture of Birch and Swinnerton-Dyer ... 268

1976–2000

A. Beauville, M. S. Narasimhan, S. Ramanan, Spectral curves and the generalised theta divisor ... 287
N. Hitchin, Spectral curves and their impact ... 298

2001–2025

C. T. McMullen, Dynamics on K3 surfaces: Salem numbers and Siegel disks ... 311
S. Cantat, Dynamical degrees, entropy, and Fatou sets ... 342

Crelle Bicentennial Edition
Preface

The first issue of the *Journal für die reine und angewandte Mathematik*, now better known as Crelle's Journal, after its founder, appeared in 1826, which makes it the oldest journal covering a wide range of mathematics. Since its beginning, Crelle's Journal has published a wealth of profound and influential works.

We are accustomed to the enduring nature of mathematical theorems and to the long-term life of mathematical articles, — that is, to their lasting relevance across different contexts and disciplinary frameworks. In the same way, through its successive editors and publishers, the Journal was able to adapt to the ever-changing conditions of the publishing landscape, allowing it to thrive for two hundred years.

This book, entitled '*Crelle Bicentennial Edition*', marks the 200th anniversary of the Journal. As a testament to its long history and to the mathematics it hosted and helped promote, this volume brings together eight articles drawn from the Journal, one for each 25-year period, illustrating the breadth and significance of its contribution to mathematics. The works, published here in English translation, are by the scholars G. Eisenstein, H.A. Schwarz, L. Kronecker, I. Schur and R. Brauer, H. Hasse, and E. Noether. In addition, there are reprints of works by B. Birch and P. Swinnerton-Dyer, as well as A. Beauville, M.S. Narasimhan and S. Ramanan, and C.T. McMullen, respectively. Each paper is followed by an extended commentary. These reviews discuss the historical background, the content, and the far-reaching influence that each article had on the development of mathematics. This collection of articles is preceded by a historical paper which aims to situate Crelle's Journal within the complex landscape of mathematical research journals over time.

Though we have tried to illustrate the variety of the topics published in the Journal, the choice could not ultimately be anything but deeply arbitrary, depending on personal preferences and available commentators. We are aware that most readers will regret the absence of their favorite piece, but we do hope that they will be willing to browse through what is here and above all to appreciate the change, not only in questions, but also in style, of mathematics through two centuries.

May 2026 Catherine Goldstein, Daniel Huybrechts, and Joachim Schwermer

 | HTTPS://DOI.ORG/10.1515/9783111260440-201

Acknowledgements

The editors wish to thank all the people who have contributed to this volume. We are very grateful to the authors, the translators, the referees, and the publication team of de Gruyter Brill. Special thanks to Barnaby Sheppard for his exceptional thorough work as a copy editor for the whole volume.

Crelle's *Journal* and its peers

By *Catherine Goldstein* at Paris

Abstract. These preliminary remarks aim to situate the development of Crelle's *Journal* within the history of mathematical publications in light of recent research on scientific journals, and to offer food for thought — if not a definitive answer— regarding the unique character of this journal.

1. Introduction

In 1989, when the 400th issue of the *Journal für die reine und angewandte Mathematik* was published, the editors, revisiting the already numerous commemorations attached to the journal, wondered whether it was appropriate or not to write another preface to the volume: "what could be said today on this subject which has not already been said or written at former occasions?", they asked. "Or what could one say that would not apply to just any mathematical journal?" [31]. It may therefore seem paradoxical to even consider writing some introductory remarks on the occasion of the journal's bicentenary, several hundred issues later. The early history of the journal and of its founder has been thoroughly documented in the pioneering work of Wolfgang Eccarius [25–27]— doubly pioneering, in fact, as, on one hand, it was devoted to a figure, August Crelle, who, while decisive in the development of 19th century-mathematics, was not otherwise a leading mathematician, and, on the other hand, it took as its very subject of study a mathematical journal, including its material and editorial aspects.[1)]

However, since the 1990s and particularly in the last two decades, there has been a keen interest in the historical study of mathematical journals, either focused on specific cases or discussing disciplinary and international comparisons.[2)] These studies invite us to revisit briefly the development of the *Journal für die reine und angewandte Mathematik* in a comparative perspective. An obvious difficulty is that the creations and transformations of the various mathematical research journals, under a change of editors or publishers or under external circumstances that interfered with the policies and the practices of mathematical publishing, be they

---

IMJ-PRG, CNRS, Sorbonne Université, Université Paris Cité.

1) Since then, several articles have also studied the editorship of the *Journal für die reine und angewandte Mathematik* at various moments, in particular [28, 34–36, 71].

2) Among the collective projects devoted to mathematical journals, let us mention [1,21,39,43,44,69,70,86]. Numerous other studies have focussed on one specific journal or country; besides the abundant bibliographies of the preceding references, see for instance [5, 7, 17, 18, 23, 24, 33, 37, 40, 48, 57, 60, 61, 63, 73, 77, 87].

 | HTTPS://DOI.ORG/10.1515/9783111260440-001

wars or institutional and educational reforms, were not synchrone: there is no global natural chronology [11]. Moreover, the available types of sources attached to a journal or a country differ widely, and, consequently, comparisons may not be accessible at all levels.[3)] Thus, we will simply use these recent historical works to shed some light on a few moments in the history of the *Journal für die reine und angewandte Mathematik*—keeping in mind the question of what "would not apply to just any mathematical journal".

2. The Crelle era I: Gergonne's and Crelle's journals

In the preface to the first issue of his journal, August Crelle only referred to two contemporary enterprises: the *Annales de mathématiques pures et appliquées*, a French journal entirely devoted to mathematics launched in 1810 by Joseph-Diez Gergonne, and the *Correspondance mathématique et physique*, a very recent initiative of Jean Guillaume Garnier and Adolphe Quetelet in Belgium. This description, however, is a bit misleading.

First of all, there already existed, of course, regular publications linked to a regional or national institution of science, such as the *Miscellanea Berolinensia ad incrementum scientiarum* of the Berlin Academy of Science (first in Latin, then replaced by the annual *Nouveaux mémoires de l'Académie royale* in French from 1745 on) and the *Philosophical Transactions of the Royal Society*, or later, to a higher-education institution, such as the *Journal de l'École polytechnique*; while multidisciplinary in principle, these publications, in particular the last one, often gave pride of place to mathematics, but they were open mostly to, or via, the members of the institution in question.

But they were not the only ones, in particular in German-speaking countries. First of all, there had already been initiatives close to our current idea of a mathematics research journal, in particular those connected with Carl Hindenburg, then professor of physics at the University of Leipzig, such as his *Leipziger Magazin für die reine und angewandte Mathematik* edited between 1786 and 1789 with Johann Bernoulli III, and his *Archiv der reinen und angewandte Mathematik* (1794–1800). We may notice how similar to that of the future Crelle's *Journal* the titles are. Among the authors of Hindenburg's *Archiv* were mathematicians such as Johann Friedrich Pfaff and Christian Kramp. Although Hindenburg wished to include amateurs as well as experts in the circle of his readers, his aim was above all to promote a specific approach to mathematics, based on combinatorial analysis— a topic on which most of the research articles of his last journal focussed.[4)] According to their printer, if Hindenburg's periodic publications were finally stopped around 1800, it was "partly because of the troubles of war, partly because of the difficulties of literary communication" [10].

Moreover, the turning of the nineteenth century witnessed in the German countries the creation of many specialized journals, including several focussing entirely on mathematics.[5)] Created at the initiative of one or two men, most often outside academia, such as mathematical practitioners of some kind or teachers, they aimed at an educated, rapidly growing, amateur audience, but also, in some cases, at the promotion of a particular topic. One finds, for

---

3) On this point, see [45]. For instance, there is no equivalent for the French *Journal de mathématiques pures et appliquées* to the rich correspondence between Crelle and his publisher Georg Reimer used in [25], as the archives of the French publishing houses, Bachelier and Gauthier-Villars, have disappeared.

4) On Hindenburg and his program, see [62, 76]. On the *Archiv der reinen und angewandte Mathematik*, see also [2, 10].

5) We are drawing here on the work of Maarten Bullynck and Thomas Morel, in particular [10, 11, 57, 58]; see also [41].

instance, *Der gemeinnützige mathematische Liebhaber* of Johann Reimers, teacher in a commercial academy in Hamburg and "Informator" of mathematics, as he called himself; the two issues of the bookkeeper Georg Friedrich Petersen's *Versuch eines Magazins für die Arithmetik*; or Heinrich Breithaupt's *Magazin für das Neueste aus der Mathematik*, which accompanied its editor's entreprise of mathematical instruments.[6] The first, short-lived, attempt by the engineer Crelle to launch a mathematical journal, his 1821–1822 *Sammlung mathematischer Aufsätze und Bemerkungen*, seems to be part of the same trend [26, 58].

All of these efforts provided the backdrop against which both Gergonne's *Annales* in 1810 and Crelle's *Journal* fifteen years later came into being. Neither Gergonne nor Crelle were members of the mathematical academia when they launched their journals; in this respect, their professional profile is close to those of the mathematics enthusiasts just mentioned. Crelle was a civil engineer in the Prussian administration, Gergonne, after a career in the army and an initiation to mathematics in an artillery school, became a high-school teacher of mathematics in Nîmes, in the south of France.

Both, however, and contrary to most practitioners, were well-informed about the mathematical research of their time and well-connected, although to varying degrees, with some of their authors. Gergonne, for instance, had met Sylvestre-François Lacroix during his studies, and through him was in contact with others of Gaspard Monge's followers. At the beginning of the 1820s, Crelle belonged to a cultural and mathematical circle in Berlin that included Niels Abel and Jakob Steiner, he had written two articles for Gergonne's *Annales*, among other works, and he had just translated the complete works of Joseph-Louis Lagrange for the publisher Georg Reimer.[7] In this respect, Crelle's framework was closer, on a more restricted scale, to that of the naturalist André Étienne d'Audebard de Férussac, whose vast library in Paris was a meeting point of many promising or established scientists, including at times Abel and Évariste Galois, and whose *Bulletin général et universel des annonces et nouvelles scientifiques*, also launched at the beginning of the 1820s, mixed original articles and reports on other publications and relevant scientific events [9].

While Gergonne's Nîmes was far from Paris and its rich scientific environment, Crelle was at the heart of a learned, lively circle in the capital of a state seeking to extend its economic and cultural importance. The creation of the Berlin University in 1809–1810, in which Wilhelm and Alexander von Humboldt played a decisive role, was to open an impressive counterpart to those of Leipzig or Göttingen, a counterpart whose research-oriented chairs and freedom of teaching would be envied by many European scholars during the 19th century.

Although he was an engineer and administrator, Crelle shared with Alexander von Humboldt, to whom he became close, a vision of the importance of a humanist culture in general and a desire to promote mathematics in particular [67]. In his plan for a mathematical institute in 1828, Crelle would argue that: "Everybody, without exception, needs pure mathematics" [75]. And as Humboldt did, Crelle would for instance support a succession of brilliant young number theorists, such as Carl Jacobi, Ernst Kummer and Gotthold Eisenstein, the publication of their works in his journal being part of this support.

In 1844, for instance, Crelle wrote to the *Kultusminister* (the minister of culture and education in Prussia), Friedrich Eichhorn, about Eisenstein [67, p. 216]:

---

6) On these, often ephemeral, journals, their owners and their public, see [11, 57, 58].

7) The important publishing house Reimer will be in charge of Crelle's *Journal* from the second year on until its replacement by De Gruyter in the 1920s. On Reimer, see [25, 68].

> For an extraordinary new talent has emerged here recently, the student H. G. Eisenstein [...] He has produced a series of papers on reciprocity laws for real and complex numbers, on quadratic and cubic forms, on biquadratic residues, on cyclotomy, on elliptic and Abelian transcendents, etc. They contain many new things, and some of them advanced on almost uncharted territory. A first series of his papers is printed in the 27th volume of the *Journal der Mathematik*, and a second series is currently in press for the 28th volume.

This quote shows that fields of science with no immediate applications could be promoted to a government; but, above all, the fact that the *Journal* was referred to as a testimony of Eisenstein's fruitfulness illustrates the prominent role it had come to play at the time in Prussia's scientific system. From the second volume on, in 1827, Crelle would mention on the first page that he benefited from the support of the authorities for his enterprise.[8] Gergonne, although he enjoyed favorable personal opinions not only from mathematicians but also in the long run from political figures, did not benefit from such an official support, nor from the help of a key publishing house; with his co-editor and colleague, Joseph Esprit Thomas-Lavernède, Gergonne had to take care of every aspect, from subscriptions to the mail to the proofreading and even engraving of the copper plates for the figures.

One thing Gergonne and Crelle had in common is the importance they give to theoretical knowledge. Gergonne would certainly concur with Crelle's defense of pure mathematics, he for whom "It is well-known that a nation which would cultivate sciences only under the sole viewpoint of their practical and immediate applications and of their material results would not see them long flourish within it" [63, pp. 48–50]. While teachers participated in the *Annales* as contributors as well as readers, Gergonne was also able to attract important articles by university or higher education professors: Augustin Louis Cauchy for instance published here his new theorems concerning residues in complex analysis, and Gergonne's *Annales* was instrumental in promoting the geometric representation of the complex numbers or the idea of duality. However, like Hindenburg and not like Crelle, Gergonne had a personal mathematics agenda and definite, even philosophical, views on mathematics, its teaching, its symbolism, and also its appropriate development. In line with the French school curriculum and his own training, he favored geometry and in particular vigorously promoted the idea and use of duality, leading him to a rather famous priority polemics with Jean-Victor Poncelet [37, 53].

In this respect, a comparison between the contents of Gergonne's *Annales* and Crelle's *Journal* is quite suggestive: analysis occupies c. 21% of the papers in Gergonne's journal, 56% of Crelle's, while geometry occupies 47% of the papers in Gergonne's journal, and 28.7% in Crelle's; both journals contain c. 7-8% of articles on mechanics and c. 7% on applied mathematics, although astronomy (which was then part of the mathematics curriculum in France) constitutes the main topic in this area in Gergonne's and less than 2% in Crelle's.[9] Let us also remark that, in the section on analysis, 10% of the papers in Crelle's were devoted to number theory, written by Jacobi, Lejeune-Dirichlet, Eisenstein and Kummer, as well as the first paper published by a woman, Sophie Germain.

---

[8] As explained by W. Eccarius, there were still a lot of financial difficulties and the *Journal für die reine und angewandte Mathematik* was partly saved thanks to another publication by Crelle, his technical *Journal für due Baukunst*, [25].

[9] The issue of classification is a bit delicate as both journals did not adopt a common one: Crelle's grouped the research articles into four sections, analysis, geometry, mechanics and applied mathematics, while Gergonne's adopted a more complex one. This issue is discussed in detail in two important studies devoted to Gergonne's *Annales* by Mario Otero and Christian Gerini [37, 63], see also [87], the conclusions of which we adopted here.

Gergonne, on the other hand, devoted 2.5% of the articles in his journal to a section called "mathematical philosophy", to which he participated actively. The views he expressed there shed light, in fact, on Gergone's numerous remarks and interventions on the articles of his authors, refering to other articles in his journal or proposing alternative proofs—to the point that Gergonne introduced a special section "mathematical polemics" to discuss his controversy with Poncelet [37]. Crelle would occasionally rewrite the articles, adding references or correcting style according to his preferences (sometimes without the approval of authors), but in general he tried to put an end to potential attacks or controversies.[10] Crelle's *Journal* also helped foster a sense of community within mathematics by publishing some obituaries and facsimiles of the writing of important mathematicians.

Both journals also proposed exercises and open questions, which led to explicit exchanges between them. For instance, Steiner published in Crelle's *Journal* a solution to a problem raised in the *Annales* [82]. Other examples, including partial reproductions of articles, are given by Norbert Verdier in [87]. Although Gergonne's *Annales* was entirely in French (with a few translations from other publications), its pages were largely open to foreign authors, often common with those of Crelle's *Journal*: 28 foreign mathematicians (c. 20%) wrote 122 articles of the *Annales*, including in particular Crelle's authors, such as Niels Abel, August Crelle himself, Julius Plücker and Jakob Steiner. As is well-known, Crelle, after a first issue of his journal entirely in German and his explicit plan to launch and develop a German-language publication for mathematics, accepted articles in various languages, including Latin, French, Italian or English. In the first 50 volumes of his journal, about a third of the authors were based outside the German countries (with a majority of French, Russian, Italian and English authors), and more than half outside Prussia [15].

3. The Crelle era II: Liouville's and Crelle's journals

In 1830, after the July Revolution and the new monarchic regime in France, Gergonne, then professor of astronomy at the University of Montpellier, again in the south of France, was appointed *recteur* in charge of the whole academic district and gave up the publishing of the *Annales*. Almost at the same time, Férussac also stopped the publication of his *Bulletin* for financial reasons. These examples showed that the still current model of publishing—one in which the burden mostly fell on a single person, even with some co-workers—was unsustainable. The following decade, during which Crelle's *Journal* thrived, witnessed several important changes in the mathematics publishing world.

The first one, in France, was the creation of *Comptes rendus hebdomadaires des séances de l'Académie des sciences* in 1836: on one hand, there had been increasing complaints about the delays of publication of the memoirs, particularly those of non-members accepted by the academicians and, on the other hand, the Academy hoped to control the public review of its meetings which were until then reported upon by journalists. As is well-known, the possibility to publish short announcements of results very quickly (within a week), before a complete publication elsewhere, would modify the habits of publishing and progressively relieve journals of the pressure to manage priorities issues.

Moreover, several journals devoted to mathematics were launched within a few years: besides Joseph Liouville's *Journal de mathématiques pures et appliquées*, on which we will

[10] See for examples on these points [26, 29].

return, one can mention *The Mathematical Miscellany* in the USA, *The Cambridge Mathematical Journal* (soon *The Cambridge and Dublin Mathematical Journal* under the direction of James Thomson) in the United Kingdom, but also, in 1841 and 1842 respectively, the *Archiv der Mathematik und Physik*, edited by Johann Grünert in Greifswald, and the *Nouvelles annales de mathématiques*, edited by Olry Terquem and Camille-Christophe Gerono in Paris.[11] The first two, which also included news of other journals or books, aimed above all at giving some space of publication to younger or isolated mathematicians and integrating them into a larger, international mathematical community. But the last two took over a part of the role of Gergonne's *Annales*, addressing primarily to teachers and students, in particular those in the various engineering schools that flourished in Europe during the nineteenth century. A new question then arose—namely, that of the distribution of authors and topics among these different types of journals; we note, for instance, that exercises as such would progressively disappear from research-oriented journals.

Let us turn now to the other important research journal founded during this period, the *Journal de mathématiques pures et appliquées*.[12] Although Liouville, its founder, established a continuity with Gergonne's *Annales* in the preface of the first issue, his model was Crelle's *Journal*, of which the young Liouville had been first of all an author. For instance, in 1833, Crelle published the report from the French Academy of Sciences on two memoirs submitted by Liouville (on integrals whose value is algebraic), which concluded that these memoirs would be printed in a volume of the Academy; to the reproduction of this report, Liouville added a note which, he said, presented his theory in a new way "to avoid duplication" [52, p. 347]. In fact, Liouville also published the same year his results in the *Journal de l'École polytechnique* (to which he had access as an alumnus), the same articles being finally printed among the memoirs of the Academy only in 1838 [55]. While the short announcements of the *Comptes rendus* were not yet available in 1833, this example shows well the importance of a journal like Crelle's for a larger diffusion in due time. Before the founding of his own journal, between 1833 and 1835, Liouville published 6 articles in Crelle, totaling 88 pages [87, partI, ch. 2]. Like Crelle, Liouville had access to an important publisher and bookseller, Charles Louis Etienne Bachelier, who already sold Crelle's *Journal* in Paris and who, unlike Reimer, was specialized in mathematics —he would also be in charge of the *Comptes rendus*, of the *Journal de l'École polytechnique* and the *Nouvelles Annales*.[13]

Liouville was also a first-order mathematician, soon himself professor at Polytechnique and Academician. After the founding of his journal, he would, like Gergonne and unlike Crelle, occupy a large place in it: he published most of his own articles in it and, conversely, 15% of the articles in his journal were signed by him [24, 87]. In the 1860s, Liouville's notes on his pet subject of the moment, sums of squares, would invade his journal, to the dismay of his colleagues [55]. But since the creation of his journal, Liouville expressed in it his mathematical priorities. Liouville's *Journal* was for instance instrumental to making Galois's work known—reproducing in 1846 all of Galois's papers first published in Gergonne's *Annales* and Férussac's *Bulletin*, as well as the famous letter to Auguste Chevalier written just before Galois's death and some hitherto unpublished pieces, with commentaries. We can draw here a parallel with Crelle's role in disseminating Eisenstein's work. Like Gergonne's and unlike Crelle's, Liouville's journal was (almost) entirely in French, which required him to ask colleagues or students

[11] On these journals, see respectively [17, 18, 48, 60, 73].

[12] Here, we draw on the significant research of Norbert Verdier on the journal [86–89]. See also [24, 54].

[13] On Bachelier, see [88].

for translations of foreign articles that Liouville considered worth republishing in his journal [14]. But an exception to the publication in French was the reproduction in the original Latin of Kummer's study of the (non-unique) decomposition into prime factors of cyclotomic numbers, a key-piece in the future development of algebraic number theory; this paper, preceded by a letter of Kummer to Liouville, and followed a few years later by a longer memoir in French, was part of a barely concealed controversy between Liouville, Guglielmo Libri, Gabriel Lamé and Augustin Louis Cauchy on Fermat's Last Theorem at the French Academy of Science.[14] These are two among many other examples of the way Liouville also used his journal to settle debates and promote specific works and directions in mathematics. In the *Journal de mathématiques pures et appliquées* under Liouville's editorship, analysis represented c. 47% of the articles, geometry 23%, mechanics 13% and applied mathematics 12.5%, a distribution consistent with Liouville's orientations and much closer to that of Crelle's *Journal.*[15] Let us also note that there was no special section devoted to exercises — in contrast to the *Nouvelles Annales* —, emphasizing a clearer distinction between research publications and those intended for teachers and students.

In this respect, a difference with Crelle (with respect to Grünert) was that Liouville had good relations with Terquem (who occasionally collaborated with Liouville, in particular as translator), and that both men, according to their respective audience, would exchange articles submitted to their respective journals if the opportunity occurred. Both journals, however, reproduced (in translation) or adapted articles from Crelle's *Journal*, highlighting its leading role in mathematical research publications. For instance, Terquem published in 1845 in his *Nouvelles Annales* a "generalization of the theory of associated numbers", following a paper of Lejeune-Dirichlet in the third volume of Crelle's *Journal*, another on the "main properties of a system of lenses", following a paper of August Ferdinand Möbius in the fifth volume of Crelle's journal, and several results from Jakob Steiner. As for Liouville, he reproduced several articles of Jacobi, for instance in 1842 an "elementary proof of a remarkable analytic formula", explicitly indicated as "extracted from the journal of M. Crelle". Indeed, c. 5% of the articles in Liouville's *Journal* were translations of articles in Crelle's *Journal*, with a few republished in the other direction.

After a few years, however, the *Journal für die reine und angewandte Mathematik* and the *Journal de mathématiques pures et appliquées* would share many authors: Jacobi, Kummer, Steiner, Eugène Catalan, Charles Hermite, Arthur Cayley, to name only a few, would publish regularly in both journals, as well as in the *Cambridge and Dublin Mathematical Journal* and its successor, the *Quarterly Journal of Mathematics*, showing that an international research scene was taking shape in mathematics.

4. The *Journal für die reine und angewandte Mathematik* after Crelle's editorship

August Crelle died in 1855. The front page of Volume 51 of the Journal, published in 1856, still mentions him as the editor, though the final pages announce both his death and the continuation of the journal. In addition to the usual articles, this volume contains several items of interest for the history of mathematical publications.

[14] For the details of this controversy and the references to the various articles involved, see [32].

[15] Liouville did not provide any classification in his journal; the figures here come from [87] to which we refer for details on the construction of these data; as pointed out by N. Verdier, mechanics in France at the time was also mostly analytic.

First, an erratum concerning previously published articles by Cayley attests to the typographical difficulties specific to mathematics [16, p. iv]:[16)]

> The reason why the distinctive symbol consisting of two intertwined arcs)(— so aptly chosen by Mr. Cayley — was not included in the printed version was that this symbol was not among those available at the printing house, and there was not enough time to have it specially cast. We were forced to indicate the difference between matrices and determinants by another means, admittedly a less suitable one.

The issue also includes a last article by Crelle himself: it adresses the construction of tables of prime numbers and of divisors —the journal's format being well-suited for their printing—, including theoretical and practical issues on their cost and the time required to produce them [16, pp. 61–99].

Finally, there is an article on "analytic factorials" by Karl Weierstrass (still listed as a high school teacher, just before his appointment as a professor in Berlin). The topic fittingly links the work of the combinatorial school and Crelle's own contributions to this school to the modern analysis of which Weierstrass would become one of the most important representatives.[17)] It is accompanied by a rather moving note from Crelle himself:

> As for my own work on the subject, I would certainly have much to say in defense of the incompleteness that Mr. Weierstrass has pointed out; but my strength, weakened by advanced age and constant illness, is no longer sufficient for such a task. I must therefore refrain from doing so; which is quite all right, since in my scientific endeavors I have never sought personal gain, never sought fame or praise, but have strived, to the best of my ability, only for the advancement of truth, and it matters not to me who comes closer to the truth—whether I or someone else—so long as a further approach to the truth is achieved at all.

At the end of the same issue, after paying tribute to Crelle, the new main editor, Carl Borchardt presents the continuation of the *Journal* as carrying on Crelle's project [16, p. 404]:

> Under the new editorial board, the journal will retain both its current format and remain faithful to the objectives it has pursued thus far. The articles that it will provide to its readers, whose subjects may be drawn from all branches of pure and applied mathematics and mathematical physics, are to contain original investigations that are novel either because of the results to which they lead or because of the reasoning behind these results.

While stressing continuity at several levels, however, the *Journal* would quietly introduce changes in two areas. First, the standard editorship consisting of one or two people in charge of most of the decisions would be replaced by a group of mathematicians: "Since my esteemed colleagues — MM. Steiner, Schellbach, Kummer, and Kronecker — had agreed to lend their kind support, I have, bolstered by this excellent collaboration, taken on the task of

[16)] A very interesting comparison from the point of view of mathematical typography between three versions of the same paper by Jacobi, in the *Journal für die reine und angewandte Mathematik*, the *Journal de mathématiques pures et appliquées* and the *Giornale arcadico di Scienze, Lettere ed arti*, an Italian, non-specialized, journal, has been done by N. Verdier in [87, part II, ch. 8].

[17)] [16, pp. 1–60]. On the topic itself and the background of this article in particular, see [84].

publishing the journal", explained Borchardt. From then on, all their names would usually appear on the front page of the Journal, preceded by the mention "Unter Mitwirkung der Herren ..." [in collaboration with Mssrs. ...].[18] This organization proved crucial to ensuring the publication's long-term survival and it was soon adopted by other journals. Camille Jordan, in France, who took over the *Journal de mathématiques pures et appliquées* from 1884 to 1921, following Liouville's withdrawal in 1874 and a transitional period overseen by Henry Résal, took a significant step by naming this group an "editorial board".[19]

Another key feature was Borchardt's desire to standardize the research articles in his journal, particularly with regard to length [16, p. 404].

> The current editorship appeals to the journal's esteemed former contributors, to all mathematicians, and in particular to those in the country, with the earnest request that they ensure the continued existence of this organ, so necessary for their professional colleagues, by submitting their works. It cannot fail to point out at the same time how desirable it is, in the interest of both the enterprise and the authors, that the contributions not exceed the length customary for articles in journals, so that space may be found for their publication all the sooner.

The issue was not entirely new. As is well known, the *Comptes rendus de l'Académie des sciences* imposed length limits on the notes which were accepted from very early on — even though the academicians themselves got around this somewhat by dividing their work into a long series of notes, published over the course of several weeks. Liouville, for his part, had adopted a quarterly rather than a monthly publication schedule for the second series of his journal, so that he could present the longer papers all at once, without dividing them into installments. Unlike Liouville, Borchardt insisted that his authors keep their submissions within a reasonable length. For example, he wrote to Jordan in 1877, before suggesting that he condense his manuscript or extract sections that could be published separately:[20]

> It is true that the length of the manuscript you have provided far exceeds the limits generally set for a journal article. I have always refused to follow the practice of many publishers, who chop up a work that they considered too long to publish it in its entirety. The natural limit to which I have always adhered (as a final limit) is the length of an issue, which is generally about 90 pages.

This reflection on the ways in which mathematical results are written and disseminated became significant in the second half of the century, in tandem with the increased structuration of the mathematical academia [44, 69]. In 1868, Hermann Amandus Schwarz, who wanted to publish quickly certain results, because he feared that "everything will soon be taken away from me" wondered where to publish it, since his work seemed too long for the *Journal* [13, pp. 20–21]. A few months later, Weierstrass suggested to write it in the form of a letter: "I believe you would feel most at ease in the form of a letter and would be able to address historical

[18] This would disappear again for short periods and at certain points in the journal's history. For instance, in a few transitional years after Weierstrass's death in 1897, only the name of the new main editor, Lazarus Fuchs, was mentioned on the front page. The *Journal für die reine und angewante mathematik* would return to an editorial board with the arrival in 1902 of Kurt Hensel as main editor, and Georg Frobenius, Georg Hettner, Johannes Knoblauch, Emil Lampe, Friedrich Schottky and Hermann Amandus Schwarz as co-editors. In 1929, a scientific advisory board is announced on the front page, instead of the mention of collaborators. And again, any mention of an editorial board, under any name, would disappear during the 1930s and 1940s.

[19] On Jordan as editor of the *Journal de mathématiques pures et appliquées*, see [7, 8].

[20] This example and the quote come from [87, p. 360].

issues with greater objectivity than in an actual memoir" [13, p. 26]. While the exercises disappeared from the *Journal*, there were indeed numerous mathematical letters addressed to one of the editors, or sometimes to another mathematician; Hermite, for example, published in the *Journal für die reine und angewandte Mathematik* more than 15 letters to Borchardt (and other letters later to Fuchs, when the latter took over as the main editor). The efficiency of the *Journal* seems at this time well recognized: when Fuchs sent to Hermite a letter (in French) intended to be published in the *Journal de mathématiques pures et appliqués*, at the time in Résal's hands, Hermite first raised doubts concerning the publication delays, then confirmed: "I was unable to ensure that your fine memoir would be published soon enough, so I am writing to let you know that I am sending it to Mr.Borchardt, who will enjoy reading it just as much as I did".[21] A year later, Hermite, again discussing the best place of publication, commented: "the German Journal, I must admit, has, given its significance, an undeniable advantage over the French one". While Fuchs's new letter was finally sent to the *Journal de mathématiques pures et appliqués*, Hermite promised to him "many offprints to make up for the more limited diffusion of the French collection". [22]

The choice of the place of publication seems nonetheless much larger in the second half of the nineteenth century. Many independent mathematical journals, separate from national academies, were established in this period, sometimes in connection with major political changes (such as independence or unification) that created new funding opportunities or triggered reforms in education or scientific research. Let us mention, without any claim to be exhaustive, the *Annali di scienze matematiche e fisiche* founded by Barnaba Tortoloni in 1850 and restructured in 1858 as the *Annali di matematica pura ed applicata* under a collective editorial board; the *American Journal of Mathematics* launched in 1878 by James Joseph Sylvester in connection with Johns Hopkins University in the US; the *Jornal de sciencias mathematicas e astronomicas* of Francisco Gomes Teixeira in 1877; and again, the *Rendiconti del Circolo Matematico di Palermo*, the *Časopis pro pěstováni matematiky a fysiky*, the *Quarterly Journal of Pure and Applied mathematics*, the *Messenger of Mathematics*, the *Mathematisk Tidsskrift*, the *Prace Matematyczno-Fizyczne*, *El Progreso matematico: periódico de matematicas puras y aplicadas*, etc.[23]

Although their editors were all aware of the model offered by Crelle's *Journal* and often mentioned it when they launched their own one, they also adopted in practice some of the other models we have encountered, in particular that of the intermediary journals intended to attract a wider audience. For example, Gaston Darboux's *Bulletin des sciences mathématiques et astronomiques* was modeled mainly after Férussac's *Bulletin*, aiming to provide a broader range of information on global research rather than original publications, even though it gradually evolved in that direction. Other journals, even those intended primarily to publish original research, also had to adapt to their readers' knowledge and interests. The links of these journals with the *Journal für die reine und angewandte Mathematik* were thus varied. *The Analyst: A Monthly Journal of Pure and Applied Mathematics*, published from 1874 in a region of the USA badly equipped in libraries and universities, included for instance commented abstracts of articles from Crelle's *Journal* by Christine Ladd, who complained about finding it "on the

[21] Letter from Hermite to Fuchs, December 13, 1876, private coll. The article appeared in 1877 in the 83th volume of Crelle's Journal.

[22] Letters from Hermite to Fuchs, November 19 and 26, 1877, private coll. Fuchs's article appeared in 1878 in the 4th volume of the third series of Resal's *Journal*.

[23] On these journals and many others, see [1, 17–19, 23, 41, 44, 48, 64, 65, 70, 80]. Some of their relations are discussed in [20, 40, 42, 44, 66, 80].

shelves of the Boston Public Library with uncut leaves" [49].

Other major launches would change the distribution of tasks among journals and their positioning. From 1870 on, reviewing publications, the *Jahrbuch über die Fortschritte der Mathematik*, followed by the *Répertoire bibliographique* — replaced during the twentieth century by *Zentralblatt für Mathematik und ihre Grenzgebiete* (now, zbMATH Open) and the *Mathematical Reviews*, now MathSciNet — would handle the systematic diffusion of abstracts of the articles published in research journals. At the very end of the century, the *Intermédiaire des mathématiques*—an ancestor of the Mathematics Stack Exchange—gathered open questions, requests for bibliographic information and answers, from its creation in 1894 until the interwar period. Journals such as the *Journal für die reine und angewandte Mathematik* would shift their focus definitively to the publication of original research articles alone, without supplementary sections for reviews or exercises.[24)]

However, two creations had a more direct impact on the scope of the *Journal*. In the German speaking world, a new research journal was launched in 1868 by Alfred Clebsch in Göttingen and Carl Neumann in Leipzig, hosted by the publishing house Teubner. While the editors primarily cited as their motivation the growing number of mathematicians — and thus of articles — and the need for more space to publish within a reasonable timeframe, the initiative also aimed at regaining some control over what was published or at least at developing a counterforce against what was perceived as Berlin's hegemony [72]. No mathematician from Berlin published in the first volume, but it did include, for example, articles by Jordan and Eugenio Beltrami. From 1876 on, Felix Klein (with Adolph Mayer at the beginning, and an editorial board) would be the driving force of the new journal, whose history closely follows the advent of Göttingen as a main (and largely international) center for mathematics.[25)]

Another important creation was the founding of *Acta Mathematica* — an idea conceived by Sophus Lie but brought to fruition primarily through the efforts of Gösta Mittag-Leffler in 1882. The aim was both to support the significant development of mathematics in the Scandinavian countries (Sweden, Denmark, Norway, and Finland were represented on the editorial board) by serving as a sounding board and a platform for disseminating their authors' work and to acknowledge an increasingly international mathematical scene [5, 22, 83]. Internationalism was therefore emphasized, while the model—a multilinguistic one — recalled in fact the one already established by Crelle.[26)] Other journals adopted other solutions to emphasize their commitment to internationalism, for instance by inviting foreigners to join the editorial board.[27)] It was also Crelle's model that Mittag-Leffler referred to in highlighting certain articles. He wrote to Poincaré in March 1882 [5, p. 143]:

> You know that it was Abel, a Norwegian, who was responsible for the success of

---

[24)] On these new types of publications, see [4, 59, 77, 81].

[25)] On various phases of the *Mathematische Annalen* and its relations to Crelle's *Journal*, see [6, 47, 70, 72, 90].

[26)] The papers in *Acta Mathematica* were mostly in French and in German. As is well-known, however, Mittag-Leffler organized the translation of Georg Cantor's memoirs on set theory into French to offer them a more international visibility. The issue of languages was seen at the time as a key one. Certain editors made efforts to launch mathematical journals in their national language in order to reach a wider local audience, even if it meant, for example, having to develop new terminology for mathematics. Others, like Mittag-Leffler, decided to adopt languages perceived as international, in his case French and German. Another example is the Polish *Fundamenta mathematica*, which also adopted French to reach an international audience, while focusing on a specialized topic, set theory and logic [23].

[27)] An example is the *Quarterly Journal of Pure and Applied Mathematics*, which included Hermite as corresponding editor [17].

> Crelle's *Journal*. Now it has occurred to Gyldén and me that you, a Frenchman, might be as generous and make a success of our journal.

Like Crelle, Mittag-Leffler was well connected to the centers of mathematical life and was able to recruit foreign authors for his journal: besides the Scandinavian authors, German and French authors were well represented, while Mittag-Leffler also secured the participation of mathematicians from Italy, Russia, Belgium, England, Austria, Holland, Switzerland and the United States. Poincaré himself contributed more than thousand pages in the first twenty volumes. The contents of *Acta mathematica* displayed a strong bias towards analysis: 89 memoirs on analysis —in fact, c. 120 if we include algebra and number theory as in Crelle's early classification—were published in the first ten volumes, against 18 in geometry.[28)] Despite Mittag-Leffler's initial concerns about the reception of his journal and the potential rivalry it might appear to create, the *Acta mathematica* was well received, both in Paris and Berlin, in particular for its "good fortune of gathering memoirs that will remain of exceptional merit for Analysis" [5, p. 160]. Signed among others by Weierstrass and Fuchs, the two editors of the *Journal für die reine und angewandte Mathematik* at the time, this 1897 praise shows that Crelle's *Journal* was then not too afraid of such new offers.

While the 1870 war between Prussia and France seemed to have had relatively little impact on the personal relations which were instrumental to strengthening the international character of mathematics research journals, the same cannot be said of the wars and political disasters of the twentieth century. Tensions that had always existed within national or international communities—whether manifested in personal or in thematic terms—also became more dramatic.[29)]

The 1915 issue of Crelle's *Journal* would be the last one for a long time to include a paper in French (or by a French mathematician, here Elie Cartan). During the First World War, there would be the occasional one or two texts in English (coming from the USA), but most of the texts were now in German, in strong contrast with the past habits of the *Journal*. The boycott of German mathematicians during part of the postwar period is well known [74,78]. The celebration of the hundredth anniversary of the Crelle's *Journal*, published in 1927, included a tribute to Crelle by Mittag-Leffler, and, besides the German contributions, mathematical texts of Godfrey Harold Hardy and John Edensor Littlewood, and of Teiji Takagi, but no participation from France or Italy.[30)]

A text in French was finally published in 1932: it is particularly moving that it came from Jacques Herbrand—and it seemed to move the editor, Helmut Hasse, too, for he accompanied the article (on the non-contradiction of arithmetic) by a note in German explaining that he had received the article for publication on the very day Herbrand died in a mountain accident at age 23 and recalling the months Herbrand had just spent in German universities with German colleagues. Herbrand's stay in Germany was part of an effort — supported in particular by funding from the Rockefeller Foundation—to rebuild ties within the European mathematical community by sending talented young mathematicians to universities abroad [79]. A year later, in the same context, Claude Chevalley would also publish in the *Journal* an article in French.

---

28) These data come from [5, pp. 156–157]. Let us note that the thematic specialization seems to increase at the time; for instance invariant theory was published abundantly in the *Quarterly Journal* and in the *Journal für die reine und angewandte Mathematik*, but not in the *Journal de mathématiques pures et appliquées*.

29) An example is the ousting of Luitzen Egbertus Jan Brouwer from the editorial board of the *Mathematische Annalen* [72,85].

30) The same is true, reciprocally, for the *Journal de mathématiques pures et appliquées*, with no German participation in the twenties and thirties.

But in general Crelle's *Journal* had become monolingual over several decades, apart from a few isolated papers.

It should be noted that during this period, the geographical and institutional division that anchored Crelle's *Journal* in Berlin was no longer in effect; although the editorial board still included Berlin professors, for instance Friedrich Hermann Schottky in the 1920s, the leading editors, Kurt Hensel, Ludwig Schlesinger and soon Helmut Hasse, connected to Berlin through their training, their mathematical preferences and their personal ties, were not based in the capital. On the other hand, thematic specialization increased sharply in this period. A detailed comparison of the contents of all the research journals in this period is still to be done, but the focus on number theory, in the continuation of Hensel's and Hasse's work, is quite striking. Both these men were prolific authors in the *Journal.* For Volume 172 in 1935, for instance, half of its 18 articles were devoted to number theory, with 5 authored by Hasse himself among them. In general, Hasse would author at least one or two articles on number theory pro issue in his journal.

Mathematics, here, is intertwined with the harshest realities of politics: both Schlesinger and Hensel, being Jewish, were ousted from the editorship in the 1930s.[31)] The arithmetical and algebraic orientation of the *Journal* was also under pressure and in 1936, a new research journal was founded in Germany, *Deutsche Mathematik,* "as a deliberate contrast to and in competition with Crelle's Journal" [71, p. 7].[32)] That journal came to a definitive end in 1942. As for Crelle's *Journal*, six years elapsed between Volume 185, published in 1943, and Volume 186, which was published in 1949, always with Hasse as unique editor. But starting in the 1950s, the *Journal für die reine und angewandte Mathematik* gradually returned to a steady publishing schedule, with international authors.

5. Conclusion

"What can one say about Crelle's *Journal* that would not also apply to just any mathematical journal?".

Throughout its long history, the role of mathematics research journals has undergone significant changes, and, as we have noted, these changes have not necessarily occurred simultaneously across all mathematical publications. The very position occupied by Crelle's *Journal* — as a model, a source, and a competitor — has also varied widely. Without claiming to be exhaustive, we have attempted here to highlight, at a few key moments, what distinguished Crelle's *Journal* from other mathematics journals — in terms of format and language, in the choice of topics, in the role of editors within the broader mathematical community, in the nature of their contributions to the journal, and in their visions of how a research journal should operate, and also in the relations between mathematics publishing and world politics.

It is not our intention here to determine which of these factors were decisive — perhaps we should leave that task to the tercentenary volume to be published in 2126. At most, we can note that amidst all these changes — and with the feeling that at certain moments in history, the survival of Crelle's *Journal* hung by a thread — its successive editors demonstrated an

31) The same happened of course in other journals, in particular *Mathematische Annalen.* The negotiations around these required withdrawals under National Socialism are described for Crelle's *Journal* from the point of view of Hasse in [71]. For various viewpoints on Hasse's role as an editor, see [30, 35, 70, 71].

32) On this journal, which wanted to champion a form of mathematics described as authentically German and more intuitive, see [50, 51, 70].

explicit commitment to continuity itself, if only as evidenced by the retention of the same title for two hundred years. This continuity stems in part both from the personal passage of the baton from one editor to the next throughout much of its history, and from the existence of an active and dynamic mathematical community around it, both for writing and for reading. It also stems from the continuity in the publishers (after the *Journal*'s first year of existence, only two successive publishers ensured its production and distribution, Reimer and Walter de Gruyter), who likewise adapted to the changing technical and mathematical conditions of typography, printing, and commercial systems. It stems finally from the important role Crelle's *Journal* played in promoting certain topics, particularly algebraic and arithmetic ones, while refusing to be confined to a total specialization.

But the most obvious element that distinguishes the *Journal für die reine und angewandte Mathematik* — and which does not apply to just any journal — is perhaps above all its very longevity and the lasting impact on the mathematical community it helped to create.

In 1844, a Note, extracted from a letter to the editor (Crelle) by Eugène Catalan, was published just above the errata of the preceding volume. It reads [12]:

> I would ask you, Sir, to include in your collection the following theorem, which I believe to be true, although I have not yet succeeded in proving it completely; perhaps others will have better luck. "Two consecutive integers, except 8 and 9, cannot be exact powers; in other words, the equation $x^m - y^n = 1$ in which the unknowns are positive integers, has only one solution."

After many partial answers, by many mathematicians, the now famous "Catalan conjecture" was finally proved by Preda Mihăilescu and published in 2004 in the *Journal* [56]. Neither Crelle nor Catalan lived to see the proof. But the *Journal für die reine und angewandte Mathematik* did.

Acknowledgement. I warmly thank Philippe Nabonnand for giving me early access to the important synthesis of the ANR Cirmath and Barnaby Sheppard and Jim Ritter for their careful reading of a first version of this text.

References

[1] *E. Ausejo* and *M. Hormigón* (eds.), Messengers of Mathematics: European Mathematical Journals (1800–1946), Siglo XXI de España Editores, Madrid 1993.

[2] *T. Archibald*, Journal Publication and Mathematical Publics in Germany, 1800–1825, in: [44], 537–560.

[3] *M. Audin*, Publier sous l'occupation, Rev. Hist. Math. **15** (2009), 7–57.

[4] *J. Auvinet*, Charles-Ange Laisant: itinéraires et engagements d'un mathématicien de la Troisième République, Hermann, Paris 2013.

[5] *J. Barrow-Green*, Gösta Mittag-Leffler and the Foundation and Administration of *Acta Mathematica*, in: [66], 139–164.

[6] *H. Behnke*, Rückblick auf die Geschichte der Mathematischen Annalen, Math. Ann. **200** (1973), i–vii.

[7] *F. Brechenmacher*, Le "journal de M. Liouville" sous la direction de Camille Jordan (1885–1922), SABIX **45** (2009), 65–71.

[8] *F. Brechenmacher*, Un journal de "rang élevé": Le *Journal de mathématiques pures et appliquées* sous la direction de Camille Jordan, in: [44], 779–851.

[9] *B. Bru* and *T. Martin*, Le baron de Férussac, la couleur de la statistique et la topologie des sciences, Electronic Journ@l for History of Probability and Statistics **1-2** (2005), 1–43.

[10] *M. Bullynck*, Stages towards a German mathematical journal (1750-1800), Arch. Int. Hist. Sci. **63** (2013), 237–251.

[11] *M. Bullynck* and *T. Morel*, Une révolution peut en cacher d'autres: Le paysage morcelé des mathématiques dans l'espace germanophone et ses reconfigurations (1750–1850), in: Sciences mathématiques 1750–1850: Continuités et ruptures, ed. *C. Gilain* and *A. Guilbaud*, CNRS Éditions, Paris 2015, 181–206.
[12] *E. Catalan*, Note, J. reine angew. Math. **270** (1844), 192.
[13] *E. Confalonieri*, Beiträge zur Geschichte der mathematischen Werke von Karl Weierstrass, III, 2013, prep.
[14] *K. Chatzis*, *T. Morel*, *T. Preveraud* and *N. Verdier*, Traduire des mathématiques "pour et par des élèves" dans la première moitié du XIXe siècle, Stud. in Book Culture **9-1** (2017), online, `https://doi.org/10.7202/1043118ar`.
[15] *A. L. Crelle*, Inhaltsverzeichnissen der Bände 1-50 dieses Journals, welche mit thätiger Beförderung hoher Königlich-Preussischer Behörden in den Jahren 1826-1855 herausgegeben wurden, J. reine angew. Math. **50** (1855), 327-392.
[16] *A. L. Crelle* (ed.), J. reine angew. Math. **51** (1856).
[17] *T. Crilly*, The Cambridge Mathematical Journal and its descendants: the linchpin of a research community in the early and mid-Victorian Age, Hist. Math. **31-4** (2004), 455–497.
[18] *S. Despeaux*, The Development of a Publication Community: Nineteenth-Century Mathematics in British Scientific Journals, Ph.D. thesis, University of Virginia, 2002.
[19] *S. Despeaux*, International Mathematical Contributions to British Scientific Journals, 1800–1900. In: [66], 61–87.
[20] *S. Despeaux*, Mathematics Sent Across the Channel and the Atlantic: Nineteenth-Century British Mathematical Contributions to International Scientific Journals, Ann. Sci. **65-1** (2008),73–99.
[21] *J. Dhombres*, Le journal professionnel au XIXe siècle: enjeux généraux d'une enquête en cours, Riv. Stor. Sci. (2) **2-2** (1994), 99-136.
[22] *Y. Domar*, On the foundation of Acta Mathematica, Acta Math. **148** (1982), 3–8.
[23] *R. Duda*, *Fundamenta Mathematica* and the Warsaw school of mathematics, in: [46], 481–500.
[24] *S. Duvina*, Le *Journal de mathématiques pures et appliquées* sous la férule de J. Liouville (1836-1874), Sciences et techniques en perspective **28** (1994), 179-217.
[25] *W. Eccarius*, Der Techniker und Mathematiker August Leopold Crelle (1780-1855) und sein Beitrag zur Förderung und Entwicklung der Mathematik in Deutschland des 19 Jahrhunderts, Dissertation, Karl-Marx-Universität Leipzig, 1974.
[26] *W. Eccarius*, August Leopold Crelle als Herausgeber wissenschaftlicher Fachzeitschriften, Ann. Sci. **33** (1976), 229–261.
[27] *W. Eccarius*, August Leopold Crelle als Herausgeber des Crelleschen Journals, J. reine angew. Math. **286-287** (1976), 5–25.
[28] *W. Eccarius*, Zur Gründungsgeschichte des Journals für reine und angewandte Mathematik, NTM **14-2** (1977), 8–28.
[29] *W. Eccarius*, Der Gegensatz zwischen Julius Plcker und Jakob Steiner im Lichte ihrer Beziehungen zu August Leopold Crelle Hintergründe eines wissenschaftlichen Meinungsstreites, Ann. Sci. **37-2** (1980), 189–213.
[30] *C. Eckes*, Organiser le recrutement de recenseurs français pour le *Zentralblatt* à l'automne 1940: les premiers liens entre Harald Geppert, Helmut Hasse et Gaston Julia sous l'occupation, Rev. Hist. Math. **24** (2018), 259–329.
[31] *The Editors*, Preface to volume 400, J. reine angew. Math. **400** (1989), 1–2.
[32] *H. Edwards*, The Background of Kummer's Proof of Fermat's Last Theorem for Regular Primes, Arch. Hist. Exact Sc. **14-3** (1975), 219–236.
[33] *H. Elkhadem*, Histoire de la Correspondance mathématique et physique d'après les lettres de Jean-Guillaume Garnier et Adolphe Quetelet, Bulletin de la Classe des lettres et des sciences morales et politiques de l'Académie royale de Belgique, **64** (1978), 316–366.
[34] *G. Frei*, Zur Geschichte des Crelleschen Journals, J. reine angew. Math. **500** (1998), 1–4.
[35] *G. Frei*, Helmut Hasse und seine Arbeiten im Crelleschen Journal, J. reine angew. Math. **500** (1998), 15–21.
[36] *J. P. Friedelmeyer*, Augustus Leopold Crelle, 1780-1855, in: Mathematics in Berlin, ed. *H. Begehr et al.*, Springer, Berlin (1998), 27–32.
[37] *C. Gérini*, Les *Annales* de Gergonne: apport scientifique et épistémologique dans l'histoire des mathématiques, Editions du Septentrion, Villeneuve d'Ascq 2003.
[38] *L. Giaccardi* and *R. Tazzioli*, The *Bolletino della Unione Matematica Italiana* et ses enjeux politiques et idéologiques (1922–1943), in: [44], 711–750.
[39] *C. Gérini* and *N. Verdier* (eds.), L'Émergence de la presse mathématique en Europe au XIXème siècle (France, Espagne, Italie et Portugal), College Publications, Oxford 2014.
[40] *H. Gispert*, Une comparaison des journaux français et italiens dans les années 1860-1875, in: [46], 391–408.
[41] *H. Gispert*, Les journaux scientifiques en Europe, in: L'Europe des sciences, constitution d'un espace scien-

tifique, ed. *M. Blay* and *E. Nicolaïdis*, Le Seuil, Paris 2001, 191–211.
[42] *H. Gispert, P. Nabonnand* and *J. Peiffer* (eds.), Interplay between mathematical journals on various scales, 1850-1950, Hist. Math. **45-4** (2018), 323–452.
[43] *H. Gispert, P. Nabonnand* and *J. Peiffer*, Les journaux mathématiques et leurs communautés de lecteurs aux XVIIIe et XIXe siècles, in: Le Monde des mathématiques, ed. *P.-M. Michel* and *P. Verschueren*, Le Seuil, Paris 2023, 497–541.
[44] *H. Gispert, P. Nabonnand* and *J. Peiffer* (eds.), Circulation des mathématiques dans et par les journaux: Histoire, territoires, publics, College Publications, London 2025.
[45] *H. Gispert* and *R. Tobies*, A Comparative Study of the French and German Mathematical Societies before 1914, in: [46], 409–430.
[46] *C. Goldstein, J. Gray* and *J. Ritter*, eds., Mathematical Europe: History, Myth, Identity, MSH, Paris 1996.
[47] *U. Hashagen*, Walther von Dyck (1856-1934): Mathematik, Technik und Wissenschaftsorganisation an der TH München, Franz Steiner, Stuttgart 2003.
[48] *D. Kent*, The Mathematical Miscellany and The Cambridge Miscellany of Mathematics: Closely connected attempts to introduce research-level mathematics in America, 1836–1843, Hist. Math. **35-2** (2008), 102–122.
[49] *D. Kent*, Publications' Places and People: Mapping 19th Century Mathematical Journals in the United States, in: [44], 171–187.
[50] *P. Kranz*, The journal "Deutsche Mathematik" (1936–1942/44), Oberwolfach Rep. **2010-3** (2010), 132–133.
[51] *H. Lindner*, "Deutsche" und "gegentypische" Mathematik: Zur Begründung einer "arteigenen" Mathematik im "Dritten Reich" durch Ludwig Bieberbach, in: Naturwissenschaft, Technik und NS-Ideologie:. Beiträge zur Wissenschaftsgeschichte des Dritten Reichs, ed. *H. Mehrtens* and *S. Richter*, Suhrkamp, Frankfurt 1980, 88–115.
[52] *J. Liouville*, Note sur la détermination des intégrales dont la valeur est algébrique, J. reine angew. Math. **10** (1833), 347–359.
[53] *J. Lorenat*, Polemics in public: Poncelet, Gergonne, Plücker, and the duality controversy, Sci. Context **28-4** (2015), 545–585.
[54] *J. Lützen*, International participation in Liouville's *Journal de mathématiques pures et appliquées*, in: [66], 89–104.
[55] *J. Lützen*, Joseph Liouville 1809-1882: Master of Pure and Applied Mathematics, Springer, New York 2013.
[56] *P. Mihăilescu*, Primary Cyclotomic Units and a Proof of Catalan's Conjecture, J. reine angew. Math. **572** (2004), 167–195.
[57] *T. Morel*, Arithmetical Periodicals in late Eighteenth-Century Germany: "Mathematics for the Use and the Pleasure of Civic Life", Arch. Int. Hist. Sci. **64** (2014), 397–427.
[58] *T. Morel*, Les publications périodiques comme miroir de l'évolution des mathématiques en langue allemande (1780-1830), in: [44], 511–536.
[59] *P. Nabonnand* and *L. Rollet*, Une bibliographie mathématique idéale ? Le Répertoire bibliographique des sciences mathématiques, Gaz. Math., Soc. Math. Fr. **92** (2002), 1–25.
[60] *P. Nabonnand* and *L. Rollet*, Les *Nouvelles annales de mathématiques: journal des candidats aux Écoles polytechnique et normale*, Conf. e Sem. dell'Associazione Subalpina Mathesis (2010–2011), 217–230.
[61] *E. Neuenschwander*, Die Edition mathematischer Zeitschriften im 19. Jahrhundert und ihr Beitrag zum wissenschaftlichen Austausch zwischen Frankreich und Deutschland, Mathematisches Institut der Universität Göttingen, Göttingen 1984.
[62] *E. Noble*, The Rise and Fall of the German Combinatorial Analysis, Birkhäuser, Boston 2022.
[63] *M. Otero*, Joseph-Diez Gergonne (1771-1859), Sciences et techniques en perspective **37** (1997), 1–270.
[64] *K. H. Parshall*, James Joseph Sylvester, Jewish Mathematician in a Victorian World, Johns Hopkins University Press, Baltimore 2006.
[65] *K. H. Parshall*, Journals in the Evolution of a National Research Community: The Case of Mathematics in the United States (1776–1940), in: [44], 647–662.
[66] *K. H. Parshall* and *A. C. Rice* (eds.), Mathematics Unbound: The Evolution of an International Mathematical Research Community 1800–1945, American Mathematical Society/London Mathematical Society, Providence, RI/London 2002.
[67] *H. Pieper*, A Network of Scientific Philanthropy: Humboldt's Relations with Number Theorists, in: The Shaping of Arithmetics after C. F. Gauss's *Disquisitiones Arithmeticae*, ed. *C. Goldstein, N. Schappacher* and *J. Schwermer*, Springer, Berlin and Heidelberg 2007, 201–233.
[68] *D. Reimer*, Passion und Kalkül: der Verleger Georg Andreas Reimer (1776–1842), Walter de Gruyter, Berlin & New-York 1999.
[69] *V. Remmert* and *U. Schneider* (eds.), Publikationsstrategien einer Disziplin: Mathematik in Kaiserreich und Weimarer Republik, Harrassowitz, Wiesbaden 2008.

[70] *V. Remmert* and *U. Schneider*, Eine Disziplin und ihre Verleger — Disziplinenkultur und Publikationswesen der Mathematik in Deutschland, 1871-1949, Transcript, Bielefeld 2010.
[71] *H. Rohrbach*, Helmut Hasse and Crelle's Journal, J. reine angew. Math. **500** (1998), 5–13.
[72] *D. Rowe*, Episodes in the Berlin-Göttingen Rivalry 1870–1930, Math. Intell. **22-1**(2000), 60–69.
[73] *P. Schreiber*, Johann August Grunert and his *Archiv der Mathematik und Physik* as an integrative factor of mathematics, in: [46], 433–446.
[74] *B. Schroeder-Gudehus*, Les scientifiques et la paix : la communauté internationale au cours des années 20, PUM, Montréal 1978.
[75] *G. Schubring*, The Conception of Pure Mathematics as an Instrument in the Professionalization of Mathematics, in: Social History of Nineteenth Century Mathematics, ed. *H. Bos*, *H. Mehrtens* and *I. Schneider*, Birkhäuser, Boston 1981, 111–134.
[76] *P. Seguin*, La recherche d'un fondement absolu des mathématiques par l'Ecole combinatoire de C. F. Hindenburg (1741-1808), Philos. Sci. **CS 5** (2005), 61–79.
[77] *R. Siegmund-Schultze*, Mathematische Berichterstattung in Hitlerdeutschland: der Niedergang des *Jahrbuchs über die Fortschritte der Mathematik*, Vandenhoeck and Ruprecht, Göttingen, 1993.
[78] *R. Siegmund-Schultze*, Opposition to the Boycott of German Mathematics in the Early 1920s, Rev. Hist. Math **17-1** (2011), 139–165.
[79] *R. Siegmund-Schultze*, Rockefeller and the Internationalization of Mathematics Between the Two World Wars, Birkhäuser, Boston 2013.
[80] *R. Siegmund-Schultze*, The interplay of various Scandinavian mathematical journals (1859–1953) and the road towards internationalization, Hist. Math. **45** (2018), 354–375.
[81] *R. Siegmund-Schultze*, The *Jahrbuch über die Fortschritte der Mathematik* (1869–1945) and the Gradual Modernization of Mathematical Reviewing, in: [44], 937–944.
[82] *J. Steiner*, Auflösung einer Aufgabe aus den Annalen der Mathematik von Herrn Gergonne, J. reine angew. Math. **2** (1827), .268–275.
[83] *L. Turner*, Cultivating Mathematics in an International Space: Roles of Gösta Mittag-Leffler in the Development and Internationalization of Mathematics in Sweden and Beyond, 1880–1920, Ph. thesis, Aarhus University, 2011.
[84] *P. Ullrich*, "A doctissimo matheseos candidato Weierstrass diligenter et accurate expositam": The First Publication of Karl Weierstrass, on Analytic Factorials, in: History of Mathematics and Its Contexts, ed. *A. Verdugo Rohrer* and *J. Zender*, Birkhäuser, Cham 2024, 209–225.
[85] *D. van Dalen*, The war of the frogs and the mice, or the crisis of the *Mathematische Annalen*, Math. Intell. **12-4** (1990), 17–31.
[86] *N. Verdier)*, Les journaux de mathématiques dans la première moitié du XIXe siècle en Europe, Philos. Sci. **13-2** (2009), 97–126.
[87] *N. Verdier*, Le Journal de Liouville et la presse de son temps: une entreprise d'édition et de circulation des mathématiques au XIXe siècle (1824–1885), Thesis, Université Paris-Sud, 2009.
[88] *N. Verdier*, Éditer puis vendre des mathématiques avec la maison Bachelier (1812–1864), Rev. Hist. Math. **19-1** (2013), 79–145.
[89] *N. Verdier*,Théodore Bailleul (1797–1875) ou le prote devenu directeur de l'Imprimerie mathématique de (Mallet)-Bachelier (1812–1864), Histoire et civilisation du livre **9** (2013), 251–269.
[90] *J. Weiss*, Erfolgreiche Alt-68er: Mathematische Annalen –Mitteilungen B. G. Teubner – Alfred Clebsch – Felix Klein – Carl Neumann, Eagle, Leipzig 2018.

Catherine Goldstein, IMJ-PRG, Case 247, 4 place Jussieu, 75252 Paris Cedex 05, France
e-mail: catherine.goldstein@imj-prg.fr

1826–1850

20.

Geometrischer Beweis des Fundamentaltheorems für die quadratischen Reste.

(Von Herrn Stud. *Gotth. Eisenstein* zu Berlin.)

Es sei p eine positive ungerade Primzahl, a der Complexus aller *geraden* Zahlen $<p$ und >0, also $a = 2, 4, 6, \ldots p-1$; q sei irgend eine durch den Modul p nicht theilbare ganze Zahl. Bezeichnet man durch r das allgemeine Glied der Reste der Vielfachen qa nach dem mod. p, so werden offenbar die Zahlen der Reihe, deren allgemeines Glied $(-1)^r . r$ ist, mit den Zahlen der Reihe a bis auf Vielfache von p übereinstimmen; also wird man die beiden Congruenzen haben:

$$q^{\frac{1}{2}(p-1)} \Pi a \equiv \Pi r \pmod{p}, \qquad \text{und} \qquad \Pi a \equiv (-1)^{\Sigma r} \Pi r \pmod{p},$$

woraus folgt

$$q^{\frac{1}{2}(p-1)} \equiv (-1)^{\Sigma r} \pmod{p}, \qquad \text{also} \qquad \left(\frac{q}{p}\right) = (-1)^{\Sigma r}.$$

Bedeutet $E\left(\frac{qa}{p}\right)$ die gröſste in dem Bruche $\frac{qa}{p}$ steckende ganze Zahl, so ist offenbar $\Sigma qa = p \Sigma E\left(\frac{qa}{p}\right) + \Sigma r$; und da alle a *gerade* sind, und $p \equiv 1 \pmod{2}$ ist, so folgt hieraus $\Sigma r \equiv \Sigma E\left(\frac{qa}{p}\right) \pmod{2}$; also ist auch

$$\left(\frac{q}{p}\right) = (-1)^{\Sigma E\left(\frac{qa}{p}\right)}.$$

Wenn $q = 2$ ist, so giebt diese Formel sogleich den Werth von $\left(\frac{2}{p}\right)$: ist dagegen q ungerade, also $q-1$ gerade, so findet man durch eine leichte Transformation

$$\Sigma E\left(\frac{qa}{p}\right) \equiv -E\left(\frac{q}{p}\right) + E\left(\frac{2q}{p}\right) - E\left(\frac{3q}{p}\right) + \ldots \pm E\left(\frac{\frac{1}{2}(q-1)a}{p}\right)$$

$$\equiv \quad E\left(\frac{a}{p}\right) + E\left(\frac{2q}{p}\right) + E\left(\frac{3a}{p}\right) + \ldots + E\left(\frac{\frac{1}{2}(q-1)a}{p}\right) \pmod{2}.$$

Wird letztere Summe durch μ bezeichnet, so hat man auch $\left(\frac{q}{p}\right) = (-1)^{\mu}$.

Man stelle sich jetzt in der Ebene ein rechtwinkliges Coordinatensystem (x, y) und die ganze Ebene durch Parallelen mit den Axen in den Abständen $= 1$ von einander in lauter Quadrate von den Dimensionen $= 1$ getheilt vor.

J. reine angew. Math. **28** (1844), 246–248
DOI 10.1515/crll.1844.28.246
Journal für die reine und angewandte Mathematik

Geometric proof of the fundamental theorem of quadratic residues

By *Gotth. Eisenstein* at Berlin

Let p be a positive odd prime, $\mathbf{a}$ the system of all *even* numbers $< p$ and > 0, i.e. $\mathbf{a} = \{2, 4, 6, \dots, p-1\}$. Let q be any integer not divisible by the modulus p. If one denotes by r the general term of the residues of the multiples qa with respect to the modulus p, then the numbers of the series whose general term is $(-1)^r \cdot r$ clearly coincide with the numbers of the series $\mathbf{a}$ up to multiples of p; therefore one has the two congruences:

$$q^{\frac{1}{2}(p-1)}\Pi a \equiv \Pi r \pmod p, \quad \text{and} \quad \Pi a \equiv (-1)^{\Sigma r}\Pi r \pmod p,$$

from which it follows that

$$q^{\frac{1}{2}(p-1)} \equiv (-1)^{\Sigma r} \pmod p, \quad \text{thus} \quad \left(\frac{q}{p}\right) = (-1)^{\Sigma r}.$$

If $E(\frac{qa}{p})$ denotes the largest integer in the fraction $\frac{qa}{p}$, then clearly $\Sigma qa = p\Sigma E(\frac{qa}{p}) + \Sigma r$; and since all a are *even*, and $p \equiv 1 \pmod 2$, it follows that $\Sigma r \equiv \Sigma E(\frac{qa}{p}) \pmod 2$; so that

$$\left(\frac{q}{p}\right) = (-1)^{\Sigma E(\frac{qa}{p})}.$$

If $q = 2$, this formula immediately gives the value of $\left(\frac{2}{p}\right)$. If, on the other hand, q is odd, so $q-1$ is even, then one finds, via an easy transformation, the identity

$$\begin{aligned}\Sigma E(\frac{qa}{p}) &\equiv -E(\frac{q}{p}) + E(\frac{2q}{p}) - E(\frac{3q}{p}) \dots \pm E(\frac{\frac{1}{2}(p-1)q}{p}) \\ &\equiv E(\frac{q}{p}) + E(\frac{2q}{p}) + E(\frac{3q}{p}) \dots + E(\frac{\frac{1}{2}(p-1)q}{p}) \pmod 2.\end{aligned}$$

If the latter sum is denoted by μ, then $\left(\frac{q}{p}\right) = (-1)^{\mu}$.

Now imagine a rectangular coordinate system (x, y) in the plane and the entire plane divided into squares of equal dimensions $= 1$ by parallels to the axes at distances $= 1$ from each other. All corner points of squares that do not lie on the two coordinate axes will be called *lattice points*. (Pl. II, Fig. 1. 2.)

If we take a point on any vertical parallel line corresponding to the coordinate y, $E(y)$ will express the number of lattice points lying between this point and the horizontal axis; and if we take a point on any horizontal parallel line corresponding to the abscissa x, then $E(x)$ will express the number of lattice points lying between this point and the vertical axis. Therefore, if any curve whose equation is $y = \varphi(x)$ is drawn in the plane (Fig. 1.), the sum

$$E\varphi(1) + E\varphi(2) + E\varphi(3) + \text{ etc.}$$

will give the number of lattice points that lie between this curve and the x-axis, including those lattice points that happen to lie on the curve itself.

Now, returning to our subject, let AB (Fig. 2.) be the straight line whose equation is $y = \frac{q}{p}x$, where p and q are both assumed to be positive odd *prime numbers*. Let $AD = FB$ be $= p$, $AF = DB = q$, $AC = EG = \frac{1}{2}(p-1)$, $AE = CG = \frac{1}{2}(q-1)$. If we denote by μ the number of lattice points between AB and AD, up to the ordinate CG incl. (which are marked with asterisks (*) in the figure), then according to what has been proven, we will have $\left(\frac{q}{p}\right) = (-1)^{\mu}$. Since the equation of our straight line can also be written as $x = \frac{p}{q}y$, we will have $\left(\frac{p}{q}\right) = (-1)^{\nu}$ in the same way, where ν denotes the number of lattice points between AB and AF up to the abscissa EG incl. (which are marked by small zeros (0)).

However, it is evident that all lattice points marked with $*$ and all those marked with 0, i.e. all lattice points to the *right* and all lattice points to the *left* of AB, taken together exhaust *all* lattice points of the rectangle $AEGC$, of which there are $\frac{1}{2}(p-1)\cdot\frac{1}{2}(q-1)$, thus, $\mu + \nu = \frac{1}{2}(p-1)\cdot\frac{1}{2}(q-1)$, and

$$\left(\frac{q}{p}\right)\left(\frac{p}{q}\right) = (-1)^{\mu+\nu} = (-1)^{\frac{1}{2}(p-1)\cdot\frac{1}{2}(q-1)};$$

which was to be proven.

Incidentally, the above transformation $\Sigma E(\frac{qa}{p}) \equiv \mu \pmod 2$ can also be proven by a very simple geometric consideration, if one considers that $\Sigma E(\frac{qa}{p})$ is nothing other than the number of lattice points lying on the *even* ordinates (which correspond to the abscissa $x = 2, 4, 6, \ldots, p-1$) between AB and AD up to BD, and that each ordinate from the axis AD to FB excl. contains $q-1$, i.e. an even number of lattice points; as well as that the two triangles BAD and ABF are congruent and that the latter relates to BF and BD in the same way as the former relates to AD and AF; the explication of which is left to the reader.

Note: There are figures for which simple formulas can be used to determine the number of lattice points contained within them. If, for example, we imagine a circle whose centre lies at the origin of the coordinates and whose radius is $= \sqrt{m}$, then the number S of lattice points enclosed by this circle, including those lying on the axes, is given by the following formula:

$$S = 1 + 4(E(m) - E(\frac{1}{3}m) + E(\frac{1}{5}m) - E(\frac{1}{7}m) + \ldots),$$

until the series itself ends.

As can be easily seen, this equation expresses a relation between the number of lattice points of a *circle* and the number of lattice points of a segment enclosed between two *hyperbolas*. If we substitute $m = \infty$ into the formula

$$\frac{1}{m}S = \frac{1}{m} + 4(\frac{1}{m}E(m) - \frac{1}{m}E(\frac{1}{3}m) + \frac{1}{m}E(\frac{1}{5}m) - \text{etc.}),$$

the left-hand side becomes π, while the right-hand side becomes $4(1 - \frac{1}{3} + \frac{1}{5} - \text{etc.})$, giving us *Leibniz*'s formula for π. There are similar formulas for the number of lattice points in a system of ellipses or sectors of hyperbolas; similar relations also occur in space and in cases involving more than three dimensions. We will return to this important subject, which is very closely related to the properties of higher forms, at a later date.

Berlin, July 1844.

Translation by Joachim Schwermer

1.

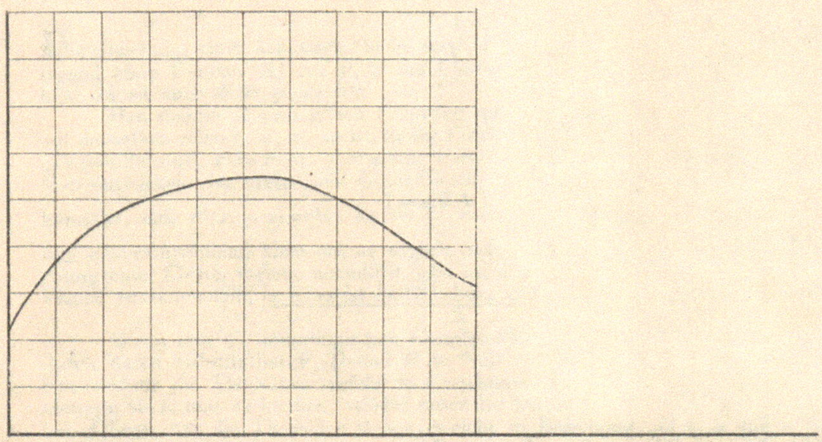

2.

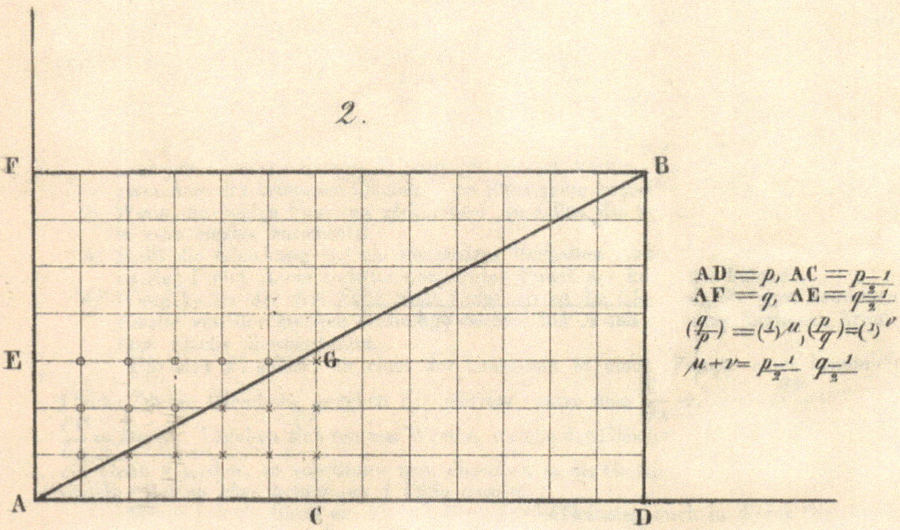
F
B
E
G
A
C
D
$AD = p$, $AC = \frac{p-1}{2}$
$AF = q$, $AE = \frac{q-1}{2}$
$\left(\frac{q}{p}\right) = (-1)^{\mu}$, $\left(\frac{p}{q}\right) = (-1)^{\nu}$
$\mu + \nu = \frac{p-1}{2} \cdot \frac{q-1}{2}$

Eisenstein's lens – a geometric view of the fundamental theorem of quadratic residues

By *Joachim Schwermer* at Vienna

1. Introduction

C.F. Gauss's treatment in the *Disquisitiones Arithmeticae* of the theory of quadratic residues, in which the *Theorema fundamentale residuorum quadraticorum* plays a central role, can be regarded as the beginning of the development of algebraic number theory. It is only in this historical process culminating in Artin's reciprocity law, class field theory and the current search for a global non-Abelian reciprocity law that this result, also known as the *loi de réciprocité* following A.-M. Legendre, reveals its clarity and significance.[1] Gauss devoted considerable time to this result, providing various proofs and thereby inspiring a great deal of scholarly work.

This paper provides a detailed analysis of the treatise "A geometric proof of the fundamental theorem of quadratic residues" by G. Eisenstein, published in 1844, and situates it within its historical context. While Eisenstein's approach is strongly inspired by Gauss's proof in *Theorematis arithmetici demonstratio nova*, it elegantly transforms the essential ingredients of the proof found there into geometric form. The concept of the integral part of a number, as introduced by Gauss, plays an important role here. It provides the background when Eisenstein considers the points of a lattice in the plane. The outcome is a brilliant proof of the fundamental theorem. This result is opened up to a new perspective, the geometric basis of which significantly expands the range of methodological tools available in number theory.

At that time, Gauss was already familiar with the idea of introducing the concept of lattices into the discussion of number-theoretical questions, in particular, in the context of the theory of positive binary or ternary integral quadratic forms. Indeed, in his review of the work of L. Seeber on ternary forms, published in 1831 [23], Gauss acknowledged the results but he also suggested to interpret these forms as lattices in space and derived in this geometric interpretation Seeber's results in a much simpler form. This review by Gauss and subsequent work by Dirichlet mark one of the sources for Minkowski's later development of the geometry of numbers (see [37, Sect. 2]).

This note is intended to be accessible to non-specialists who are interested in reading Eisenstein's paper.

[1] We refer to [1], [2] and [5].

© 2026 THE AUTHOR(S) | HTTPS://DOI.ORG/10.1515/9783111260440-003

2. Congruences

2.1. Congruence in $\mathbb{Z}$. Suppose that r, s are integers and that $m \neq 0$ is a natural number. One says that r is congruent to s modulo m if m divides $s - r$. This relation is denoted $r \equiv s \pmod m$. One easily verifies that the congruence relation modulo m is an equivalence relation on the set $\mathbb{Z}$ of integers. The equivalence classes are called residue classes or congruence classes modulo m.

If $r \equiv r' \pmod m$ and $s \equiv s' \pmod m$, then clearly $r + s \equiv r' + s' \pmod m$ and $r - s \equiv r' - s' \pmod m$. Furthermore, since m divides $r's' - rs = r'(s' - s) + s(r' - r)$ one has $rs \equiv r's' \pmod m$. Consequently, given any polynomial F with integer coefficients, $F(r) \equiv F(r') \pmod m$.

Given an integer $d \geq 1$, one denotes by $\phi(d)$ the Euler ϕ-function defined as the number of integers z with $1 \leq z \leq d$ which are prime to d. One easily verifies $\phi(p^i) = p^i - p^{i-1}$ for all powers p^i of a prime number p.

We call a set $\{r_1, \ldots, r_m\}$ of m integers, one from each residue class modulo m, a complete set of residues modulo m. Observe that, given any complete set $\{r_1, \ldots, r_m\}$ of residues $(\operatorname{mod} m)$, then also $\{nr_1, \ldots, nr_m\}$ is a complete set of residues $(\operatorname{mod} m)$ for each natural number n prime to m.

By a reduced set of residues $(\operatorname{mod} m)$ one means a set of $\phi(m)$ numbers, one from each of the $\phi(m)$ many residue classes modulo m that consist of numbers relatively prime to m. For example, the numbers r with $1 \leq r < m$ and r, m relatively prime form a reduced set of residues $(\operatorname{mod} m)$.

We remark that, if m and m' are two natural numbers relatively prime with one another and r and r' run through reduced sets of residues $(\operatorname{mod} m)$ and $\pmod{m'}$ respectively, then the elements $rm' + r'm$ run through a reduced set of residues $\pmod{mm'}$. In particular, it follows that the ϕ-function is multiplicative.

Given a reduced set $\{s_1, \cdots, s_{\phi(m)}\}$ of residues $(\operatorname{mod} m)$, then also $\{as_1, \cdots, as_{\phi(m)}\}$ forms a reduced set of residues $(\operatorname{mod} m)$ for any natural number a relatively prime with m. Therefore, $\prod_i s_i = \prod_i as_i \pmod m$. By cancelling out the factor $\prod_i s_i$ on both sides one obtains Euler's theorem, which states that $a^{\phi(m)} \equiv 1 \pmod m$. As a particular case, if p is a prime number not dividing a, this reads as $a^{p-1} \equiv 1 \pmod p$, already known to P. Fermat.

First, consider a linear congruence. Given natural numbers a, m and an integer b, the congruence $ax \equiv b \pmod m$ is solvable for some integer x if and only if the greatest common divisor $d > 0$ of a and m divides b. Second, given a system of linear congruences

$$x \equiv b_1 \pmod{m_1}, \ldots\ldots, x \equiv b_k \pmod{m_k}$$

where $m_1, \ldots, m_k$ are natural numbers which are coprime in pairs and $b_1, \ldots, b_k$ are any integers, the congruences are solvable simultaneously for some integer x. Moreover, there is a unique solution with respect to the modulus $m = m_1 \cdots m_k$. This result is known as the Chinese remainder theorem.

2.2. Lagrange's theorem. An easy inductive argument proves the validity of the following result, known as Lagrange's theorem: Given a prime p and a polynomial f of degree n with integer coefficients and whose leading coefficient is not divisible by p, the congruence $f(x) \equiv 0 \pmod p$ has at most n solutions modulo p. Note that the analogous statement fails for moduli which are not square free.

3. Quadratic residues and the "Theorema fundamentale"

3.1. Quadratic forms and quadratic residues. Gauss's "Theorema fundamentale" in the theory of quadratic residues has its origins in L. Euler's, J.L. Lagrange's and, finally, in Gauss's investigations of the arithmetic of quadratic forms, a central part of Diophantine analysis, which concerns at its basis the study of the solutions of equations in integers. In specific cases, Fermat and Euler had studied the nature of integral solutions of equations of the form $g(X, Y) = aX^2 + cY^2 = m$, where a, c, m are integers. In particular, the question of whether a given integer m can be represented by g (i.e. whether there exist integers x, y such $g(x, y) = m$) played a decisive role in the formation of the theory. By unifying these studies, Lagrange laid the foundations of the general theory of binary quadratic forms $f(X, Y) = aX^2 + bXY + cY^2$, where a, b, c are integers, and X, Y variables. Gauss gave a systematic treatment of this theory in articles 171–175 of his *Disquisitiones arithmeticae.*

The quadratic form $Q_a(X, Y) = X^2 - aY^2, a \in \mathbb{Z}$, is closely linked to the study of the properties of quadratic residues and the derivation of the fundamental theorem. Indeed, we say that an integer m divides the form Q_a if there exist integers x, y, relatively prime to one another, such that m divides $x^2 - ay^2$. In such a case, m and y are relatively prime. Therefore there exists an integer w relatively prime to m such that $wy \equiv 1 \pmod{m}$. Hence

$$w^2(x^2 - ay^2) = (wx)^2 - a(wy)^2 \equiv (wx)^2 - a \pmod{m}$$

holds. It follows that any divisor of the form $Q_a(X, Y) = X^2 - aY^2$ also divides the form $X^2 - a$. The reverse is obviously also true. The task of specifying all numbers represented by the quadratic form $Q_a(X, Y)$ and determining the divisors of $Q_a(X, Y)$ thus leads to the consideration of the congruence

$$X^2 - a \equiv 0 \pmod{m}$$

and the question for which m, given a, it is solvable in the integers.

Following Euler, given any integer a and a natural number m, the integer a is called a quadratic residue ("residuum quadraticum") modulo m if the congruence $X^2 - a \equiv 0 \pmod{m}$ is solvable and a quadratic non-residue ("non residuum quadraticum") modulo m if the congruence is not solvable. Gauss used the notation aRm to denote a quadratic residue modulo m and aNm to denote a quadratic non-residue modulo m.

In this context, two fundamental questions arise:

(I) Given a natural number m, which integers are quadratic residues modulo m?

(II) Given an integer a, for which natural numbers m is a a quadratic residue modulo m?

In discussing these questions we may restrict our considerations to odd primes $m = p$. Indeed, let the modulus have the prime factorisation $m = p_1^{e_1} \cdots p_k^{e_k}$, with distinct primes $p_i \neq p_j$, $i, j = 1, \ldots, k$, and positive integral exponents e_i, $i = 1, \ldots, k$. The Chinese remainder theorem implies that the congruence $X^2 - a \equiv 0 \pmod{m}$ is equivalent to the system of simultaneous congruences $X^2 - a \equiv 0 \pmod{p_j^{e_j}}$, $j = 1, \ldots, k$. Therefore, we can limit ourselves to considering prime powers p^e. If p is an odd prime that does not divide a, then from a solution x of the congruence $X^2 - a \equiv 0 \pmod{p^e}$ one obtains a solution y, given as $y = x + p^e z$, of $Y^2 - a \equiv 0 \pmod{p^{e+1}}$, by solving the linear congruence $2xZ + (x^2 - a)/p^e \equiv 0 \pmod{p}$ with z. Since $2x$ and p are coprime, this is possible. It follows that a is a quadratic residue modulo p if and only if a is one modulo p^e for any natural number e. Consider $p = 2$, and suppose that a is not divisible by 2. Each a is a quadratic residue modulo 2. With regard to the modulus 2^2, any odd number of the form $4k+1$ is a quadratic residue, the other ones of the form $4k+3$ are quadratic non-residues. In the remaining cases, a is a quadratic residue modulo 2^e, $e \geq 3$, if and only if $a \equiv 1 \pmod 8$.

A simple answer to question I can be given by listing all squares $1^2, 2^2, \dots, (\frac{1}{2}(p-1))^2$ Then all numbers that are congruent to one of these modulo p are quadratic residues modulo p, whereas all numbers that are not congruent to any of these squares are non-residues modulo p.

Beyond this simple approach, the following result, usually called Euler's criterion (see [12], [13]) proves to be very helpful.

Proposition 3.1. *If p is an odd prime number and a is an integer which is not divisible by p, then*

$$a^{\frac{1}{2}(p-1)} \equiv \begin{cases} 1 \pmod p & \text{if } aRp, \\ -1 \pmod p & \text{if } aNp. \end{cases} \tag{3.1}$$

Proof. By the result of Fermat, one of the two congruences as given always holds. In particular, given an integer a with aRp, there exists a natural number x with $x^2 \equiv a \pmod p$. It follows that $a^{\frac{1}{2}(p-1)} \equiv x^{p-1} \equiv 1 \pmod p$. The numbers $1^2, 2^2, \dots, \frac{1}{2}(p-1)$ are pairwise not congruent $\pmod p$ to one another. Numbers a which belong to one of their residue classes are solutions of the congruence $x^{\frac{1}{2}(p-1)} \equiv 1 \pmod p$. By Lagrange's theorem, the congruence has at most $\frac{1}{2}(p-1)$ solutions $\pmod p$. Therefore, if aNp then a is not a solution of the congruence, and it follows that $a^{\frac{1}{2}(p-1)} \equiv -1 \pmod p$. □

3.2. Law of quadratic reciprocity. One of the most significant findings in number theory is certainly the insight that the answers to questions I and II are related. More precisely, the solution to the problem of whether a prime number p is a quadratic residue or non-residue modulo a prime q depends on the inverse question of whether q is quadratic residue or non-residue modulo p. In 1744 Euler had already conjectured an early form of this relationship on the basis of extensive sample material, obtained in his work on divisors of quadratic forms (see [14]). It was only towards the end of his life that Euler was able to make a fully precise and complete conjecture about this problem [15], but he was unable to prove it.

A.M. Legendre dealt with the same subject in his excellent treatise *Recherches d'analyse indéterminée,* [32], completed later on in his *Essai sur la theorie des nombres*, published in 1798. In it, he arrived at a statement which is complete. He called it the "*loi de réciprocité*". Legendre also attempted a proof but it was incomplete because he assumed many unknown mathematical assertions. Finally, it was just Gauss who provided a complete proof and called it the fundamental theorem, in his Disquisitiones Arithmeticae, published in 1801.

The law of quadratic reciprocity states that, given two distinct odd primes p, q, if p or q is congruent to $1 \pmod 4$ then pRq if and only if qRp. If both primes are congruent to 3 $\pmod 4$ then pRq if and only if qNp.

The considerations involved in this first proof of the theorem are elementary. In 1854, G.L. Dirichlet notes "the consideration never leaves the field of second-degree congruences, to which the theorem essentially belongs" [6, p 139]). However, the proof is not concise, as it has to present extensive relations between different terms. Given the numerous distinctions between cases required in Gauss's argument, Dirichlet [6] revisits the proof using the notational convenience introduced by Legendre [33, p. 186] which is today called the Legendre symbol. This approach and some slight modifications in the argument significantly shorten the proof.

Already for Legendre, this symbol was very helpful. In the further development, the formulation of the fundamental theorem takes on a very elegant form through its use. Given a prime number p and an integer a not divisible by p, the Legendre symbol $\left(\frac{a}{p}\right)$ is defined to be 1 if a is a quadratic residue modulo p and to be -1 if a is a quadratic non-residue modulo p. Clearly, we have $\left(\frac{a}{p}\right) = \left(\frac{a'}{p}\right)$ if $a \equiv a' \pmod{p}$. In terms of the Legendre symbol Euler's criterion takes the form $a^{\frac{1}{2}(p-1)} = \left(\frac{a}{p}\right)$. Therefore one obtains as an immediate consequence of Euler's criterion the multiplicative property $\left(\frac{a}{p}\right) \cdot \left(\frac{b}{p}\right) = \left(\frac{ab}{p}\right)$ of the Legendre symbol. Since every integer has a prime factorisation $a = (\pm 1) \cdot 2^{n_0} \cdot q_1^{n_1} \cdots q_k^{n_k}$, $n_i \geq 0$, with positive odd prime numbers q_i, when dealing with questions I and II , one can therefore limit oneself to the cases $a = -1, 2$ or a is an odd prime number p.

In terms of the Legendre symbol, the law of quadratic reciprocity reads as the formula $\left(\frac{p}{q}\right) \cdot \left(\frac{q}{p}\right) = (-1)^{\frac{1}{4}(p-1)(q-1)}$.

Beyond the initial proof, Gauss provided seven further proofs of what he termed his fundamental theorem. These are characterised by very different methodological approaches: on the one hand, they elaborate on the core content of the theorem; on the other hand, they were also created in an effort to find a generalisation for congruences of the form $X^n \equiv a \pmod{m}$. Of particular significance is the second proof of the fundamental theorem published in the Disquisitiones Arithmeticae [19, Art. 262]. It is based on the theory of binary quadratic forms, of which Gauss provides a groundbreaking treatment.

For the discussion of Eisenstein's publication "A geometric proof of the fundamental theorem for quadratic residues", the proof of the fundamental theorem set out by Gauss in [21], usually labelled as his third proof, is of crucial importance as a point of reference.

4. Gauss's publication *Theorematis arithmetici demonstratio nova*

This approach to the proof of the fundamental theorem has two essential components. On the one hand, there is what is now known as Gauss's lemma. On the other hand, there is the introduction of the concept of the integral part of a real number, which is used rigorously in the argument.

4.1. Gauss's lemma. In our discussion of this result we closely follow Gauss's text. For any integer a and any natural number n we define the least residue of $a \pmod{n}$ as that integer a' for which $a \equiv a' \pmod{n}$ and $0 < a' \leq n$.

Lemma 4.1. *Let p be an odd prime, and let k be any natural number not divisible by p. We write the system of numbers* $1, 2, \ldots, p-2, p-1$ *as the disjoint union of the two systems A and B of numbers where*

$$A = \{1, 2, \ldots, (1/2)(p-1)\} \quad \textit{resp.} \quad B = \{(1/2)(p+1), \ldots, p-1\}.$$

Let k_j *be the least residue of* $jk \pmod{p}$ *for* $j = 1, 2, \ldots, \frac{p-1}{2}$*, and let* μ *denote the number of least positive integers* k_j *which belong to the system* B*. Then* k *is a quadratic residue (resp. non-residue)* $(\bmod\, p)$ *if and only if* μ *is even (resp. odd).*

Proof. Let $a_1, \dots, a_\lambda$ (resp. $b_1, \dots, b_\mu$) denote the least residues k_j, $j = 1, \dots, \frac{p-1}{2}$, which belong to the system A (resp. B). Then we have $\lambda + \mu = \frac{p-1}{2}$, and we note that the numbers $a_1, \dots, a_\lambda, p - b_1, \dots, p - b_\mu$ are pairwise distinct. Therefore, they range over the system of numbers $1, \dots, \frac{p-1}{2}$. Consequently, we have

$$1 \cdots \frac{p-1}{2} = a_1 \cdots a_\lambda \cdot (p - b_1) \cdots (p - b_\mu).$$

The right-hand side of this identity satisfies modulo p the congruence

$$\begin{aligned} a_1 \cdots a_\lambda \cdot (p - b_1) \cdots (p - b_\mu) &\equiv (-1)^\mu a_1 \cdots a_\lambda \cdot b_1 \cdots b_\mu \\ &\equiv (-1)^\mu k \cdot 2k \cdots \frac{p-1}{2} k \\ &\equiv (-1)^\mu k^{\frac{p-1}{2}} 1 \cdots \frac{p-1}{2} \qquad \pmod{p}. \end{aligned}$$

Therefore we derive the congruence $1 \equiv (-1)^\mu k^{\frac{p-1}{2}} \pmod{p}$ from the initial identity above. Taking into account Euler's criterion, this proves the claim. □

Gauss repeats this lemma and its proof at the beginning of his later publication [22].

4.2. The integral part of a real number. Let p be a prime, and let k be any natural number not divisible by p. Following Gauss, we denote by (k, p) the number of least residues k_j of $jk \pmod{p}$ for $j = 1, 2, \dots, \frac{p-1}{2}$ for which $k_j > \frac{1}{2}p$. For any non-integral real number x, let $[x]$ denote the unique largest integer such that $x - 1 < [x] \le x$. One calls $[x]$ the *integral part* of x, and one calls the difference $x - [x]$ the *fractional part* of x. It is written $\{x\}$ and satisfies $0 < \{x\} < 1$. Among the relations established by Gauss, we highlight the following, labelled III, $[x] + [h - x] = h - 1$ for any integer h. One also has

$$\text{IV} \quad \text{If } x - [x] \lneq \frac{1}{2}, \text{ then } [2x] - 2[x] = 0, \text{ and, if } x - [x] \ge \frac{1}{2}, \text{ then } [2x] - 2[x] = 1.$$

Now, if we take the prime number p into account in this consideration, these relations result in the formula

$$\text{VI } (k, p) = [\frac{2k}{p}] + [\frac{4k}{p}] + \cdots + [\frac{(p-1)k}{p}] - 2[\frac{k}{p}] - 2[\frac{2k}{p}] + \cdots + 2[\frac{\frac{1}{2}(p-1)k}{p}].$$

Subject to a wealth of transformations enforced by case-by-case distinctions, it forms the methodological tool used to prove the fundamental theorem. In 1884, by avoiding some of the technicalities, L. Kronecker published a simplified version [30] of this proof.

5. The young Gotthold Eisenstein

In July 1844, when Eisenstein wrote the work in question, he was 21 years old. He had been a student at Friedrich-Wilhelm-University in Berlin since 21 October 1843. Behind him lay an eventful and sometimes difficult childhood and time as a pupil, characterised by medical challenges and frequent changes of location.

Eisenstein was born on 16 April 1823 in Berlin, the first of six children in a family of merchants. His parents, Johann and Helene Eisenstein, had converted from Judaism to Protestantism prior to his birth. While his siblings died at a young age, he survived meningitis and other illnesses. However, Eisenstein suffered from poor health throughout his life. While he was still at secondary school, his teachers noticed his talent for mathematics and encouraged him to pursue his studies. From 1840 onwards, while still visiting the Friedrich-Wilhelms-Gymnasium, he attended lectures given by Dirichlet and others. Eisenstein's father moved to England in 1840 in search of better career opportunities. However, in the long run, these did not materialise. In the summer of 1842, Helene Eisenstein followed with her son, and they were forced to move between different parts of the country, including Wales and Ireland. According to his later biographical writings [35], he devoted himself intensively to mathematical questions during these months, focusing particularly on Gauss's Disquisitiones Arithmeticae.[2)]

In June 1843, Eisenstein returned to Berlin with his mother, and, on 22 September, he graduated as an external candidate from the Friedrich-Wilhelms-Gymnasium.[3)]

During his time in Dublin, Eisenstein encountered W.R. Hamilton, who entrusted him with a reprint [26] of an extensive treatise on a work by Abel, intended for the Berlin Academy. Eisenstein consigned this work, but included his own work about cubic forms in two variables, written in December 1843, in an accompanying letter, requesting permission to submit it to the Academy. The Academy agreed to this request and commissioned A.L. Crelle on 20 February 1844 to inform Eisenstein. Immediately afterwards, this work [7] appeared, along with others written by Eisenstein in December 1843, in the first issue of volume 27 of the *Journal für die reine und angewandte Mathematik*, published by Crelle. This first encounter marked the beginning of a long-lasting collaboration between the young student and scholar Eisenstein and Crelle. As an editor, Crelle paved the way for Eisenstein to publish his mathematical treatises.

Crelle's appreciation for Eisenstein is evident in his letter to the Prussian Minister of Culture, J.A.F. Eichhorn, dated 6 July 1844.

> "Recently, a new and extraordinary mathematical talent has emerged here in the person of the student Mr G. Eisenstein, whose abilities are so exceptional that it is perhaps not too much to hope that this young man will more or less replace Abel, who passed away at the age of 27 and to whom current mathematics owes so much, and who promised to become a mathematician of the first order. Although only just over 20 years old, Mr Eisenstein has already made several genuine discoveries in analysis, particularly in number theory, and has tackled the most difficult problems with an energy that is truly astonishing. He has published a series of works on the laws of reciprocity in real and complex numbers, on quadratic and cubic forms, on biquadratic residues, on cyclotomy, on elliptic and Abelian transcendents, etc., which contain many new things and, in some cases, break new ground." [4)]

[2)] In the years 1840 up to the summer of 1842 while he was still a pupil in the prima of the Friedrich-Werdersche-Gymnasium, Eisenstein came into contact with the Disquisitiones Arithmeticae and he later purchased the French edition *Recherches Arithmétiques* in 1842. This copy is now held in the archives of the University Library in Giessen. It is available at https://digisam.ub.uni-giessen.de/urn/urn:nbn:de:hebis:26-digisam-189157. In [39], Ullrich meticulously reviews the notes and annotations that Eisenstein wrote in his copy.

[3)] Eisenstein enclosed his curriculum vitae [35], written in August 1843, with his application for admission. We refer to [4] for a comprehensive biographical overview.

[4)] "Es hat sich nämlich so eben in der letzten Zeit hier in der Person des Studiosus Herrn G. Eisenstein ein neues ausserordentliches mathematisches Talent hervorgethan, von solcher ungewöhnlichen Kraft, dass es vielleicht selbst nicht zu viel ist, zu hoffen, dass dieser junge Mann werde den unvergesslichen, leider! schon in seinem 27.

At the same time, Crelle also advocated personally for financial support for Eisenstein.

> "However, Mr Eisenstein certainly needs external support, as his family is in very poor circumstances. His father has had to seek employment abroad for some time now, and his son lives here with his mother in poverty. His parents have done everything in their power to support him. Born here, he received his early education at the local secondary school and is now continuing his studies at the local university." [5)]

This later request was successful, as were others, such as those made by A. von Humboldt.

Thanks to the intervention of Gauss's student, J.F. Encke, and also Humboldt's recommendation, Eisenstein managed to visit Gauss in Göttingen in June 1844. Gauss was impressed by the young scholar, also by having examined the works published by Eisenstein in the first issues of volume 27 of Crelle's Journal. Gauss wrote to H.C. Schumacher on 7 July 1844 [24, vol. IV, pp. 265–266]:

> "The 27th volume of Crelle's Journal contains a number of essays by him, mostly on higher arithmetic, which, as far as I have been able to ascertain, are of varying value, but even the weakest among them, considering that they originate from a 21-year-old young man, give rise to favourable expectations, while some are of such a nature that a master of the first rank would not have to disavow them. I expect great things from him." [6)]

The extent to which the discussions with Gauss influenced Eisenstein's works, including the one treated here, is unclear. At the very least, however, Gauss's publication *Theorematis arithmetici demonstratio nova* is of great significance for Eisenstein's work.

6. The paper by G. Eisenstein

Eisenstein's approach to proving the fundamental theorem is geometric in nature. However, it also bears a strong formal resemblance to the line of argument taken by Gauss in his work *Theorematis arithmetici demonstratio nova.*

Lebensjahre dahingeschiedenen Abel, dem die neuere Mathematik schon so Bedeutendes verdankt und der ein Mathematiker erster Grösse zu werden versprach, mehr oder weniger ersetzen. Herr Eisenstein hat schon, obgleich erst wenig über 20 Jahre alt, mehrere wirkliche Entdeckungen in der Analysis gemacht, besonders in der Theorie der Zahlen, und ist in die schwierigsten Aufgaben mit einer Energie gedrungen, welche Staunen erregt. Er hat über die Gesetze der Reciprocität bei reellen und complexen Zahlen, über die quadratischen und cubischen Formen, über die biquadratischen Reste, über die Kreistheilung, über die elliptischen und Abelschen Transcendenten u.s.w. eine Reihe von Abhandlungen geliefert, die gar manches Neue enthalten und zum Theil auf fast unbetretenen Wegen vordringen." (see [3])

5) "Nun bedarf aber auch Herr Eisenstein ganz sicher der äussern Unterstützung: denn seine Familie befindet sich in sehr dürfthigen Umständen. Sein Vater hat müssen schon seit länger im Auslande seinen Unterhalt suchen, und der Sohn lebt hier mit seiner Mutter im Mangel. Seine Eltern haben nach ihren Kräften an ihn gewendet. Hier geboren, hat er seine Vorbildung auch am hiesigen Gymnasium erhalten und setzt jetzt seine Studien auf der hiesigen Universität fort." (see [3])

6) "Der 27. Band des Crelle'schen Journals enthält eine Menge von Aufsätzen von ihm, grösstentheils zur höhern Arithmetik angehörig, die allerdings, so weit ich bisher sie habe beachten können, von ungleichem Wert sind, aber auch die schwächsten darunter, würden in Erwägung, dass sie von einem 21jährigen jungen Menschen herrühren, günstige Erwartungen erregen, während manches der Art ist, dass der erste Meister es nicht zu desavouiren haben würde. Ich verspreche mir grosses von ihm."

The first step involves reducing the determination of the Legendre symbol $\left(\frac{q}{p}\right)$, q, p odd primes, $q \neq p$, to the calculation of a sum of quantities modulo 2. This sum is given as $\sum[\frac{qa}{p}]$ where a ranges over the even numbers $a = 2, 4, 6, \dots, p-1$. This is similar to the Gauss Lemma but, as Eisenstein remarks in a letter, probably written in July 1844, to M.A. Stern:

> "I did not rest until I had freed my geometric proof of the law of reciprocity, which you enjoyed so much and which, incidentally, Jacobi also finds extremely pleasing, from the lemma on which it was still dependent, and it is now so simple that it can be communicated in a few lines. The main difference between my approach and Gauss's is that I do not divide the numbers $< p$ into those $< \frac{p}{2}$, and those $\geq \frac{p}{2}$ as Gauss does, but into *even* and *odd* numbers." [7)]

The second step in his approach, also briefly outlined in the aforementioned letter to Stern, is achieved by transforming the problem of determining $\sum[\frac{qa}{p}]$ into one of counting lattice points in the standard grid in the plane. Eisenstein writes:

> "Now imagine a rectangular coordinate system (x, y) in the plane and the entire plane divided into squares of equal dimensions $= 1$ by parallels to the axes at distances $= 1$ from each other. All corner points of squares that do not lie on the two coordinate axes will be called *lattice points.* (Pl. II, Fig. 1.2.)" [8)]

The crucial question of whether r, the remainder in the equation $z = [z] + r$ where $z = \frac{qa}{p}$, $a = 2, 4, 6, \dots, p-1$, is greater than or less than $\frac{1}{2}$ becomes a question about the position of z in the lattice relative to the straight line whose equation is $y = \frac{q}{p}x$.

We closely follow Eisenstein's notation and its occasional abuse. In particular, in contrast to Gauss in [21], he writes, given any real number x, $E(x)$ to denote the largest integer such that $x - 1 < E(x) \leq x$. This notation has been on occasion used by later writers, see e.g. [38].

6.1. Eisenstein's lemma.

Lemma 6.1. *Let p be a positive odd prime,* **a** *the system of all even numbers* $< p$ *and* > 0, *i.e.* $\mathbf{a} = \{2, 4, 6, \dots, p-1\}$. *Let q be any integer not divisible by the modulus p. Then we have* $\left(\frac{q}{p}\right) = (-1)^{\sum_{a \in \mathbf{a}} E(\frac{qa}{p})}$.

Proof. Given $a \in \mathbf{a}$, we have $qa = pE(\frac{qa}{p}) + r_a$ where r_a denotes the least positive residue of $qa \pmod p$. Then the quantities $(-1)^{r_a} r_a$, $a \in \mathbf{a}$, have even least positive residue, and they are all distinct. Indeed, suppose $(-1)^{r_a} r_a \equiv (-1)^{r'_a} r_{a'} \pmod p$ for $a, a' \in \mathbf{a}$, $a \neq a'$. Then we have $(-1)^{r_a} qa \equiv (-1)^{r'_a} qa' \pmod p$, hence, $a \equiv \pm a' \pmod p$. Since $a \neq a'$, the congruence $a + a' \equiv 0 \pmod p$ has to hold. However, this contradicts the fact that

7) "Ich habe nicht eher geruht, als bis ich meinen geometrischen Beweis des Reciprocitätsgesetzes, der Ihnen so viel Spass gemacht hat, und der auch, beiläufig gesagt, Jacobi ausserordentlich gefällt, von dem Lemma befreit habe, von dem er noch abhängig war, und er ist jetzt so einfach, dass er sich in ein paar Zeilen mittheilen lässt. Der Hauptunterschied zwischen meinem Gange und dem Gaussischen besteht darin, dass ich nicht wie Gauss die Zahlen $< p$, in solche $< \frac{p}{2}$, und in solche $\geq \frac{p}{2}$ theile, sondern in *gerade* und *ungerade*." [27, pp. 173–174]

8) "Man stelle sich jetzt in der Ebene ein rechtwinkliges Coordinattensystem (x, y) und die ganze Ebene durch Parallelen mit den Axen in den Abständen $= 1$ von einander in lauter Quadrate von den Dimensionen $= 1$ getheilt vor. *Gitterpuncte* sollen alle Eckpuncte von Quadraten heissen, welche nicht in den beiden Coordinaten-Axen liegen (Taf. II.Fig. 1.2.)" [9]

$a, a' \lneq p$. Consequently, up to multiples of p, the quantities $(-1)^{r_a} r_a, a \in \mathbf{a}$, coincide with the numbers in $\mathbf{a}$. Therefore one has the two congruences

$$q^{\frac{1}{2}(p-1)} \Pi_{a \in \mathbf{a}} a \equiv \Pi_{a \in \mathbf{a}} r_a \pmod{p}, \quad \text{and} \quad \Pi_{a \in \mathbf{a}} a \equiv (-1)^{\Sigma r_a} \Pi_{a \in \mathbf{a}} r_a \pmod{p},$$

from which it follows that

$$q^{\frac{1}{2}(p-1)} \equiv (-1)^{\Sigma_{a \in \mathbf{a}} r_a} \pmod{p}, \quad \text{thus} \quad \left(\frac{q}{p}\right) = (-1)^{\Sigma_{a \in \mathbf{a}} r_a}.$$

The final step of the conclusion is based on Euler's criterion. Since $p \equiv 1 \pmod 2$, and $a \equiv 0 \pmod 2$, the identity $qa = pE(\frac{qa}{p}) + r_a$ implies that we can replace the exponent in the last formula by $\Sigma_{a \in \mathbf{a}} E(\frac{qa}{p})$. □

As Eisenstein points out, this result allows one to decide whether 2 is a quadratic residue $\pmod p$ or not. One obtains that 2 is a quadratic residue modulo all primes $\equiv \pm 1 \pmod 8$ and a quadratic non-residue modulo all primes $\equiv \pm 3 \pmod 8$. To arrive at this result one has to determine the number, say κ, of a's in the system $\mathbf{a}$ such that $2a > p - 1$. Since a is even it suffices to count the number of quantities $2 \cdot 1, 2 \cdot 2, \ldots, \frac{1}{2}(p-1)$ that exceed $\frac{1}{2}(p-1)$. For $1 \leq j \leq [\frac{1}{4}p]$ we have $2j \leq \frac{1}{2}(p-1)$, and for $[\frac{1}{4}p] \lneq j \leq \frac{1}{2}(p-1)$ we have $2j \gneq \frac{1}{2}(p-1)$. Therefore we see $\Sigma_{a \in \mathbf{a}} E(\frac{2a}{p}) = \kappa = \frac{1}{2}(p-1) - [\frac{1}{4}p]$. One readily checks the claim.

6.2. An easy transformation. In dealing with the final formula in Lemma 6.1, in the case of an odd prime number q, Eisenstein modifies the exponent modulo 2 by means of an easy transformation as follows, without giving any evidence, however:

$$\begin{aligned} \Sigma_{a \in \mathbf{a}} E(\frac{qa}{p}) &\equiv -E(\frac{q}{p}) + E(\frac{2q}{p}) - E(\frac{3q}{p}) \cdots \pm E(\frac{\frac{1}{2}(p-1)q}{p}) \\ &\equiv E(\frac{q}{p}) + E(\frac{2q}{p}) + E(\frac{3q}{p}) \cdots + E(\frac{\frac{1}{2}(p-1)q}{p}) \pmod 2. \end{aligned}$$

If the latter sum is denoted by μ, then $\left(\frac{q}{p}\right) = (-1)^{\mu}$. Eisenstein reverses the previous use of system $\mathbf{a}$ – the even numbers $2, 4, 6, \ldots, p-1$ – and reverts to the parameterisation, given as $1, 2, \ldots, \frac{1}{2}(p-1)$, as in Gauss.

The following could be the argument for this crucial conclusion: The identity III for the largest integer function $E(x)$, denoted $[x]$ by Gauss [21], states that for any integer h, $E(x) + E(h-x) = h - 1$. In particular, if h is an odd integer, we have $E(x) \equiv E(h-x) \pmod 2$. Therefore we have for any pair of odd primes $p, q, p \neq q$

$$E(\frac{2hq}{p}) \equiv E(\frac{(p-2h)q}{p}) \pmod 2 \text{ where } h \text{ is any integer.} \tag{6.1}$$

Now we let h range over the system $1, 2, \ldots, \frac{p-1}{2}$. If $h \lneq \frac{p}{4}$, the numbers $2h$ run through all even integers $\lneq \frac{p}{2}$, and if $h \gneq \frac{p}{4}$, thus, $2h \gneq \frac{p}{2}$, the numbers $p - 2h$ range over all odd integers $\lneq \frac{p}{2}$. Since $a \in \mathbf{a}$ runs through the even numbers, we use the left side of the congruence for $h \lneq \frac{p}{4}$ and the right side for $h \gneq \frac{p}{4}$. This gives us the congruence[9)]

$$\Sigma_{a \in \mathbf{a}} E(\frac{aq}{p}) = \Sigma_{h=1}^{\frac{p-1}{2}} E(\frac{2hq}{p}) \equiv \Sigma_{h=1}^{\frac{p-1}{2}} E(\frac{hq}{p}) \pmod 2.$$

9) In [31], however, this crucial transformation is only discussed in terms of its geometric disguise, which only becomes visible once the lattice points have been counted.

6.3. The geometric approach. To determine the exponent $\mu = \Sigma_{h=1}^{\frac{p-1}{2}} E(\frac{hq}{p}) \pmod 2$ in the formula $\left(\frac{q}{p}\right) \equiv (-1)^{\mu}$, Eisenstein now adopts a geometrically inspired interpretation of the quantity $E(\frac{hq}{p})$, which is defined as the integral part of $\frac{hq}{p}$ where $h = 1, 2, \ldots, \frac{p-1}{2}$. As the latter depend on the integral parameter h, it is natural to view them as points in the standard lattice in the plane formed by all pairs (r, s) with integers r, s. Using current notation, we denote this lattice by $\mathbb{Z}^2$.

Given any curve $y = \phi(x)$ in the plane, parametrised by the coordinates x, y, the number of lattice points between a given integral point x_0 and the point $y_0 = \phi(x_0)$ on the curve is given by $E(y_0)$. By summing over all integer coordinates, Eisenstein obtains an expression which specifies the number of lattice points that lie between this curve and the x-axis, including those lattice points that happen to lie on the curve itself.

Now, this idea is applied to the straight line ϕ, given by the equation $y = \frac{q}{p}x$. In order to determine the quantity $E(\frac{qh}{p})$ we must restrict ourselves to the parameters $h = 1, 2, \ldots, \frac{p-1}{2}$. From our geometric perspective, this corresponds to counting the lattice points in the standard lattice within the region enclosed by the sides AB and AD, up to and including the ordinate CG. Note that, since p and q are relatively prime there are no lattice points on the straight line ϕ. The result is the expression $\Sigma_{h=1}^{\frac{p-1}{2}} E(\frac{hq}{p})$ which coincides, as proved before, with $\mu \pmod 2$. Therefore, one obtains $\left(\frac{q}{p}\right) = (-1)^{\mu}$.

Next, if we change our perspective on the geometric interpretation by considering the straight line $x = \frac{p}{q}y$, that is, if we swap the roles of the coordinates, we obtain $\left(\frac{p}{q}\right) = (-1)^{\nu}$, where ν is the number of lattice points in the the region enclosed by the sides AB and AF, up to and including the abscissa EG. Now, as Eisenstein writes in conclusion of his line of arguments,

> "However, it is evident that all lattice points marked with $*$ and all those marked with 0, i.e. all lattice points to the *right* and all lattice points to the *left* of AB, taken together exhaust *all* lattice points of the rectangle $AEGC$, of which there are $\frac{1}{2}(p-1) \cdot \frac{1}{2}(q-1)$, thus, $\mu + \nu = \frac{1}{2}(p-1) \cdot \frac{1}{2}(q-1)$, and
>
> $$\left(\frac{q}{p}\right)\left(\frac{p}{q}\right) = (-1)^{\mu+\nu} = (-1)^{\frac{1}{2}(p-1)\cdot\frac{1}{2}(q-1)};$$
>
> which was to be proven".

The congruence modulo 2 of the exponents underlying this result was announced by Eisenstein in a generalised form for arbitrary integers M, N, $(M, N) = 1$ some time earlier as item V in a list of exercises and theorems [8]. A non-geometric proof was already given in [21, Sect. 6].

Within the context of Eisenstein's geometric approach, the following paragraph indicates an intuitive explanation of the congruence $\Sigma_{a \in \mathbf{a}} E(\frac{aq}{p}) = \Sigma_{h=1}^{\frac{p-1}{2}} E(\frac{2hq}{p}) \equiv \Sigma_{h=1}^{\frac{p-1}{2}} E(\frac{hq}{p}) \pmod 2$, called "an easy transformation" by Eisenstein. The more formal details assigned to the reader by him can be found here in Section 6.2.

Indeed, as Eisenstein points out, given $a \in \mathbf{a} = \{2, 4, 6, \ldots, p-1\}$, the term $E(\frac{aq}{p})$ stands for the number of lattice points in the column above the ordinate a between AB and AD. Since $p \neq q$ are unequal primes, $E(\frac{aq}{p})$ cannot lie on the straight line AB. The total number of lattice points in the column above $a \in \mathbf{a}$ between the lines AD and FB is $q - 1$, thus, an even number. Consequently, the number of lattice points on this column below the straight line AB has the same parity as the number of lattice points on this column above AB.

The rectangle $ADBF$ has centre $Z = (\frac{p}{2}, \frac{q}{2})$, which lies on the straight line AB. We choose the points $G = (\frac{p}{2}, 0)$ and $H = (\frac{p}{2}, q)$. Rotating the triangle AGZ, say Δ, clockwise around the point Z by 180 degrees results in a congruent right-angled triangle ZHB, denoted Δ' with hypotenuse ZB. Given $a \in \mathbf{a}$ with $a \gneq \frac{p}{2}$, we see $p - a$ is an odd number with $p - a \lneq \frac{p}{2}$. The lattice points in the column above $p - a$ that lie in Δ correspond one-to-one with the lattice points in the column above a that lie in the triangle Δ'. In particular, the parities in question coincide. This is the geometric explanation for the "easy transformation".

6.4. Eisenstein's note. Eisenstein concludes his treatise with a note pointing out that simple formulas can be used to determine the number of lattice points in figures. As an example, he selects the question of how many lattice points are enclosed by a circle with a radius of $\sqrt{m}$, including those that lie on the axes. He states without proof that this number is given as

$$S = 1 + 4(E(m) - E(\frac{1}{3}m) + E(\frac{1}{5}m) - E(\frac{1}{7}m) + \cdots).$$

The formula can be explained as follows: To begin with, given a natural number n, we have to decide whether it can be expressed in the form $n = a^2 + b^2$ for some integers a, b. If $n \gneq 1$, as already known to Fermat for n a prime, this is the case if and only if every prime divisor p of n with $p \equiv 3 \pmod 4$ only occurs to an even power in the prime decomposition of n. Note that a number n may have different representations as a sum of two squares. In particular, representations that only differ in the order of summands or in the signs of a and b have to be viewed as different.

In other words, we have to know the number $\rho_2(n)$ of lattice points (x, y) in the standard lattice $\mathbb{Z}^2$ in the plane whose coordinates satisfy $x^2 + y^2 = n$. As proved by Jacobi [28][10)], $\rho_2(n) = 4(d_1(n) - d_3(n))$ where $d_1(n)$ (resp. $d_3(n)$) denotes the number of divisors of n which are congruent to 1 (resp. 3) $\pmod 4$. Next, we have to form the sum over all values $r_2(n)$ where n ranges over all $n \leq m$. We rearrange this summation: we first determine the number, say $p_1(n)$, of all prime divisors of the form $4k+1$ of all numbers $n \leq m$, then we subtract the number $p_3(n)$ of all prime divisors of the form $4k+3$ of all numbers $n \leq m$. In the first case, we get the finite sum,

$$p_1(n) = E(m) + E(\frac{1}{5}m) + E(\frac{1}{9}m) + \ldots \tag{6.2}$$

and in the second case, we have the finite sum

$$p_3(n) = E(\frac{1}{3}m) + E(\frac{1}{7}m) + E(\frac{1}{11}m) + \ldots. \tag{6.3}$$

Therefore we obtain as the number of lattice points which are enclosed by a circle with a radius of $\sqrt{m}$, including those that lie on the axes, the value $1 + 4(p_1(n) - p_3(n))$. This implies the formula for S as stated.

Via multiplication by the factor $\frac{1}{m}$ and letting the parameter m go to infinity, one derives an infinite alternating series, namely $\frac{\pi}{4} = 1 - \frac{1}{3} + \frac{1}{5} - \frac{1}{7} + \frac{1}{9} - \cdots$, known as the Leibniz formula for π.

10) Eisenstein noted this result in the margin of page 157 in his copy of the Disquisitiones Arithmeticae (see [39])

7. Two concluding remarks

Volumes 27 and 28 of Crelle's Journal, published in 1844, contain 28 of Eisenstein's publications. The quality of these publications varies, and their topics range from cubic forms and the fundamental theorem to reciprocity laws for cubic and biquadratic residues, cyclotomy and elliptic functions. This already indicates Eisenstein's main interests in the years that followed. Although Eisenstein was now considered a promising scientist, his career as a scholar never really took off. Despite receiving numerous honours from the scientific community, he did not succeed in obtaining a professorship. From August 1845 onwards, mental and health problems became increasingly prevalent and hampered his work. In retrospect, Eisenstein cannot point to a rich treasure trove of great mathematical results, nor is he the creator of far-reaching structural theories. His significance lies in examining questions and known results from new perspectives. By doing so, he paved the way for wide-ranging developments.

Loneliness, disputes over priorities and increasing health problems characterised the last years of Eisenstein's life. The number of his publications was decreasing, but their content was of a high quality. As well as addressing questions in number theory and the theory of algebraic functions they also covered reduction theory of positive ternary forms. On 11 October 1852, Eisenstein passed away from a progressive pulmonary disease, probably tuberculosis. He had already suffered a pulmonary haemorrhage during the summer.

The short note by Eisenstein "A geometric proof of the fundamental theorem of quadratic residues" does not present any new deep mathematical results or develop any far-reaching theories. Instead, it provides an ingenious geometric proof of the *Theorema fundamentale residuorum quadraticorum* whose proof had been published by Gauss in 1801. Gauss's original proof in [21], characterised by the use of the largest integer function and laborious case distinctions, is simplified to a mere count of integral points in a lattice. Eisenstein's work is one of many attempts to get to the heart of what the law of reciprocity is all about. It was highly influential and served as an impetus for further papers by others.

One of these later papers is "Über das quadratische Reziprozitätsgesetz II", by G. Frobenius, published in 1914 [17], a sequel to [16] which also concerns quadratic reciprocity. Frobenius was inspired to pursue Eisenstein's geometric approach to the fundamental theorem after reading a highly admirable method of counting points due to Pastor Zeller, presented to the Academy by L. Kronecker on 16 December 1874 [40]. This method was characterised by a principle of symmetry which ultimately helped to simplify the proof. Frobenius writes:

> "Proofs of the reciprocity law based on Gauss's Lemma require the counting of lattice points, which can be achieved by cutting and reassembling figures in a manner similar to the numerous proofs of Pythagoras' Theorem. Zeller's (Monatsber. 1874, p. 846) direct approach to this counting has rightly aroused the greatest admiration. However, his conclusions can be further simplified by applying this method consistently and making more use of his principle of symmetry." [11)]

[11)] "Die vom Gaussischen Lemma ausgehenden Beweise des Reziprozitätsgesetzes erfordern eine Abzählung von Gitterpunkten, die sich durch Zerschneiden und Zusammensetzen von Figuren in ähnlicher Weise ausführen lässt wie die zahlreichen Beweise des pythagoreischen Lehrsatzes. Die grösste Bewunderung hat mit Recht die Art erregt, wie Zeller (Monatsber. 1874, S. 846) diese Abzählung ganz direkt ausführt. Aber auch seine Schlüsse lassen sich noch durch passende Anwendung jenes Verfahrens und konsequentere Benutzung seines Symmetrieprinzips vereinfachen." [16, p. 628]

Frobenius succeeded in clearly identifying the symmetrical aspects of Zeller's approach and presenting them in a geometric form. However, it should not be overlooked that a certain idea of symmetry is already contained in Eisenstein's geometric approach.

References

[1] *E. Artin*, Beweis des allgemeinen Reziprozitätsgesetzes, Abh. Math. Semin. Univ. Hambg. **5** (1927), 353–363.

[2] *E. Artin* and *J. Tate*, Class Field Theory, Benjamin, New York-Amsterdam, 1967.

[3] *K.R. Biermann*, A.L. Crelles Verhältnis zu G. Eisenstein, Mon.Ber.Dt. Akad. Wiss. Berlin, **1** (1959), 67–72.

[4] *K.R. Biermann*, Gotthold Eisenstein–Die wichtigsten Daten seines Lebens, J. reine angew. Math. **214/15** (1964), 19–30;

[5] *J.W. Cogdell*, L-functions and non-abelian class field theory, from Artin to Langlands, In: D. Dumbaugh, J. Schwermer, Emil Artin and beyond–class field theory and L-functions, Zürich: European Math. Society 2015, 127–161.

[6] *G.L. Dirichlet*, Über den ersten der von Gauss gegebenen Beweise des Reciprocitätsgesetzes in der Theorie der quadratischen Reste, J. reine angew. Math. **47** (1854), 139–150.

[7] *G. Eisenstein*, Théoremes sur les formes cubiques et solution d'une équation du quatrième degré à quatre indéterminées, J. reine angew. Math. **27** (1844), 75–79.

[8] *G. Eisenstein*, Théorèmes arithmetiques, J. reine angew. Math. **27** (1844), 281.

[9] *G. Eisenstein*, Geometrischer Beweis des Fundamentaltheorems für die quadratischen Reste, J. reine angew. Math. **28** (1844), 246–248; Math. Werke I, 164-166; Engl. Transl. Quart. J. Math. 1 (1857), 186-191.

[10] *G. Eisenstein*, Mathematische Werke. AMS Chelsea Publishing, Providence 1975, 2nd Edition 1989..

[11] *L. Euler*, Opera Omnia / Series prima, Birkhäuser, Basel 1992.

[12] *L. Euler*, Theoremata circa divisores numerorum, Opera Omnia I-2, 62–85.

[13] *L. Euler*, Theoremata circa residua ex divisione potestatum relicta, Opera Omnia I-2, 493–518.

[14] *L. Euler*, Theoremata circa divisores numerorum in hac forma $pa^2 \pm qb^2$ contentorum, Comm. Acad. Sc. Petersburg 14 (1744/46), 151–181; Opera Omnia I-2, 194–222.

[15] *L. Euler*, Observationes circa divisionem quadratorum per numeros primes, Opera Omnia I - 3 (1783), 477–512.

[16] *G. Frobenius*, Über das quadratische Reziprozitätsgesetz, Sitzungsberichte Kgl. Preuss. Akad. Wiss. Berlin, Berlin (1914), 335–349; Ges. Abhandl. III, 628–642.

[17] *G. Frobenius*, Über das quadratische Reziprozitätsgesetz II, Sitzungsberichte Kgl. Preuss. Akad. Wiss. Berlin, Berlin (1914), 484–488; Ges. Abhandl. III, 643–647.

[18] *G. Frobenius*, Gesammelte Abhandlungen, ed. J-P. Serre, Springer, Berlin–Heidelberg, 1968.

[19] *C.F. Gauss*, Disquisitiones Arithmeticae, Leipzig 1801.

[20] *C.F. Gauss*, Werke, Kgl. Gesell. Wiss. Göttingen, Vol. II, Göttingen 1863.

[21] *C.F. Gauss*, Theorematis arithmetici demonstratio nova, Comment. Soc. regiae sci. Göttingen XVI (1808), 69; Werke II, pp. 1–8.

[22] *C.F. Gauss*, Theorematis fundamentalis in doctrina de residuis quadraticis demonstrationes et ampliationes novae. Commentationes societatis regiae scientiarium Gottingensis recentiores IV (1818),; Werke II, pp. 47–64.

[23] *C.F. Gauss*, Recension der "Untersuchungen über die Eigenschaften der positiven ternären quadratischen Formen von Ludwig August Seeber", Göttingsche Gelehrte Anzeigen, Juli 9, 1831, pp. 1065. Repr: J. reine angew. Math. **20** (1840), 312–320,; Werke II, pp. 188–196

[24] *C.F. Gauss* and *H.C. Schuhmacher*, Briefwechsel zwischen C. F. Gauss und H. C. Schumacher, ed. by C.A.F. Peters, Gustav Esch: Altona 1860–1865; reprint: Georg Olms: Hildesheim, New York 1975.

[25] *C. Goldstein, N. Schappacher* and *J. Schwermer*, The Shaping of Arithmetic after C.F. Gauss's Disquisitiones Arithmeticae, Springer, Berlin-Heidelberg 2007.

[26] *W.R. Hamilton*, On the argument of Abel, respecting the impossibility of expressing a root of any general equation above the fourth degree, by any finite combination of radicals and rational functions, Transact. of the Royal Irish Academy, **18**, Part II, (1839), 171–259.

[27] *A. Hurwitz* and *F. Rudio*, Briefe von G. Eisenstein an M.A. Stern, Zeitschrift für die Mathematik und Physik **40** (1895), Supplement: Abh. Geschichte d. Mathematik, issue VII, 169–204.

[28] *C.G. Jacobi*, Fundamenta nova theoriae functionum ellipticarum, Regiomonti, Borntaeger 1829, In: Ges. Werke, vol.1, 49–239.

[29] *C.G. Jacobi*, Gesammelte Werke, Cambridge U. Press, Cambridge, 2013.

[30] *L. Kronecker*, Der dritte Gausssche Beweis des Reciprocitätsgesetzes für die quadratischen Reste, in vereinfachter Darstellung, J. reine angew. Math. **97** (1884), 93–94.

[31] *R.C. Laubenbacher* and *D.J. Pengelley* Eisenstein's misunderstood geometric proof of the quadratic reciprocity theorem. The College Maths. J. **25** (1994), 29–34.

[32] *A.M. Legendre*, Recherches d'analyse indeterminee, Histoire de l'Academie Royale des Sciences de Paris (1785), 465-559, Paris 1788.

[33] *A.M. Legendre*, Essai sur la theorie des nombres, 1st ed. Paris 1798, 2nd ed. Paris 1808, 3rd ed. Paris 1830; German transl. (Zahlentheorie) of the 3rd. ed., Leipzig 1886; Quatrieme edition conforme ala troisieme, nouveau tirage corrige, Paris 1955.

[34] *F. Lemmermeyer*, Reciprocity Laws: from Euler to Eisenstein, Springer Monographs in Maths., Berlin-Heidelberg 2000.

[35] *F. Rudio*, Eine Autobiographie von Gotthold Eisenstein. Mit ergänzenden biographischen Notizen. Zeitschrift für die Mathematik und Physik **40** (1895), Supplement: Abh. Geschichte d. Mathematik, VII, 143–168.

[36] *N. Schappacher*, Gotthold Eisenstein, 16 April 1823–11 October 1852, In: H.G.W. Begehr, H. Koch, J. Kramer, N. Schappacher, E.-J. Thiele (Ed.) Mathematics in Berlin, Berlin-Basel-Boston: Birkhäuser 1998, pp. 55–60.

[37] *J. Schwermer*, Reduction theory of quadratic forms: toward *Räumliche Anschauung* in Minkowki's early work. In: [25], 483–504.

[38] *M.A. Stern*, Ueber einige Eigenschaften der Function Ex, J. reine. angew. Math. **59** (1861), 146–162.

[39] *P. Ullrich*, Über das Exemplar der "Disquisitiones Arithmeticae" aus dem Besitz von Gotthold Eisenstein, Mitt. Math. Ges. Hamb. **21** (2002), 35–58.

[40] *Ch. Zeller* Untitled, Monatsberichte Kgl. Preuss. Akad. Wiss. Berlin, Berlin (1872), 846–847.

Joachim Schwermer, Faculty of Mathematics, University of Vienna, Oskar-Morgenstern-Platz 1, A-1090 Vienna, Austria
e-mail: Joachim.Schwermer@univie.ac.at

1851–1875

Ueber diejenigen Fälle, in welchen die *Gauss*ische hypergeometrische Reihe eine *algebraische* Function ihres vierten Elementes darstellt.

Nebst zwei Figurentafeln.

(Von Herrn *H. A. Schwarz* in Zürich.)

Die nachfolgende Abhandlung beschäftigt sich mit der Aufgabe:

Alle Fälle zu ermitteln, in denen der linearen Differentialgleichung zweiter Ordnung

$$(\mathrm{A.})\qquad \frac{d^2y}{dx^2}+\frac{\gamma-(\alpha+\beta+1)x}{x(1-x)}\cdot\frac{dy}{dx}-\frac{\alpha\cdot\beta}{x(1-x)}\cdot y = 0,$$

von welcher die Gaussische hypergeometrische Reihe $F(\alpha,\beta,\gamma,x)$, als Function ihres vierten Elementes betrachtet, ein particuläres Integral ist, durch eine algebraische Function von x genügt werden kann.

Die Einleitung der *Gauss*ischen Abhandlung „Disquisitiones generales circa seriem infinitam" enthält die Angabe einiger Fälle, in denen für specielle Werthe von α, β, γ die Function $F(\alpha,\beta,\gamma,x)$ eine algebraische Function von x ist. (Werke Bd. III., pag. 127, Formel I.—V.)

Eine Programm-Abhandlung des Herrn *Kummer*, Osterprogramm 1834 des Gymnasiums zu Liegnitz: „De generali quadam aequatione differentiali tertii ordinis" enthält die Entwickelung einer Methode, hypergeometrische Reihen mit einander zu vergleichen, deren letzte Elemente durch eine gewisse Differentialgleichung dritter Ordnung von einander abhängen. Die Bedeutung dieser in mehr als einer Hinsicht bemerkenswerthen Differentialgleichung für die allgemeine Theorie der hypergeometrischen Reihe ist bekannt. Ein specieller Fall dieser Differentialgleichung, zu welchem eine von dem Grundgedanken jener Methode etwas verschiedene Gedankenverbindung führt, ist für die Lösung des hier aufgestellten Problems von wesentlichem Einflusse.

Die erwähnte *Gauss*ische Abhandlung, die *Kummer*sche Abhandlung: „Ueber die hypergeometrische Reihe", Bd. 15 dieses Journals, die Abhandlung *Riemanns*: „Beiträge zur Theorie der durch die *Gauss*ische Reihe $F(\alpha,\beta,\gamma,x)$ darstellbaren Functionen" (Abhandlungen der Königl. Gesellschaft der Wissenschaften zu Göttingen, Bd. 7) und die nachgelassene Abhandlung von *Gauss* (Werke Bd. III., pag. 207) bieten verschiedene Mittel dar, um eine grössere

J. reine angew. Math. **75** (1873), 292–335
DOI 10.1515/crll.1873.75.292

Journal für die reine und angewandte Mathematik

On the cases when the Gaussian hypergeometric series represents an algebraic function of its fourth element

With two plates

By *H. A. Schwarz* at Zürich

The following memoir deals with the question:

To determine all cases in which the second order linear differential equation

$$(A.)\quad \frac{d^2y}{dx^2} + \frac{\gamma-(\alpha+\beta+1)x}{x(1-x)}\cdot\frac{dy}{dx} - \frac{\alpha\cdot\beta}{x(1-x)}\cdot y = 0,$$

of which the Gaussian hypergeometrical series $F(\alpha,\beta,\gamma,x)$, considered as a function of its fourth element, is a particular integral, can be satisfied by an algebraic function of x.

The introduction of *Gauss*'s memoir, "Disquisitiones generales circa seriem infinitam", contains the indication of some cases in which, for special values of α,β,γ, the function $F(\alpha,\beta,\gamma,x)$ is an algebraic function of x. (Werke, vol. III, p. 127, Formulas L–V.)

A program paper of Mr. *Kummer* (Easter-Program 1834 of the Gymnasium at Liegnitz): "De generali quadam aequatione differentiali tertii ordinis" contains the development of a method for comparing hypergeometric series, whose last arguments are related through a certain third-order differential equation. The importance of this differential equation, which is remarkable in more than one respect for the general theory of hypergeometric series, is well known. A connection of ideas somewhat different from the basic idea of that method leads to a special case of this differential equation, which is of essential importance for the solution of the problem posed here.

Gauss's memoir mentioned above, *Kummer*'s memoir: "Über die hypergeometrische Reihe", vol. 15 of this journal, *Riemann*'s memoir "Beiträge zur Theorie der durch die Gaussche Reihe $F(\alpha,\beta,\gamma,x)$ darstellbaren Functionen" (Abhandlungen der Königl. Gesellschaft der Wissenschaften zu Göttingen, vol. 7) and the posthumous *Gauss*'s memoir (Werke, vol. III., p. 207) offer different means to find a larger number of special cases, in which $F(\alpha,\beta,\gamma,x)$ represents an algebraic function of x; the path of investigation taken in these memoirs, however, does not seem to provide a general method by which all cases of this kind could be determined.

In the task in question two cases must be distinguished: Firstly: the differential equation (A) has *one* particular integral which is an algebraic function of x; but this algebraic function is such that itself or its logarithmic derivative is a rational function of x. In this case, the quotient of two different branches of this algebraic function is a quantity which is independent of x, thus constant, and the existence of a second algebraic integral of the differential equation which is linearly independent from the first does not follow from the existence of such a particular algebraic integral.

Secondly: The differential equation has two particular algebraic integrals whose quotient s is not constant. This case includes that where there exists a particular algebraic integral of the differential equation whose logarithmic derivative is not a rational function of x, because in this case each branch of the algebraic integral is a particular integral of the differential equation and, thus, there are always two such branches whose quotient is not a constant.

The method of investigation is different for each of these two cases. Art.I of the following memoir deals with the first case; the remaining articles, with the exception of Article III, concern the second case.

The main content of Articles I to VI was the object of a communication in August 1871 at the session of the mathematical section of the Swiss Society of Natural Sciences, a summary of which has been published in the proceedings of this society, year 1871, pp. 75–77.

I.

It is assumed here that the differential equation of the *Gauss*ian series has a particular algebraic integral whose logarithmic derivative is a rational function of x.

It follows from the general theory of integration of linear differential equations with variable coefficients (see Mr. *Fuchs*'s memoir: "Zur Theorie der linearen Differentialgleichungen mit veränderlichen Coefficienten", this journal, vol. 66, p. 121) that each integral of the differential equation (A) has for all values of x, except for $x = 0$, $x = 1$ and $x = \infty$, the character of an entire function and that, moreover, when convergent series of the form

$$x^{\mathfrak{a}}(1+c_1x+c_2x^2+\cdots),\ (\frac{1}{x})^{\mathfrak{b}}(1+\frac{c_1}{x}+\frac{c_2}{x^2}+\cdots),\ (1-x)^{\mathfrak{c}}(1+c_1(1-x)+c_2(1-x)^2+\cdots)$$

are particular integrals of this differential equation, the exponents

$$\mathfrak{a},\qquad \mathfrak{b},\qquad \mathfrak{c}$$

can only have, respectively, the values

$$0 \text{ or } 1-\gamma,\qquad \alpha \text{ or } \beta,\qquad 0 \text{ or } \gamma-\alpha-\beta.$$

If the differential equation (A) has a particular algebraic integral y_1 whose logarithmic derivative is a rational function of x, this integral must have the form

$$y_1 = x^{\mathfrak{a}}(1-x)^{\mathfrak{c}}g(x),$$

where $\mathfrak{a}$ and $\mathfrak{c}$ are rational numbers and $g(x)$ is an entire function of x.

For it turns out that if $\mathfrak{a}$ and $\mathfrak{c}$ are correctly determined, the product

$$x^{-\mathfrak{a}}(1-x)^{-\mathfrak{c}}\cdot y_1$$

has the character of an entire function for all finite values of x; since this product is an algebraic function of x, it is thus an entire function of this variable.

For the sake of simplicity, one can set $\mathfrak{b} = \alpha$. Since it is always possible to reduce the case $\mathfrak{b} = \beta$ to the preceding one by exchanging α and β, there is no loss of generality in the investigation if one sets $\mathfrak{b} = \alpha$.

Let n be the degree of the entire function $g(x)$, then $\mathfrak{b} = \alpha = -(\mathfrak{a} + \mathfrak{c} + n)$, and one obtains the following table:

	$\mathfrak{a}$	$\mathfrak{c}$	$\mathfrak{b} = \alpha$	β	γ
1	0	0	$-n$	β	γ
2	0	$\gamma - \alpha - \beta$	$-n - \mathfrak{c}$	$n + \gamma$	γ
3	$1 - \gamma$	0	$-n - \mathfrak{a}$	β	$1 - \mathfrak{a}$
4	$1 - \gamma$	$\gamma - \alpha - \beta$	$-n - \mathfrak{a} - \mathfrak{c}$	$n + 1$	$1 - \mathfrak{a}$

It will now be shown that in each of the four cases characterized in this table the differential equation (A) possesses a particular integral whose logarithmic derivative is a rational function of x.

1^a. If γ is not an integer between 0 and $-n$ (0 included, $-n$ excluded), then

$$F(\alpha, \beta, \gamma, x) = F(-n, \beta, \gamma, x)$$

is an algebraic, and in fact an entire, function of x of degree n.

1^b. If on the contrary $\gamma = 1 - n'$, where n' is an integer with $1 \leq n' \leq n$, and if β is not at the same time a negative integer $\beta = -n''$, with the condition $0 \leq n'' < n'$, then the expression $F(\alpha, \beta, \gamma, x)$ loses its meaning; in this case, however (cf. Mr. *Kummer*'s memoir, this journal, vol. 15, formula 3 of the table on the page 52) the integral

$$x^{1-\gamma} F(\alpha - \gamma + 1, \beta - \gamma + 1, 2 - \gamma, x) = x^{n'} F(-n + n', \beta + n', 1 + n', x)$$

represents an algebraic, and in fact an entire, function of x; for it, however, the number $\mathfrak{a}$ is not equal to 0, but to n' (cf. 3^a.)

1^c. If, as before, the numbers n, n', n'' designate integers and if one has at the same time

$$\alpha = -n; \qquad \gamma = 1 - n', \quad 1 \leq n' \leq n; \qquad \beta = -n'', \quad 0 \leq n'' < n',$$

then the expression given in 1^a as well as that given in 1^b represent an algebraic integral of the differential equation (A) and in fact both integrals are linearly independent as entire functions of x of degrees n'' and n.

Remark: The case, specified in 1^c, of two particular integrals of the differential equation (A) which can be expanded as a power series in x that are not deduced from each other by a constant factor seems to contradict a theorem of Mr. *Kummer* on such integrals, mentioned on p. 53 of the memoir quoted above:

> "If there exist several integrals of the equation (A) which can be expanded as power series of x, these integrals can differ only by constant factors."

This contradiction is, however, only apparent, because this theorem is only derived under the assumption that the three elements α, β, γ are totally arbitrary; in fact, it then suffers one exception, when γ is a negative integer, and also only under the assumption that the value $x = 0$

is an accidental singular value[1)] for the differential equation. The derivation of the condition for the occurence of this case is to be found in Art. III and Art. VII.

2^a. If γ is not a negative integer $1-n'$, for $1 \leq n' \leq n$, then, under the assumption that the number $\gamma - \alpha - \beta = \mathfrak{c}$ is a rational number, the integral

$$[2.]^{2)} \qquad (1-x)^{\gamma-\alpha-\beta}F(\gamma-\alpha,\gamma-\beta,\gamma,x) = (1-x)^{\mathfrak{c}}F(n+\gamma+\mathfrak{c},-n,\gamma,x)$$

represents an algebraic function of x.

2^b. If, on the contrary, $\gamma = 1-n'$, where $1 \leq n' \leq n$, and if at the same time $n+\mathfrak{c}$ is not an integer n'' satisfying the condition $0 \leq n' < n'$, the expression above loses its meaning. In this case, again under the assumption that $\gamma-\alpha-\beta = \mathfrak{c}$ is a rational number, the integral

$$\begin{aligned}[4.] \quad & x^{1-\gamma}(1-x)^{\gamma-\alpha-\beta}F(1-\alpha,1-\beta,2-\gamma,x) \\ & \qquad = x^{n'}(1-x)^{\mathfrak{c}}F(1+n+\mathfrak{c},-n+n',1+n',x)\end{aligned}$$

represents an algebraic function of x. (cf. 4^a.)

2^c. But if $\gamma = 1-n'$, while $1 \leq n' \leq n$ and also $n+\mathfrak{c} = n''$, where n'' denotes an integer satisfying the condition $0 \leq n'' < n'$, then the expression in 2^a and 2^b represents an algebraic function of x. See the remark after 1^c.

3^a. If $\gamma = 1-\mathfrak{a}$ is not equal to a positive integer $1+n'$, where n' satisfies the condition $1 \leq n' \leq n$, and if $\mathfrak{a}$ denotes a rational number, the integral

$$[3.] \quad x^{1-\gamma}F(\alpha-\gamma+1,\beta-\gamma+1,2-\gamma,x) = x^{\mathfrak{a}}F(-n,\beta+\mathfrak{a},1+\mathfrak{a},x)$$

represents an algebraic function of x.

3^b. If, on the contrary, $\gamma = 1+n'$, $\mathfrak{a} = -n'$, $1 \leq n' \leq n$, and $\beta - n'$ is not at the same time equal to 0 or equal to a negative integer $-n''$, where $0 \leq n'' < n'$, the preceding expression loses its meaning; however, in this case, $F(\alpha,\beta,\gamma,x) = F(-n+n',\beta,1+n',x)$ represents an algebraic, and indeed an entire, function, of x (cf. 1^a.)

3^c. But if at the same time $\gamma = 1+n'$, $\mathfrak{a} = -n'$, $1 \leq n' \leq n$, $\beta - n' = -n''$, $0 \leq n'' < n'$ (thus $0 \leq n'' < n' \leq n$), then the expression given in 3^a and that given in 3^b represent an algebraic function of x. From this there follow in this case the two particular algebraic integrals

$$x^{-n'}F(-n,-n'',1-n',x) \qquad \text{and} \qquad F(-n+n',n'-n'',1+n',x).$$

4^a. If γ is not a positive integer, for which, if one sets $\gamma = 1+n'$, $\mathfrak{a} = -n'$, the condition $1 \leq n' \leq n$ is satisfied, then

$$[4.] \quad x^{1-\gamma}(1-x)^{\gamma-\alpha-\beta}F(1-\alpha,1-\beta,2-\gamma,x) = x^{\mathfrak{a}}(1-x)^{\mathfrak{c}}F(1+n+\mathfrak{a}+\mathfrak{c},-n,1+\mathfrak{a},x),$$

under the assumption that $\mathfrak{a}$ and $\mathfrak{c}$ are rational numbers, represents an algebraic function of x.

[1)] Tr.'s note: Schwarz, following Weierstrass, uses the terminology "ausserwesentlich singulär", inessential singular, when a singularity of a coefficient of the differential equation does not give rise to a singularity of the solutions. Several translations were used at the time, such as "apparently singular" or "accidental singularity". We are using here the latter, following *A. R. Forsyth*, Theory of Differential Equations, Cambridge UP, Cambridge 1890–1906.

[2)] The numbers in brackets denote the numbers of the corresponding formulas in Mr. *Kummer*'s table, this journal, vol. 15, pp. 52 and 53.

4^b. If, on the contrary, $\mathfrak{a} = -n'$, $1 \leq n' \leq n$, the preceding expression loses its meaning, if one does not have at the same time $1 + n - n' + \mathfrak{c} = -n''$, $0 \leq n'' < n'$; if one disregards this exception, then in this case

$$[2.] \quad (1-x)^{\gamma-\alpha-\beta} F(\gamma-\alpha, \gamma-\beta, \gamma, x) = (1-x)^{\mathfrak{c}} F(1+n+\mathfrak{c}, -n+n', 1+n', x),$$

represents an algebraic function of x, if $\mathfrak{c}$ is a rational number. (cf. 2^a.)

4^c. Finally, if at the same time $\mathfrak{a} = -n'$, $1 \leq n' \leq n$, $1+n-n'+\mathfrak{c} = -n''$, $0 \leq n'' < n'$, the expression given in 4^a and that given in 4^b represent algebraic functions of x.

With this we have completed all the cases in which the differential equation (A.) has a particular algebraic integral whose logarithmic derivative is a rational function of x.

II.

We will now assume that the differential equation (A.) presented above has *two* particular algebraic integrals whose quotient is not a constant.

In this case each particular integral of the differential equation is an algebraic function of x; thus the *quotient* of any two particular solutions is an algebraic function of x.

Now the following theorems apply:

1. If

$$(B.) \quad \frac{d^2y}{dx^2} + p \cdot \frac{dy}{dx} + q \cdot y = 0$$

is a linear differential equation of the second order and y_1 and y_2 are two linearly independent particular integrals of it, then, as *Abel* has shown,

$$(C.) \quad y_2 \cdot \frac{dy_1}{dx} - y_1 \cdot \frac{dy_2}{dx} = C \cdot e^{-\int p dx}.$$

(*Abel*: Œuvres T. 1, p. 93; this journal, vol. 2, pp. 22–25.) In the special case of the differential equation (A.), one has

$$p = \frac{\gamma - (\alpha+\beta+1)x}{x(1-x)} = \frac{\gamma}{x} - \frac{\alpha+\beta-\gamma+1}{1-x},$$

and it results that

$$(D.) \quad y_2 \cdot \frac{dy_1}{dx} - y_1 \cdot \frac{dy_2}{dx} = C \cdot x^{-\gamma}(1-x)^{\gamma-\alpha-\beta-1}.$$

(*Gauss*'s Werke III, p. 222. — *Kummer*, this journal, vol. 15, p. 62. – *Riemann*'s Abhandlungen Art. IV.)

If one now assumes that y_1 and y_2 are algebraic functions, it immediately follows that the numbers γ and $\alpha + \beta$ should be real, and in fact rational, numbers.

From formulas 9 and 10 in *Kummer*'s table, that has been mentioned several times, one draws the conclusion that the numbers α and β ought to be individually rational.

For the general integral of the differential equation of the hypergeometric series to be an algebraic function of x, it is necessary that the three numbers α, β, and γ are real numbers and in fact rational numbers.

This hypothesis will hold for the rest of this article.

2. If y_1 and y_2 are algebraic functions of x, then $\frac{y_1}{y_2}$ is also an algebraic function of x. Mr. *Heine* has drawn my attention to the fact that this theorem also has a converse.

If the quotient of two particular solutions y_1 and y_2 of a linear differential equation of the second order

$$\frac{d^2y}{dx^2}+p\cdot\frac{dy}{dx}+q\cdot y=0$$

is an algebraic function of x, without being a constant, and if, at the same time, $e^{-\int pdx}$ is an algebraic function of x, then the general integral of the differential equation is likewise an algebraic function of x.

Proof: According to the hypothesis, $\frac{y_1}{y_2}=f(x)$ is an algebraic function of x, thus

$$\frac{d}{dx}(\frac{y_1}{y_2})=f'(x)=(y_2\cdot\frac{dy_1}{dx}-y_1\cdot\frac{dy_2}{dx})\cdot\frac{1}{y_2^2}$$

is also an algebraic function of x; with the help of formula (C.), it follows that

$$(E.)\quad y_2^2=\frac{C}{f'(x)}\cdot e^{-\int pdx},$$

or, in other words: y_2 and also $y_1=f(x)\cdot y_2$ are algebraic functions of x, as is the general integral Ay_1+By_2, as the above lemma states.

A differential equation will be now derived for the quotient of two particular solutions y_2 and y_1 of the differential equation (B.).

If $C_1y_1+C_2y_2$ and $C_3y_1+C_4y_2$ are two particular solutions, let us put:

$$s=\frac{C_1y_1+C_2y_2}{C_3y_1+C_4y_2}.$$

This expression contains three independent constants; the differential equation which the quotient s satisfies, and for which these constants are constants of integration, should thus be of the *third* order; and in fact this differential equation can be obtained either by eliminating the ratios $C_1:C_2:C_3:C_4$ through repeated differentiation, or by putting $y_2=sy_1$ and then eliminating the particular integrals y_1 and y_2 with the help of (B.).

That is

$$\begin{aligned}\frac{d^2y_2}{dx^2}&=s\cdot\frac{d^2y_1}{dx^2}+2\frac{ds}{dx}\cdot\frac{dy_1}{dx}+\frac{d^2s}{dx^2}\cdot y_1\\ p\cdot\frac{dy_2}{dx}&=p\cdot s\cdot\frac{dy_1}{dx}+p\cdot\frac{ds}{dx}\cdot y_1\\ q\cdot y_2&=q\cdot s\cdot y_1\\ \hline 0&=2\frac{ds}{dx}\cdot\frac{dy_1}{dx}+y_1(\frac{d^2s}{dx^2}+p\frac{ds}{dx})\end{aligned}$$

$$\frac{1}{2}\cdot\frac{\frac{d^2s}{dx^2}}{\frac{ds}{dx}}+\frac{\frac{dy_1}{dx}}{y_1}+\frac{1}{2}p=\frac{1}{2}\cdot\frac{d}{dx}\log(\frac{ds}{dx})+\frac{d}{dx}\log y_1+\frac{1}{2}p=0$$

Repeating the differentiation gives

$$\frac{1}{2}\cdot\frac{d^2}{dx^2}\log(\frac{ds}{dx})+\frac{y_1\cdot\frac{d^2y_1}{dx^2}-\left(\frac{dy_1}{dx}\right)^2}{y_1^2}+\frac{1}{2}\cdot\frac{dp}{dx}=0,$$

and by elimination of y_1

$$\frac{1}{2}\cdot\frac{d^2}{dx^2}\log(\frac{ds}{dx}) - \frac{y_1 p\cdot\frac{dy_1}{dx} + y_1^2\cdot q + (\frac{dy_1}{dx})^2}{y_1^2} + \frac{1}{2}\cdot\frac{dp}{dx} = 0,$$

$$\frac{1}{2}\cdot\frac{d^2}{dx^2}\log(\frac{ds}{dx}) - (\frac{d}{dx}\log y_1 + \frac{1}{2}p)^2 + \frac{1}{4}p^2 - q + \frac{1}{2}\cdot\frac{dp}{dx} = 0,$$

$$\frac{d^2}{dx^2}\log(\frac{ds}{dx}) - \frac{1}{2}(\frac{d}{dx}\log\frac{ds}{dx})^2 = 2q - \frac{1}{2}p^2 - \frac{dp}{dx}.$$

In this equation, which does not contain y_1 anymore, let the expression on the left side be denoted by $\Psi(s,x)$, the expression on the right side by $F(x)$. One then has:

$$(F.)\quad \Psi(s,x) = \frac{2\frac{ds}{dx}\cdot\frac{d^3s}{dx^3} - 3\left(\frac{d^2s}{dx^2}\right)^2}{2(\frac{ds}{dx})^2} = 2q - \frac{1}{2}p^2 - \frac{dp}{dx} = F(x).$$

This differential equation is a special case of the one considered by Mr. *Kummer* in the program paper mentioned above:

$$2\frac{d^3z}{dzdx^2} - 3(\frac{d^2z}{dzdx})^2 - Z\frac{dz^2}{dx^2} + X = 0.$$

Namely one has to set:

$$x = x,$$
$$z = s,$$
$$Z = 0,$$
$$X = 2\frac{dp}{dx} + p^2 - 4q.$$

In this case one has

$$q = -\frac{\alpha\cdot\beta}{x(1-x)},\qquad p = \frac{\gamma-(\alpha+\beta+1)x}{x(1-x)} = \frac{\gamma}{x} + \frac{\gamma-(\alpha+\beta+1)}{1-x},$$

$$\frac{dp}{dx} = -\frac{\gamma}{x^2} + \frac{\gamma-(\alpha+\beta+1)}{(1-x)^2},$$

$$(G.)\left\{\begin{aligned} F(x) &= 2q - \tfrac{1}{2}p^2 - \frac{dp}{dx} \\ &= \frac{1-(1-\gamma)^2}{2x^2} + \frac{1-(\gamma-\alpha-\beta)^2}{2(1-x)^2} \\ &\qquad - \frac{(1-\gamma)^2 - (\alpha-\beta)^2 + (\gamma-\alpha-\beta)^2 - 1}{2x(1-x)}. \end{aligned}\right.$$

(Regarding the function $\Psi(s,x)$, cf. this journal, vol. 70, p. 116, along with the improvement, vol. 71, p. iv; Monatsberichte der Berliner Akademie 1870, p. 767, as well as the memoir of the author "Bestimmung einer speciellen Minimalfläche", published by the same Academy, p. 10.)

III

Under the hypothesis that the three magnitudes α, β, γ have *real* values, let us now integrate the differential equation (F.) of Art. II, in which the value found for the function $F(x)$ ought to be inserted (see equation (G.)). (cf. "Bestimmung einer speciellen Minimalfläche", pp. 17–21.)

When the three squares

$$(1-\gamma)^2, \qquad (\alpha-\beta)^2, \qquad (\gamma-\alpha-\beta)^2$$

are respectively denoted by

$$\lambda^2, \qquad \mu^2, \qquad \nu^2,$$

the mentioned differential equation becomes

$$(H.) \quad \Psi(s,x) = \frac{1-\lambda^2}{2x^2} + \frac{1-\nu^2}{2(1-x)^2} - \frac{\lambda^2-\mu^2+\nu^2-1}{2x(1-x)},$$

and thus only contains the squares of the magnitudes λ, μ, ν. In order to have something well determined, it may be assumed that the numbers λ, μ, ν, which are real under the mentioned hypothesis, denote the *positive* values of the square roots of λ^2, μ^2, ν^2.

For the general integration of this differential equation, it is sufficient to discover one particular integral, because, when $s = \sigma$ is a particular integral, $s = \dfrac{C_1\sigma + C_2}{C_3\sigma + C_4}$ is likewise an integral of the differential equation, which contains the appropriate number of arbitrary constants.

For the argument x, the values $0, \infty, 1$ are the only singular ones. If, instead of the variable x, one introduces a rational function of it of the first degree $z = \dfrac{c_1x + c_2}{c_3x + c_4}$ as a new independent variable, one then has the following identity:

$$\Psi(s,x) = (\frac{dz}{dx})^2 \cdot \Psi(s,z).$$

If one now denotes the general integral of the differential equation above by $s(\lambda, \mu, \nu, x)$ for short, the following formulas are deduced from the last given identity:

$$\begin{aligned} s(\lambda,\mu,\nu,z) &= s(\lambda,\mu,\nu,x) \quad \text{when } z = x, \\ &= s(\nu,\mu,\lambda,x) \quad \text{for } z = 1-x, \\ &= s(\mu,\lambda,\nu,x) \quad \text{for } z = \frac{1}{x}, \\ &= s(\nu,\lambda,\mu,x) \quad \text{for } z = \frac{1}{1-x}, \\ &= s(\lambda,\nu,\mu,x) \quad \text{for } z = \frac{x}{x-1}, \\ &= s(\mu,\nu,\lambda,x) \quad \text{for } z = \frac{x-1}{x}, \end{aligned}$$

in complete agreement with *Riemann*'s theorem concerning the function denoted $P(\lambda, \mu, \nu, x)$ by him. (*Riemann*'s Abhandlungen, Art. V.)

In fact, the function $s(\lambda, \mu, \nu, x)$ turns out to be the quotient of two linearly independent Riemann functions $P(\lambda, \mu, \nu, x)$.

Let now x_0 be a non-singular value of the argument and

$$F(x) = a_0 + a_1(x - x_0) + \cdots + a_{n-1}(x - x_0)^{n-1} + \cdots$$

the expansion of $F(x)$ as a series of ascending powers of $x - x_0$, convergent when the absolute value of $x - x_0$ is small enough. If s' denotes a particular integral, let us write:

$$\frac{d}{dx} \log \frac{ds'}{dx} = r = \frac{-2}{x - x_0} + \star + b_1(x - x_0) + b_2(x - x_0)^2 + \cdots + b_n(x - x_0)^n + \cdots,$$

where the coefficients b and the domain of convergence are still to be determined. Then

$$\frac{dr}{dx} = \frac{2}{(x - x_0)^2} + b_1 + 2b_2(x - x_0) + \cdots + n \cdot b_n(x - x_0)^{n-1} + \cdots$$

$$\begin{aligned}\frac{dr}{dx} - \frac{1}{2}r^2 = {}& b_1 + 2b_2(x - x_0) + \cdots + n \cdot b_n(x - x_0)^{n-1} + \cdots \\ & + 2b_1 + 2b_2(x - x_0) + \cdots + 2b_n(x - x_0)^{n-1} + \cdots \\ & - \frac{1}{2}(b_1(x - x_0) + b_2(x - x_0)^2 + \cdots + b_{n-2}(x - x_0)^{n-2} + \cdots)^2 \\ = {}& a_0 + a_1(x - x_0) + \cdots + a_{n-1}(x - x_0)^{n-1} + \cdots .\end{aligned}$$

If one denotes by $G(b_1, \cdots, b_{n-2})$ the coefficient of $(x - x_0)^{n-1}$ in the expansion of

$$(b_1(x - x_0) + b_2(x - x_0)^2 + \cdots + b_{n-2}(x - x_0)^{n-2})^2,$$

which is an entire function of $b_1, b_2, \cdots, b_{n-2}$ with positive integer coefficients, it follows from the comparison of the terms with the same power of $x - x_0$ that

$$(J.) \quad b_n = \frac{1}{n+2}(a_{n-1} + \frac{1}{2}G(b_1, \cdots, b_{n-2})).$$

Thus the coefficient b_n is a entire function of the quantities $a_0, a_1, \cdots, a_{n-1}$ with *positive* rational coefficients and thus is uniquely determined by these quantities a. It is thus now only a question of proving that the series for r, when the coefficients b in it are determined in the way just explained, is convergent for all values of $x - x_0$ whose absolute value does not exceed a certain bound. For this purpose, let us put for comparison

$$r = \frac{-2}{x - x_0} + \star + \beta_1(x - x_0) + \cdots + \beta_n(x - x_0)^n + \cdots,$$

where $\beta_n = 2(\frac{1}{\xi})^{n+1}$; one thus has

$$r = \frac{-2}{x - x_0} + \frac{2}{\xi} \cdot \frac{x - x_0}{\xi - (x - x_0)};$$

here ξ denotes a positive quantity, to be determined more precisely later. From this hypothesis, it results that

$$\frac{dr}{dx} - \frac{1}{2}r^2 = \alpha_0 + \alpha_1(x - x_0) + \cdots + \alpha_{n-1}(x - x_0)^{n-1} + \cdots,$$

where $\alpha_0 = \frac{6}{\xi^2}$, while, for all the other coefficients, one has the equation:

$$\alpha_{n-1} = 8 \cdot (\frac{1}{\xi})^{n+1}, \qquad n > 1.$$

If one now sets $a_n = \alpha_n$ in the above equation, it would thus follow, when b_n is computed with the formula (J.), that $b_n = \beta_n 2(\frac{1}{\xi})^{n+1}$ and then the resulting series for r will unconditionally converge when $[x - x_0] < \xi$.[3)] If the law of the coefficients a_n is then another one and if one only assumes that the series

$$a_0 + a_1(x - x_0) + \cdots + a_n(x - x_0)^n + \cdots$$

converges for all values of $x - x_0$ whose absolute value is sufficiently small, it will then be possible in any case to determine a strictly positive number ξ such that the absolute value $[a_n]$ of a_n is smaller than α_n for all values of n. Because the convergence is assumed, there should exist a non-zero value $x_1 - x_0$ of the difference $x - x_0$ for which all the absolute values of the terms of the series are individually smaller than a finite quantity g; thus:

$$[a_n] \cdot [x_1 - x_0]^n < g.$$

Let us now choose ξ small enough, such that $\xi < [x_1 - x_0]$ and $g \cdot \xi^2 < 6$. If one then computes the coefficients b_n thanks to the formula (J.), it follows that the absolute value of b_n is smaller than that of β_n, and thus the series obtained for r certainly converges when $[x - x_0] < \xi$.

This proof is based on the same principles of comparison between series expansions that Mr. *Weierstrass* has been seeking to develop for several years in his lectures, when founding certain fundamental theorems in the theory of ordinary differential equations, and on which the proof of the general lemma expressed in the memoir of the same author, "Ueber die Theorie der analytischen Facultäten", this journal, vol. 51, p. 43, is based.

From the series obtained, it follows by integration and passing from logarithms to numbers, with an appropriate determination of the constants of integration

$$\begin{aligned}
\log \frac{ds'}{dx} &= -2\log(x - x_0) + \frac{1}{2}b_1(x - x_0)^2 + \frac{1}{3}b_2(x - x_0)^3 + \cdots + \log(-1), \\
\frac{ds'}{dx} &= \frac{-1}{(x - x_0)^2} \cdot e^{\frac{1}{2}b_1(x - x_0)^2 + \frac{1}{3}b_2(x - x_0)^3 + \cdots} \\
&= \frac{-1}{(x - x_0)^2} + b'_0 + b'_1(x - x_0) + \cdots
\end{aligned}$$

and by another integration

$$s' = \frac{1}{x - x_0} + b'_0(x - x_0) + \frac{1}{2}b'_1(x - x_0)^2 + \cdots,$$

where the quantities b' are entire functions of the coefficients b with rational coefficients and the series all converge unconditionally when the absolute value of $x - x_0$ is smaller than ξ.

3) Tr.'s note: Karl Weierstrass defined an "unconditionally convergent series" ("*unbedingt convergent Reihe*") as one which remains convergent upon reordering of its terms, see *K. Weierstrass*, Über die Theorie der analytischen Facultäten, J. reine angew. Math. (1856), 1–60, p. 41. At the time, it was regarded as the same as an "absolutely convergent series".

The particular integral s' of the differential equation $\Psi(s,x) = F(x)$ obtained in this way, expressed in a certain neighborhood of the non-singular point $x = x_0$, is characterized by the property that the value $s = \infty$ corresponds to the value $x = x_0$; now, this integral is *unramified* in the neighborhood of the value $x = x_0$ and, when x gets close to the value x_0, it becomes infinite in the same way as $\frac{1}{x-x_0}$, thus, to one loop of x around the value x_0, there corresponds likewise one loop of the value s' around the value ∞. Let us thus define a simply connected domain X in the neighborhood of the value $x = x_0$, entirely lying inside the domain $[x - x_0] < \xi$; if the region of all the values that can be taken on the points of X by the function element denoted by s' is represented geometrically by a part S' of the plane s, then the point at ∞ lies inside this domain S', which is also simply connected and, indeed, the point at infinity is a *simple* point on the surface S'.

The general integral arises from the particular integral s' by means of the equation $s = \dfrac{C_1 s' + C_2}{C_3 s' + C_4}$. If S designates the region corresponding to the points of the domain X and represents geometrically the values of the integral s when the values of the constants C are fixed, it then follows that the two regions S and S' are linked together by the well-known relation of a Möbius transformation. The point of the domain S which uniquely corresponds to the point x_0 is thus a *simple* point of S. Since the point x_0 is any non-singular point and the domain X can be extended arbitrarily up to the singular points, the following theorem follows: *If in the plane of the argument x one delimits an arbitrary simply-connected domain X whose interior does not contain any singular point, there corresponds to it by means of a particular integral of the differential equation $\Psi(s) = F(x)$ a likewise simply-connected domain S whose interior may contain the point at infinity once or several times, but no winding point.*

In particular if the value x_0 is a non-singular *real* value and if the quantities λ^2, μ^2, ν^2 occurring in the differential equation have *real* values, as assumed, then real values of s' correspond to real values of x near the point $x = x_0$ by means of the particular integral s', since all the quantities b' have real values. Thus, to the segment of the real axis containing the point x_0 in the plane x, there corresponds a *straight line* on the surface S' and on the surface S a circular arc, generally speaking.

We now have to investigate more closely the behavior of a particular integral of the differential equation in the neighborhood of a *singular* value of the argument. According to the formulas above, it is sufficient to carry out this investigation for the singular value $x = 0$.

Let

$$F(x) = \frac{1-\lambda^2}{2x^2} + \frac{a_0}{x} + a_1 + a_2 x + \cdots + a_n x^{n-1} + \cdots$$

be the expansion as a series of ascending powers, which converges unconditionally for $[x] \leq \xi_0$. One sets, with s' denoting again a particular integral:

$$\frac{d}{dx} \log \frac{ds'}{dx} = r = -\frac{1+\lambda}{x} + b_0 + b_1 x + b_2 x^2 + \cdots + b_n x^n + \cdots,$$

thus

$$\frac{dr}{dx} = \frac{1+\lambda}{x^2} + b_1 + 2b_2 x + \cdots + n b_n x^{n-1} + \cdots,$$

and

$$\frac{dr}{dx} - \frac{1}{2}r^2 = \frac{1-\lambda^2}{2x^2} + \frac{(1+\lambda)b_0}{x} + (1+\lambda)(b_1 + b_2 x + \cdots + b_n x^{n-1} + \cdots)$$
$$+ (b_1 + 2b_2 x + \cdots + n b_n x^{n-1} + \cdots)$$
$$- \frac{1}{2}(b_0 + b_1 x + b_2 x^2 + \cdots + b_{n-1}x^{n-1} + \cdots)^2.$$

It follows that

$$b_0 = \frac{a_0}{1+\lambda}, \qquad b_1 = \frac{a_1 + \frac{1}{2}b_0^2}{2+\lambda}, \qquad \text{etc.,}$$

and generally, if $G(b_0, b_1, \ldots, b_{n-1})$ denotes the coefficient of x^{n-1} in the expansion of

$$(b_0 + b_1 x + b_2 x^2 + \cdots + b_{n-1}x^{n-1})^2$$

according to the powers of x, which is an entire function of the quantities $b_0, b_1, \ldots, b_{n-1}$ with positive coefficients,

$$(K.) \quad b_n = \frac{1}{n+1+\lambda} \cdot (a_n + \frac{1}{2}G(b_0, b_1, \ldots, b_{n-1})).$$

Thus b_n is an entire function of $a_0, \ldots, a_n$ with positive coefficients.

The convergence follows from the comparison with the series

$$r = -\frac{1+\lambda}{x} + \beta_0 + \beta_1 x + \beta_2 x^2 + \cdots + \beta_n x^n + \cdots,$$

if one sets $\beta_n = 2(\frac{1}{\xi})^{n+1}$, thus $r = -\frac{1+\lambda}{x} + \frac{2}{\xi - x}$. Then it follows that

$$\frac{dr}{dx} - \frac{1}{2}r^2 = \frac{1-\lambda^2}{2x^2} + \frac{\alpha_0}{x} + \alpha_1 + \alpha_2 x + \cdots + \alpha_n x^{n-1} + \cdots,$$

$$\alpha_n = 2(1+\lambda)(\frac{1}{\xi})^{n+1}.$$

From the hypothesis that the series

$$\frac{a_0}{x} + a_1 + a_2 x + \cdots + a_n x^{n-1} + \cdots$$

converges for $[x] = \xi_0$, there is a positive value g such that for all values of n, $[a_n] \cdot \xi_0^{n-1} < g$. Let us now choose ξ small enough such that $\xi < \xi_0$ and also $g\xi\xi_0 < 2(1+\lambda)$; then $[b_n] < \beta_n$, and the series thus obtained certainly converges for all values of x such that $[x] < \xi$.

Through integration and passing from logarithms to numbers, one obtains, by an appropriate choice of the constants introduced in the integration

$$\int r \cdot dx = \log \frac{ds'}{dx} = \log C - (1+\lambda)\log x + b_0 x + \frac{1}{2}b_1 x^2 + \frac{1}{3}b_2 x^3 + \cdots,$$
$$\frac{ds'}{dx} = C \cdot x^{-(1+\lambda)} \cdot e^{b_0 x + \frac{1}{2}b_1 x^2 + \frac{1}{3}b_2 x^3 + \cdots}$$
$$= C \cdot x^{-(1+\lambda)}(1 + b_1' x + b_2' x^2 + \cdots),$$

where the quantities b' are entire functions of the quantities b with positive coefficients and the series inside the parenthesis surely converges, as soon as $[x] < \xi$. From this it follows by integration, if λ is not equal to 0 nor to an integer, that

$$s' = -Cx^{-\lambda}(\frac{1}{\lambda} - \frac{b'_1 x}{1-\lambda} - \frac{b'_2 x^2}{2-\lambda} - \cdots) + C'$$

or, if one denotes $-\frac{\lambda}{C} \cdot s'$ by σ,

$$(L.) \quad \sigma = x^{-\lambda}(1 + b''_1 x + b''_2 x^2 + \cdots) + C''.$$

If on the contrary λ is equal to 0 or to an integer m, there occurs in the expansion of $\frac{1}{C} \cdot \frac{ds'}{dx}$ a term $\frac{b'_m}{x}$, which by integration provides $b'_m \log x$, such that one obtains in this case a particular integral of the form

$$(L^\star.) \quad \sigma = x^{-m}(1 + b''_1 x + b''_2 x^2 + \cdots \text{to inf.}) + B_m \log x + C''.$$

The coefficients b'', among which only b''_m remains undefined, are entire functions of the quantities a.

When m is different from 0, the factor B_m is an entire function of $a_0, a_1, \ldots, a_{m-1}$. The following method can serve to compute it.

If, in the differential equation (B.), one sets $p = -\frac{m-1}{x}$ and q is determined by the condition

$$2q - \frac{1}{2}p^2 - \frac{dp}{dx} = F(x),$$

from which one has

$$q = \frac{1}{2}\left[F(x) - \frac{1-m^2}{2x^2}\right] = \frac{1}{2x}(a_0 + a_1 x + a_2 x^2 + \cdots),$$

then, as follows from formula (E.) of Article II.,

$$y_1 = x^{\frac{m-1}{2}} \cdot (-\frac{1}{m} \cdot \frac{ds}{dx})^{-\frac{1}{2}} \quad \text{and} \quad y_2 = s \cdot y_1$$

are two integrals of this differential equation which can now be written in the form

$$(M.) \quad 2x \cdot \frac{d^2y}{dx^2} - 2(m-1)\frac{dy}{dx} + (a_0 + a_1 x + a_2 x^2 + \cdots + a_{m-1}x^{m-1} + \cdots) \cdot y = 0.$$

If one sets $s = \sigma$, where σ is taken from equation ($L^\star$), the integral y_1 has the character of an entire function in the neighbourhood of x_0, and its expansion in this domain takes the form

$$y_1 = x^m(1 + c_1 x + c_2 x^2 + \cdots).$$

The integral y_2, on the other hand, has the expansion:

$$y_2 = (1 + b''_1 x + b''_2 x^2 + \cdots) \cdot (1 + c_1 x + c_2 x^2 + \cdots) + B_m \log x \cdot y_1 + C'' \cdot y_1$$

or

$$y_2 = 1 + A_1 x + A_2 x^2 + \cdots + A_{m-1}x^{m-1} + \star + A_{m+1}x^{m+1} + \cdots$$
$$+ B_m \log x \cdot x^m(1 + c_1 x + c_2 x^2 + \cdots) + C_1 \cdot y_1.$$

Without restricting the generality of the investigation, it can be assumed that A_m has the value 0, since this can be achieved by an appropriate determination of C_1. It thus provides surely a particular integral of the differential equation (M.) of the form $y_2 - C_1 y_1$ and, indeed, the coefficients $A_1, A_2, \ldots, A_{m-1}$ and the coefficient B_m can be computed by the method of indeterminate coefficients, since the existence of this expansion is not in question. Hereby, when one proceeds up to the terms with the power x^{m-1} — the terms containing the factor $\log x$ cancel each other out — one obtains the following equations:

$$\begin{aligned}
0 &= a_0 - 2\cdot 1\cdot(m-1)A_1,\\
0 &= a_1 + \quad a_0A_1 \quad - 2\cdot 2\cdot(m-2)A_2,\\
0 &= a_2 + \quad a_1A_1 \quad + a_0A_2 \quad - 2\cdot 3\cdot(m-3)A_3,\\
&\ldots\ldots\ldots\ldots\ldots\\
0 &= a_{m-2} + a_{m-3}A_1 + a_{m-4}A_2 + a_{m-5}A_3\cdots - 2\cdot(m-1)\cdot 1A_{m-1},\\
0 &= a_{m-1} + a_{m-2}A_1 + a_{m-3}A_2 + a_{m-4}A_3\cdots + a_0A_{m-1} + 2mB_m
\end{aligned}$$

These equations allow a step-by-step resolution. Let us denote by a_{gh} a quantity in which both indices g and h can take all the values $0, 1, 2, \ldots m-1$ independently of each other and put

$$\begin{aligned}
a_{gh} &= a_{g-h}, && \text{if} \quad h \le g,\\
&= -2h(m-h), && \text{if} \quad h = g+1,\\
&= 0, && \text{if} \quad h > g+1.
\end{aligned}$$

Following Mr. *Kronecker*'s procedure, let us denote by $\mid a_{gh} \mid$ the determinant constructed from the m^2 quantities a_{gh}; one obtains:

$$\mid a_{gh} \mid_{(g,h=0,1\cdots,m-1)} = \begin{vmatrix}
a_0 & -2\cdot 1\cdot(m-1) & 0 & \cdots & 0 & 0\\
a_1 & a_0 & -2\cdot 2\cdot(m-2) & \cdots & 0 & 0\\
\cdot & \cdot & \cdot & \cdots & \cdot\cdot & \\
\cdot & \cdot & \cdot & \cdots & \cdot\cdot & \\
a_{m-3} & a_{m-4} & a_{m-5} & \cdots & -2\cdot(m-2)\cdot 2 & 0\\
a_{m-2} & a_{m-3} & a_{m-3} & \cdots & a_0 & -2\cdot(m-1)\cdot 1\\
a_{m-1} & a_{m-3} & a_{m-3} & \cdots & a_1 & a_0
\end{vmatrix}.$$

The coefficient B_m thus has the value

$$(N.) \quad B_m = -\frac{1}{2^m}\cdot\frac{1}{m!}\cdot\frac{1}{(m-1)!}\cdot \mid a_{gh} \mid .$$

The necessary and sufficient condition that $\mid a_{gh} \mid$ be zero is that the singular point $x = 0$ under consideration be an accidental singular point in Mr. *Weierstrass*'s terminology. (cf. Mr. *Fuchs*'s memoir, this journal, vol. 68, p. 378.)

For $m = 1, \quad 2, \quad 3, \quad 4,$ it follows that
$| a_{gh} |= a_0,\ a_0^2 + 2a_1, a_0^3 + 8a_0a_1 + 16a_2, a_0^4 + 20a_0^2a_1 + 96a_0a_2 + 36a_1^2 + 288a_3.$

When $\lambda = 0$, the singular point $x = 0$ is never accidental ; when $\lambda = 1$, a necessary and sufficient condition for $x = 0$ to be an accidental singular value is that $a_0 = \frac{1}{2}(\mu + \nu)(\mu - \nu) = 0$. One finds likewise, when one actually does the computation by which the values of the coefficients a are determined, the respective conditions for $\lambda = 2, 3, 4$ (cf. Art. VII):

for $\lambda = 2$: $(\mu + \nu + 1)(\mu + \nu - 1)(\mu - \nu + 1)(\mu - \nu - 1) = 0,$

for $\lambda = 3$: $(\mu + \nu)(\mu - \nu)(\mu + \nu + 2)(\mu + \nu - 2)(\mu - \nu + 2)(\mu - \nu - 2) = 0,$

for $\lambda = 4$:

$$(\mu + \nu + 1)(\mu + \nu - 1)(\mu - \nu + 1)(\mu - \nu - 1)(\mu + \nu + 3) \\ (\mu + \nu - 3)(\mu - \nu + 3)(\mu - \nu - 3) = 0.$$

Because, in the case considered here, real values of $F(x)$ always correspond to real values of x, all the quantities a as well as the quantities b and c, as well as the quantity B_m have exclusively real values; it follows that real values of σ always correspond to small real positive values of x — when one gives, for positive values of x, its real value to the power $x^{-\lambda}$ and, in the case when a logarithmic term intervenes, its real value to the logarithm for positive value of x, and the value 0 to the constant C'' —; the integrals σ and s' thus describe *straight lines* when x describes a part of the segment $0 \cdots 1$.

If x now describes a half-circle clockwise around the point $x = 0$ and if then to the variables x are assigned only negative values $x = -z$, $x^{-\lambda}$ thus contains the factor $e^{-\lambda\pi i}$ and one has $\sigma = e^{-\lambda\pi i}z^{-\lambda}(1 - b_1''z + \cdots)$; thus σ and s' describe straight lines also for *negative* values of x.

If λ is equal to m and B_m is not equal to 0, $\log x$ increases by πi and σ again describes, for small real negative values of x, a line which is parallel to the preceding one and is at a distance $B_m\pi$ from it.

The particular integral $s = \frac{1}{\sigma}$ serves as a conformal mapping of a part of the plane x situated on the positive side of the real axis in a neighborhood of the point $x = 0$ (Pl. I. Fig. 1^a) onto a region whose boundary contains in general *two circular arcs* which correspond to the two segments of the real axis of the plane x intersecting at the point $x = 0$. (Plate I. Fig. 1^b and 1^c). The tangents of these circular arcs enclose an angle $\lambda\pi$, in the sense that, when one cuts from the region a part of it in the neighborhood of the vertex by a circular line with a sufficiently small radius ρ, this cut-off part possesses either exactly or approximatively the shape of a circular sector with a central angle $\lambda\pi$. If λ is an integer and B_m is not 0, the two circular arcs touch each other; when B_m is 0, both arcs belong to the same circle; if λ is neither 0 nor an integer, the circles to which these circular arcs belong intersect in two distinct points.

As a similar investigation leads to similar results for the singular points $x = 1, x = \infty$, one has the following theorem:

The half-plane E situated on the positive side of the real axis in the plane of the argument x is conformally mapped by a particular integral of the differential equation

$$\Psi(s, x) = \frac{1 - \lambda^2}{2x^2} + \frac{1 - \nu^2}{2(1 - x)^2} - \frac{\lambda^2 - \mu^2 + \nu^2 - 1}{2x(1 - x)},$$

when the three quantities λ, μ, ν have real values, onto a simply-connected domain S, which *does not contain any winding point in its interior, and whose boundary in general consists of three circular arcs forming a triangle.*

The *angles* of this *circular triangle* S, whose vertices correspond to the singular values $x = 0, x = 1, x = \infty$, are respectively $\lambda\pi, \mu\pi, \nu\pi$.

Or: *The quotient of two linearly independent particular solutions of the differential equation of the hypergeometric series provides, when the quantities* α, β, γ *have real values, the conformal mapping of a half-plane into a circular triangle with the angles* $[1-\gamma]\pi$, $[\alpha-\beta]\pi$, $[\gamma-\alpha-\beta]\pi$. From what precedes, this circular triangle does not contain any winding point in its interior; its vertices can all be winding points only when, among the angles of the circular triangle, there are some which are bigger than 2π (cf. this journal, vol. 70, p. 117.)

Remark. Also for the case when one, two or all three among the real quantities λ^2, μ^2, ν^2 have *negative* values, the various sheets of the domain S corresponding to the half-plane E, which then cover parts of the plane s infinitely often, in general, are limited by three circles, while, correspondingly, one, two, or three of the proper vertices of the circular triangle are lost.

IV

The relative situation of the circles to which the three sides of the circular triangle S belong obviously depends on the numbers λ, μ, ν. This dependence must now be investigated more closely. Let us consider one of the three circular arcs and let us denote by K_1 the circle to which it belongs. Let us characterize this circular arc by a point moving along it in such a way that the interior of the circular triangle lies on the *left-hand* side with respect to the direction of the movement. In this way one uniquely determines a simply-connected, one-sheeted surface which is completely bounded by the circle K_1 and which also lies on the left-hand side of this circular arc; one may denote it by (1), whether or not it contains the point at infinity in its interior. In the same way, one defines two other circular surfaces (2) and (3) which are bounded by the circles K_2 and K_3.

It may now be assumed that the three numbers λ, μ, ν have positive values and, firstly, that none of them is equal to an integer.

The vertex L corresponding to the point $x = 0$ on the boundary of S is a point common to two circles, say K_2 and K_3. The angle of S at the point L has the value $\lambda\pi$.

The two circular surfaces (2) and (3) have in common a circular digon $(2, 3)$. If its angle is denoted by $\lambda'\pi$, λ' is a positive number which is smaller than 1 and is indeed either

$$\lambda' \equiv \lambda \pmod 2, \quad \text{or} \quad 2 - \lambda' \equiv \lambda \pmod 2.$$

The number λ' is thus uniquely determined by these two properties, namely λ' is the absolute value of the smallest residue of λ modulo 2.

Likewise μ' and ν' are uniquely determined as the absolute values of the smallest residues modulo 2 of μ and ν.

A circle K_1 cuts the plane on which it is situated into *two* parts, the *interior* and the *exterior* of the circle K_1.

By the *two* circles K_1 and K_2 intersecting in two distinct points, the plane is divided into *four* circular digons which have two common vertices.

If one adds to the two circles K_1 and K_2 a *third* circle K_3, which *cuts* the circles K_1 and K_2 and does not meet any of the two intersection points of K_1 and K_2, two cases are to be distinguished.

Namely, either 1) the two intersection points of K_1 and K_2 are situated on the *same* side of K_3, that is, both in the interior or both in the exterior of K_3, or 2) the two intersection points of K_1 and K_2 are situated on *different* sides of K_3, that is, one of the two intersection points is situated inside, the other outside, K_3.

For this classification, the order in which the three circles are denoted by K_1, K_2, K_3 is indifferent. In the first case there is a circle, which may also be a straight line, cutting the three circles orthogonally, in the other case, there is no such orthogonal circle.

In the first case the plane is divided by the three circles into three circular digons, two circular triangles and three circular quadrangles; in the second case, on the contrary, the plane is divided into eight circular triangles whose angles are together smaller than π.

One can nonetheless in the first case as well as in the second one distinguish eight different circular triangles, whose angles together are smaller than π. Indeed, if one adds to one of the aforementioned circular triangles one of the three contiguous circular quadrangles, a new circular triangle arises, and there are six of this sort; by counting the two circular triangles already mentioned, there are thus for the first case eight circular triangles altogether, which are different from each other. In the first case as well as in the second, these eight circular triangles correspond pairwise by the equality of their angles — apart from their different order in regard to the interior of the circular triangle —, such that they form altogether *four* pairs, whose angles, when $\lambda''\pi$, $\mu''\pi$, $\nu''\pi$ denote the angles of any of the eight triangles, can be represented by the following scheme, with π omitted:

$$\begin{matrix} \lambda'' & \mu'' & \nu'' \\ \lambda'' & 1-\mu'' & 1-\nu'' \\ 1-\lambda'' & \mu'' & 1-\nu'' \\ 1-\lambda'' & 1-\mu'' & \nu''. \end{matrix}$$

The two triangles belonging to the same pair are either completely separated, such that their surfaces have no common point, or they have in common one of the three circular quadrangles of the first case.

As now one may also decide on the interior or the exterior of K_1, K_2, K_3 with regard to the domains (1), (2), (3), there are, among the eight triangles determined by these three circles when λ', μ', ν' have the meaning mentioned above, always two (and in general only two) triangles belonging to the same pair, with the angles λ', μ', ν'; or in other terms:

The circles, to which the sides of the circular triangle S with the angles $\lambda\pi$, $\mu\pi$, $\nu\pi$ belong, also determine in the same plane a circular triangle S' with the angles $\lambda'\pi$, $\mu'\pi$, $\nu'\pi$.

The problem we are dealing with will now be simplified to some extent by the fact that the numbers λ, μ, ν can be replaced by the absolute values of the smallest residues modulo 2, λ', μ', ν', without detracting from the generality of the investigation.

Save the mentioned pair of the circular triangles, three other pairs of circular triangles are also determined by the same circles. The ratios to the number π of the angles of these four pairs of circular triangles are represented by the scheme

$$\begin{matrix} \lambda' & \mu' & \nu' \\ \lambda' & 1-\mu' & 1-\nu' \\ 1-\lambda' & \mu' & 1-\nu' \\ 1-\lambda' & 1-\mu' & \nu' \end{matrix}$$

For the sake of further simplification, it is now useful, among these four pairs determined together by these three circles, to single out one pair, namely the one for which the sum of the angles is a *minimum.* A circular triangle belonging to this pair may be denoted by S'' and its angles respectively by $\lambda''\pi$, $\mu''\pi$, $\nu''\pi$.

In the first of the cases mentioned above, there is always one and only one pair for which the sum of the angles is smaller than π, namely the pair of the two circular triangles first mentioned, of which one is situated totally inside, the other totally outside, the orthogonal circle.

In the second case, the sum of the angles is bigger than π for each of the four pairs and if one does not have $\lambda = \mu = \nu = \frac{1}{2}$, thus for one of the four pairs at least the sum of the angles is smaller than $\frac{3}{2}\pi$.

As a limit case, we have to handle the case when, among the triangles, there is one in which the sum of the angles is equal to π, as this case can only occur when the circle K_3 passes through one of the two intersection points of K_1 and K_2.

The circular triangle S'', for which the sum of the angles is a minimum, may be called the *reduced* circular triangle associated with the circular triangle S. For some questions, it is sufficient to replace the circular triangle S by the associated reduced circular triangle S''.

Plate I. Fig. 2 shows the construction of a circular triangle LMN with two straight-lined sides LM and LN, whose angles are respectively $\lambda\pi$, $\mu\pi$, $\nu\pi$, in the case when $\lambda < 1$, $\mu < 1$, $\nu < 1$; $\lambda + \mu + \nu < \frac{3}{2}$. The arcs AB, BC, CD of the circle with the center O correspond respectively to the central angles $\mu\pi$, $\lambda\pi$, $\nu\pi$. The arcs AM and DN are quadrants, the sides ML and NL are respectively parallel to the radii OB and OC. (In the figure, the factor π is omitted everywhere in the description of the angles.)

When $\lambda + \mu + \nu = 1$, a triangle with three straight-lined sides and angles $\lambda\pi, \mu\pi, \nu\pi$ replaces the circular triangle.

In order to be in the second of the two aforementioned cases 1) and 2), it is necessary and sufficient, as a more precise investigation shows, that the following four conditions be satisfied simultaneously:

$$\begin{aligned} \lambda + \mu + \nu &> 1, \\ -\lambda + \mu + \nu &< 1, \\ \lambda - \mu + \nu &< 1, \\ \lambda + \mu - \nu &< 1. \end{aligned}$$

When these conditions are satisfied, and only under this assumption, there exists a circle with center L and radius $\sqrt{OM^2 - OL^2}$, which is cut by each side of the circular triangle LMN in two diametrically opposed points. In this case, it is thus possible to transform conformally by reciprocal radii the plane of the circular triangle onto a *sphere*, such that to the circular triangle LMN corresponds a *spherical* triangle formed by three arcs of great circles.

The circular triangle S corresponds to the half-plane E situated on the positive side of the real axis of the plane x. One can now *analytically continue* the conformal map on each of the segments of the real axis $-\infty \cdots 0$, $0 \cdots 1$, $1 \cdots +\infty$ and thus extend the domain of the argument of the function s to the half-plane E_1 situated on the negative side of the real axis.

For each of these analytic continuations a very simple law applies, because to the *straight-lined* parts of the boundaries of E corresponds a *circular* boundary of S.

By the continuation of the transformation on one of the three segments of the real axis, there corresponds indeed to the half-plane E_1, exactly as for the half-plane E, a circular triangle which has a common side with S and which may be denoted by S_1. If now the points of the two half-planes E and E_1 are matched one to one through the symmetry with respect to the real axis, a point-by-point correspondence between the domains S and S_1 arises from this matching, and this correspondence is in fact that which is known under the name of *Möbius* circle-transformation or transformation by reciprocal radii. The circle to which the common side of the domains S and S_1 belongs is the directrix of this transformation (cf. Monatsberichte der Berl. Ak. 1870, p. 774.)

According to the aforementioned law, which may well be called the "Law of symmetry with respect to a circle", the domain S can also be analytically continued over each of its three sides and each of the three domains S_1 thus arising can be called a "symmetric repetition" of the domain S.

Obviously the same considerations about the analytic continuation are now also valid, according to the laws of symmetry, for the symmetric repetitions, as they are likewise bounded by circular arcs or straight-lined segments. From this, the idea arises first of all of an infinite number of domains S and S_1, of infinitely many conformal mappings of the half-planes E and E_1, which are all conveyed by the analytic function s and its analytic continuations.

In all the cases in which the function s is an *algebraic* function of x, the number of the domains S and S_1 arising in the given way and which are *different from one another* will be *finite*.

As this theorem can be inverted, in the sense that, when the number of the domains arising from the symmetric repetition of the domain S and different from each other is *finite*, the function s is an *algebraic* function of its argument x, the examination of a question which originally belonged to function theory is transformed into a geometrical problem, namely:

To find all circular triangles which only give rise, when reproduced according to the law of symmetry, to a finite number of circular triangles which are different by their situation and their form.

The consideration of the circles to which the sides of the circular triangle S and its symmetric repetitions belong now shows that the same circular lines which occur in the formation of the symmetric repetitions of the circular triangle S are also obtained for the associated reduced circular triangle S'' and its symmetric repetitions. Thus, it is permitted, when none of the three numbers λ, μ, ν is an integer, to limit oneself for the present aim of the research to the consideration of the associated reduced circular triangle.

The possibility of such a converse depends on the lemma that there is an identity between three hypergeometric series whose first three elements only differ by integers, which makes it possible in general to express each of the three hypergeometric series by the two others with rational functions of x as coefficients.

V

When the sum of the angles of the reduced circular triangle S'' (explained in Art. IV) is *smaller* than π, the first case occurs among the two aforementioned cases and there is thus a real circular line which cuts orthogonally the three circles to which the sides of S'' belong. One may assume that this circle does not reduce to a straight line, as this case can be avoided by an appropriate transformation by reciprocal radii. As already mentioned, it then occurs that one of the two reduced circular triangle is inside, the other outside, the orthogonal circle.

If one now constructs the symmetric repetitions of the domain S'' situated inside the orthogonal circle, it follows that they are also situated totally *inside* the orthogonal circle, moreover, that the circles to which the boundaries of the symmetric repetitions belong all cut the orthogonal circle at a right angle, and finally, that the vertices of the symmetric repetitions of S'' arrive arbitrarily close to the circumference of the orthogonal circle, but cannot reach it truly. Plate II. shows a figure designed under the assumption that $\lambda'' = \frac{1}{5}, \mu'' = \frac{1}{4}, \nu'' = \frac{1}{2}$.

By a finite number of symmetric repetitions, performed one after the other, the vertices always remain at a finite distance from the circumference and the repetition is never achieved.

In this case, the quantity s always has infinitely many determinations, thus is a transcendental function of x. On the contrary, the case may occur where, inversely, x is a uniform function of s; this happens when $\lambda = \lambda''$, $\mu = \mu''$, $\nu'' = \nu$ and when, at the same time, $\frac{1}{\lambda''}, \frac{1}{\mu''}, \frac{1}{\nu''}$ are three integers. However, there arises here the remarkable circumstance that the domain of the variable s is limited to the interior of a circle and that the periphery of this circle forms *a natural boundary* for each domain, over which there is no possible analytic continuation of the relation of dependence between function and argument in the ordinary sense.

In 1863, in his lectures on the theory of analytic functions, Mr. *Weierstrass* has drawn the attention on the possibility of such circumstance. From the same time also dates the origin of the example

$$\sqrt{\frac{2K}{\pi}} = 1 + 2q + 2q^4 + 2q^9 + \cdots,$$

in which the domain of q is always restricted to the interior of a circle; an example which I know of thanks to Mr. *Kronecker*.

I owe to an oral communication of Mr. *Weierstrass* in 1867 an example in which a natural boundary exists for the domain of the integral of an *algebraic* differential equation, while the argument varies without a limit, and which also comes from the theory of the elliptic functions; this example is closely related to the limit case $\lambda = 0, \mu = 0, \nu = 0$ which will be considered immediately.

One can also note that, in the above case, the situation of the natural boundary, apart from the numbers λ'', μ'', ν'', depends on the values which have been fixed for the constants of integration arising in the output element.

As far as I know, one finds the first printed communication on the occurrence of natural boundaries in a memoir of Mr. *Weierstrass*, Monatsberichte der Berl. Aka. 1866, p. 17. Mr. *Hankel*'s program paper "Ueber die unendlich oft oscillirenden und unstetigen Functionen", Tübingen 1870 contains several relevant examples.

One can connect to what precedes the observation of the two limit cases $\lambda'' + \mu'' + \nu'' = 0$ and $\lambda'' + \mu'' + \nu'' = 1$.

In the first case, the circular triangle S'' is a circular triangle with three cusps.

If one sets with Jacobi

$$K = \int_0^1 \frac{d\xi}{\sqrt{(1-\xi^2)(1-k^2\xi^2)}}, \qquad K' = \int_0^1 \frac{d\xi}{\sqrt{(1-\xi^2)(1-k'^2\xi^2)}},$$

where $k^2 + k'^2 = 1$, it thus follows for $\lambda = 0, \mu = 0, \nu = 0$

$$s = \frac{a'K + b'K'}{aK + bK'}$$

as a function of $x = k^2$. For real values of k and k' the quantities K and K' have a well-defined meaning; for complex values, they can be defined by simultaneous analytic continuations. (Cf. *Jacobi*, Fundamenta, p. 74; this journal, vol. 3, p. 194.)

In this case, the domain of the variable s is also bounded, while the quantity k^2 is an unbounded variable quantity. Conversely, k^2 and k'^2 as well as all the powers of these quantities arise as uniform functions of s, as soon as a, b, a', b' are given.

In the second limit case, $\lambda'' + \mu'' + \nu'' = 1$, the circular triangle S'' can be replaced by a straight-lined triangle with the angles $\lambda''\pi$, $\mu''\pi$, $\nu''\pi$ and indeed, in this case, one can imagine that the radius of the orthogonal circle has become infinitely big. Here, one must assume that among the three numbers λ'', μ'', ν'', none has the value 0.

The symmetric repetitions of a straight-lined triangle are again straight-lined triangles and the number of the symmetric repetitions which are different from each other is unlimited. In this limit case, s is thus also an infinitely multivalued function of x.

The special cases contained in this limit case, in which x is a *uniform* function of s, deserve a special mention. For this to happen, it is necessary and sufficient that $\frac{1}{\lambda''}, \frac{1}{\mu''}, \frac{1}{\nu''}$ be integers and that $\lambda = \lambda'', \mu = \mu'', \nu = \nu''$. Then either $\lambda'' = \mu'' = \nu'' = \frac{1}{3}$ or one of the numbers λ'', μ'', ν'', say λ'', is bigger than $\frac{1}{3}$. As $\frac{1}{\lambda''}$ should be an integer, this integer can only be equal to 2; it thus follows that $\lambda'' = \frac{1}{2}$; $\lambda'' + \nu'' = \frac{1}{2}$; one has either $\mu'' = \nu'' = \frac{1}{4}$, and, with this, one has found a second case $\lambda'' = \frac{1}{2}, \mu'' = \frac{1}{4}, \nu'' = \frac{1}{4}$, or one of the two numbers, say μ'', is bigger than $\frac{1}{4}$. As $\frac{1}{\mu''}$ should be an integer, it follows that $\mu'' = \frac{1}{3}$ and $\nu'' = \frac{1}{6}$.

In these three cases, in which the straight-lined triangle under consideration is equilateral, respectively right-angled isosceles, or has the angles 30°, 60°, 90°, x is a *uniform elliptic* function of the complex quantity s represented geometrically by a point situated in the plane of the straight-lined triangle.

One would be mistaken to believe that these are the only cases of the conformal mapping of a half-plane (x) on the surface of a plane straight-lined triangle (s), in which the dependence obtained under the hypothesis $\lambda+\mu+\nu = 1$ between x and s can be inverted without restriction, such that to each value of x correspond only a *finite* number of values of x. If the problem is formulated in this way, one case should be added to the ones already introduced, namely the case $\lambda'' = \frac{2}{3}, \mu'' = \frac{1}{6}, \nu'' = \frac{1}{6}$; because in this case x is a *double-valued elliptic* function of an argument u, depending on s by a bilinear equation, which can be expressed by the integral:

$$u = \int_0^x \frac{dx}{\sqrt[6]{x^2(1-x)^3}}.$$

With this are exhausted all the cases in which the conformal mapping of the surface of a straight-lined triangle on the surface of a half-plane is obtained by an analytic function, by which to each value of the unlimited variable argument correspond only a *finite* number of values of the function. (See this journal, vol. 70, p. 118.)

VI

The investigation for the second of the cases considered in Art.IV, in which the sum of the angles of the associated reduced circular triangle S'' is bigger than π, is different.

There is then no circumference orthogonally cutting the three circles K_1, K_2, K_3: on the contrary, there is a circle, which is cut by the three named circles in diametrically opposed points and whose center is the power center of the three circles. Let us now imagine that a sphere is constructed, for which this circle is a great circle, and that one projects from one of its two poles the two circular triangles S and S'' on the spherical surface, a conformal mapping of these circular triangles is then generated on the surface of the sphere and in fact great circles correspond to the boundaries of their sides. It is now advantageous to consider, instead of the plane circular triangle S'', the corresponding spherical triangle with the angles $\lambda''\pi$, $\mu''\pi$, $\nu''\pi$, which may be denoted by $S^\star$. For this triangle, the symmetric repetitions are also symmetric or congruent figures in the usual narrower sense, while the planes of the bounding sides are this time the planes of symmetry.

The problem is therefore reduced to the following:

To find all spherical triangles whose symmetric and congruent repetitions on the surface of the sphere only give rise to a finite number of spherical triangles differing by their situation.

When the spherical triangle $S^\star$ only leads to a finite number of symmetric and congruent repetitions differing by their situation, the function s, by which the conformal mapping of a half-plane E on this spherical triangle $S^\star$ is conveyed, is always an *algebraic* function. Indeed, the two *Riemann* surfaces which represent geometrically the domain of the argument x and the domain of the function s by unbounded change of the two variables, are in this case *closed* surfaces with a finite number of sheets. It follows from this (cf. Art. 20 of *Riemann*'s inaugural dissertation) that, under the given hypothesis, the two variables x and s are linked to each other by an *algebraic* equation.

The same reasoning that is indicated in what precedes is also mentioned at the beginning of Art. 13 in *Riemann*'s memoir "Ueber die Fläche vom kleinsten Inhalt bei gegebener Begrenzung" (Abhandlungen der Königlichen Gesellschaft der Wissenschaften zu Göttingen, vol. 13.). One reads there: "The minimal surface is determined as soon as one of the quantities u, η, X, Y, Z has been expressed by one of the remaining ones. This succeeds in many cases. Among them, particular consideration is deserved for those in which $\frac{du}{d\log\eta}$ is an algebraic function of η. For this it is necessary and sufficient that the mapping on the sphere and its symmetric and congruent continuations form a closed surface, which covers once or several times the whole sphere."

This place and the content of Article 18 in the same memoir are also related to the content of a little note of the author (Monatsberichte der Berliner Akademie, 1865, p. 149), in which it is mentioned among other things that with the conformal mapping of the total surface of the regular polyhedra on the sphere are associated a number of minimal surfaces, on which infinitely many straight lines are situated.

In general, for each of the cases considered by *Riemann*, in which the function $\frac{du}{d\log\eta}$ is an algebraic function of η, a relation can be proved with one of the regular polyhedra, among which, in the present case, the regular double pyramids must also be counted, inasmuch as a spherical polygon together with its symmetric and congruent repetitions on the spherical surface can form a closed *Riemann* surface only when the planes to which the sides of this polygon belong are the planes of symmetry of a regular polyhedron.

The problem, to which the investigation has led here, is already indicated in *Steiner*'s article: "Einfache Beweise der isoperimetrischen Hauptsätze" in this journal, vol. 18, p. 295, but only partially solved. One can find the complete solution in a remark in another memoir of the same geometer, this journal vol. 24, p. 247, in which place, however, the relation to the regular polyhedra is not mentioned.

Apart from *Steiner*'s memoir, one can also mention a memoir of Mr. *C. Jordan*, "Recherches sur les polyèdres", this journal, vol. 66.

Through geometrical considerations of, by the way, a very simple nature, into which we will not enter closer here, one gets the insight that a spherical triangle leads to a finite number of symmetric or congruent repetitions different from each other only when the three planes in which the sides of the triangle are situated are the three planes of symmetry of a regular polyhedron. Apart from this, however, there is also the case in which two angles of the spherical triangle are right angles. Then the necessary and sufficient condition for the spherical triangle to offer only a finite number of symmetrical repetitions is that the third angle be commensurable with π.

When one excludes this case, which corresponds to a regular double pyramid, and sets oneself the task to discover all the remaining possible cases, it is only necessary to consider the sections of the planes of symmetry of the regular bodies with a concentric sphere and to seek out the spherical triangles occurring here.

From what precedes, it is sufficient to consider only the reduced circular triangles S'', different from one another, that occur here.

One then obtains for the numbers λ'', μ'', ν'' the following table, which is organized such that $\lambda'' \geq \mu'' \geq \nu''$ and that, inside each group, the sum of the angles $(\lambda'' + \mu'' + \nu'')\pi$ for a case is bigger than the preceding one or at least equal.

Table

containing, apart from the common factor π, the value of the angles and the area of the surface of the reduced spherical triangles, which are determined on a sphere of radius 1 by the planes of symmetry of a concentric regular double pyramid or of a concentric regular polyhedron.

No.	λ''	μ''	ν''	$\frac{\text{Area}}{\pi}$	Polyhedron
I.	$\frac{1}{2}$	$\frac{1}{2}$	ν	ν	Regular double pyramid
II.	$\frac{1}{2}$	$\frac{1}{3}$	$\frac{1}{3}$	$\frac{1}{6} = A$	Tetrahedron
III.	$\frac{2}{3}$	$\frac{1}{3}$	$\frac{1}{3}$	$\frac{1}{3} = 2A$	
IV.	$\frac{1}{2}$	$\frac{1}{3}$	$\frac{1}{4}$	$\frac{1}{12} = B$	Cube and Octohedron
V.	$\frac{2}{3}$	$\frac{1}{4}$	$\frac{1}{4}$	$\frac{1}{6} = 2B$	
VI.	$\frac{1}{2}$	$\frac{1}{3}$	$\frac{1}{5}$	$\frac{1}{30} = C$	Dodecahedron and Icosahedron
VII.	$\frac{2}{5}$	$\frac{1}{3}$	$\frac{1}{3}$	$\frac{1}{15} = 2C$	
VIII.	$\frac{2}{3}$	$\frac{1}{5}$	$\frac{1}{5}$	$\frac{1}{15} = 2C$	
IX.	$\frac{1}{2}$	$\frac{2}{5}$	$\frac{1}{5}$	$\frac{1}{10} = 3C$	
X.	$\frac{3}{5}$	$\frac{1}{3}$	$\frac{1}{5}$	$\frac{2}{15} = 4C$	
XI.	$\frac{2}{5}$	$\frac{2}{5}$	$\frac{2}{5}$	$\frac{1}{5} = 6C$	
XII.	$\frac{2}{3}$	$\frac{1}{3}$	$\frac{1}{5}$	$\frac{1}{5} = 6C$	
XIII.	$\frac{4}{5}$	$\frac{1}{5}$	$\frac{1}{5}$	$\frac{1}{5} = 6C$	
XIV.	$\frac{1}{2}$	$\frac{2}{5}$	$\frac{1}{3}$	$\frac{7}{30} = 7C$	
XV.	$\frac{3}{5}$	$\frac{2}{5}$	$\frac{1}{3}$	$\frac{1}{3} = 10C$	

This results in the rule:

Let us consider the three numbers λ, μ, ν which are assumed to be rational, none of which being an integer, reduce them to their lowest terms and then conclude as follows.

1. If two of the three denominators are equal to 2, then s is an *algebraic* function of x (No. I of the table.)

2. If one of the three denominators is bigger than 5, while the two other denominators are not equal to 2 at the same time, then s is a *transcendantal* function of x.

3. If all the three denominators are smaller than 6, let us take the absolute value λ', μ', ν' of the smallest residues of λ, μ, ν modulo 2 and choose from the 4 systems

$$\begin{array}{ccc} \lambda' & \mu' & \nu' \\ \lambda' & 1-\mu' & 1-\nu' \\ 1-\lambda' & \mu' & 1-\nu' \\ 1-\lambda' & 1-\mu' & \nu' \end{array}$$

the system of three numbers which gives the smallest possible sum. Let us denote the three numbers of this system by λ'', μ'', ν'' according to their magnitude, such that $\lambda'' \geq \mu'' \geq \nu''$; then three cases are to be distinguished:

a. If the sum $\lambda'' + \mu'' + \nu''$ is smaller than 1 or equal to 1, s is a *transcendental* function of x.

b. If the sum is bigger than 1, while, however, the system λ'', μ'', ν'' does not coincide with any of the 15 systems of values in the table given above, s is a *transcendental* function of x.

c. But if the system λ'', μ'', ν'' coincides with one of the 15 systems of the table, s is an *algebraic* function of x. —

In the case $\lambda = \frac{1}{2}$, $\mu = \frac{1}{2}$, one obtains $\alpha = \frac{1-\nu}{2}$, $\beta = -\frac{\nu}{2} = \alpha - \frac{1}{2}, \gamma = \frac{1}{2}$;

$$F(\alpha, \alpha - \frac{1}{2}, \frac{1}{2}, x) = \frac{1}{2}[(1+\sqrt{x})^\nu + (1-\sqrt{x})^\nu].$$

Cf. the formulas II. and IV. of *Gauss*'s tables (Werke, vol. III, p. 127). A second integral of the differential equation is

$$x^{\frac{1}{2}} F(\alpha, \alpha + \frac{1}{2}, \frac{3}{2}, x) = \frac{1}{2\nu}[(1+\sqrt{x})^\nu - (1-\sqrt{x})^\nu].$$

It follows that the quantity s is a rational function of the first degree of $(\frac{1+\sqrt{x}}{1-\sqrt{x}})^\nu$ and, for the limit case $\nu = 0$, of $\frac{1}{2}\log\frac{1+\sqrt{x}}{1-\sqrt{x}}$.

Less simple is the algebraic relation between the two variable quantities s and x for the case designated by II. in the above table. It seems appropriate to give for this case the results with some detail.

Example. Investigation of the special case $\lambda = \frac{1}{3}, \mu = \frac{1}{3}, \nu = \frac{1}{2}$.

Let ξ, η, ζ be Cartesian coordinates — think of a regular tetrahedron in the situation that two opposite edges, one of which is the ξ-axis, and the other is parallel to the η-axis, are cut by the ζ-axis. Let the origin of the coordinates coincide with the center of the tetrahedron. Around it let us describe a sphere with radius 1 and let us project on it, from the center point, the vertices, the mid-points of the faces and of the edges of the tetrahedron. The planes of symmetry of the tetrahedron, in which these points are situated, cut the surface of the sphere in 24 congruent spherical triangles, whose angles are $60°$, $60°$, $90°$.

From the point $\xi = 0, \eta = 0, \zeta = 1$, let us project stereographically the surface of the sphere on the plane $\zeta = 0$ and let us consider a point of this projection, whose coordinates are respectively $\xi_1, \eta_1, 0$, as the geometrical representative of the complex quantity $\xi_1 + \eta_1 i = s$.

The tetrahedron may have, among the two situations satisfying the given conditions, the one whose vertices correspond to the roots of the equation $1 - 2\sqrt{3}s^2 - s^4 = 0$; the mid-points of the faces then correspond to the roots of the equation $1+2\sqrt{3}s^2 - s^4 = 0$ and the mid-points of the edges correspond to the roots of the equation $s(1 + s^4) = 0$, to which $s = \infty$ is added.

The conformal mapping of one of the 24 aforementioned spherical triangles on a half-plane, in such a way that by the transformation of the total surface of the sphere, the point $x = 0$ corresponds to the vertices of the tetrahedron, the point $x = \infty$ to the mid-points of the faces and the point $x = 1$ to the mid-points of the edges, gives the equation

$$(\frac{1 - 2\sqrt{3}s^2 - s^4}{1 + 2\sqrt{3}s^2 - s^4})^3 = x.$$

The consideration of the value $x = 1$ leads to the identity:

$$(1 + 2\sqrt{3}s^2 - s^4)^3 - (1 - 2\sqrt{3}s^2 - s^4)^3 = 12\sqrt{3}(s(1 + s^4))^2.$$

When the equation between x and s is solved as a function of s and, for brevity,

$$e^{i\frac{\pi}{12}} = \frac{\sqrt{3}+1}{2\sqrt{2}} + \frac{\sqrt{3}-1}{2\sqrt{2}}i$$

is denoted by δ, $e^{\frac{2}{3}\pi i} = \delta^8$ by ϵ, one of the twelve values of s is given by the formula

$$s = -\sqrt{-i} \cdot \sqrt{\frac{\delta\sqrt{1 - \epsilon\sqrt[3]{x}} - \delta^{-1}\sqrt{1 - \epsilon^2\sqrt[3]{x}}}{\delta\sqrt{1 - \epsilon\sqrt[3]{x}} + \delta^{-1}\sqrt{1 - \epsilon^2\sqrt[3]{x}}}},$$

where s has the value $-\frac{\sqrt{3}-1}{\sqrt{2}}$ for $x = 0$. From this, one obtains, when one sets $\alpha = \frac{1}{4}$, $\beta = -\frac{1}{12}$, $\gamma = \frac{2}{3}$, while using the formulas communicated in Art. II, the two following particular integrals of the differential equation of the hypergeometric series:

$$y_1 = \sqrt{\delta\sqrt{1 - \epsilon\sqrt[3]{x}} - \delta^{-1}\sqrt{1 - \epsilon^2\sqrt[3]{x}}}, \qquad y_2 = \sqrt{\delta\sqrt{1 - \epsilon\sqrt[3]{x}} + \delta^{-1}\sqrt{1 - \epsilon^2\sqrt[3]{x}}}.$$

As each of the two branches is exchanged with the other, when $\sqrt[3]{x}$ describes a loop around the point $\sqrt[3]{x} = \epsilon$, y_1 and y_2 are two different branches of the same algebraic function y.

When $\sqrt[3]{x}$ describes a loop around the point $\sqrt[3]{x} = \epsilon^2$, y_1^2 becomes $-y_2^2$, and y_2^2 becomes $-y_1^2$. It follows that y^4 is a double-valued function of $\sqrt[3]{x}$ and that y is a 24-valued function of x. If now $\sqrt[3]{x}$ describes a loop twice around the point $x = 0$, the two branches y_1, y_2 become two new branches y_3, y_4 after the first loop and, after the second loop, two new branches y_5, y_6:

$$y_3 = \sqrt{\delta\sqrt{1 - \epsilon^2\sqrt[3]{x}} - \delta^{-1}\sqrt{1 - \sqrt[3]{x}}}, \qquad y_4 = \sqrt{\delta\sqrt{1 - \epsilon^2\sqrt[3]{x}} + \delta^{-1}\sqrt{1 - \sqrt[3]{x}}},$$

$$y_5 = \sqrt{\delta\sqrt{1 - \sqrt[3]{x}} - \delta^{-1}\sqrt{1 - \epsilon\sqrt[3]{x}}}, \qquad y_6 = \sqrt{\delta\sqrt{1 - \sqrt[3]{x}} + \delta^{-1}\sqrt{1 - \epsilon\sqrt[3]{x}}}.$$

The function y satisfies a linear differential equation of the second order, and thus $y_3, \ldots, y_6$ can be expressed linearly and with constant coefficients by y_1 and y_2; in fact, one gets:

$$y_3 = \frac{1}{\sqrt{2}} \cdot \delta(+iy_1 + y_2), \qquad y_4 = \frac{1}{\sqrt{2}} \cdot \delta(-iy_1 - y_2)$$
$$y_5 = \frac{1}{\sqrt{2}} \cdot \delta^{-1}(-iy_1 + iy_2), \qquad y_6 = \frac{1}{\sqrt{2}} \cdot \delta^{-1}(y_1 + y_2).$$

Now one has:

$$\begin{aligned} y_1 + y_2 + y_3 =& \frac{1}{\sqrt{2}}(\sqrt{2} + i\delta - i\delta^{-1} i)y_1 + \frac{1}{\sqrt{2}}(\delta + i\delta^{-1})y_2 \\ =& 3\sqrt{i} \cdot \sqrt[4]{\frac{\delta - \delta^{-1}}{i(\delta + \delta^{-1})}} \cdot \{\frac{1}{\sqrt{3}}\sqrt{\frac{\delta - \delta^{-1}}{2i}} \cdot \frac{y_1}{\sqrt{i}} + \frac{1}{\sqrt{3}}\sqrt{\frac{\delta + \delta^{-1}}{2}} \cdot y_2\}. \end{aligned}$$

From this it follows that:

$$F(\frac{1}{4}, -\frac{1}{12}, \frac{2}{3}, x) = \frac{1}{\sqrt{3}}\sqrt{\frac{\delta - \delta^{-1}}{2i}} \cdot \frac{y_1}{\sqrt{i}} + \frac{1}{\sqrt{3}}\sqrt{\frac{\delta + \delta^{-1}}{2}} \cdot y_2.$$

The conformal mapping of the surface of the sphere on the surface of the regular tetrahedron is given by the elliptic integral

$$u = C \cdot \int \frac{ds}{\sqrt{1 - 2\sqrt{3} \cdot s^2 - s^4}}.$$

For it, the invariant that Mr. *Weierstrass* has denoted g_2 has the value 0. For the multiplication of this integral by a third root of unity ϵ, an algebraic multiplication theorem occurs. One can now ask to which of *Riemann*'s classes the algebraic functions considered here belong, or, respectively, the equations by which they are expressed.

The equation between x and s belongs to the *first* (rational) class ($p = 0$), since x can be expressed rationally by s.

The equation between x and y_1 belongs to the *second* class ($p = 1$), the class from which the elliptic integrals come; for x can be expressed rationally by y_1 and $\sqrt{y_1^4 - \sqrt{3}}$,

$$x = [1 - \sqrt{3} \cdot y_1^4 + 2y_1^2\sqrt{y_1^4 - \sqrt{3}}]^3.$$

The elliptic integrals arising from this equation are thus in particular *lemniscate* integrals.

On the other hand, the equation that exists between y_1 and s belongs to the *fourth* of *Riemann*'s classes ($p = 3$), since three of the integrals belonging to this equation, linearly independent from each other, are transformed elliptic integrals of the first kind. Cf. an article by Mr. *Röthig*, this journal, vol. 56, p. 197.

It indeed follows that:

$$y_1 = \frac{-\sqrt[8]{3}(1+i)s}{\sqrt[4]{1 + 2\sqrt{3}s^2 - s^4}}, y_2 = \frac{\sqrt[8]{3} \cdot \sqrt{2}}{\sqrt[4]{1 + 2\sqrt{3}s^2 - s^4}}, \sqrt{y_1^4 - \sqrt{3}} = i\sqrt[4]{3} \cdot \frac{1 + \sqrt{3}s^2}{\sqrt[4]{1 + 2\sqrt{3}s^2 - s^4}}.$$

The three integrals of the first kind are the following:

$$\int \frac{ds}{\sqrt[4]{(1+2\sqrt{3}s^2-s^4)^3}}, \quad \int \frac{ds}{\sqrt[4]{(1+2\sqrt{3}s^2-s^4)}}, \quad \int \frac{sds}{\sqrt[4]{(1+2\sqrt{3}s^2-s^4)^3}}.$$

The first and the last of these three integrals are transformed lemniscate integrals, as follows from the equations:

$$\frac{dy_1}{\sqrt{y_1^4-\sqrt{3}}} = \frac{-1+i}{\sqrt[8]{3}} \cdot \frac{ds}{\sqrt[4]{(1+2\sqrt{3}s^2-s^4)^3}}, \qquad \frac{dy_2}{\sqrt{y_1^4-\sqrt{3}}} = -\frac{\sqrt{2}}{\sqrt[8]{3}} \cdot \frac{sds}{\sqrt[4]{(1+2\sqrt{3}s^2-s^4)^3}}.$$

The second conveys the conformal mapping of the surface of the sphere onto the surface of a regular tetrahedron; an algebraic multiplication theorem occurs for the multiplication by a fourth root of unity in the case of the first and the last integrals, for the multiplication by a third root of unity in the case of the second integral.

The cases denoted by IV. and VI. in the table above can be treated in a similar way, only with some more computational effort.

Here are the equations relating x and s for these cases, with a choice of the available constants which provides the greatest possible simplicity to the final result. On this occasion, the formulas by which the conformal mapping of the surface of the regular polyhedron into the surface of the sphere is expressed may also find a place here.

If one imagines a regular octahedron in the position where its vertices lie on the axes of coordinates, then, by the projection of the vertices on the sphere and by the projection of the sphere on the plane $\zeta = 0$, the roots of the equation $s(1-s^4) = 0$ and $s = \infty$ correspond to the vertices of the octahedron, the mid-points of the faces to the roots of the equation $1 + 14s^4 + s^8 = 0$, and the mid-points of the edges to the roots of the equation $1 - 33s^4 - 33s^8 + s^{12} = 0$.

If one constructs the polar figure of the octahedron, that is, the cube, one obtains the same equations, only up to the exchange of the equations relative to the vertices and those relative to the midpoints of the faces.

The conformal mapping of the surface of the sphere on the surface of the octahedron is conveyed by the integral

$$u = C \cdot \int \frac{ds}{\sqrt[3]{s(1-s^4)}}.$$

This integral belongs to these integrals of algebraic functions which can be reduced by an *algebraic* transformation to elliptic integrals. The substitution is the following: $s = t^{\frac{3}{2}}$ leads to $u = \frac{3}{2}C \cdot \int \frac{dt}{\sqrt[3]{1-t^6}}$, through which the surface of the circle $[t] \leq 1$ is conformally mapped onto the interior of a regular hexagon in the plane u. If one sets $t^3 - \frac{1}{t^3} = -2v$, one has:

$$t^3 = -v + \sqrt{v^2+1}, \quad 1 - t^6 = 2v \cdot t^3, \quad dt = -\frac{1}{3}t \cdot \frac{dv}{\sqrt{v^2+1}},$$

$$u = -\frac{C}{2\sqrt[3]{2}} \int \frac{dv}{\sqrt[3]{v} \cdot \sqrt{v^2+1}}; \quad v = w^{\frac{3}{2}}, \quad u = -\frac{3C}{4\sqrt[3]{2}} \int \frac{dw}{\sqrt{w^3+1}},$$

$$(s^2 - \frac{1}{s^2})^2 = 4w^3.$$

The doctoral thesis of Mr. *Amstein*, which is printed in the Vierteljahrsschrift der Naturforschenden Gesellschaft in Zürich, XVI. Jahrgang, 1871, pp. 297–341, contains more details on the conformal mapping of the surface of the regular octahedron on the surface of the sphere.

The conformal map of the sphere on the surface of a cube is conveyed by the integral $u = C \cdot \int \frac{ds}{\sqrt[4]{1+14s^4+s^8}}$. This integral is also reducible by an algebraic transformation to an elliptic one, and, in fact, to a lemniscate one.

The function

$$\frac{(1+14s^4+s^8)^3}{4\cdot 27(s(1-s^4))^4} = x$$

conveys the conformal mapping of a spherical triangle with the angles $\frac{\pi}{3}, \frac{\pi}{4}, \frac{\pi}{2}$ $(60°, 45°, 90°)$, whose vertices are formed respectively by a vertex of the inscribed cube, by a vertex of the inscribed regular octahedron and by the projection of a mid-point of an edge, into a half-plane such that the points $0, \infty, 1$ to correspond to these points. If one considers the value $x = 1$, one obtains the identity:

$$(1+14s^4+s^8)^3 - 4\cdot 27(s(1-s^4))^4 = (1-33s^4-33s^8+s^{12})^2,$$

which may serve as the solution of the problem: to solve in rational numbers the equation $p^3 - Aq^4 = r^2$, when A is an integer.

If one imagine a regular icosahedron and a regular dodecahedron inscribed in a sphere, such that the two surfaces are polar figures with respect to a concentric sphere, the roots of the equation $\phi_{12}(s) = 0$, where $\phi_{12}(s) = s(1-11s^5-s^{10})$, correspond to the vertices of the icosahedron and to the mid-points of the faces of the dodecahedron, with a suitable position of them with respect to the system of coordinates except for the value $s = \infty$, the roots of the equation $\phi_{20}(s) = 0$, where $\phi_{20}(s) = 1 + 228s^5 + 494s^{10} - 228s^{15} + s^{20}$, correspond to the mid-points of the faces of the icosahedron and the vertices of the dodecahedron, and the roots of the equation $\phi_{30}(s) = 0$, where

$$\phi_{30}(s) = 1 - 522s^5 - 10005s^{10} - 10005s^{20} + 522s^{25} + s^{30},$$

correspond to the mid-points of the edges of both surfaces. The conformal mapping of the surface of the sphere on the surface of a regular icosahedron is given by the integral

$$u = \int \frac{ds}{\sqrt[6]{s(1-11s^5-s^{10})}} = \int \frac{ds}{\sqrt[6]{\phi_{12}(s)}},$$

which, by an appropriate algebraic substitution, can be transformed into an *elliptic* integral of the first kind.

The conformal mapping of the surface of the sphere on the surface of a regular dodecahedron is conveyed by the integral

$$u = \int \frac{ds}{\sqrt[10]{\phi_{20}(s)}}$$

whereby it should be noted that this integral is not, contrarily to those considered until now, a transformed elliptic one.

[The integral

$$u = \int \frac{ds}{\sqrt[15]{\phi_{30}(s)}}$$

conveys the conformal mapping of the surface of a sphere onto the surface of a half-regular polyhedron, which is formed by twelve regular pentagons and twenty equilateral triangles.]

The conformal mapping of the surface of a spherical triangle with the angles $\frac{\pi}{3}, \frac{\pi}{5}, \frac{\pi}{2}$ on a half-plane x such that the values $x = 0, x = \infty, x = 1$ correspond respectively to its vertices is conveyed by the function

$$\frac{[\phi_{20}(s)]^3}{4^3 \cdot 3^3 [\phi_{12}(s)]^5} = x.$$

Taking into account the value $x = 1$, one obtains the following identity:

$$[\phi_{20}(s)]^3 - 4^3 \cdot 3^3 \cdot [\phi_{12}(s)]^5 = [\phi_{30}(s)]^2.$$

VII

The case when one or two of the three numbers λ, μ, ν are *integers* has been ruled out in the investigations until now. This case should now be discussed.

When the number λ has the value 0, the integral of the differential equation

$$\Psi(s, x) = F(x)$$

is a transcendental function of x, because the particular integral σ found earlier in this case always contains a logarithmic term which does not disappear. (See equation (L*) in Art. III.)

When one of the three numbers, say λ, is equal to an *integer* m, it is necessary, for the integral s to have the character of an algebraic function in the neighborhood of the corresponding singular value $x = 0$, that the coefficient B_m of this logarithmic term have the value 0 and thus that the value $x = 0$ be an accidental singular value for the differential equation, according to Mr. *Weierstrass*'s terminology.

In Art. III the condition for the cancellation of this coefficient is generally reduced to the cancellation of a certain determinant $\mid a_{gh} \mid$, which is an entire function with integer values of the m-th degree with respect to the coefficients a occurring in it; only the coefficient a_0 intervenes with the highest power and, in fact, a_0^m has the factor 1; this term is also the only term contained in $\mid a_{gh} \mid$ of the m-th dimension.

We now have to examine in particular the condition $\mid a_{gh} \mid = 0$ in the case of the hypergeometric series.

Following the notation introduced in Art. III, one has:

$$a_0 + a_1 x + a_2 x^2 + \cdots = x[F(x) - \frac{1 - m^2}{2x^2}] = \frac{1 - \nu^2 + \mu^2 - m^2}{2(1 - x)} + \frac{(1 - \nu^2) \cdot x}{2(1 - x)^2}.$$

It follows that

$$a_n = \frac{1}{2}(1 - \nu^2 + \mu^2 - m^2) + \frac{1}{2} n (1 - \nu^2).$$

Thus $2^m \cdot \mid a_{gh} \mid$ is an entire function with integer coefficients of the $2m$-th degree of μ and ν and the coefficient of μ^{2m} in it is always equal to 1.

One can now show that this function is decomposable in $2m$ linear factors built from μ and ν with integer coefficients and different from each other, which have the form

$$\mu \pm \nu \pm (m - m')$$

where m' runs over all odd positive integers not bigger than m. Since the number of different expressions of this form, when all possible combinations of signs and all possible values of m' are taken into account, is exactly equal to $2m$, it is only necessary to prove that the general integral of the differential equation of the hypergeometric series is always unramified in the neighborhood of the value $x = 0$, when there is a relation between μ and ν such that one of the $2m$ linear expressions just explained has the value 0. It then follows that each of the $2m$ expressions has to be a factor of $2^{m}\cdot \mid a_{gh} \mid$, and that thus

$$2^{m}\cdot \mid a_{gh} \mid = \prod [\mu \pm \nu \pm (m - m')].$$

For the objective of the present investigation one can, when μ and ν are variables, write $-\mu$ instead of μ and $-\nu$ instead of ν, without prejudice to the generality, since only μ^2 and ν^2 occur in the differential equation $\Psi(s, x) = F(x)$. Therefore one can set $\mu = \nu - (m - m')$.

Let m'' denote a positive integer, which is smaller than $\frac{1}{2}m$, 0 not excluded. One can set $m' = 2m'' + 1$, and thus $\mu = \nu - m + 2m'' + 1$.

When one now sets

$$\alpha = \frac{-\lambda + \mu - \nu + 1}{2}, \quad \beta = \frac{-\lambda - \mu - \nu + 1}{2}, \quad \gamma = 1 - \lambda, \quad \lambda = m,$$

thus

$$\alpha = -(m - m'' - 1), \quad \beta = -(m'' + \nu), \quad \gamma = -(m - 1),$$

one gets the following linearly independent integrals of the differential equation of the hypergeometric series

$$\text{[1.]} \quad F(\alpha, \beta, \gamma, x) = F(-(m - 1 - m''), -m'' - \nu, -(m - 1), x),$$

$$\text{[3.]} \quad x^{1-\gamma} F(\alpha - \gamma + 1, \beta - \gamma + 1, 2 - \gamma, x) = x^{m} F(m'' + 1, m - m'' - \nu, 1 + m, x).$$

Among these two integrals the first represents an entire function of x of degree $m - m'' - 1$ and the second also possesses the character of an entire function in the neighborhood of the value $x = 0$. It follows that the value $x = 0$ for the differential equation of the hypergeometric series as well as for the differential equation $\Psi(s, x) = F(x)$ is *accidentally singular*; consequently, $\mu - \nu + (m - m'')$ is a factor of $2^{m}\cdot \mid a_{gh} \mid$.

Likewise all the other expressions of the form $\mu \pm \nu \pm (m - m')$ must also be factors of $2^{m}\cdot \mid a_{gh} \mid$. It thus follows that

$$2^{m}\cdot \mid a_{gh} \mid = \prod [\mu \pm \nu \pm (m - m')].$$

With the condition of the cancellation of one of the $2m$ factors, we have therefore exhausted all the possibilities in which the value $x = 0$ is an accidental singular point of the differential equation. In order to investigate if the point $x = 0$ is an accidental singular point or not for the differential equation (A.), let us construct the quantities $\lambda = 1 - \gamma$, $\mu = \alpha - \beta$, $\nu = \gamma - \alpha - \beta$ and denote the absolute values of λ, $\mu + \nu$, $\mu - \nu$ by $[\lambda]$, $[\mu + \nu]$, $[\mu - \nu]$. It is then necessary and sufficient, in order that $x = 0$ be an accidental singular point, that

1. λ be an integer different from 0, and that
2. one of the two differences $[\lambda] - [\mu + \nu]$ and $[\lambda] - [\mu - \nu]$ be a positive odd integer.

The question now arises: when $x = 0$ is an accidental singular point of the differential equation, as henceforth assumed, which condition must be added for the general integral of the differential equation to be an algebraic function of x ?

When the number ν does not denote an integer, the two integrals $F(\alpha, \beta, \gamma, x)$ and

$$[7.] \quad (1-x)^{\gamma-\alpha-\beta} F(\gamma-\alpha, \gamma-\beta, \gamma-\alpha-\beta+1, 1-x)$$

are linearly independent from each other; the last integral is equal to

$$(1-x)^{\nu} F(-m'', -m+m''+1+\nu, 1+\nu, 1-x);$$

the hypergeometric series F intervening here represents an entire function of $1-x$ of degree m'' and the differential equation (A.) therefore has two particular integrals whose logarithmic derivatives are rational functions of x; — the case where ν is equal to a negative integer is however to be excluded here.

Since, in what precedes, the change of ν to $-\nu$ was allowed, the case in which ν is equal to an integer has to be treated specifically.

When ν is a rational number (not an integer), there is, besides the integral $F(\alpha, \beta, \gamma, x)$, still a second particular integral [7.] which is linearly independent from the first and an algebraic function of x, provided that the point $x = 0$ is an accidental singular point.

A remark may find a place here, concerning two special cases in which the conformal mapping conveyed by the function s offers some interest:

Let ν be a positive number which is smaller than 1 and let us put $\lambda = m = 2$, $\mu = 1 - \nu$, $\alpha = 0$, $\beta = \nu - 1$, $\gamma = -1$. It follows that $s = (1 - \nu \cdot x)(1-x)^{-\nu}$. This function conveys the conformal mapping of the half-plane x situated on the positive side of the real axis onto a domain whose boundary consists of two straight lines. Plate I. Fig. 3^a.

By a transformation by reciprocal radii, this domain can be reduced to the interior of a circle, where the surface of this circle along a circular arc forming with the circumference the angle $\nu\pi$ appears to be cut up to an interior point of the surface of the circle. Plate I. Fig. 3^b.

The second case is obtained from the preceding one, when ν is changed into $-\nu$. Plate I. Fig. 4^a and 4^b.

It now still remains to investigate the case in which, besides λ, ν is also put equal to an integer, and when then, from what precedes, μ is also equal to an integer. With this hypothesis, for the general integral of the differential equation of the hypergeometric series to be an algebraic function of x, it is necessary that the point $x = 1$ not be a ramification point for it; when this condition is satisfied, the point $x = \infty$ too is not a ramification point.

It is sufficient to assume that the three numbers λ, μ, ν are positive, since only their squares enter into the equation $\Psi(s, x) = F(x)$.

A condition for the point $x = 0$ to be an accidental singular point has been found:

Either

$$\lambda - \mu - \nu \qquad \text{or} \qquad \lambda - [\mu - \nu],$$

where $[\mu - \nu]$ denotes the absolute value of $\mu - \nu$, has to be a odd positive number. From this, it follows that only two cases are permitted: either from the three integers λ, μ, ν one is odd, while the two others are even, or all three are odd.

For the point $x = 1$ to be a ramification point, it is necessary that either

$$\nu - \mu - \lambda \quad \text{or} \quad \nu - [\lambda - \mu]$$

be an odd positive number.

These two requirements are then only compatible when at the same time $\lambda - [\mu - \nu]$ and $\nu - [\lambda - \mu]$ are positive. It follows that

$$\lambda + \nu > \mu, \qquad \lambda + \mu > \nu, \qquad \mu + \nu > \lambda.$$

Thus the three quantities λ, μ, ν should have the same properties as the measures of the three sides of a straight-lined triangle: the sum of any two of them must be bigger than the third one.

From this, the necessary and sufficient conditions for the integral of the differential equation

$$\Psi(s, x) = \frac{1-\lambda^2}{2x^2} + \frac{1-\nu^2}{2(1-x)^2} - \frac{\lambda^2 - \mu^2 + \nu^2 - 1}{2x(1-x)}$$

to be a *rational* function of x are the following:

1. The three numbers λ, μ, ν must be real integers different from 0.
2. The sum of them must be an odd number.
3. The sum of the absolute values of any two of the three numbers must be bigger than the absolute value of the third one.

If these conditions are satisfied, the general integral of the differential equation (A.) is also a rational function of x and indeed the cases resulting here completely coincide with those considered in Art. I under 1^c and 3^c, as well as under 2^c and 4^c.

Zürich, 1872.

Translation by Catherine Goldstein

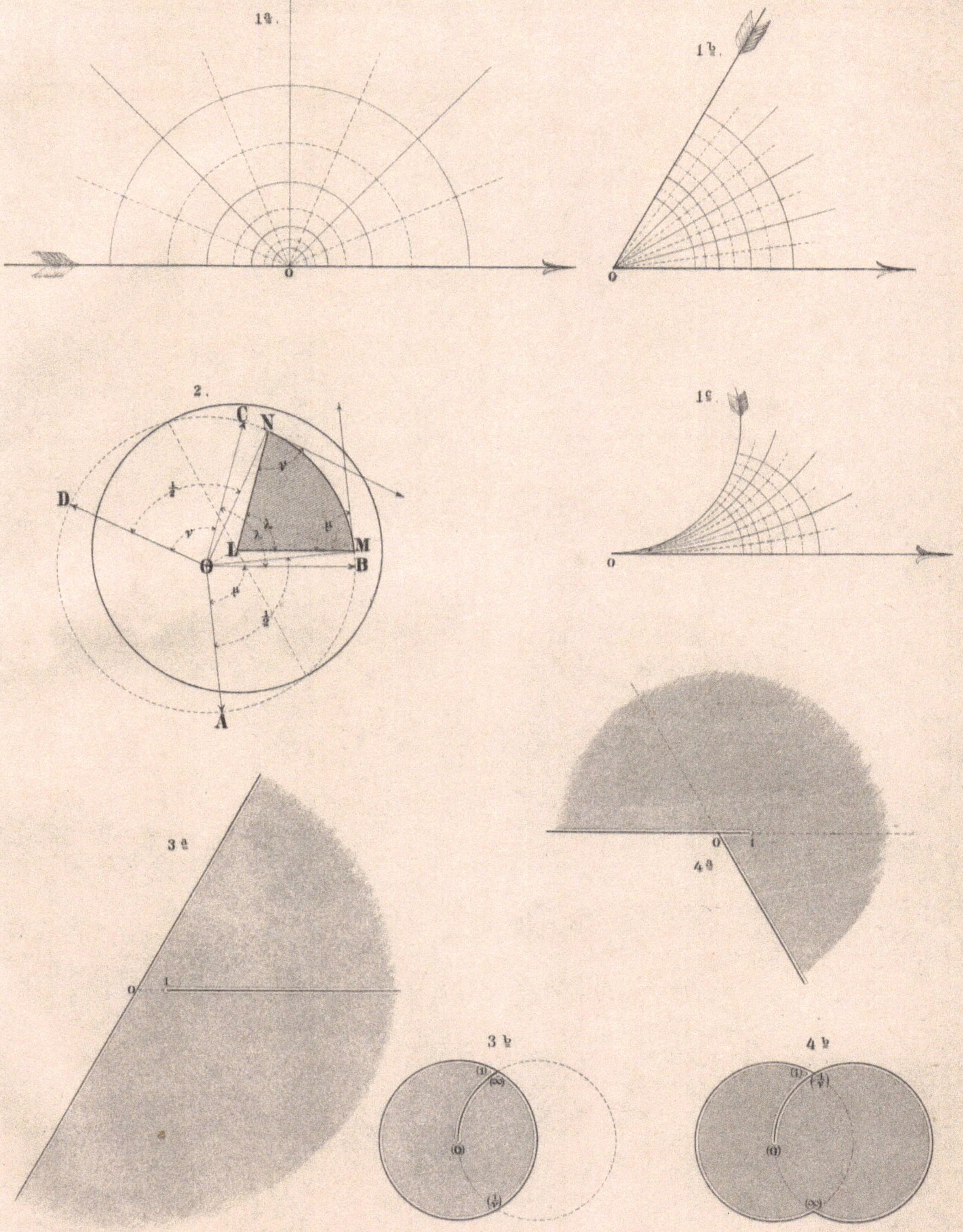
1a.
0
1b.
0
2.
C
N
D
O
L
M
B
A
1c.
0
3a
0
1
4a
0
1
3b
(1)
(∞)
(0)
4b
(1)
(0)
(∞)

Lith. Anst v. Leopold Kraatz in Berlin

At the crossroads: H. A. Schwarz on linear differential equations

By *Catherine Goldstein* at Paris

Abstract. This brief commentary on Hermann Amandus Schwarz's article "On the cases when the Gaussian hypergeometric series represents an algebraic function of its fourth element", whose subject matter seems somewhat isolated within Schwarz's body of work, aims to explain its context, to summarize some of its main features, and to discuss briefly some aspects of its reception in both the short and longer term.

1. Introduction

On August 22, 1871, a special session on mathematics was held as part of the 45th annual meeting of the *Schweizerische Naturforschende Gesellschaft*, the Swiss Society of Natural Sciences. After a communication by Carl Friedrich Geiser on wave surfaces, Hermann Amandus Schwarz, his colleague at the Federal Polytechnic School (now ETH) in Zurich who chaired the session, introduced his own results on the algebraic solutions of a particular second-order linear differential equation, the hypergeometric differential equation—the complete presentation of these results, published in the *Journal für die reine und angewandte Mathematik*, is the article reproduced above in English translation [74]. As Schwarz explained at the beginning of his communication [73]:

> The problem of determining whether an (ordinary) algebraic differential equation possesses a particular algebraic integral and, when this is the case, of finding all its particular algebraic integrals, still belongs today to the most difficult problems in analysis. As it seems, this problem must, given the present state of science, be attacked in each single case with the help of methods which are proper to the particular case in consideration.

This modest approach to tackling a major issue by starting with what Schwarz presented as a very specific case—we will see that this is not exactly the situation—was quite characteristic of the author's style.

IMJ-PRG, CNRS, Sorbonne Université, Université Paris Cité.

 | HTTPS://DOI.ORG/10.1515/9783111260440-005

2. Hermann Amandus Schwarz

Schwarz (1843–1921) had been appointed full professor at the Federal Polytechnic School two years before this address to the Swiss Society, at the age of 26, following an initial position in Halle. A former student of Karl Weierstrass and Ernst Eduard Kummer, it was under the latter that he defended his thesis, with the rare mention *eximia cum laude*, in August 1864; since its beginning, he had also been a regular participant in the famous Berlin mathematics seminar, an activity for which he even received several bonuses.[1)] Schwarz had married one of Kummer's daughters, Marie, in 1868 and, in 1890, he would dedicate his own complete works to his father-in-law "with admiration and gratitude" [75]. He also remained close to Weierstrass throughout the latter's life, as their correspondence attests, and Weierstrass would play a decisive role in getting Schwarz hired in Berlin as his successor in 1892.[2)] But by 1871, Schwarz had already made a name for himself through his work on two topics that would remain his favorites throughout his career: minimal surfaces (i.e.surfaces that locally minimize area among those with the same prescribed boundary) and explicit conformal mappings in two and three dimensions.

At first glance, these topics may seem quite distant from the one we are focusing on here, linear differential equations, but that is not the case; in fact, Schwarz's early research is closely linked in several ways to his work on the hypergeometric differential equation. First of all, a number of consistent features can be found throughout Schwarz's work, in particular the combination of geometric and analytical considerations and a preference for explicit results. He would explain in 1893, in his acceptance speech at the Berlin Academy of Sciences [76, p. 624]:

> Higher geometry and the theory of analytic functions have held the greatest appeal for me from the very beginning of my studies. It has always given me particular satisfaction whenever I succeeded in bringing a difficult scientific problem closer to its final solution by combining these two disciplines.

In more florid terms, Ludwig Bieberbach, in his obituary of Schwarz for the Berlin Mathematical Society, would later link this approach to a dual legacy [10, p. 48]:

> In Schwarz's work, alongside Weierstrass's stimuli, Riemann's thoughts also become operational. In his work, one can first discern a kind of reconciliation between the opposing schools. Generally speaking, each school tended to see in the other only what was different, what was alien to its own nature, which repelled it and made it deaf to the fact that rigor and mildness produce the right harmony only when combined. Thanks to Riemann's impulsion, his wealth of ideas, and the bold, forward-driving spirit of his school, Schwarz managed to solve many of his problems and did not rest until he had tamed the unruly with the means of Weierstrass's rigor.

It has become common to contrast Riemann's mathematical characteristics with those of Weierstrass—a contrast sometimes underpinned by institutional or personal tensions between

[1)] The lectures at the Friedrich Wilhelm University of Berlin and the organization of the Berlin mathematics seminar by Kummer and Weierstrass are described in Kurt Biermann's major book [11]. On Schwarz's *Diplomarbeit*, in particular, see [11, pp. 112–3, and doc. 5, pp. 276–7].

[2)] At this time, Schwarz was a professor in Göttingen. The succession of Weierstrass (and Kronecker) in Berlin is discussed in [11, pp. 150–152]. On Weierstrass's circle, see [14]. The correspondence between Schwarz and Weierstrass is reproduced in [20].

their direct respective successors, and by the growing opposition between Göttingen and Berlin as competing mathematical centers at the end of the nineteenth century [36]. However, their research themes often overlapped closely[3] and Berlin mathematicians, at least in the 1860s, also favored a certain kind of assimilation of Riemann's approach: in his report on Friedrich Prym's thesis in 1863, Kummer wrote for instance [11, p. 94]:

> The Riemann method is still not widely used. Studying it from Riemann's treatises is very difficult because it is not explained there with the desired level of detail; therefore, its elaborated application to a specific and limited problem given by the candidate is a useful work for many mathematicians.

Weierstrass also advised Schwarz in 1868 to write in detail a mapping of the ellipse on a disk, because

> however trivial the subject may seem to you now, perhaps it is precisely in a task like this that we can see how advantageous it is when someone with specialized analytical knowledge has also made the tools of pure geometry his own [20, p. 25].

The similarities between Schwarz's early works, however, go much beyond a mere convergence of two approaches; they operate on a more technical level, the same key ingredients being at play in all the topics Schwarz addressed at the time, the same techniques being mobilized and adapted by Schwarz to address what thus appears to be a common set of questions. Let us briefly illustrate this with a few elements of Schwarz's first papers that would reappear in his 1873 article.

One of Schwarz's first results concerns the theorem of conformal representation — now also known as the Riemann mapping theorem: it states that for any non-empty simply connected open strict subset U of the complex plane, there exists a biholomorphic map from it to the unit disk D (or, equivalently, to the upper half-plane), which can be extended as a homeomorphism between their respective closures if the boundary of U is regular enough; it is even unique if one fixes the image of an interior point and that of a point on the boundary. In Riemann's and Schwarz's time, its usual formulation was that two simply connected plane surfaces can be mapped, one onto the other, "similarly in their smallest parts". Riemann had justified it with what he called the Dirichlet principle, which asserts the existence of a function minimizing a certain integral. This principle, as is well-known, was criticized by Weierstrass and his successors (eventually with counterexamples), and Schwarz, later in his career, would develop specific tools to avoid it and prove the theorem of conformal representation without it for a large class of surfaces.[4] According to Schwarz himself, however, his incentive to study the question at first was not the theoretical uncertainty regarding the validity of the Dirichlet principle, but a remark by Franz Mertens, another Berlin student at the same time, during a lecture by Weierstrass in 1823-24, that no *explicit* mapping was known for some of the simplest cases, such as the mapping of a triangle to the unit disk, because of the singularities created by the vertices. In this respect, this early work also illustrates another defining feature of Schwarz's approach to mathematics: his preference for concrete, effective solutions.

Schwarz tackled Mertens's question not only for triangles, but also for squares (and to a certain degree, for polygons), as well as for some surfaces delimited by simple curves, such

[3] Weierstrass and Riemann both worked on minimal surfaces in the mid-1860s, for instance, with quite similar results and even similar choices of examples [66, 81].

[4] On Schwarz's work on the theorem of conformal representation in general, see [77]. The development of this theorem and its variants, in particular the uniformization theorem—every simply connected Riemann surface is conformally equivalent to the open disk, the complex plane or the Riemann sphere— is beautifully told in [67].

as circular arcs [70].[5] The first step is local, dealing with the situation at a vertex: Schwarz used the simple function $t(u) = u^{\frac{1}{\alpha}}$, α a real number, $0 < \alpha < 2$, to represent conformally the (punctured) angular sector in the complex plane

$$\{u = re^{i\phi} \in \mathbb{C} \mid 0 < r < r_0, 0 \leq \phi \leq \pi\alpha\}$$

on the upper half-plane, more specifically onto a punctured half-disk centered at the origin $\{t = \rho e^{i\psi} \in \mathbb{C} \mid 0 < \rho < \rho_0, 0 \leq \psi \leq \pi\}$. The function takes real values for real values of the argument and maps the sides of the angular sector (that is, the points for which $\phi = 0$ or $\phi = \alpha\pi$) to the two real radii of the half-disk (that is, to the points for which, respectively, $\psi = 0$ or $\psi = \pi$).

Schwarz's second step was to use a symmetry principle: if the boundary of a domain U in the upper half-plane includes a segment I of the real axis, a function f which is holomorphic inside the domain, continuous on its boundary and which takes real values on the segment I can be extended as a holomorphic function to the reflection of U with respect to the real axis, by setting $f(\overline{z}) = \overline{f(z)}$. In particular, the function $u^{\frac{1}{\alpha}}$, as a function of t, can be extended to an entire disk centered at the origin (the origin $t = 0$ being sent to the origin $u = 0$), and thus can be expanded into a Taylor series. Schwarz finally obtained $u = C^{\alpha}t^{\alpha}(1 + c_1 t + c_2 t^2 + \cdots)$, for a strictly positive real constant C and with real coefficients c_i. Analogous expansions (with t replaced by $t - t_0$ or $-1/t$) would be obtained if the vertex of the angular sector is mapped to a point t_0 of the plane or to ∞ respectively.

The third step involved putting all this local information together. Schwarz remarked that his local mappings are defined up to a change of axes or a resizing, that is, up to an affine transformation, which introduces arbitrary constants. In order to eliminate them, he searched for his global function in the form of a solution of a differential equation whose differential term on the left side, $\frac{d}{dt}\log\frac{du}{dt}$, is invariant under affine transformations and whose right side gathers the local information. For instance, for the angular sector of vertex 0, the local solution found above provided for the right side a term $\frac{\alpha-1}{t} + d_1 + d_2 t + \cdots$, with real d_i. It is rather easy to prove, by studying what happens near ∞, that the global expression on the right side of the differential equation is in fact a rational function. In the case of a square, for instance, the angles $\alpha\pi$ are all equal to $\frac{\pi}{2}$ and it is possible, for reasons of symmetry, to choose the four values $-1, 0, 1, \infty$ as the values of the function $t(u)$ at the four vertices. Schwarz's final step was thus to solve the differential equation

$$\frac{d}{dt}\log\frac{du}{dt} = -\frac{1}{2}\left(\frac{1}{t+1} + \frac{1}{t} + \frac{1}{t-1}\right).$$

Its general solution is $u(t) = C_1 \int_0^t \frac{dt'}{\sqrt{4t'(1-t'^2)}} + C_2$, with two arbitrary constants C_1 and C_2. In other terms, the solution is given in this case by a lemniscate integral.[6] It is noticeable that Schwarz then took the pain to compute the values of the lemniscate integral with four decimals and to provide precise drawings of the mapping of the square on the half-plane and on the disk, "to make this representation intuitive by means of figures" [70, pp. 112–3].

[5] The general case of polygons leads to what is now known as the Christoffel–Schwarz theorem, but it involves accessory parameters which cannot be determined a priori. For a slightly modernized explanation of Schwarz's proof and the Christoffel–Schwarz theorem, see [67, ch. IV.2].

[6] Schwarz had already noticed it directly, because of the symmetries with respect to the four sides of the square.

As a prelude to his 1873 article, it is useful to mention another case handled by Schwarz, that of a "circular triangle", that is, a triangle whose sides are arcs of circles, with respective angles $\alpha\pi$, $\beta\pi$, $\gamma\pi$. In this case, the three vertices can be sent to $0, 1, \infty$ and the argumentation follows the same lines. The symmetry principle applies by using inversions with respect to the circles defining the sides of the "triangle", instead of complex conjugation (that is, instead of reflection with respect to a straight line, the real axis). This time, however, the local mappings can be changed by Möbius transformations, instead of affine transformations, and thus a more complicated differential equation should be used to ensure the correct number of constants in the solution (three instead of two) and the required invariance. Schwarz thus introduced another differential term, which, this time, is invariant under a homographic change of variables:

$$\frac{d^2}{dt^2}\log\frac{du}{dt} - \frac{1}{2}\left(\frac{d}{dt}\log\frac{du}{dt}\right)^2.$$

Schwarz denotes it by $\Psi(u,t)$, and is thus led to solve a differential equation

$$\Psi(u,t) = F(t),$$

for an appropriate rational function F gathering the local information. The differential expression $\Psi(u,t)$, which can also be written as $\frac{u'''}{u'} - \frac{3}{2}\left(\frac{u''}{u'}\right)^2$, was not new with Schwarz — in fact, Schwarz would provide for it a chronology dating back to Joseph-Louis Lagrange in the comments he later added to this paper for his complete works [75, vol. 2, pp. 351–355]—, but it would soon be known as the 'Schwarzian derivative", following Arthur Cayley's suggestion [19], and designated by the symbol $\{u,t\}$. Schwarz would then characteristically comment [75, vol. 2, p. 353]:

> As for the notation $\Psi(u,t)$, one might object that, with regard to this expression, we are not dealing with a function of the two arguments u and t, but rather with a differential expression formed using the function u, by differentiating it in a certain way with respect to t. Thus, the notation and terminology adopted by M. *Cayley* undoubtedly have an advantage over those used by the author.

This Schwarzian derivative will play an important role in the 1873 article. It is worth noting that, years before, Schwarz had been made aware by Weierstrass of its direct connection to the hypergeometric differential equation, which would occupy him a few years later [70, pp. 79–80]:

> It is easy to prove that the general integral of the differential equation $\Psi(u,t) = F(t)$ can be represented as a quotient of two solutions of the same second-order differential equation with rational functions as coefficients. I owe this observation to a kind communication of Mr. Weierstrass. If the figure to be represented is a triangle whose sides are circular arcs, the differential equation is that of the hypergeometric series.

Very similar constructions were used by Schwarz in his early work on minimal surfaces, whose link with conformal mappings was only emerging at the time [21, 59]. Although Schwarz's publications on minimal surfaces officially predate those on conformal mappings, he presented the latter quite early at the Berlin seminar and used a conformal mapping of the cube onto the sphere to construct his first minimal surfaces, accompanied with models that were displayed at the Berlin Academy of Sciences in 1865, when Kummer presented Schwarz's work [69]. Two years later, a prize at the Berlin Academy of sciences gave rise to another construction. The theme for this prize was: "Any significant problem whose subject matter belongs to algebra, number theory, integral calculus, geometry, mechanics, and mathematical physics is to be completely solved with the aid of elliptic or Abelian transcendents" [72].

In his memoir, Schwarz studied in detail a minimal surface leaning on a skew quadrilateral formed by four edges A, B, C, D of a regular tetrahedron, with a plane of symmetry through AC.[7)] Using the property that a minimal surface has a null mean curvature, Weierstrass had recently proved that the points on a simply connected, open, minimal surface can be parametrized:

$$x(u) = x_0 + \Re \int_{u_0} (G^2(u) - H^2(u))du,$$
$$y(u) = y_0 + \Re \int_{u_0} i(G^2(u) + H^2(u))du,$$
$$z(u) = z_0 + \Re \int_{u_0} 2G(u)H(u)du,$$

where G and H are single-valued complex functions with some extra conditions, u_0 is the point in the complex plane corresponding to a point x_0, y_0, z_0 of the surface and $\Re$ designates as usual the real part of a complex number. Putting $s = \frac{H}{G}$, and choosing $0, 1, \infty, -1$ for the points corresponding to A, B, C, D (which is possible because of the symmetry of the quadrilateral), Schwarz showed that

$$\frac{d^2}{du^2}\log\frac{ds}{du} - \frac{1}{2}\left(\frac{d}{du}\log\frac{ds}{du}\right)^2 = \frac{a + bu^2 + cu^4}{u^2(u^2-1)^2}$$

where the constants a, b, c can be expressed as functions of the angles of the quadrilateral. A solution of the differential equation describes a segment or a circular arc when the argument u describes a segment of the real axis which does not contain any singular point. A meticulous study of the behavior of the solutions around the singular points then allowed Schwarz to give an explicit parametrization of this minimal surface bounded by the four sides of the quadrilateral and without interior singularity [72]:

$$x = \Re \int_0 (1 - s^2)\mathcal{F}(s)ds$$
$$y = \Re \int_0 i(1 + s^2)\mathcal{F}(s)ds$$
$$x = \Re \int_0 2s\mathcal{F}(s)ds,$$

where $\mathcal{F}(s) = \dfrac{1}{\sqrt{1 - 14s^4 + s^8}}$, $\mathcal{F}(0) = 1$.[8)]

7) This very example also appeared in Riemann's posthumous paper on minimal surfaces [66], that Karl Hattendorff completed and published.

8) The zeros of the denominator of the function $\mathcal{F}$ correspond to the projections of the vertices and other special points of the tetrahedron, see for example [59, pp. 233–8].

This surface inherits the symmetries of the quadrilateral and can be extended infinitely by reflections; Schwarz thus obtained the first examples of periodic minimal surfaces. The function $\mathcal{F}$ is linked to elliptic integrals and moreover, Schwarz showed that the equation of the surface can be expressed rationally by elliptic functions of the coordinates. He again pointed out the connection with the hypergeometric equation [75, vol.I, p. 25]:

> The differential equation has four singular points. But an algebraic transformation (for example, u^2 = t) can reduce the number of singular points to three. The function s is obtained in terms of this new variable as the quotient of two solutions to a second-order differential equation, which coincides with the well-known differential equation of the hypergeometric series.

Several tools that would be crucial to Schwarz's 1873 article were therefore already at his disposal in this earlier work: a differential expression invariant under Möbius transformations (the future "Schwarzian derivative"); the relation between the solutions of a third-order differential equation expressed with the Schwarzian and of the hypergeometric differential equation; conformal mappings obtained as solutions of relevant differential equations; symmetries expressed both analytically and geometrically, as needed; expansions in power series; and finally elliptic functions and polyhedra. All this prepared him to explore the question of determining under what conditions a particular second-order linear differential equation has algebraic solutions, to which we will now turn.

3. The hypergeometric differential equation

At the center of Hermann Amandus Schwarz's 1873 article is a second-order linear ordinary differential equation with rational functions as coefficients:

$$\frac{d^2y}{dx^2} + \frac{\gamma - (\alpha + \beta + 1)x}{x(1-x)} \cdot \frac{dy}{dx} - \frac{\alpha \cdot \beta}{x(1-x)} \cdot y = 0 \tag{1}$$

where α, β, γ are complex numbers. Its only possible singular values are $0, 1, \infty$. The main question raised in his paper is to determine for which values α, β, γ it can have algebraic solutions and to find these solutions. A solution u is algebraic if it satisfies an equation $A_m u^m + A_{m-1} u^{m-1} + \cdots + A_1 u + A_0 = 0$, for an integer m and rational functions A_i in $\mathbb{C}(x)$ as coefficients.

This differential equation — called the hypergeometric differential equation — was well-known in the 19th century, but the real heroine, so to speak, was first a series satisfying it: the hypergeometric series $F(\alpha, \beta, \gamma, x)$, defined as

$$F(\alpha,\beta,\gamma,x) = 1+\frac{\alpha\beta}{1\cdot\gamma}x+\frac{\alpha(\alpha+1)\beta(\beta+1)}{1\cdot 2\cdot\gamma(\gamma+1)}xx+\frac{\alpha(\alpha+1)(\alpha+2)\beta(\beta+1)(\beta+2)}{1\cdot 2\cdot 3\cdot\gamma(\gamma+1)(\gamma+2)}x^3+\cdots \tag{2}$$

Here γ is not a negative or null integer. If α or β is equal to a negative or null integer, the series reduces to a polynomial in x. In general, it converges for $\mid x \mid < 1$.[9)]

Although he was not the first to study it,[10)] Gauss drew attention to this series in a paper presented to the Academy of Sciences in 1812, demonstrating that it was related to most of

[9)] The literature on the hypergeometric function and the hypergeometric differential equation is huge. For different presentations at different moments, we refer, for instance, to [7,25,33,43,50,58,82].

[10)] On the early history of the hypergeometric function, see [24].

the functions known at the time [29]. For instance, $(x+u)^n = u^n F(-n,\beta,\beta,-\frac{x}{u})$ (for an arbitrary choice of β), $\log(1+x) = xF(1,1,2,-x)$ and $\arcsin(x) = xF(\frac{1}{2},\frac{1}{2},\frac{3}{2},x^2)$. He also showed that there exist linear relations between $F(\alpha,\beta,\gamma,x)$, $F(\alpha',\beta',\gamma',x)$ and $F(\alpha'',\beta'',\gamma'',x)$ when all the differences $\alpha-\alpha'$, $\alpha'-\alpha''$, $\beta-\beta'$, etc., are integers. For instance,

$$\gamma(1-x)F(\alpha,\beta,\gamma,x) - \gamma F(\alpha,\beta-1,\gamma,x) + (\gamma-\alpha)xF(\alpha,\beta,\gamma+1,x) = 0$$

In a last section, he focussed on the value $x=1$, linking the hypergeometric series, when it converges, to the Γ-function which interpolates the values $n!$. Gauss seems to have been particularly interested in the practical uses of the hypergeometric function and its expansions for approximation purposes, and more generally, as a prime candidate for expressing many functions or relations relevant to physics and astronomy [30]. Although Gauss never published it during his lifetime, a paper in his *Nachlass* shows that he had also found a differential equation of the second order satisfied by the hypergeometric function:

$$0 = \alpha\beta P - (\gamma - (\alpha+\beta+1)x)\frac{dP}{dx} - (x-x^2)\frac{d^2P}{dx^2},$$

the general solution of which can be written as $MF(\alpha,\beta,\gamma,x)+NF(\alpha,\beta,\alpha+\beta+1-\gamma,1-x)$, with two constants M and N (we recognize the differential equation (1)). Again, Gauss emphasized its relation with the establishment of astronomical ephemerides [31].

But it was Kummer, then still a high school teacher and at the very beginning of his career, who particularly drew the attention of mathematicians to the differential equations associated with the hypergeometric series. His own interest had been sparked by a remark in Carl Jacobi's *Fundamenta nova theoriae functionum ellipticarum*, in the context of the transformations between elliptic integrals. A modulus k being given, it is a matter of finding a rational function $y(x)$ (with co-prime numerator of odd degree n and denominator of degree $n-1$) and modulus $\lambda(k)$, such that.

$$\frac{dy}{\sqrt{1-y^2}\sqrt{1-\lambda^2y^2}} = \frac{1}{M}\frac{dx}{\sqrt{1-x^2}\sqrt{1-k^2x^2}}$$

for a constant $M(\lambda,k)$. For each degree n, there exists an algebraic equation (the modular equation) of degree $n+1$ between $\sqrt[4]{k}$ and the corresponding $\sqrt[4]{\lambda}$, but Jacobi also showed that, independently of the degree, λ as a function of k satisfies a third-order differential equation (of the Riccati type). This equation has thus infinitely many algebraic solutions; however, its complete integration requires transcendental elliptic functions —a phenomenon that Jacobi called "the most memorable and singular thing" [40, p. 32].

Kummer then began to study transformations directly on differential equations, first in a short program paper in 1834 [52], then in a longer article in Crelle's *Journal*. More specifically, he searched for when a solution $v(z)$ of a second-order linear ordinary differential equation

$$\frac{d^2v}{dz^2} + P(z)\frac{dv}{dz} + Q(z)v = 0$$

can be transformed into a solution of another such differential equation

$$\frac{d^2y}{dx^2} + p(z)\frac{dy}{dx} + Q(x)y = 0$$

with $z = z(x)$ and $y(x) = w(x)v(z(x))$. These conditions lead to the resolution of a third-order differential equation:

$$2\frac{d^3z}{dzdx^2} - 3(\frac{d^2z}{dzdx})^2 - Z(z)(\frac{dz}{dx})^2 + X(x) = 0 \tag{3}$$

for appropriate functions $Z(z)$ and $X(x)$. The "multiplier" $w(x)$ is easily expressed in function of $P(z)$ and $p(x)$. This setting encompasses that studied by Jacobi and the transformations of the hypergeometric differential equation (1). One also recognizes the "Schwarzian derivative" on the left-hand side of (3). In 1836, Kummer used equation (3) to complete Gauss's results by studying the transformations of a hypergeometric function (as a solution of a hypergeometric differential equation) under changes in its fourth element, that is, Kummer searched for when a function $w(x)F(\alpha', \beta', \gamma', z(x))$ is a solution of the hypergeometric differential equation (1), for α', β', γ' affine functions of α, β, γ. He found that z and x are linked by a homographic relation—more precisely that z could be x, $1-x$, $\frac{1}{x}$, $\frac{1}{1-x}$, $\frac{x}{x-1}$ or $1-\frac{1}{x}$— and that the α', β', γ' could take four triples of values. This provided 23 other solutions of the equation (1) from $F(\alpha, \beta, \gamma, x)$. For instance, for $z(x) = 1-x$, Kummer found that $F(\alpha, \beta, \alpha+\beta-\gamma+1, 1-x)$, $(1-x)^{\gamma-\alpha-\beta}F(\gamma-\alpha, \gamma-\beta, \gamma-\alpha-\beta+1, 1-x)$, $x^{1-\gamma}F(\alpha-\gamma+1, \beta-\gamma+1, \alpha+\beta-\gamma+1, 1-x)$ and $x^{1-\gamma}(1-x)^{\gamma-\alpha-\beta}F(1-\alpha, 1-\beta, \gamma-\alpha-\beta+1, 1-x)$ are also solutions of (1).[11)]

Kummer also raised the question of the algebraic integrals of (3), in terms that would be echoed by those of Schwarz in 1871 [53, p. 44]:

> The problem of finding all particular algebraic integrals of a given differential equation is undoubtedly one of the most difficult problems posed in recent times; and especially for differential equations of higher orders, [...] the solution to this problem generally still does not seem to be feasible.

As pointed out by Bernhard Riemann a few years later, the fact that the hypergeometric series has only radius of convergence 1 means that alternative methods should be found to study on the entire complex domain the function it represents on the unit disk. The techniques used by Gauss and Kummer relied mainly on extensive algebraic and differential computations. It is these calculations that Riemann aimed to eliminate by reframing the study of $F(\alpha, \beta, \gamma, x)$ within his theory of analytic functions [65]. He examined functions of a complex variable satisfying what we now call a Riemann scheme, denoted by

$$P\begin{Bmatrix} a & b & c & \\ \alpha & \beta & \gamma & x \\ \alpha' & \beta' & \gamma' & \end{Bmatrix},$$

and satisfying the following properties: the function is single-valued and analytic for all complex x, except at a, b, c; all its analytic continuations (its "branches", according to Riemann's terminology) are linear combinations of two of them; in a neighborhood of a, the function can be written as $c_\alpha P^{(\alpha)} + c_{\alpha'} P^{(\alpha')}$, where c_α and $c_{\alpha'}$ are constants, and the branches $P^{(\alpha)}$ and $P^{(\alpha')}$ are such that $P^{(\alpha)}(x-a)^{-\alpha}$ and $P^{(\alpha')}(x-a)^{-\alpha'}$ are single-valued functions that are neither zero nor infinite at $x = a$ (and the same conditions for b and c *mutatis mutandis*). In other words, the P-functions have three branch points at a, b, c with a given behavior at the

11) [53, p. 52]. Kummer also found other solutions when α, β, γ satisfy extra relations and computed numerous particular values of the series. On these questions, see also [7, 58].

branch points, described by α, α', $\dots \gamma$, γ', that Riemann called the exponents.[12] With two branches of a function P whose quotient is not constant, it is possible to express linearly every P-function with the same exponents.

The hypergeometric series of course served as a model for the functions P; in fact:

$$F(\alpha,\beta,\gamma,x) = P^{(0)}\left\{\begin{matrix} 0 & \infty & 1 & \\ 0 & \alpha & 0 & x \\ 1-\gamma & \beta & \gamma-\alpha-\beta & \end{matrix}\right\}.$$

Conversely, it is possible to transform any Riemann P-function into a hypergeometric one by homographic transformations. Riemann also studied the transformations of a P-function when the *complex* variable x follows a circuit around one of the branch points a, b or c: for instance $P^{(\alpha)}$ is transformed into $P^{(\alpha)}e^{2\pi\alpha}$ when the variable x describes a positive circuit around a. Riemann also showed that the P-functions satisfy second-order linear differential equations with rational functions whose coefficients depend on the exponents and he recovered many relations obtained by Gauss and Kummer for the hypergeometric function. The main point, of course, was that the hypergeometric function then entered the realm of complex analysis, which was quite active at the time.

Riemann's work was continued, in particular, by Lazarus Fuchs, in a setting that is reminiscent of the one we discussed in connection with conformal mappings. Fuchs, then at the beginning of his career and still a teacher at the Friedrichs-Werdersche Gewerbeschule (a Berlin vocational school), had been, like Schwarz, a student of Kummer and Weierstrass and he was motivated to study differential equations by a course the latter taught in 1863–64. In a series of articles [26, 27], Fuchs identified and highlighted a broad class of linear ordinary differential equations of order n, with a finite number of singular points:

$$\frac{d^m y}{dx^m} = \frac{F_{\rho-1}(x)}{\psi}\frac{d^{m-1}y}{dx^{m-1}} + \cdots + \frac{F_{(m-1)(\rho-1)}(x)}{\psi^{m-1}}\frac{dy}{dx} + \frac{F_{m(\rho-1)}(x)}{\psi^m}y$$

with $\psi = (x-a_1)\cdots(x-a_\rho)$, the a_i are complex numbers and the F_s are polynomials of degree less than or equal to s [26, p. 146]. With these conditions, the singular points are what will be called "regular". The singular points of the solutions are included into those of the equation, that is, at most, the a_i and the point at ∞.[13] The hypergeometric differential equation belongs to this class, for $m = 2$, $\rho = 2$ and three regular singular points $a_0 = 0, a_1 = 1, \infty$. It was already known that any solution of such a linear differential equation can be written as a linear combination, with constant coefficients, of a fundamental system of n independent solutions, say $y_1, \cdots, y_n$.

Fuchs did two important things. First of all, following Riemann, he studied the analytic continuation of a solution when the variable describes a loop around a singular point. For any a_i, each y_k becomes a new branch $\tilde{y_k}$ which is again a solution of the differential equation, thus can be expressed as a linear combination of the y_k. Fuchs attached to each singular point a $n \times n$ (monodromy) matrix encapsulating the behavior of the fundamental system of solutions when the variable describes a loop around the point and he studied what we now call its eigenvalues—for him, the roots of what he designated as the "fundamental equation" belonging to the corresponding singular point— and its eigenvectors. Fuchs showed in particular that any

[12] Riemann also assumed that $\alpha-\alpha'$, $\beta-\beta'$, $\gamma-\gamma'$, are not integers and that $\alpha+\alpha'+\beta+\beta'+\gamma+\gamma' = 1$.

[13] Fuchs developed some of his theory under less restrictive hypotheses. For a detailed presentation of his work, and of its relations with those of Riemann, Weierstrass and others, see [34, 35].

eigenvalue $\omega_{i,s}$, pour $1 \leq s \leq m$ associated to the singular point a_i can be written as $e^{2\pi i r_{i,s}}$ and that there exists a solution of the differential equation $(x-a_i)^{r_{i,s}}\phi_{i,s}(x-a_i)$, where $\phi_{i,s}$ is single-valued, analytic in a neighborhood of a_i and $\phi_{i,s}(a_i) \neq 0$.[14]

Then, conversely, Fuchs looked for solutions

$$y(x) = (x-a)^r(e_0 + e_1(x-a) + \cdots + e_n(x-a)^n + \cdots)$$

in the neighborhood of a singular point a. By substituting this expression back into the equation, it is easy to see that r must be a solution to an equation of degree m, now called the indicial equation.[15] Fuchs also showed that $e^{2\pi i r}$ is an eigenvalue of the corresponding monodromy matrix. In the case of the hypergeometric equation, for instance, the indicial equation for $a = 0$ is $r^2 - (1-\gamma)r = 0$, with the roots 0 and $1-\gamma$: if γ is not an integer, Kummer's solutions $F(\alpha, \beta, \gamma, x)$ and $x^{1-\gamma}F(\alpha-\gamma+1, \beta-\gamma+1, 2-\gamma, x)$ are solutions near 0 which correspond, respectively, to $r = 0$ (and the eigenvalue 1) and $r = 1-\gamma$ (and the eigenvalue $e^{2\pi i(1-\gamma)}$. Similarly, the roots of the indicial equations for $a_1 = 1$ (resp. for ∞) are 0 and $\gamma - \alpha - \beta$ (resp. α and β).

Fuchs also remarked that if a n-th order linear differential equation has only algebraic solutions, it belongs to his special class. However, he did not succeed then to characterize such equations inside the class. Weierstrass, too, tried unsuccessfully to solve this question. On April 24, 1871, he wrote to Schwarz [20, p. 32]:

> I have sought in vain to answer in general the question of under what conditions such a D[ifferential] equation has algebraic integrals; however, it can be resolved in the case where the equation is of the second order and the number of essential singular points of the argument is no greater than three. Nevertheless, your treatment, based on geometric considerations, should be no less welcome for that reason. To my knowledge, nothing has yet been published on the subject.

The result alluded to is that which Schwarz presented a few months later to the Swiss Society and to which we now turn.

4. Outline of Schwarz's paper

The paper is divided into seven sections, the first handling the case where one (and, up to a constant, only one) solution of the hypergeometric differential equation (1) is an algebraic function, the following sections the case where all solutions are algebraic. In this second case, Schwarz used the fact, already noticed in his work on conformal representation, that the quotient s of two independent solutions of the hypergeometric differential equation (which satisfies a third-order non linear equation and is algebraic) defines a conformal mapping from the complex plane onto a "circular" plane triangle, and finally onto a spherical triangle; its analytic continuations, or branches, give rise to other spherical triangles which are symmetric with respect to the first one and cover the sphere. The solution s is algebraic if and only if there are only finitely many of them, and the problem then is to determine for which α, β, γ, this procedure provides a finite covering of the sphere by such triangles. As will be seen, however, the paper is not strictly limited to this determination and has an even larger scope than the mere search for algebraic solutions.

[14] Solutions with a logarithmic term may also occur, in particular in the case of multiple eigenvalues.

[15] For the singular point at ∞, he replaced x by $\frac{1}{t}$ and applied the previous reasoning for $t = 0$.

4.1. Section I. In the first section, Schwarz discusses the situation where only one solution of the hypergeometric equation (1), up to a constant, is algebraic. In this case, the various analytic branches of this solution are multiples of one of them (if not, two of these branches, which are then all algebraic, would be independent and generate the whole set of solutions) and the logarithmic derivative of the solution is a rational function (satisfying a first-order differential equation). Writing it as a sum of simple elements, it is in fact easy to find by comparison that its poles are of degree 1, that the term corresponding to the pole 0 can only be 0 or $\frac{1-\gamma}{x}$, the term corresponding to the pole 1 only 0 or $\frac{\gamma-\alpha-\beta}{x-1}$;[16] a computation at ∞ shows that one of α or β or $\gamma-\alpha$ or $\gamma-\beta$ is necessarily an integer (equivalently, that one of the $\pm\lambda\pm\mu\pm\nu$ is an odd integer).

In more modern terms, this case is the one in which the global monodromy group of the equation, generated for instance by two circuits around 0 and ∞, is reducible. The group is reducible when the two generators have a common eigenvector, that is, when they have a common eigenvalue, and, as the eigenvalues corresponding to these two circuits are respectively $1, \exp 2\pi i(1-\gamma)$ for 0 and $\exp 2\pi i\alpha, \exp 2\pi i\beta$ for ∞, we find again the conditions mentioned above [7].

Schwarz states that the solution has the form $y = x^{\mathfrak{a}}(1-x)^{\mathfrak{c}}g(x)$, where g is a polynomial of degree n, such that $g(0) \neq 0$, $g(1) \neq 0$. The exponents $\mathfrak{a} = 0$ or $1-\gamma$ and $\mathfrak{c} = 0$ or $\gamma-\alpha-\beta$ should be rational numbers, and n is in each case expressed as a linear function of α, β, γ. For all the relevant cases, Schwarz expresses the solution by means of Kummer's solutions for the hypergeometric equation, one of the intervening hypergeometric series being here a polynomial [58].

For instance, if α is a negative integer, $\mathfrak{a} = \mathfrak{c} = 0$, $n = -\alpha$ and $F(\alpha, \beta, \gamma, x)$ is a polynomial, and thus an algebraic solution (case 1^a in Schwarz's article). If β a strictly positive integer $n+1$, α and γ are rational numbers, and $2-\gamma$ is not an integer such that $-n < 2-\gamma \leq 0$, then $\mathfrak{a} = 1-\gamma$, $\mathfrak{c} = \gamma-\alpha-\beta$, $n = \beta - 1$ and

$$x^{1-\gamma}(1-x)^{\gamma-\alpha-\beta}F(1-\alpha, 1-\beta, 2-\gamma, x)$$

is an algebraic solution (case 4^a for Schwarz).

4.2. Section II. From now on, Schwarz discusses the case where all the solutions of the hypergeometric differential equation are algebraic. The numbers α, β, γ should necessarily be all rational in this case. Schwarz considers in this section the (slightly) more general differential equation:[17]

$$\frac{d^2y}{dx^2} + p \cdot \frac{dy}{dx} + q \cdot y = 0.$$

The quotient $s = \frac{y_2}{y_1}$ of two linearly independent algebraic solutions is obviously also algebraic. But, conversely, if this quotient is algebraic and not a constant, and if $e^{-\int p dx}$ is algebraic, all the solutions are algebraic; it is the case for the hypergeometric equation. This is where Schwarz introduces a third-order differential equation (F) satisfied by s:

[16] The possible coefficients correspond to the roots of the indicial equations for the poles 0 and 1.

[17] There is no explicit hypothesis on the coefficients p and q in Schwarz's paper, no more than in Niels Abel's paper to which Schwarz here refers [1]. Given the context, one may assume that p and q are single-valued analytic functions of x.

$$(F.) \quad \Psi(s,x) = \frac{2\dfrac{ds}{dx}\cdot\dfrac{d^3s}{dx^3} - 3\left(\dfrac{d^2s}{dx^2}\right)^2}{2\left(\dfrac{ds}{dx}\right)^2} = 2q - \frac{1}{2}p^2 - \frac{dp}{dx} = F(x).$$

We recognize here the differential equation met in the context of conformal representations; it highlights the differential expression $\Psi(s,x)$ (the "Schwarzian derivative") which, as Schwarz himself pointed out, already appeared in Kummer's articles [52, 53] (cf. equation (3) above). While the solutions of a linear second-order equation are linear combinations of two of them, the solutions of this third-order differential equation depend homographically on one of them.

4.3. Section III. The third section is entirely devoted to the integration of (F), in the case of the hypergeometric differential equation and when α, β, γ are real numbers.

As just explained, it is enough to find one integral, as every homographic transformation of s then also gives a solution. Such a transformation contains three independent constants, as needed for the set of solutions of the third-order differential equation (F).

Schwarz provides the solution through series, applying Weierstrass's theory there in full to establish their convergence. Given a non-singular point x_0, and a simply connected neighborhood of x_0 without any singularity of the differential equation, Schwarz shows that an integral of F transforms this neighborhood into a simply connected domain which does not contain any branching point. Moreover, when α, β, γ and x_0 are real numbers, a segment of the real axis containing x_0 is transformed into a segment or a circular arc.

When x_0 is a singular point, that is, here, $x_0 = 0$ (or, respectively, $x_0 = 1$ or $x_0 = \infty$), it is situated on the real axis (completed by the point at ∞), and the integral s transfers two real half segments on each side of x_0 into two circular arcs, whose tangents form an angle $\lambda\pi$ (resp. $\mu\pi$, resp. $\nu\pi$), with $\lambda^2 = (1-\gamma)^2, \quad \mu^2 = (\alpha-\beta)^2, \quad \nu^2 = (\gamma-\alpha-\beta)^2$ (Schwarz chooses λ, μ, ν positive).

Putting together these local informations, Schwarz concludes that the half-plane (say, the upper one) is conformally sent by a solution s of (F) into a simply connected domain, without interior branching points, whose boundary is a circular triangle, with angles $\lambda\pi, \mu\pi, \nu\pi$. This line of argumentation exactly mimics (up to the computation of the series) the procedure used by Schwarz for constructing conformal maps. Let us notice that branching points may occur at the vertices, when the corresponding angle is greater that 2π.

4.4. Section IV. Schwarz now begins to examine more closely the relative disposition of the circles whose arcs bound one of the triangles obtained in the preceding section.

After normalizing the angles (replacing λ by its absolute value modulo 2, and the same for the other angles, which gives the same geometrical triangles) and choosing an orientation of each circle, Schwarz distinguishes two main cases:[18)]

— in the first one, the sum of the angles of the original triangle is less than π. The figure formed by the three circles displays two circular triangles (including the original one), three circular biangles and three circular quadrangles, and there exists a circle (which may be a straight line) orthogonal to the three original circles;

18) These two cases correspond respectively to cases III and I in [50, pp. 214–215].

— in the second case, the sum of the angles of the original triangle is greater than π, the figure formed by the three circles displays eight circular triangles (including the original one) and no orthogonal circle. In this second case, the triangles can be grouped into pairs, the two grouped together having the same angles. The angles of the other triangles are obtained when one changes two of the λ, μ, ν into $1-\lambda, 1-\mu, 1-\nu$, which allows Schwarz to select a privileged pair, for which the sum of the angles is minimal and less than $\frac{3}{2}\pi$.

In this second case, it is then possible through a transformation by reciprocal radii to map the plane of the circular original triangle onto a sphere: the original triangle becomes a spherical triangle whose sides belong to great circles. Let us mention that this construction also appeared in Schwarz's work on minimal surfaces.

The analytic continuation $\tilde{s}$ of a solution s across one of the segments $]-\infty,0[,]0,1[$ or $]1,\infty[$ can then be interpreted very simply: if $s(I)$ is the side of the circular triangle which is the image by s of the chosen real segment I, the image by $\tilde{s}$ of a point symmetric with respect to I of a point x_0 of the upper half-plane corresponds to the transformed of $s(x_0)$ by a Möbius transformation with respect to the side $s(I)$; in this way, a new circular triangle is created. Schwarz baptizes the procedure (already used in his work on minimal surfaces and conformal representation) "the law of symmetry with respect to a circular arc". When s is algebraic, the number of different domains obtained in this way is finite, and conversely.

Schwarz has thus transformed the initial analytic question into a purely geometrical one: to find all circular triangles which, reproduced through this law of symmetry, give rise only to a finite number of circular triangles differing by size or place.

4.5. Section V. This section is devoted to Schwarz's first case, where the sum of the angles is less than π. Recalling the figure formed by the three circles delimiting the circular triangle and selecting among the two circular triangles of the figure the one which is inside the orthogonal circle, Schwarz shows that all the triangles obtained by symmetry from it as explained above remain inside this orthogonal circle: the circular sides of the symmetric triangles are orthogonal to this circle, while their vertices are closer and closer to it without reaching it. There are thus infinitely many different triangles obtained by symmetry from the original one. In this case, the function s is a transcendental function of x, while x may be a holomorphic function of s in certain cases (when $\frac{1}{\lambda}, \frac{1}{\mu}, \frac{1}{\nu}$ are integers). The values of s remain inside the orthogonal circle which appears as a natural boundary (without possible analytic continuation beyond it). This question of a natural boundary, which seemed quite remarkable to the mathematicians of the time, had been emphasized by Weierstrass. Schwarz mentions several examples of this phenomenon, due to Weierstrass and Kronecker.

Schwarz also studies the two limit cases, when the sum of the angles is exactly 0 or π. In the first one, the triangle offers three cusps at its vertices and one is led to an elliptic situation. Let k and k' be such that $k^2 + k'^2 = 1$, the corresponding elliptic periods are

$$K = \int_0^1 \frac{d\xi}{\sqrt{(1-\xi^2)(1-k^2\xi^2)}}, \quad K' = \int_0^1 \frac{d\xi}{\sqrt{(1-\xi^2)(1-k'^2\xi^2)}},$$

$s = \dfrac{a'K + b'K'}{aK + bK'}$, $x = k^2$ is a uniform function of s.

In the second case, the triangle is a standard plane triangle, the orthogonal circle being a straight line, and there are infinitely many symmetric repetitions of the triangle. However, x may be a uniform function of s, when $\lambda = \mu = \nu = \frac{1}{3}$ (equilateral triangle),

$\lambda = \frac{1}{2}, \mu = \nu = \frac{1}{4}$, $\lambda = \frac{1}{2}, \mu = \frac{1}{3}, \nu = \frac{1}{6}$ (right-angled triangle) and in these cases, x is a uniform elliptic function of s. In the case $\lambda = \frac{2}{3}, \mu = \nu = \frac{1}{6}$, x is an elliptic function with two branches of $\displaystyle\int_0^x \frac{dx}{\sqrt[6]{x^2(1+x)^3}}$.

4.6. Section VI. Here, Schwarz studies his second case, mapped onto the sphere. That is, he looks for spherical triangles whose symmetric repetitions provide only a finite number of different triangles and cover the sphere. In such a case, s is algebraic. Piecing together results by Riemann and himself on minimal surfaces and studies on polyhedra by Camille Jordan and Jakob Steiner, Schwarz states that a spherical polygon and its symmetric repetitions form a closed Riemann surface only if the plane of the sides of the polygon are symmetry planes of a regular polyhedron or the double pyramid. In our present case, either two angles of the spherical triangle are right angles, so a necessary and sufficient condition for the finiteness of the repetitions is that the third angle be commensurable with π (this is the case of the double pyramid) or the sides of the triangle should be in three symmetry planes of a regular polyhedron.

This provides to Schwarz a table of the various possibilities (now known as "Schwarz's list"). In the same way he had explicitly constructed his minimal surfaces, Schwarz uses the algebraic equations satisfied by particular points of the polyhedra (vertices, mid-points of the faces and of the sides), as well as the conformal transformations between the polyhedron and the surface of the sphere, to finally provide a procedure in order to determine which is the case according to the rational, non-integral, values of λ, μ, ν. In each of them, Schwarz also gives a fundamental system of algebraic integrals for the hypergeometric differential equation. For instance, when $\lambda = \mu = \frac{1}{2}$ and $\nu \neq 0$, he proposes as two independent integrals:

$$F(\alpha, \alpha - \frac{1}{2}, \frac{1}{2}, x) = \frac{1}{2}[(1+\sqrt{x})^\nu + (1-\sqrt{x})^\nu], \quad \frac{1}{2\nu}[(1+\sqrt{x})^\nu + (1-\sqrt{x})^\nu].$$

4.7. Section VII. This last section is devoted to the cases left aside until now, those for which one or two of the λ, μ, ν are integers. If λ is a non-zero integer, for instance, $x = 0$ should be an accidental singular point for the solutions to be algebraic, and conversely, if ν is not an integer, the differential equation has two independent integrals whose logarithmic derivatives are rational.

This, as Schwarz said, completes the determination of the values α, β, γ for which the hypergeometric equation has an algebraic solution. Let us just add that his memoir was accompanied by a set of beautiful plates, including one of the first representations, if not the first, of a hyperbolic tesselation of the unit disk (plate II).[19)] They were drawn by the lithographic establishment of Leopold Kraatz in Berlin, a specialist in atlases and maps, who worked in particular for Dietrich Reimer, son and brother of the publishers of Crelle's *Journal* [63, pp. 157–158].

5. On the reception of Schwarz's paper

To fully appreciate the reception Schwarz's article first received,[20)] one must recall the close thematic confluence of important figures in the European mathematical community we

19) There is no indication, however, that Schwarz had considered a non-Euclidean interpretation of this case at the time.

20) On the issue of reception, see [57].

already alluded too, a confluence that gave rise to numerous chronological and diplomatic clarifications attempting to establish priority and independence for the results. Regarding the minimal surfaces mentioned above, Weierstrass had written in 1868 to Schwarz [20, p. 24]:

> The beginning of my work bears such a strong resemblance to that of yours that this should indeed be a sign for both of us that we are embarking on the necessary and natural path, but for new friends it would be something very striking.

Friends, and also foes, displayed different opinions on what constituted a "necessary and natural path" and therefore on what to take from Schwarz's work. Let us first mention that several tools in Schwarz's proof, in particular its reflection principle and the Schwarzian derivative were (and are still) used in a variety of mathematical contexts. Two main issues, however, gave rise to key developments in the last decades of the nineteenth century and remain important today.

The first one, somewhat off-topic compared to the main question of the paper, is linked to section V of Schwarz's memoir, dealing with the (in general non-algebraic) case where the images of the circular triangles by the conformal mapping s remain inside the unit disk, and with the inversion of s. This is related to the question of uniformization, a question at the center of Henri Poincaré's and Felix Klein's work around 1880, and of their famous polemic about Poincaré's use of the term "Fuchsian functions" for the new transcendents he constructed in his own study of differential equations. This story has been abundantly examined and recounted in detail and we will not go over it again.[21)] Of interest here is the fact that, according to the correspondence between Klein and Poincaré, it was the former who introduced Schwarz's work to the latter (and perhaps as well to his friend and colleague Émile Picard).[22)] Although Charles Hermite, who generally played a decisive role in introducing German texts to his younger colleagues in France — especially after the 1870 war [5]—, had known of Schwarz's text quite early, he did not seem to have fully appreciated its significance first, partly, as he said, because of his difficulties with German language, but also, we may add, because of his own distaste for geometry. Almost ten years after the publication of the article, we have evidence that he made amends, writing to its author:

> Since I am mentioning my son-in-law [Émile Picard], I feel obliged to share with you a conversation we had regarding your memoir on volume 75 of Borchardt's Journal—a memoir I had read through but did not fully understand, due to the German language. Mr. Picard has authorized me to inform you that, in his judgment, the results you obtained led to the concept of the transcendents that Mr. Poincaré has called Fuchsian functions. Mr. Poincaré is a fine and excellent young man, who seeks nothing but truth and justice; I shall bring your work to his attention, as it is of the utmost importance, and none of us, rest assured, will fail in our duty to do you full justice.[23)]

21) On this development, see in particular [2, 35, 67] and their bibliographies, as well as [18, 78].

22) "My own evolution" wrote Klein in June 1881, "was strongly influenced by Schwarz's lines of thought, closely linked to Riemann's and of great significance, which were published in Vol. 75 of Borchardt's Journal (and which I can strongly recommend if you are unaware of them)" [67, p. 389] (rev. tr.) (Carl Borchardt was the editor of the *Journal für die reine und angewandte Mathematik* at the time, when it was still usual to refer to a journal by the name of his main editor.) Klein also mentioned Schwarz three times in his important text on the influence of Riemann, reproduced in the French translation of Riemann's work [49].

23) Letter from Hermite to Schwarz, November 1, 1882, Nachlass Schwarz 924, Archiv, Berlin Brandenburgische Akademie der Wissenschaften. For Picard's use of Schwarz's paper, see for instance his work on hypergeometric series in two variables [61].

The second line of development is that linked to the main question of Schwarz's paper, that of the algebraicity of the solutions of linear differential equations. The second half of the 1870s saw the publication of several articles that either directly addressed Fuchs's 1866 paper in their own terms or responded to Schwarz's memoir, trying to simplify its proofs, to establish its results by other means, and, in doing so, to eventually extend its scope.[24)]

In France, Camille Jordan, who had worked on both polyhedra and algebraic equations, in particular those equations linked to geometry,[25)] by means of the new group theory he promoted, had already intervened in 1871 on linear differential equations with constant coefficients and soon extended his approach to linear n-order differential equations with single-valued functions of one variable as coefficients: he first recalled that if the variable describes a closed path (including no singularity), a solution is changed into another which can be represented by a linear transformation of dimension n, all these transformations forming a group (in other terms, he explains the representation of the monodromy group of the differential equation in the linear group). He then set up a program explicitly analogous to that for algebraic equations: given a differential equation, first of all, determine its group, then find the conditions on the group to satisfy the requirements of a problem concerning the equation, and lastly, verify, through techniques of group theory, if the conditions are filled. Although he gave as an example the fact that for the solutions to be algebraic, the group should be finite, what he tackled in detail in his memoir was another question, that of the reducibility of the equation (raised by Georg Frobenius in Crelle's *Journal*, also in 1873) [41].[26)] However, between 1876 and 1880, Jordan returned to the topic and worked on the problem of finding finite groups which can be represented as linear groups of dimension n. With the help of the then recent Sylow theorems, he successfully found the answer for $n = 2$ (recovering Schwarz's list) and for the new case $n = 3$.[27)]

Fuchs himself sought to regain control of the subject by considering the general case of a second-order differential equation, and eliminating the geometric aspects of Schwarz's approach thanks to the leading algebraic theory of the time: the theory of invariants [28]. Fuchs himself described his approach as "algebrico-analytic". An algebraic solution u of the differential equation satisfies an irreducible equation

$$A_m u^m + A_{m-1} u^{m-1} + \cdots + A_0 = 0, \tag{4}$$

where the A_i are rational functions. All the roots u of this algebraic equation are also solutions of the differential equation and when the variable describes a loop around a singular point, the new branch $\tilde{u}$ is again a root of (4). If the quotient of two solutions is a rational function, it is easy to see that, because of the irreducibility, this rational function should be a root of unity; the algebraic equation is then a binomial one and the algebraic solution u is a root of a rational function. Whether a given second-order differential equation has such an algebraic solution, as Fuchs noticed, can be in fact directly determined by a system of linear equations.

If the differential equation has an algebraic solution which is not the root of a rational function, there exists another algebraic solution which is linearly independent of the first and then all the solutions are algebraic. Fuchs then reduced the original differential equation to the

24) These early works have been presented with more detail by Jeremy Gray in [35, ch. III] and we just summarize here a few distinctive features that shed light on the reception of Schwarz's paper.

25) On this issue, see [56].

26) On the context of Jordan's work, in particular with respect to matrices, see [17, 38, 51].

27) [42]. This, in particular Jordan's initial omission and correction, is explained in more detail in [35, sec. 3.5].

form $\frac{d^2y}{dz^2} = Py$, where P is a rational function with the same singular points as the coefficients of the original differential equation; under an easy-to-check condition on the coefficients, the solutions of the two differential equations are algebraic at the same time. Fuchs then showed that the roots of (4) can be organized in the form $y_i, y_i j, \cdots, y_i j^{l-1}$, for $0 \leq i \leq n-1$ $(nl = m)$, where j is a l-th root of unity, such that the ratio of y_i and y_k is not a constant when $i \neq k$ and that y_i is multiplied by an appropriate root of unity when the variable describes a circuit around a singularity [28, p. 112]. In effect, Fuchs described the algebraic extension generated by an algebraic solution u of the differential equation over the field of rational functions. Fuchs calls "a reduced system of roots for (4)" the set of the y_i, $0 \leq i \leq n-1$. Let us remark that if $n = 1$, y_0 is a root of a rational function. As all the y_i are linear combinations of two fixed particular solutions, say u_1, u_2, their product is a binary form $f(u_1, u_2)$ of degree n. When the variable describes a loop around a singular point, this binary form is multiplied by a root of unity and thus an appropriate power of it remains the same, that is, is a rational function. To summarize, Fuchs had proved that there always exists an algebraic solution of the differential equation, or a finite product of such solutions, which is a root of a rational function.

Fuchs called such a binary form of the smallest possible degree, say N, a "prime form". He showed that every solution of the differential equation which is a root of a rational function is a product of such prime forms, and that all the covariants of a binary form that is a root of a rational function are also roots of rational functions.[28] In particular, the Hessian $\frac{\partial^2 f}{\partial u_1^2} \cdot \frac{\partial^2 f}{\partial u_2^2} - \frac{\partial^2 f}{\partial u_1 \partial u_2} \cdot \frac{\partial^2 f}{\partial u_2 \partial u_1}$ of a prime form f is a product of prime forms and, being of degree $2N - 4$ in the variables, it is itself a prime form. Using both invariant theory and considerations of monodromy, Fuchs proved that $N \leq 12$ and finally provided a finite list of possible N, with their associated binary forms; this list contained in particular that of Schwarz for the hypergeometric differential equation. Moreover, Fuchs also tried to find a complete procedure to decide in a finite number of steps whether a given differential equation has algebraic solutions: he replaced the original differential equation by another one of higher order, with the same singular points, such that the original one has algebraic solutions if and only if this new one is satisfied by the root of a rational function — a condition that again depends on a system of linear equations. However, his procedure did not seem truly practical [34].

Reinforcing what we said earlier, Hermite who, besides his dislike of projective geometry (and even more of the non-Euclidean geometries then in full development), looked at Jordan's group theory with some suspicion, was quite pleased with Fuchs's approach, once presented to him in French by its author:

> Your memoir [on second-order equations that admit algebraic integrals] is a monumental work; studying it requires the utmost effort ... I was quickly enlightened upon reading in your translation that the results concerning the hypergeometric series, which had been obtained by the eminent mathematician Mr. Schwarz, follow directly and as a very specific consequence from your principles.[29]

More generally, Hermite encouraged several of his students — Jules Tannery, Georges-Henri Halphen, Gaston Floquet, among others who defended a PhD thesis in France at the end

---

[28] A covariant of a binary form $f(u_1, u_2)$ is a polynomial in the coefficients of the form and the variables u_1, u_2 which remains invariant under a linear change of variables of determinant 1. Fuchs spoke of "degree" for the degree of the covariants as polynomials with respect to the variables, a terminology which clashes with the usual one, but which we follow here. On invariant theory at the time, see [60].

[29] Unpub. letter from Hermite to Fuchs, April 14, 1876, pr. coll.

of the 1870s — to study and extend Fuchs's "algebrico-analytic" approach. Schwarz's article became a major reference in France only from the 1880s on, partly under the incentive of Klein, as seen above, and also that of Darboux, and of his student Edouard Goursat, who were more open to geometric arguments [21, 32].

The list of possible cases provided by Fuchs actually contained some irrelevant values, which were corrected by Felix Klein and Francesco Brioschi. Brioschi, in particular, refined Fuchs's study by using other invariants and covariants of the prime forms. The specialist of invariant theory Paul Gordan, who had been Klein's colleague for a year in Erlangen and might have been the one who introduced him to Schwarz's 1873 paper, finally established Fuchs's (corrected) list by arguments based purely on invariant theory [35, p. 88].

Klein himself devoted several papers to simplifying Schwarz's presentation. Using a characteristic mixture of group theory, invariant theory and geometrical arguments, he used (algebraic) solutions of the Schwarzian equation (F.) to construct forms which are invariant under certain groups of transformations. He would emphasize that "M. Fuchs's 'prime' forms are my binary forms which are invariant under these linear transformations" [45–47]. He also showed that a linear second-order differential equation with only algebraic solutions can be obtained from the hypergeometric ones on Schwarz's list with an appropriate change of variables by a rational function (the so-called Klein pullback method). That is, the solutions are of the form $e^{\int g}H(f)$, for rational functions f and g and an hypergeometric series H with parameters α, β, γ in Schwarz's list [79]. In Riemann's footsteps, the hypergeometric equation therefore no longer appeared to be a particular case requiring methods specially tailored to it, but a central case and a model. Some of Klein's students would also complete his work on the geometrical interpretation of binary forms [80] or Schwarz's s-function [68], for example.

The central role thus played by hypergeometric functions and the associated differential equations also spurred research into multivariable analogues and the partial differential equations associated to them [3, 55, 61]. A number-theoretical approach to the algebraicity of the hypergeometric series itself, modeled on Eisenstein's criteria, is due to Edmund Landau (completed by Eric Stridsberg and Alfred Errera a few years later) [54]. This, again, had been extended to other hypergeometric-type series.[30)]

The problem of deciding if and when certain types of linear differential equations have algebraic solutions resurfaced in the 1970s, framed this time in the language of the new algebraic geometry; it is partly associated to what is known as Grothendieck's conjecture, linking the nature of the solutions of a linear ordinary differential equation to that of its reductions modulo p, for almost all prime numbers p [22, 44]. It gave rise to rewritings of Klein's, Fuchs's and Schwarz's results on the hypergeometric equation using more modern techniques; in particular Francesco Baldassari and Bernard Dwork obtained a decision procedure that had been Fuchs's dream [6]. Certain features of the hypergeometric case and its variants—whether regarding the number and nature of singularities or the question of rigidity, which in the hypergeometric case is linked to the fact that the monodromy matrices around the singular points, up to conjugation, determine the global monodromy—have also been highlighted and examined within these new frameworks, in particular with the aim of addressing the question of algebraicity for other types of differential equations, whether they be linear differential equations defined, for instance, over algebraic curves in general (and not only the projective line as in the classical case) or other types of differential equations, such as Painlevé or Lamé equations [6, 13, 23].

30) For some examples among many, see [9, 12]. For a renewal of an arithmetical and effective approach to the algebraicity of the solutions of linear differential equations, see [16].

In his tribute to Schwarz, Georg Hamel described him as "not only a sharp-witted analyst [but] also a geometer, gifted with a vivid imagination and intuitive insight", also emphasizing how, for Schwarz, "the task [at hand] was [...] painstakingly carried out, thoroughly calculated, and, where possible, the result was drawn, modeled, and possibly even verified experimentally" [37, p. 11]. His work on the hypergeometric equation exemplified these multiple aspects that he aimed at intertwining. As mentioned earlier, a number of texts have been devoted to a presentation of Schwarz's results since the end of the nineteenth century; they are interesting witnesses of the changes in the priorities and writing of mathematics. For instance, Schwarz's geometrical approach to his spherical triangles was replaced by group theory quite early on, but this in quite different versions [7,33,43,50,58]. As for Schwarz's ideal of synthesis between opposite tendencies, it would be more and more explicitly challenged for a while by one of purity of methods [4].

But Schwarz's 1873 article also exemplified the central role played by the *Journal für die reine und angewandte Mathematik* in creating such a confluence of trends: except for references to complete works and to a few communications to learned societies, all the references mentioned and used by Schwarz in his memoir were to articles published in this Journal. They illustrate the mathematical variety of its authors, such as Abel, Jacobi, Kummer, Fuchs, Steiner, Weierstrass, Jordan. The 1870s were precisely a turning point. Indeed, the articles which extended or completed that of Schwarz in the following decade were hosted by a variety of journals: besides *Crelle's Journal* itself, the recently created, and soon-to-be rival, *Mathematische Annalen* edited by Klein, the *Zeitschrift für Mathematik und Physik*, the *Annales de l'École normale supérieure*, and soon *Acta mathematica*. New forms of institutional, personal, and sometimes disciplinary solidarities would reshape the landscape of mathematical publications and their uses.

References

[1] *N. Abel*, Ueber einige bestimmte Integrale, J. reine angew. Math. **2** (1827), 22–30.

[2] *W. Abikoff*, The Uniformization Theorem, Am. Math. Mon. **88** (1981), 574–592.

[3] *P. Appell*, Sur les fonctions hypergéométriques de deux variables, J. Math. Pures Appl. (3) **8** (1882), 173–216.

[4] *A. Arana*, Elements of Purity, Cambridge University Press, Cambridge 2024.

[5] *T. Archibald*, Charles Hermite and German Mathematics in France, in: Mathematics Unbound: The Evolution of an International Mathematical Research Community, 1800-1945, ed. *K. Hunger Parshall* and *A. C. Rice*, American Math. Soc. and London Math. Soc., Providence 2002, 123–138.

[6] *F. Baldassarri* and *B. Dwork*, On second order linear differential equations with algebraic solutions, Am. J. Math. **101-1** (1979) 42–76.

[7] *F. Beukers*, Gauss' Hypergeometric Function, in: Arithmetic and Geometry Around Hypergeometric Functions, ed. *R.-P. Holzapfel*, *A. M. Uludağ* and *M. Yoshida*, Birkhäuser, Basel, Boston, Berlin, 2007, 23–42.

[8] *F. Beukers* and *G. Heckman*, Monodromy for the hypergeometric function ${}_nF_{n-1}$, Invent. Math. **95** (1989), 325–354.

[9] *F. Beukers* and *A. van der Waall*, Lamé equations with algebraic solutions, J. Differ. Equations **197-1** (2004), 1–25.

[10] *L. Bieberbach*, H. A. Schwarz, Berl. Ber. **21** (1922), 47–52.

[11] *K.-R. Biermann*, Die Mathematik und ihre Dozenten an der Berliner Universität 1810–1933, Akademie-Verlag, Berlin 1988.

[12] *P. Beazley Cohen* and *J. Wolfart*, Fonctions hypergéométriques en plusieurs variables et espaces des modules de variétés abéliennes, Ann. Sci. Éc. Norm. Supér. (4) **26-6** (1993), 665–690.

[13] *P. Boalch*, Towards a nonlinear Schwarz's list, in: The many facets of geometry: a tribute to Nigel Hitchin, ed. *O. Garcia-Prada*, *J.-P. Bourguignon* and *S. Salamon*, Oxford University Press, Oxford 2010, 210–236.

[14] *R. Bölling*, Weierstrass and some members of his circle: Kovalevskaia, Fuchs, Schwarz, Schottky, in: Mathematics in Berlin, ed. *H. Begehr* and al., Springer, Basel 1998, 71-82.

[15] *O. Borůvka*, Sur la transformation des intégrales des équations différentielles linéaires ordinaires du second ordre, Annali di Mat. **41** (1956), 325-342.

[16] *A. Bostan*, *X. Caruso* and *J.Roques*, Algebraic solutions of linear differential equations: An arithmetic approach, Bull. Am. Math. Soc., New Ser. **61** **(4)** (2024), 609–658.

[17] *F. Brechenmacher*, La controverse de 1874 entre Camille Jordan et Leopold Kronecker, Rev. Hist. Math. **13** (2007), 187–257.

[18] *F. Brechenmacher*, Autour de pratiques algébriques de Poincaré: héritages de la réduction de Jordan, prepr., `hal-00630959v3`.

[19] *A. Cayley*, On the Schwarzian Derivative and the Polyhedral Functions, Trans. Camb. Philos. Soc. **13-1** (1881), 5–68.

[20] *E. Confalonieri*, Beiträge zur Geschichte der mathematischen Werke von Karl Weierstrass, III, 2013, prep.

[21] *G. Darboux*, Leçons sur la théorie générale des surfaces et les applications géométriques du calcul infinitésimal, vol. 1, Gauthier-Villars, Paris 1887.

[22] *P. Deligne*, Equations différentielles à points singuliers réguliers, LNM 163, Springer, Berlin Heidelberg 1970.

[23] *P. Deligne* and *G. D. Mostow*, Monodromy of hypergeometric functions and non-lattice integral monodromy, Publ. Math. Inst. Hautes Étud. Sci. **63** (1986), 5–89.

[24] *J. Dutka*, The early history of the hypergeometric function, Arch. Hist. Exact Sci. **31** (1984), 15–34.

[25] *D. Duverney*, An Introduction to Hypergeometric Functions, Birkhäuser, Cham 2024.

[26] *L. Fuchs*, Zur Theorie der linearen Differentialgleichungen mit veränderlichen Coefficienten, J. reine angew. Math. **66** (1866), 121–160.

[27] *L. Fuchs*, Zur Theorie der linearen Differentialgleichungen mit veränderlichen Coefficienten, J. reine angew. Math. **68** (1868), 354–385.

[28] *L. Fuchs*, Ueber die linearen Differentialgleichungen zweiter Ordnung, welche algebraische Integrale besitzen und eine neue Anwendung der Invariantentheorie, J. reine angew. Math. **81** (1876), 97–142, **85** (1878),1–25.

[29] *C. F. Gauss*, Disquisitiones generales circa seriem infinitam $1 + \frac{\alpha\beta}{1\cdot\gamma}x + \frac{\alpha(\alpha+1)\beta(\beta+1)}{1\cdot2\cdot\gamma(\gamma+1)}xx+$ etc. (Pars prior), Societatis regiae scientiarum Gottingensis Commentationes – Classis mathematicae **2** (1812), 3–46. Rep. in Gauss's *Werke*, vol. III, Akademie der Wissenschaften, Göttingen 1866, 123–162.

[30] *C. F. Gauss*, Methodus nova integralium valores per approximationem inveniendi, Societatis regiae scientiarum Gottingensis Commentationes – Classis mathematicae **3** (1814). Rep. in Gauss's *Werke*, vol. III, Akademie der Wissenschaften, Göttingen 1866, 163–196.

[31] *C. F. Gauss*, Determinatio serieri nostrae per aequationem differentialem secundi ordinis, in: Gauss's *Werke*, vol. III, Akademie der Wissenschaften, Göttingen 1866, 207–229.

[32] *É. Goursat*, Sur l'équation différentielle linéaire, qui admet pour intégrale la série hypergéométrique, Ann. Sci. Éc. Norm. Supér. (2) **10** (1881), 3–142 (supp.).

[33] *É. Goursat*, Leçons sur les séries hypergéométriques et sur quelques fonctions qui s'y rattachent, Hermann, Paris 1936.

[34] *J. Gray*, Fuchs and the Theory of Differential Equations, Bull. Am. Math. Soc. New Ser. **10-1** (1984), 1–26.

[35] *J. Gray*, Linear Differential Equations and Group Theory from Riemann to Poincaré, Birkhäuser, Boston 1986.

[36] *J. Gray* and *D. Rowe*, Episodes in the Berlin-Göttingen Rivalry, 1870–1930, Math. Intell. **22** (2000), 60–69.

[37] *G.G. Hamel*, Zum Gedächtnis an Hermann Amandus Schwarz, Jahresber. Dtsch. Math.-Ver. **32** (1923), 6–13.

[38] *T. Hawkins*, Thomas Weierstrass and the theory of matrices, Arch. Hist. Exact Sci., **17** (1977), 119–163.

[39] *C. Houzel*, L'équation générale du cinquième degré, in La Géométrie algébrique: recherches historiques, Paris: Blanchard 2002, 73–80.

[40] *C. G. J. Jacobi*, Fundamenta nova theoriae functionum ellipticarum, Gebrüder Borntraeger, Königsberg 1829.

[41] *C. Jordan*, Sur une application de la théorie des substitutions à l'étude des équations différentielles linéaires, Bull. Soc. Math. Fr. **2** (1873–1874), 100–127.

[42] *C. Jordan*, Mémoire sur les équations différentielles linéaires à intégrale algébrique, J. reine angew. Math. **84** (1877), 89–215.

[43] *J. Kampé de Fériet*, La fonction hypergéométrique, Gauthier-Villars, Paris 1937.

[44] *N. Katz*, Algebraic solutions of differential equations, Invent. Math. **18** (1972), 1–118.

[45] *F. Klein*, Ueber binäre Formen mit linearen Transformationen in sich selbst, Math. Ann. **9** (1875), 183–208.

[46] *F. Klein*, Ueber lineare Differentialgleichungen, Math. Ann. **11** (1876), 115–118.

[47] *F. Klein*, Ueber lineare Differentialgleichungen, Math. Ann. **12** (1877), 167–180.

[48] *F. Klein*, Vorlesungen über das Ikosaeder und die Auflösung der Gleichungen vom fünften Grade, Teubner, Leipzig 1884.

[49] *F. Klein*, Riemann et son influence sur les mathématiques modernes, in: Œuvres mathématiques de Riemann, Gauthier-Villars, Paris 1898, xiii-xxxv.

[50] *F. Klein*, Vorlesungen über die hypergeometrische Funktion gehalten an der Universität Göttingen im Wintersemester 1893/94, Springer, Berlin 1906.

[51] *R. Krömer*, The set of paths in a space and its algebraic structure: A historical account, Ann. Fac. Sci. Toulouse, Math. (6) **22** (2013), 915–968.

[52] *E. E. Kummer*, De generali quadam aequatione differentiali tertii ordinis, Program des evangelischen Königl. und Stadtgymnasiums in Liegnitz (1834). Repr. in J. reine angew. Math. **100** (1887), 1–9.

[53] *E. E. Kummer*, Über die hypergeometrische Reihe $1 + \frac{\alpha\beta}{1\cdot\gamma}x + \frac{\alpha(\alpha+1)\beta(\beta+1)}{1\cdot 2\cdot\gamma(\gamma+1)}x^2 + \cdots$, J. reine angew. Math. **15** (1836), 39–83, 127–172.

[54] *E. Landau*, Eine Anwendung des Eisensteinsches Satz auf die Theorie der Gaussche Differentialgleichung, J. Reine Angew. Math. **127** (1904), 92–102, 1904.

[55] *G. Lauricella*, Sulle funzioni ipergeometriche a più variabili, Rend. Circ. Mat. Palermo (2) **7** (1893), 111–158.

[56] *F. Le*, "Geometrical Equations": Forgotten Premises of Felix Klein's Erlanger Program, Hist. Math. **42-3** (2015), 315–342.

[57] *J. McCleary*, A Theory of Reception for the History of Mathematics in: The History of Modern Mathematics, ed. *D. Rowe* and *J. McCleary*, vol. I, Academic Press, Boston 1989, 3-14.

[58] *M. Matsuda*, Lectures on Algebraic Solutions of Hypergeometric Differential Equations, Kinokuniya, Kyoto 1985.

[59] *J. Nitsche*, Lectures on Minimal Surfaces, Cambridge University Press, Cambridge, New York, Melbourne 1989.

[60] *K. Parshall*, Toward a History of Nineteenth-Century Invariant Theory, in: The History of Modern Mathematics, ed. *D. Rowe* and *J. McCleary*, vol. I, Academic Press, Boston 1989, 157–206.

[61] *E. Picard*, Sur les fonctions hyperfuchsiennes provenant des séries hypergéométriques de deux variables, Ann. Sci. Éc. Norm. Supér. **2** (1885), 357–384.

[62] *E. G. C. Poole*, Introduction to the Theory of Linear Differential Equations, Clarendon Press, Oxford 1936.

[63] *D. Reimer*, Passion und Kalkül: Der Verleger Georg Andreas Reimer, Walter de Gruyter, Berlin and New York 1999.

[64] *B. Riemann*, Grundlagen für eine allgemeinen Theorie der Functionen einer verändlicher komplexer Grösse, Inauguraldissertation, Göttingen, 1851. Engl. transl.: Foundations for a general theory of functions of a complex variable, in: Collected Papers, Kendrick Press, Heber (2004), 1–42.

[65] *B. Riemann*, Beiträge zur Theorie der durch die Gauss'sche Reihe $F(\alpha, \beta, \gamma, x)$ darstellbaren Functionen, Abh. Königl. Ges. d. Wissensch. Göttingen, Mathem. Cl (1857), 3–22.

[66] *B. Riemann* (and *K. Hattendorff*, ed.), Über die Fläche vom kleinsten Inhalt bei gegebener Begrenzung, Abh. Königl. Ges. d. Wissensch. Göttingen, Mathem. Cl. (1868), 3–52.

[67] *H. P de Saint-Gervais*, Uniformization of Riemann Surfaces: Revisiting a Hundred-Year-Old Theorem, EMS Press, Zürich 2011.

[68] *F. Schilling*, Beiträge zur geometrischen Theorie der Schwarzschen s-Funktion, Math. Annalen **44** (1894), 161–260.

[69] *H. A. Schwarz*, Ueber die Minimalsfläche, deren Begrenzung als ein von vier Kanten eines regulären Tetraeders gebildetes, windschiefes Viereck gegeben ist, Monatsber. der könig. preussischen Akad. d. Wissensch. Berlin (1865), 149–153.

[70] *H. A. Schwarz*, Ueber einige Abbildungsaufgaben, J. reine angew. Math. **70** (1869), 105–120. Berichtigung, J. reine angew. Math. **71** (1869), iv.

[71] *H. A. Schwarz*, Conforme Abbildung der Oberfläche eines Tetraeders auf di Oberfläche einer Kugel, J. reine angew. Math. **70** (1869), 121–136.

[72] *H. A. Schwarz*, Bestimmung einer speciellen Minimalfläche, Akad. Wiss. Berlin (G. Vogt), Berlin 1871.

[73] *H. A. Schwarz*, Über diejenigen Fälle, in denen die Gauss'sche hypergeometrische Reihe $F(\alpha, \beta, \gamma, x)$ eine algebraische Funktion ihres vierten Elementes darstellt, Verh. Schweiz. Naturforsch. Ges. (1871), 74–77.

[74] *H. A. Schwarz*, Über diejenigen Fälle, in denen die Gauss'sche hypergeometrische Reihe $F(\alpha, \beta, \gamma, x)$ eine algebraische Funktion ihres vierten Elementes darstellt, J. reine angew. Math. **75** (1873), 292–335.

[75] *H. A. Schwarz*, Gesammelte Abhandlungen, 2 vols., Springer, Berlin 1890.

[76] *H. A. Schwarz*, Antrittsrede, Berl. Ber. (1893), 623–626.

[77] *R. Tazzioli*, Il Teorema di rappresentazione di Riemann: critica et interpretazione di Schwarz, Suppl. Rend. Circ. Mat. Palermo (2) **34** (1994), 95–132.

[78] *R. Tazzioli*, Fonctions fuchsiennes ou schwarziennes ?, Images des mathématiques (2010), online `10.60868/t51a-9w75`, `10.60868/v1s1-gv07`.

[79] *M. van Hoeij* and *J.-A. Weil*, Solving second order differential equations with Klein's theorem, in: ISSAC05, ACM, New York 2005, 340–347.

[80] *L. Wedekind*, Beiträge zur geometrische Interpretation binärer Formen, Math. Annalen **9** (1875), 209–217.

[81] *K. Weierstrass*, Über die Flächen, deren mittlere Krümmung überall gleich Null ist, Monatsber. König. Preuss. Akad. Wiss. Berlin (1866), 387, 612–625, 855–856.
[82] *M. Yoshida*, Fuchsian Differential Equations, With Special Emphasis on the Gauss-Schwarz Theory, Springer, Wiesbaden 1987.
[83] *M. Yoshida*, Hypergeometric Functions, My Love: Modular Interpretations of Configuration Spaces, Vieweg and Sohn, Braunschweig and Wiesbaden 1997.
[84] *G. Zappa*, Storia della risoluzione delle equazioni du quinto e sesto grado, Rendiconti Semin. Mat. Fis. Milano **65** (1995), 89–107.

Catherine Goldstein, IMJ-PRG, Case 247, 4 place Jussieu, 75252 Paris Cedex 05, France
e-mail: catherine.goldstein@imj-prg.fr

1876–1900

Zur Theorie der Abelschen Gleichungen.

Bemerkungen zum vorstehenden Aufsatze des Herrn *Schwering*.

(Von *Kronecker*.)

I.

Die von Herrn *Schwering* gefundene und nachgewiesene Eigenschaft der complexen Zahlen $\psi(\alpha)$ lässt sich auch aus allgemeineren Eigenschaften der bei den Abelschen Gleichungen auftretenden Ausdrücke herleiten.

Beschränkt man sich auf *einfache* Abelsche Gleichungen und bezeichnet man wie in meinem Aufsatze „Ueber Abelsche Gleichungen" (Monatsbericht der Akademie der Wissenschaften zu Berlin vom December 1877, S. 845 u. folg.) mit

$$x_h \qquad (h=0, 1, \dots n-1)$$

n unbestimmte Grössen, mit ω eine primitive n^{te} Wurzel der Einheit und mit ϖ_h die Summe

$$\sum_r \omega^{hr} x_r \qquad (r=0, 1, \dots n-1),$$

so wird der Quotient

$$\frac{\varpi_{ht}\varpi_{kt}}{\varpi_{(h+k)t}}$$

eine ganze Function von ω^t, deren Coefficienten rationale cyklische Functionen der n Grössen x sind, und welche mit $\psi_{h,k}(\omega^t)$ bezeichnet werden soll. Setzt man nämlich, wie schon a. a. O. angegeben worden, zuvörderst eine *Variable* w an Stelle der Wurzel der Einheit ω, so dass an die Stelle der mit ϖ_h bezeichneten Summe der Ausdruck

$$\sum_r w^{hr} x_r \qquad (r=0, 1, \dots n-1)$$

tritt, so findet offenbar die Congruenz

$$\sum_r w^{mr} x_r \equiv w^{ms} \sum_r w^{mr} x_{r+s} (\text{mod.}(w^n-1)) \qquad (r=0, 1, \dots n-1)$$

statt, wenn m, s irgend welche nicht negative ganze Zahlen bedeuten und unter x_k für den Fall, dass $k > n-1$ ist, der kleinste nicht negative Rest modulo n verstanden wird. Hieraus folgt die Congruenz

J. reine angew. Math. **93** (1882), 338–364
DOI 10.1515/crll.1882.93.338

Journal für die reine und angewandte Mathematik
© De Gruyter 1882

On the theory of Abelian equations

By *Leopold Kronecker* at Berlin

Preliminary Note[1)]

In an article entitled "On the Theory of the Arithmetic Functions that are called $\psi(\alpha)$ by *Jacobi*", [10][2)], Herr Schwering develops the theorem to the effect that "$1 + \psi(\alpha)$ is always divisible by $(1 - \alpha)^3$." The functions $\psi(\alpha)$ were originally defined by *Jacobi* (cf. Vol. 30, p. 166 of the Journal für Mathematik [3]) via the equation

$$\psi_{\mu,\nu}(\alpha) \sum_r \alpha^{(\mu+\nu)r} x^{g^r} = \sum_r \alpha^{\mu r} x^{g^r} \cdot \sum_r \alpha^{\nu r} x^{g^r} \qquad (r = 0, 1, \ldots, p-2)$$

where α is a λ-th, and x a p-th root of unity, g is a primitive congruence root modulo p, and p a prime number of the form $2m\lambda + 1$, with λ a prime number greater than three. This defining equation immediately implies—as Jacobi has indicated in the above-mentioned note ([3], p. 170 / 259), and explained in more detail in his lectures courses—the representation of the functions $\psi(\alpha)$:

$$\psi_{\mu,\nu}(\alpha) = \sum_m \alpha^{\mu \text{ ind. } m \, + \, \nu \text{ ind. } (m+1)} \qquad (m = 1, 2, \ldots, p-2),$$

which is labelled (6.) in Herr *Schwering*'s paper (cf. also p. 93 of Bachmann's book on cyclotomy [1]). This representation of the functions $\psi(\alpha)$ is what Herr Schwering bases his proof of the congruence

$$\psi(\alpha) \equiv -1 \ \big(\text{mod. } (1-\alpha)^3\big)$$

on, and from which he also deduces in his addendum at the end— [10], p. 337— the reduction of the number of the ψ-functions to the sixth part of them. This reduction had already been mentioned by Jacobi on p. 170 of Vol. 30 of the Journal für Mathematik [3], p. 170 / 259; cf. Part II of the following paper.

Berlin, December 1882. *Kronecker.*

---

1) *Translator's note*: In the 93rd volume of *Crelle's Journal*, Kronecker's article *On the Theory of Abelian Equations* follows Schwering's four-page paper *On the Theory of the Arithmetic Functions that are called $\psi(\alpha)$ by Jacobi* immediately, and its subtitle and content allude explicitly to it. However, the offprints of his article, which Kronecker sent to colleagues, did not include a copy of Schwering's note. They were instead preceded by an opening page which we translate here from Vol. IV of Kronecker's Collected Works, p. 133.

2) *Translator's note*: In this translation of Kronecker's paper we take the liberty to help today's readers by including references to a bibliography at the end, which we have added to Kronecker's text.

On the Theory of Abelian Equations.

Remarks on the preceding paper by Herr Schwering.

(By *Kronecker.*)

I.

The property of the complex numbers $\psi(\alpha)$ that was found and proved by Herr *Schwering* can also be deduced from more general properties of the expressions that occur in the context of Abelian equations.

If we restrict to *simple* Abelian equations[3)], and denote, as in my 1877 paper "On Abelian Equations" [6], by

$$x_h \qquad (h = 0, 1, \dots, n-1)$$

n indeterminate quantities, by ω a primitive n-th root of unity, and by $\tilde{\omega}_h$ the sum

$$\sum_r \omega^{hr} x_r \qquad (r = 0, 1, \dots, n-1),$$

then the quotient

$$\frac{\tilde{\omega}_{ht}\,\tilde{\omega}_{kt}}{\tilde{\omega}_{(h+k)t}}$$

turns out to be an entire function of ω^t whose coefficients are rational cyclic functions of the n quantities x. This polynomial will be written $\psi_{h,k}(\omega^t)$. Indeed, if one first takes an indeterminate w instead of the root of unity ω, so that the sum $\tilde{\omega}_h$ is replaced with the expression

$$\sum_r w^{hr} x_r \qquad (r = 0, 1, \dots, n-1),$$

then one obviously has the congruence

$$\sum_r w^{mr} x_r \equiv w^{ms} \sum_r w^{mr} x_{r+s} \quad (\text{mod. } (w^n - 1)) \qquad (r = 0, 1, \dots, n-1),$$

where m, s are any non-negative integers and where, for $k > n-1$, we read the index of x_k as its smallest residue modulo n. This implies the congruence

$$\prod_m \sum_r w^{mr} x_r \equiv \prod_m \sum_r w^{mr} x_{r+s} \quad (\text{mod. } (w^n - 1)) \qquad (r = 0, 1, \dots, n-1),$$

if the products are taken over a sequence of numbers m whose sum is divisible by n. Since the product

$$\prod_m \sum_r w^{mr} x_{r+s} \qquad (r = 0, 1, \dots, n-1; m = m_0, m_1, m_2, \dots),$$

does not change modulo $(w^n - 1)$ when s is replaced by one of the numbers $0, 1, \dots, n-1$, provided that $m_0 + m_1 + m_2 + \cdots \equiv 0$ (mod. n), this yields the congruence

$$\prod_m \sum_r w^{mr} x_r \equiv f(w) \quad (\text{mod. } (w^n - 1)) \qquad (r = 0, 1, \dots, n-1; m = m_0, m_1, m_2, \dots),$$

[3)] *Translator's Note.* Kronecker refers here to equations with a cyclic Galois group.

for an entire function $f(w)$ of degree $(n-1)$, whose coefficients are entire *cyclic* functions of the n quantities x. Linking this to the congruence

$$\prod_m \sum_r w^{mr} x_r \equiv \bar{f}(w) \quad (\text{mod. } (w^n - 1)) \qquad (r = 0, 1, \ldots, n-1; m = m_0, \bar{m}_1, \bar{m}_2, \ldots),$$

where we suppose that $m_0 + \bar{m}_1 + \bar{m}_2 + \cdots \equiv 0$ (mod. n), and where $\bar{f}(w)$ is a function of the same type as $f(w)$, we arrive at the general congruence

$$\bar{f}(w) \prod_m \sum_r w^{mr} x_r \equiv f(w) \prod_{\bar{m}} \sum_r w^{\bar{m}r} x_r \quad (\text{mod. } (w^n - 1)) \qquad (r = 0, 1, \ldots, n-1),$$

in which the numbers m, resp. $\bar{m}$, over which the products are taken, are only subject to the condition that their respective sums are congruent to each other according to the modulus n. This general congruence may also be stated like this:

$$(A.) \qquad \Big(\sum_r w^{mr} x_r\Big)^{\ell} \cdot \Big(\sum_r w^{m'r} x_r\Big)^{\ell'} \cdot \Big(\sum_r w^{m''r} x_r\Big)^{\ell''} \cdots \equiv \varphi(w) \quad (\text{mod. } (w^n - 1))$$

$$\big(r = 0, 1, \ldots, n-1;\ lm + l'm' + l''m'' + \cdots \equiv 0 \ (\text{mod. } n)\big),$$

where $\varphi(w)$ is an entire function in w of degree $(n-1)$, whose coefficients are *cyclic*, rational (entire or fractional) functions of $x_1, x_2, \ldots, x_n$.[4] Substituting in the congruence (*A.*), instead of the variable w, the n-th root of unity ω^t yields the *equation*:

$$(B.) \quad \tilde{\omega}_{mt}^{\ell} \cdot \tilde{\omega}_{m't}^{\ell'} \cdot \tilde{\omega}_{m''t}^{\ell''} \cdots = \varphi(\omega^t) \qquad \big(\ell m + \ell' m' + \ell'' m'' + \cdots \equiv 0 \ (\text{mod. } (n))\big).$$

For the special case $\ell = \ell' = 1, \ell'' = -1$, this does indeed give the relation mentioned above

$$(C.) \qquad \frac{\tilde{\omega}_{ht}\, \tilde{\omega}_{kt}}{\tilde{\omega}_{(h+k)t}} = \psi_{h,k}(\omega^t),$$

that is, the quotient on the left-hand side is an entire function of ω^t, whose coefficients are rational cyclic functions of $x_0, x_1, \ldots, x_{n-1}$. If x is, as in Herr *Schwering*'s paper, a primitive p-th root of unity, g a primitive congruence root of p, and if one puts $n = p - 1$ and

$$x_0 = x, x_1 = x^g, x_2 = x^{g^2}, \ldots, x_{n-1} = x^{g^{p-2}},$$

then the ψ-expressions defined by the equation (*C.*) coincide with *Jacobi*'s.

Now let again $x_0, x_1, \ldots, x_{n-1}$ stand for *indeterminate* quantities. Then expanding the product

$$2 \sum_r w^{hr} x_r \cdot \sum_r w^{kr} x_r \qquad (r = 0, 1, \ldots, n-1)$$

according to increasing powers of $(w-1)$, discarding all terms that contain a higher power of $(w-1)$ than the second, and using the abbreviations

$$\sum_r x_r = X, \ \sum_r r x_r = X_1, \ \sum_r r^2 x_r = X_2 \qquad (r = 0, 1, \ldots, n-1)$$

---

[4] *Translator's Note.* This should be $x_0, x_1, \ldots, x_{n-1}$.

yields the following expression (where $r, s = 0, 1, \dots, n-1$):

$$2X^2 - (h+k)(w-1)(w-3)XX_1 + (h+k)^2(w-1)^2XX_2 - hk(w-1)^2\sum_{r,s}(r-s)^2x_rx_s.$$

On the other hand, expanding $2\sum w^{(h+k)r}x_r$ and disregarding terms that start with $(w-1)^3$ yields the aggregate:

$$2X - (h+k)(w-1)(w-3)XX_1 + (h+k)^2(w-1)^2X_2.$$

We thus obtain the congruence, where again r, s vary over $0, 1, \dots, n-1$:

$$2\sum_r w^{hr}x_r \cdot \sum_r w^{kr}x_r \equiv 2X\sum_r w^{(h+k)r}x_r - hk(w-1)^2\sum_{r,s}(r-s)^2x_rx_s,$$

which is to be understood in the sense that, dividing each side by $(w-1)^3$, yields as remainders one and the same entire function with integral coefficients of the quantities

$$w, x_0, x_1, \dots, x_{n-1}.$$

Writing now, for short,

$$\sum_r x_rx_{r+h} = \Xi_h, \qquad (r = 0, 1, \dots, n-1)$$

we have the congruence

$$\sum_{r,s}(r-s)^2\, x_rx_s \equiv \sum_r r^2\, \Xi_r \pmod{n}, \qquad (r, s = 0, 1, \dots, n-1)$$

where $\Xi_1, \Xi_2, \dots$ are cyclic entire functions of the n quantities x. This implies that the congruence (where again $r = 0, 1, \dots, n-1$)

$$(D.) \qquad 2\sum_r w^{hr}x_r \cdot \sum_r w^{kr}x_r \equiv 2\sum_r x_r \cdot \sum_r w^{(h+k)r}x_r - hk(w-1)^2\sum_r r^2\, \Xi_r$$

holds, in the sense explained on p. 77 of the 92nd volume of this Journal ([7], p. 77 / 335), with respect to the modular system

$$\big(n(w-1)^2, (w-1)^3\big).$$

Putting, according to the congruence $(A.)$:

$$(\bar{A}.) \qquad \sum_r w^{hr}x_r \cdot \sum_r w^{kr}x_r \cdot \sum_r w^{(h+k)(n-1)r}x_r \equiv \varphi_{h,k}(w) \quad (\text{mod. } (w^n-1)),$$

and multiplying the congruence $(D.)$ by

$$\sum_r w^{(h+k)(n-1)r}x_r,$$

transforms this latter congruence into the congruence

$$(E.)\ 2\varphi_{h,k}(w) \equiv 2\varphi_{0,h+k}(w) - hk(w-1)^2\sum_r r^2\, \Xi_r \quad (\text{modd. } n(w-1)^2, (w-1)^3, w^n-1),$$

where the terms $\varphi(w)$ are entire functions in w of degree $(n-1)$, whose coefficients are *cyclic entire functions with integral coefficients* of the n quantitites x. This congruence $(E.)$ contains as a special case the result presented by Herr *Schwering*.

To begin with, taking as above for w the n-th root of unity ω^t, where t is any divisor of n, the modular system of the congruence $(E.)$ reduces to the system of the *two* moduli

$$n(\omega^t-1)^2, (\omega^t-1)^3.$$

Furthermore, since $\omega^t - 1$ is a complex *unit* as soon as the quotient $\frac{n}{t}$ is divisible by *distinct* prime numbers, the only case to be studied is the one where this quotient is a prime power ν^m. In that case, however, ν, and therefore also $t\nu^m$, i.e., n, is divisible by $\omega^t - 1$, so that the modular system in $(E.)$ reduces to the simple modulus

$$(\omega^t-1)^3.$$

The terms $\varphi(w)$ occurring in congruence $(E.)$ are determined by the equation

$$(F.)\qquad \tilde{\omega}_{ht}\,\tilde{\omega}_{kt}\,\tilde{\omega}_{-(h+k)t} = \varphi_{h,k}(\omega^t),$$

which follows from congruence $(A.)$ by setting $w = \omega^t$. The expression $\sum_r r^2\,\Xi_r$ in $(E.)$ transforms into

$$\frac{1}{n}\sum_{h,r} r^2\omega^{hr}\tilde{\omega}_h\tilde{\omega}_{-h}m, \qquad (h,r=0,1,\ldots,n-1)$$

when one inserts the values

$$x_s = \frac{1}{n}\sum_h \omega^{-hs}\tilde{\omega}_h,\, x_{s+r} = \frac{1}{n}\sum_h \omega^{-h(s+r)}\tilde{\omega}_h \qquad (h=0,1,\ldots,n-1)$$

in

$$\Xi_r = \sum_s x_s x_{s+r}. \qquad (s=0,1,\ldots,n-1)$$

For an arbitrary quantity w (and $r = 0,1,\ldots,n-1$), one has:

$$(w-1)^2\sum_r r^2 w^r = n^2w^n(w-1) - 2nw^{n+1} + w(w+1)\cdot\frac{w^n-1}{w-1}.$$

It follows that

$$(G.)\qquad \sum_r r^2\,\Xi_r = \frac{1}{n}\sum_{h,r} r^2\omega^{hr}\tilde{\omega}_h\tilde{\omega}_{-h} = \chi, \qquad (h,r=0,1,\ldots,n-1)$$

where χ is defined by the equation

$$(\bar{G}.)\quad \chi = \frac{1}{6}(n-1)(2n-1)\tilde{\omega}_0^2 - \sum_k \frac{n+(2-n)\omega^k}{(1-\omega^k)^2}\tilde{\omega}_k\tilde{\omega}_{-k}. \qquad (k=1,2,\ldots,n-1)$$

In this way one gets from congruence $(E.)$ to the congruence

$$(\bar{E}.)\qquad 2\varphi_{h,k}(\omega^t) \equiv 2\varphi_{0,h+k}(\omega^t) - hk(\omega^t-1)^2\chi \pmod{(\omega^t-1)^3},$$

where the terms φ, χ have the meaning introduced in (F.) and (G.).

Specialising now the quantities $x_0, x_1, \dots, x_{n-1}$ as before, for an odd prime number p:

$$x_0 = x, x_1 = x^g, x_2 = x^{g^2}, \dots, x_{n-1} = x^{g^{p-2}}, n = p-1,$$

where g is a primitive congruence root of p, and x a primitive p-th root of unity, one finds:

$$(H.) \qquad \tilde{\omega}_0 = -1, \;\; \tilde{\omega}_k\, \tilde{\omega}_{-k} = \omega^{\frac{1}{2}kn} p \qquad (k = 1, 2, \dots, n-1).$$

Thus

$$\chi = \frac{1}{n} \sum_{h,r} r^2 \omega^{hr} \tilde{\omega}_h \tilde{\omega}_{-h} = \frac{1}{4} p(p-1)^2 - \sum_r r^2, \quad (h = 0, 1, \dots, n-1; r = 1, 2, \dots, n-1)$$

or

$$\chi = -\frac{1}{12}(p-1)(p^2 - 11p + 12) = -\frac{1}{12} n(n^2 - 9n + 2),$$

and therefore

$$\chi \equiv -\frac{1}{12} n(n-2)(n+5) \;\; (\text{mod. } n).$$

Since $n = t\nu^m$, so that $\chi \equiv 0$ (mod. ν)—if the prime number $\nu \geq 3$, and if in case $\nu = 3$ we do also not have $t \equiv \pm 1$ (mod. 3)—the congruence ($\bar{E}$.) transforms into the simpler

$$(J.) \qquad \varphi_{h,k}(\omega^t) \equiv \varphi_{0,h+k}(\omega^t) \quad \big(\text{mod. } (\omega^t - 1)^3\big),$$

which in turn, thanks to the equations (C.), (F.), and (H.), can be brought into the form

$$(\bar{J}.) \qquad p\psi_{h,k}(\omega^t) \equiv -p \quad \text{and thus} \quad \psi_{h,k}(\omega^t) \equiv -1 \quad \big(\text{mod. } (\omega^t - 1)^3\big).$$

Here ω^t means a primitive root of unity whose exponent is any odd prime power dividing $p-1$, the only exception being the case where one has both $t = \frac{1}{3}(p-1)$ and t not divisible by 3.

The simplifaction of the congruence ($\bar{E}$.) is not at all limited to the cyclotomic equations. Indeed, for an arbitrary odd number of quantities $x_0, x_1, \dots, x_{n-1}$, it is possible to determine an equal number of entire rational functions of them

$$x'_0, x'_1, \dots, x'_{n-1}$$

in such a way that the resulting expressions $\tilde{\omega}'$ have precisely the property which, in the case of the cyclotomic equations, made the last term of the congruence ($\bar{E}$.) disappear. This reflects the fact that the product $\tilde{\omega}_h \tilde{\omega}_{-h}$ has the same value for all the $n-1$ nonzero values of the index h. In order to find such quantities x', put (for $h, i = 0, 1, \dots, n-1; k = 1, 2, , \dots, n-1$):

$$(K.) \qquad \frac{\tilde{\omega}_h}{\tilde{\omega}_{-h}} \prod_k \tilde{\omega}_k = \tilde{\omega}'_h, \quad x'_i = \frac{1}{n} \sum_h \omega^{-hi}\, \tilde{\omega}'_h.$$

These n quantities x' are entire integral functions of the n quantities x, and each of the n quantities x' can thus be written as an entire function of one of the quantities x in such a way that the coefficients are rational cyclic functions of $x_0, x_1, \dots, x_{n-1}$. Within the domain of rationality formed by the symmetric and cyclic functions of $x_0, x_1, \dots, x_{n-1}$, the quantity x_0

thus represents a genus [*Gattung*] of algebraic functions of the n-th order to which belong both all the n quantities x and all the n quantities x'. The expressions

$$\tilde{\omega}'_h = \sum_r \omega_{hr} x'_r \qquad (h, r = 0, 1, \ldots, n-1)$$

obtained from the x' satisfy, thanks to the equations $(K.)$, the condition

$$(L.) \qquad \tilde{\omega}'_h \tilde{\omega}'_{-h} = \tilde{\omega}'_0 \tilde{\omega}'_0 \qquad (h = 1, 2, \ldots, n-1).$$

Therefore the term

$$\frac{1}{n} \sum_{h,r} r^2 \omega^{-hr}\, \tilde{\omega}'_h \tilde{\omega}'_{-h}, \qquad (h, r = 0, 1, \ldots, n-1)$$

which corresponds to what we have called χ above, is equal to *zero*. The congruence $(\bar{E}.)$ thus takes on the simple form:

$$(E.') \qquad \varphi'_{h,k}(\omega) \equiv \varphi'_{0,h+k}(\omega) \quad (\text{mod. } (\omega - 1)^3),$$

where we have put $t = 1$ and omitted the factor 2. This congruence makes sense only when n is a prime power, because otherwise $\omega - 1$ is a complex unit. However, if $n = \nu^m$, with an odd prime number ν, then one may replace the modulus $(\omega - 1)^3$ by $\nu^{\frac{3}{\sigma}}$, where $\sigma = \nu^{m-1}(\nu - 1)$. Introducing instead of the terms φ' the terms ψ' defined by

$$\tilde{\omega}'_h \tilde{\omega}'_k = \tilde{\omega}'_{h+k} \psi'_{h,k}$$

transforms the congruence $(E.')$ into

$$(E.'') \qquad \psi'_{h,k}(\omega) \equiv \sum_r x'_r \quad (\text{mod. } (\omega - 1)^3) \qquad (r = 0, 1, \ldots, n-1).$$

As a consequence, we see that *in each genus* [Gattung] *of abelian equations in an arbitrary domain of rationality, there exist such equations for which one has a congruence $(E.'')$ that is as simple as the one for the cyclotomic equations.*

If one introduces the expressions $\tilde{\omega}'$ themselves, then one obtains instead of $(E.'')$ the congruence:

$$\tilde{\omega}'_h\, \tilde{\omega}'_k \equiv \tilde{\omega}'_0\, \tilde{\omega}'_{h+k} \quad (\text{mod. } (\omega - 1)^3),$$

and substituting the definitions for the $\tilde{\omega}'$ from the equation $(K.)$:

$$\frac{\tilde{\omega}_h\, \tilde{\omega}_k}{\tilde{\omega}_{-h}\, \tilde{\omega}_{-k}}\, \tilde{\omega}'_0\, \tilde{\omega}'_0 \equiv \frac{\tilde{\omega}_{h+k}}{\tilde{\omega}_{-h-k}}\, \tilde{\omega}'_0\, \tilde{\omega}'_0 \quad (\text{mod. } (\omega - 1)^3).$$

The factors $\tilde{\omega}'_0$ may be omitted from this congruence because

$$\tilde{\omega}'_0 = \prod_k \tilde{\omega}_k \qquad (k = 1, 2, \ldots, n-1)$$

and therefore obviously not divisible by $\omega - 1$. The congruence

$$(M.) \qquad \tilde{\omega}_h\, \tilde{\omega}_k\, \tilde{\omega}_{-h-k} \equiv \tilde{\omega}_{-h}\, \tilde{\omega}_{-k}\, \tilde{\omega}_{h+k} \quad (\text{mod. } (\omega - 1)^3)$$

thus emerges as being equivalent to the congruence ($E.''$). This congruence ($M.$) can be verified immediately. For, if one substitutes on each side the values of the terms $\tilde{\omega}$, and if one replaces the indices $-h,\ -k$ by $n-h,\ n-k$, in order to have but positive exponents, and if one takes at the end a variable w in the place of ω, then the left-hand side becomes (for $r,s,t=0,1,\ldots,n-1$)

$$(N.)\qquad \sum_{r,s,t} w^{hr+ks+(2n-h-k)t}\, x_r\, x_s\, x_t$$

and the right-hand side

$$(\bar{N}.)\qquad \sum_{r,s,t} w^{(n-h)r+(n-k)s+(h+k)t}\, x_r\, x_s\, x_t.$$

Expanding these summations according to increasing powers of $w-1$ the first term on either side is the same one, to wit:

$$\Big(\sum_r x_r\Big)^3 \qquad (r=0,1,\ldots,n-1);$$

the second term on either side—the one that contains the first power of $(w-1)$—will be divisible by n. Indeed, the coefficient of $(w-1)$ will be congruent, with respect to the modulus n, to the following sum, taken either positively or negatively:

$$\sum_{r,s,t}\big(hr+ks-(h+k)t\big)\, x_r\, x_s\, x_t, \qquad (r,s,t=0,1,\ldots,n-1)$$

which is obviously zero, since the product

$$\sum_r r x_r \sum_r x_r \sum_r x_r$$

occurs in it multiplied first by h, then by k, and finally by $-(h+k)$. In the third term, finally, the coefficient of $(w-1)^2$, up to multiples of the number n, equals ($r,s,t=0,1,\ldots,n-1$)

$$\frac{1}{2}\sum_{r,s,t}\big(hr+ks-(h+k)t\big)^2\, x_r\, x_s\, x_t \mp \frac{1}{2}\sum_{r,s,t}\big(hr+ks-(h+k)t\big)\, x_r\, x_s\, x_t.$$

Here the upper sign of the second term occurs for the left-hand side, the lower one for the right-hand side of the congruence. But this second term is identically equal to zero. Thus the two sums ($N.$) and ($\bar{N}.$) are congruent to each other with respect to the modular system $\big(n, (w-1)^3\big)$. And for $w=\omega$, this modular system reduces to the simple modulus $(\omega-1)^3$, as n is divisible by $(\omega-1)^3$.

II.

Just as we have here deduced the property of the complex numbers $\psi(\omega)$ that is encoded in the congruence ($\bar{J}.$) from their *original* definition by the equation ($C.$)—and not, as in Herr *Schwering's* work, from that second definition that is contained in equation (6.) of the preceding paper [10]—the reduction of all the functions $\psi_{h,k}$ to the sixth part of them also follows immediately from that equation ($C.$). And just as the simplification of the congruence ($\bar{E}.$), this reduction is based on the property that we have stressed above: that the product $\tilde{\omega}_h\,\tilde{\omega}_{-h}$ has the same value for all the $n-1$ indices h. It therefore applies to all abelian equations having that property. Indeed, putting as above

$$(P.)\qquad \psi'_{h,k} = \frac{\tilde{\omega}'_h\,\tilde{\omega}'_k}{\tilde{\omega}'_{h+k}},\quad \tilde{\omega}'_h\,\tilde{\omega}'_{-h} = \tilde{\omega}'_k\,\tilde{\omega}'_{-k},\qquad (h,k=1,2,\ldots,n-1)$$

it is immediately obvious that the six expressions

$$\psi'_{h,k},\ \psi'_{k,h},\ \psi'_{h,-h-k},\ \psi'_{k,-h-k},\ \psi'_{-h-k,h},\ \psi'_{-h-k,k}$$

are all equal to each other. This observation can already be found in the notes of *Jacobi*'s 1836 lecture course, which is mentioned in the preface to Bachmann's treatise on cyclotomy [1]. *Jacobi* there uses the following notations (with the indices $h = 0,1,2,\ldots,\lambda-2$ and $k=0,1,2,\ldots,p-2$):

$$(\alpha,x) = \sum_r \alpha^k x^{g^k},\quad \psi_h(\alpha) = \frac{(\alpha,x)(\alpha^h,x)}{(\alpha^{h+1},x)},$$

where x is a primitive p-th root of unity, α a primitive λ-th root of unity, g a primitive congruence root of the prime number p, and λ a prime factor of $p-1$. He discusses in detail the conditions under which some of the six different index systems of the ψ-functions that have the same values are identical, and formulates the final result like this:

> "The number of the functions (ψ) that have to be formed is equal to the integer that exceeds $\frac{1}{6}\lambda$ but by a true fraction."[5)]

Here λ denotes the number that we have written n above. The fact that the functions ψ can "always be reduced to the sixth part of them", *Jacobi* has also stated in [3], p. 170 / 259, footnote. In this footnote, *Jacobi* adds: "I have even found by an induction going up to $\mu = 31$ that all the functions ψ can always be expressed by the values of a single one of them." (Here μ denotes the number for which *Jacobi* used the letter λ in his lecture courses.) As for this, the copy in my possession of the notes of *Jacobi's* lecture no. XVII contains the following *more detailed* explanation[6)]:

> We have seen in the XVth lecture that one may always reduce the number of functions ψ to be formed to the sixth part of them. These reductions are the only ones that reduce one function to the other *immediately* by simply exchanging α with another root. However, if one just poses the problem to express one function in terms of the others, one may obtain very different relations by which one obtains one function from another, by replacing α with various other roots and multiplying several of these expressions. In general, one obtains from one function $\psi(\alpha)$ the $\lambda-1$ functions
>
> $$\psi(\alpha),\psi(\alpha^2),\psi(\alpha^3),\ldots,\psi(\alpha^{\lambda-1}),$$
>
> and multiplying several of these functions will lead one again to functions ψ. However, it is very difficult to formulate the general procedure, because every prime

5) *Translator's Note.* See [2], p. 161 of the pdf file; cf. [4], p. 109.

6) *Translator's Note.* Cf. [2], pp. 170–172 of the pdf file. There is word for word agreement with Kronecker's quote at the beginning, but slight divergences in the latter part of the quote. Cf. [4], p. 116, where the whole passage has a slightly different wording.

> number λ enters here as an individual case, which requires very special procedures. This is why I have, a while ago, engaged into an induction which I have pushed pretty far, proving the theorem not in general, but for all the prime numbers up to 31, one by one. Doing this one encounters very curious expressions under the radical sign, which turn into products of different values of a function ψ that is determined by the nature of the prime number. A useful formula here is for instance the one by which the product of two functions ψ may be expressed as the product of two others.

Jacobi then derives a proposition, which in terms of the notation defined by $(C.)$ above can be expressed by saying that the product

$$\psi_{h,k}\;\psi_{h+k,\ell}$$

remains unchanged under any permutation of h, k, ℓ. The meaning of this proposition becomes evident once the product $\psi_{h,k}\;\psi_{h+k,\ell}$ is transformed by means of the expressions (α, x) or $\tilde{\omega}$ into the quotients

$$\frac{(\alpha^h, x)(\alpha^k, x)(\alpha^\ell, x)}{(\alpha^{h+k+\ell}, x)} \quad \text{or} \quad \frac{\tilde{\omega}_h\,\tilde{\omega}_k\,\tilde{\omega}_\ell}{\tilde{\omega}_{h+k+\ell}}.$$

Jacobi uses this result in the sequel of his lectures to obtain the desired representation of the term $\tilde{\omega}$ by a single one of the functions ψ, for a series of prime numbers λ. He introduces the argument by noting that the general theory is laden with great difficulties, and that it is therefore advisable to first test the representations in a number of examples. Combining relations that hold between the various ψ-functions he then arrives at the formulae[7]:

$$\begin{aligned}
(\alpha, x)^3 &= p\,\psi_1(\alpha^2),\\
(\alpha, x)^5 &= p\,\psi_1(\alpha^2)\psi_1^2(\alpha),\\
(\alpha, x)^7 &= \psi_1(\alpha^4)\psi_1^2(\alpha^2)\psi_1^4(\alpha),\\
(\alpha, x)^{11} &= \tfrac{1}{p^2}\,\psi_2(\alpha^2)\psi_2^2(\alpha)\psi_2^3(\alpha^8)\psi_2^4(\alpha^6)\psi_2^5(\alpha^7),\\
(\alpha, x)^{13} &= \tfrac{1}{p^4}\,\psi_2(\alpha^3)\psi_2^2(\alpha^8)\psi_2^3(\alpha)\psi_2^4(\alpha^4)\psi_2^5(\alpha^{11})\psi_2^6(\alpha^7),\\
(\alpha, x)^{17} &= \tfrac{1}{p^{11}}\,\psi_3(\alpha^6)\psi_3^2(\alpha^3)\psi_3^3(\alpha^2)\psi_3^4(\alpha^{10})\psi_3^5(\alpha^8)\psi_3^6(\alpha)\psi_3^{10}(\alpha^4)\psi_3^8(\alpha^5),\\
(\alpha, x)^{19} &= \tfrac{1}{p^{13}}\,\psi_2(\alpha^4)\psi_2^2(\alpha^4)\psi_2^3(\alpha^{14})\psi_2^4(\alpha)\psi_2^5(\alpha^{16})\psi_2^6(\alpha^7)\psi_2^7(\alpha^6)\psi_2^8(\alpha^{10})\psi_2^9(\alpha^{11}),\\
(\alpha, x)^{23} &= \tfrac{1}{p^{23}}\,\psi_3(\alpha^3)\psi_3^2(\alpha^{13})\psi_3^3(\alpha)\psi_3^4(\alpha^{18})\psi_3^5(\alpha^{19})\psi_3^6(\alpha^{12})\psi_3^7(\alpha^7)\psi_3^8(\alpha^9)\psi_3^9(\alpha^8)\psi_3^{11}(\alpha^{17})\psi_3^{13}(\alpha^2).
\end{aligned}$$

For the first eight odd prime numbers λ, *Jacobi* thus has indeed expressed the value of $(\alpha, x)^\lambda$ in terms of products

$$\prod \psi^p(\alpha^q)$$

(where $\psi^p(\alpha^q)$ denotes the p-th power of $\psi(\alpha^q)$), i.e., in terms of products of *conjugate* ψ-functions. He stresses as remarkable properties of these products that, for the different factors $\psi^p(\alpha^q)$, the congruence value of $p \cdot q$ (mod. λ) remains the same, and that the arrangement of the products makes it possible to recognize how the different radical quantities that occur when computing the conjugate values $(\alpha, x), (\alpha^2, x), (\alpha^3, x), \ldots$ are determined by one another. What we mean is this: For instance, if $\lambda = 7$, the radical expressions

$$(\alpha, x) = \sqrt[7]{\psi_1(\alpha^4)\psi_1^2(\alpha^2)\psi_1^4(\alpha)}, \quad (\alpha^2, x) = \sqrt[7]{\psi_1(\alpha)\psi_1^2(\alpha^4)\psi_1^4(\alpha^2)}$$

---

7) *Translator's Note.* In Kronecker's Crelle paper, the exponent of the term $\psi_3(\alpha^{17})$ in the last identity is printed as 10 instead of 11. This mistake is tacitly corrected in [9], p. 144. See also [11], p. 62.

are arranged such that the proportion $(\alpha, x)^2 : (\alpha^2, x)$ equals $\sqrt[7]{\psi_1^7(\alpha)}$, so that the root is determined by the condition: $(\alpha, x)^2 = \psi_1(\alpha)\,(\alpha^2, x)$. Furthermore, *Jacobi* notes that for the numbers $\lambda = 3, 4, 11, 13, 19$, the exponents of the conjugate functions ψ form the series of natural numbers from 1 through $\frac{1}{2}(\lambda - 1)$, and he develops the conditions under which *such* a representation of $(\alpha, x)^\lambda$ is possible. However, he fails to investigate in general the question of whether the expression $(\alpha, x)^\lambda$ can be written as a product of conjugate ψ-functions for all prime numbers λ—even though he seems to have conjectured this.

In my opinion it is these further developments concerning the reduction of the ψ-functions in *Jacobi*'s lectures—and not just to the simple and short remarks about reducing the number of ψ-functions to their sixth part—that Bachmann refers to in the passage from [1], p. 92, to which Herr *Schwering* alludes ([10], p. 334). Bachmann writes there: "the ψ-functions can be reduced to an even smaller number, which simplifies the calculation. However, I do not think it fitting to reproduce *Jacobi*'s finer investigations of this here."

The question of reducing the ψ-functions, which *Jacobi* has dealt with in detail, can be treated in a very simple and elegant way by means of the methods that I have used in my dissertation about the complex units (cf. p. 3 of this volume [5], or p. 125 of my *Festschrift* for Kummer's doctorate anniversary). Indeed, if n is a prime number and γ a primitive congruence root of it, then *Jacobi* looks for one of the $n - 1$ functions

$$\psi_{1,\gamma^m} \qquad (m = 0, 1, \ldots, n-2),$$

for which $\tilde{\omega}^n$ may be written as a product of powers of the conjugate functions

$$\psi_{t,t\gamma^m} \qquad (t = 1, 2, \ldots, n-1),$$

and of the $n - 1$ expressions

$$\tilde{\omega}_t\,\tilde{\omega}_{-t} \qquad (t = 1, 2, \ldots, n-1).$$

He thus looks for an equation (with $h, k = 0, 1, \ldots, n-2; t = 1, 2, \ldots, n-1$)

$$(Q.) \qquad \tilde{\omega}_t^n = \prod_h \psi_{t\gamma^h, t\gamma^{h+m}}^{r_h} \prod_k \tilde{\omega}_{t\gamma^k}^{s_k}\,\tilde{\omega}_{-t\gamma^k}^{s_k},$$

where the exponents r and s are integers. Here it does not make any difference whether $\tilde{\omega}^k$ stands for the *general* expression

$$\sum_r \omega^{kr}\,x_r \qquad (r = 0, 1, \ldots, n-1),$$

or—like in *Jacobi*'s question—for the particular expression that arises from the substitution

$$x_r = x^{g^r}, \qquad (r = 0, 1, \ldots, n-1)$$

with a primitive p-the root of unity x; it is just that in this special case the second product on the right-hand side of equation $(Q.)$ reduces to an integral (positive or negative) power of the prime number p. As we explained above, *Jacobi* also requires from the representation of $\tilde{\omega}^n$ given by the equation $(Q.)$ that it "shows" which of the n-th roots belong to the various conjugate values $\tilde{\omega}_t$. In other words, in such a presentation the proportion $\tilde{\omega}_{t\gamma}^n : \tilde{\omega}_t^{n\gamma}$ has to appear in terms of complete n-th powers of ψ-functions. This means, since the first product on the right-hand side of $\tilde{\omega}_{t\gamma}^n$ becomes:

$$\prod_h \psi_{t\gamma^{h+1},t\gamma^{h+m+1}}^{r_h} \quad \text{or} \quad \prod_h \psi_{t\gamma^h,t\gamma^{h+m}}^{r_{h-1}} \qquad (h = 0, 1, \ldots, n-2),$$

the exponents r_k have to satisfy the congruence conditions

$$(R^0.) \qquad \gamma r_h \equiv r_{h-1} \ (\text{mod. } n) \qquad (h = 0, 1, \ldots, n-2).$$

Replacing in the equation $(Q.)$ the functions ψ by their expressions in terms of the functions $\tilde{\omega}$, we find (with $h, k = 0, 1, \ldots, n-2; t = 1, 2, \ldots, n-1$):

$$(Q'.) \qquad \tilde{\omega}_t^n = \prod_h \tilde{\omega}_{t\gamma^h}^{r_h} \, \tilde{\omega}_{t\gamma^{h+m}}^{r_h} \, \tilde{\omega}_{t\gamma^{h+\ell}}^{-r_h} \prod_k \tilde{\omega}_{t\gamma^k}^{s_k} \, \tilde{\omega}_{-t\gamma^k}^{s_k},$$

where the number ℓ is determined by the congruence

$$\gamma^\ell \equiv 1 + \gamma^m \ (\text{mod. } n) \qquad \text{or} \qquad \ell \equiv \text{ind. } (1 + \gamma^m) \ \big(\text{mod. } (n-1)\big).$$

The exponents r and s are determined, apart from the congruence conditions $(R^0.)$, only by the conditions:

$$(R.) \qquad r_h - r_{h-\ell} + r_{h-m} + s_h + s_{h-\nu} = n\, \delta_{0k} \qquad (h = 0, 1, \ldots, n-2),$$

where

$$\nu = \frac{1}{2}(n-1), \quad \delta_{0k} = 0 \text{ for } h > 0, \ \delta_{00} = 1,$$

and where both the exponents r and the exponents s whose indices differ but by integral multiples of n are taken to be the same, so that for each number k one has

$$r_k = r_{k+n}, \ s_k = s_{k+n}.$$

If now ζ is any primitive root $(n-1)$-st of unity, then multiplying the equation $(R.)$ by ζ^{hk} and summing subsequently over all numbers $h = 0, 1, \ldots, n-2$ yields the equation

$$\sum_h (r_h - r_{h-\ell} + r_{h-m})\, \zeta^{hk} + \sum_h (s_h + s_{h-\nu})\, \zeta^{hk} = n \qquad (h, k = 0, 1, \ldots, n-2).$$

This immediately transforms into

$$(1 - \zeta^{k\ell} + \zeta^{km}) \sum_h r_h \zeta^{hk} + (1 + \zeta^{\nu k}) \sum_h s_h \zeta^{hk} = n \qquad (h, k = 0, 1, \ldots, n-2).$$

Since $\zeta^\nu = -1$, this in turn yields for odd numbers k the conditional equation

$$(1 - \zeta^{k\ell} + \zeta^{km}) \sum_h r_h \zeta^{hk} = n \qquad (h = 0, 1, \ldots, n-2; k = 1, 3, 5, \ldots, n-2).$$

As a consequence, if z is a variable, we get the congruence condition

$$(1 - z^\ell + z^m) \sum_h r_h z^h + (1 + z^\nu) \sum_h s_h z^h \equiv n \ \big(\text{mod. } (z^{n-1} - 1)\big) \qquad (h = 0, 1, \ldots, n-2).$$

Since $n-1=2\nu$, this congruence condition contains the following

$$(R''.) \qquad (1-z^\ell+z^m)\sum_h r_h z^h \equiv n \ \big(\text{mod. } (z^\nu+1)\big) \qquad (h=0,1,\dots,n-2).$$

But it is also *sufficient*, for if $\sum_h r_h z^h$ satisfies this congruence, and thus an equation of the form

$$(1-z^\ell+z^m)\sum_h r_h z^h = n+(1+z^\nu)\Phi(z) \qquad (h=0,1,\dots,n-2),$$

then the numbers s are determined by the congruence

$$\sum_h s_h z^h \equiv -\Phi(z) \ \big(\text{mod. } (z^\nu-1)\big) \qquad (h=0,1,\dots,n-2).$$

Therefore the equation $(R'.)$, or the perfectly equivalent congruence $(R''.)$, joined to the congruences $(R^0.)$, express the necessary and sufficient condition for the solvability of *Jacobi*'s problem discussed above.

The congruence conditions $(R^0.)$ can be transformed like this:

$$r_h \equiv \gamma^{-h} r_0 \ (\text{mod. } n) \qquad (h=1,2,\dots,n-2).$$

This shows that one has to have

$$\sum_h r_h \zeta^h \equiv r_0 \sum_h \gamma^{-h}\zeta^h \ (\text{mod. } n) \qquad (h=0,1,\dots,n-2).$$

Since therefore the following congruence modulo the system of divisors $(n,\gamma-\zeta^k)$—in the sense of [7], p. 72 / 329—has to hold:

$$\sum_h r_h \zeta^h \equiv r_0 \sum_h \zeta^{h(1-k)} \qquad (h=0,1,\dots,n-2),$$

and since for $k>1$ this gives the congruence

$$\sum_h r_h \zeta^h \equiv 0 \qquad (h=0,1,\dots,n-2),$$

it follows that the complex number $\sum_h r_h \zeta^h$ has to be divisible by each one of the "algebraic divisors"[8)]

$$\text{div}[nu+\gamma-\zeta^k] \qquad (k=2,3,\dots,n-2).$$

If one considers only numbers k that are smaller than, and relatively prime to $n-1$, the different algebraic divisors $\text{div}[nu+\gamma-\zeta^k]$ represent all the conjugate, or linked, prime divisors of n in the theory of the numbers formed from *primitive* roots of unity. The product

$$\prod_k \text{div}[nu+\gamma-\zeta^k],$$

extended over all numbers k, is therefore absolutely equivalent to the prime number p. Since for $k>1$ we have the congruence

$$\sum_h r_h \zeta^h \equiv 0 \ (\text{mod}[nu+\gamma-\zeta^k]) \qquad (h=0,1,\dots,n-2),$$

8) Cf. §15 of my paper in the 92th volume of this Journal [7].

so that for *all* values of k:

$$\mathrm{div}[nu+\gamma-\zeta]\cdot\sum_h r_h\zeta^h \equiv 0 \ \ (\mathrm{mod}[nu+\gamma-\zeta^k]) \qquad (h=0,1,\ldots,n-2),$$

the product on the left-hand side of this congruence has to be divisible by the product of all conjugate prime divisors $\mathrm{div}[nu+\gamma-\zeta^k]$, and thus by the prime number n. Thanks to the equation $(R'.)$ we thus obtain:

$$\mathrm{div}[nu+\gamma-\zeta]\cdot\sum_h r_h\zeta^h \equiv 0 \ \ \big(\mathrm{mod.}\ (1-\zeta^\ell+\zeta^m)\sum_h r_h\zeta^h\big) \qquad (h=0,1,\ldots,n-2),$$

or

$$\mathrm{div}[nu+\gamma-\zeta] \equiv 0 \ \ \big(\mathrm{mod.}\ (1-\zeta^\ell+\zeta^m)\big).$$

On the other hand we have:

$$1-\zeta^\ell+\zeta^m \equiv 1-\gamma^\ell+\gamma^m \big(\mathrm{mod.}\ (\gamma-\zeta)\big),$$

$$1-\gamma^\ell+\gamma^m \equiv 0 \ \ (\mathrm{mod.}\ n)$$

and thus

$$1-\zeta^\ell+\zeta^m \equiv 0 \ (\mathrm{mod}[nu+\gamma-\zeta]).$$

We therefore have to have

$$1-\zeta^\ell+\zeta^m \sim \mathrm{div}[nu+\gamma-\zeta].$$

In other words, the complex number $1-\zeta^\ell+\zeta^m$ is itself a prime divisor of n, so that

$$(S.) \qquad \mathrm{Nm}(1-\zeta^\ell+\zeta^m)=n,$$

where the norm is defined to be the product of all the complex numbers $1-\zeta^\ell+\zeta^m$, with ζ varying over all *primitive* $(n-1)$-th roots of unity.

The condition $(S.)$ is necessary, but not sufficient for *Jacobi*'s problem to be solvable. It is verified for all prime numbers $n=3,5,7,\ldots$ up to 43; but *not* for $n=47$. Indeed, in this last case, $-\zeta$ is a primitive 23rd root of unity, and the number 47 cannot be written as the norm of a complex number made up from 23rd roots of unity; there exists absolutely no complex number that would be equivalent to the *prime form* $nu-2-\zeta$, or to the algebraic prime divisor $\mathrm{div}[nu-2-\zeta]$, and would thus satisfy the condition imposed for one of the complex numbers $1-\zeta^\ell+\zeta^m$.[9)]

If, however, we disregard *Jacobi*'s *second* postulate expressed by the congruence condition $(R^0.)$, then the problem can be solved also for $n=47$. That is, even in this case, $\tilde{\omega}^n$ can still be represented as a product of powers of conjugate ψ-functions. In fact, for this to be possible, only the condition expressed in the equation $(R'.)$ or in the congruence $(R''.)$ matters. Its meaning will now be explained in more detail.

The congruence $(R''.)$ taken for itself—i.e., disregarding the congruence conditions $(R^0.)$, which will now be put aside—allows for totally arbitrary values of the coefficients r. It can

9) Here we take the primitive root γ to be -2.

therefore be rewritten in terms of the divisor systems of higher levels introduced in my work (in the 92nd volume of this Journal; [7], p. 72 / 329), as a simple congruence:

$$(T.) \qquad n \equiv 0 \pmod{\text{modd. } 1 - z^\ell + z^m, 1 + z^\nu},$$

where $(1-z^\ell+z^m, 1+z^\nu)$ is a divisor system of level 2. This divisor system can be decomposed into factors that correspond to the irreducible factors of $1 + z^\nu$. Indeed, if $F(z)$ and $G(z)$ are two divisors of $1 + z^\nu$, which do not have a common divisor (of level one), then the modular system

$$\big(1 - z^\ell + z^m, F(z), G(z)\big)$$

is the composition[10)] of the two modular systems

$$\big(1 - z^\ell + z^m, F(z)\big), \ \big(1 - z^\ell + z^m, G(z)\big).$$

Indeed, multiplying the elements of these two modular systems produces at first the system

$$\big((1 - z^\ell + z^m)^2, (1 - z^\ell + z^m)F(z), (1 - z^\ell + z^m)G(z), F(z) \cdot G(z)\big).$$

But here the functions that are multiplied by $(1 - z^\ell + z^m)$ in the first three terms:

$$(1 - z^\ell + z^m), F(z), G(z)$$

form a modular system equivalent to *One*, i.e., it is not a proper modular system. In order to explain this more in detail, I first emphasize that the elimination resultant of $F(z)$ and $G(z)$, as a homogeneous linear function of the two, may be added to the modular system $\big(F(z), G(z)\big)$ as a further element. This resultant is represented by the product of all the factors $(\zeta^h - \zeta^k)$, which one obtains by letting ζ^h vary over all the zeros of the equation $F(z) = 0$, and ζ^k over all the zeros of the equation $G(z) = 0$. Its value is an integer d, and since it is given by a product

$$\pm \prod_{h,k} (1 - \zeta^{h-k}) \quad \text{or} \quad \pm \prod_{h,k} (1 - e^{\frac{(h-k)\pi i}{\nu}}),$$

it can only be a divisor of ν. Let us note in passing that, if this divisor d is not One, then the functions $F(z)$ and $G(z)$ have a divisor system of level *two* in common. For instance,

$$1 + z^3 + z^6 = (1 + z + z^2)(-2 + 2z - z^3 + z^4) + 3,$$

and therefore

$$1 + z^3 + z^6 \equiv 0 \ (\text{modd. } 1 + z + z^2, 3),$$

and thus the modular system of level *two* $(1 + z + z^2, 3)$ that occurs here is contained in *both* functions $1 + z + z^2, 1 + z^3 + z^6$, even though these have no divisor of level *one* in common.

---

10) Cf. Vol. 92 of this Journal, p. 78, V ([7], p. 78 / 336, V). At this place one may add that a modular system $(M \cdot M', M_1, M_2, \dots)$ is the composition of the two systems $(M, M_1, M_2, \dots)$ and $(M', M_1, M_2, \dots)$ even in the case when $(M, M', M_1, M_2, \dots) \sim 1$. In fact, the composed system is built from the elements

$$M \cdot M', M \cdot M_1, M \cdot M_2, \dots, M' \cdot M_1, M' \cdot M_2, \dots, M_1 \cdot M_1, M_1 \cdot M_2, \dots,$$

and for each $h = 1, 2, \dots$, the elements $M \cdot M_h, M' \cdot M_h, M_1 \cdot M_h, M_2 \cdot M_h, \dots$ may be replaced by the sole element M_h, because the system constituted by the elements that are multiplied here with M_h, is equivalent to 1 according to our hypothesis.

From what has been said we now have the equivalence

$$(1 - z^\ell + z^m, F(z), G(z)) \sim (1 - z^\ell + z^m, F(z), G(z), d).$$

According to the congruence ($T.$), the number n must contain *each of* the modular systems

$$(1 - z^\ell + z^m, F(z)), \quad (1 - z^\ell + z^m, G(z)),$$

and therefore also the modular system which is common to both of them:

$$(1 - z^\ell + z^m, F(z), G(z)),$$

and consequently also the following, which is equivalent to it:

$$(1 - z^\ell + z^m, F(z), G(z), d).$$

We see that the number n may be added to this modular system as an element.[11)] Then, however, having the two numbers d and n as elements, since d divides ν and hence also $n - 1$, already the modular system (d, n) is equivalent to 1, so that we also find

$$(1 - z^\ell + z^m, F(z), G(z)) \sim 1.$$

The upshot is that, as claimed above, the system

$$(1 - z^\ell + z^m, F(z) \cdot G(z))$$

is the composition of the two modular systems

$$(1 - z^\ell + z^m, F(z)), \quad (1 - z^\ell + z^m, G(z)).$$

If one introduces, as in §22 of my paper published in the 92nd volume of this Journal, instead of the divisor systems the *'forms'* with indeterminates $u, u', \ldots$, then the congruence ($T.$) tells us that

> the form of level two $(1 - z^\ell + z^m)u + 1 + z^\nu$ must be contained in the number n.[12)]

Consequently, if $F(z)$ denotes, as above, a divisor of $1 + z\nu$, the form

$$(1 - z^\ell + z^m)u + F(z)$$

must be contained in the number n as well. On the other hand, if the two forms

$$(1 - z^\ell + z^m)u + F(z), \ (1 - z^\ell + z^m)u + G(z)$$

are contained in n, then the same necessarily holds also for the composed form

$$(1 - z^\ell + z^m)u + F(z) \cdot G(z).$$

---

11) Cf. Vol. 92 of this Journal, p. 77, III ([7], p. 77 / 335, III).

12) Cf. Vol. 92 of this Journal, p. 85, I, and p. 91, IX^0 ([7], p. 85 / 343, I and p. 91 / 351, IX^0).

Indeed, since the modular system

$$\big(1 - z^\ell + z^m, F(z), G(z)\big)$$

is equivalent to One, as explained above, those two forms have no other one in common, and the fact that each of them is contained in n implies the same for the product. To prove this here without recourse to my paper that was mentioned several times, let $P(z), Q(z), R(z)$ be integral functions satisfying the equation

$$(1 - z^\ell + z^m)P(z) + F(z)\cdot Q(z) + G(z)\cdot R(z) = 1,$$

which is required for the equivalence

$$\big(1 - z^\ell + z^m, F(z), G(z)\big) \sim 1$$

to hold. Furthermore, let $\Phi(z), \Phi_1(z), \Psi(z), \Psi_1(z)$ be integral functions satisfying the equations

$$n = (1 - z^\ell + z^m)\Phi(z) + F(z))\Phi_1(z), \quad n = (1 - z^\ell + z^m)\Psi(z) + G(z))\Psi_1(z),$$

that correspond to the congruences

$$n \equiv 0 \;\big(\text{modd. } 1 - z^\ell + z^m, F(z)\big), \quad n \equiv 0 \;\big(\text{modd. } 1 - z^\ell + z^m, G(z)\big).$$

Then we have the equation

$$\begin{aligned} n = \big(1 - z^\ell + z^m\big)\big(G(z)R(z)\Phi(z) + F(z)Q(z)\Psi(z) + nP(z)\big) \\ + F(z)G(z)\big)\big(R(z)\Phi_1(z) + Q(z)\Psi_1(z)\big), \end{aligned}$$

which may be expressed by the congruence

$$n \equiv 0 \;\big(\text{modd. } 1 - z^\ell + z^m, F(z)\cdot G(z)\big).$$

This shows that the form

$$(1 - z^\ell + z^m)u + F(z)\cdot G(z)$$

is indeed contained in the number n. The congruence (T.) may thus be replaced by the congruence

$$(T^0.) \qquad n \equiv 0 \;\big(\text{modd. } 1 - z^\ell + z^m, f_h(z)\big) \qquad (h = 1, 2, \ldots),$$

where $f_h(z)$ represents all distinct irreducible divisors of $z^\nu + 1$. The necessary and sufficient conditions for the fact that $\tilde{\omega}^n$ can be written in the form (Q'.), with *arbitrary* exponents r—that is, irrespective of the restrictive congruences (R^0.)—may therefore be formulated like this: There have to exist numbers ℓ, m,

> for which all the forms of level two $(1 - z^\ell + z^m)u + f_h(z)$ are contained in the prime number n; or else: for which the prime number n is divisible by each one of the complex numbers
>
> $$1 - \zeta^{kl} + \zeta^{km} \qquad (k = 1, 3, 5, \ldots, n-2).$$

Indeed, the values $\zeta, \zeta^3, \zeta^5, \ldots, \zeta^{n-2}$ are the ν different roots of all the equations $f_h(z) = 0$. As a consequence of all this, we thus see that the congruences

$$(T'.) \qquad n \equiv 0 \ \big(\text{modd.}\ (1 - \zeta^{kl} + \zeta^{km})\big) \qquad (k = 1, 3, 5, \ldots, n-2),$$

which immediately emerge from the equation $(R'.)$ as *necessary* conditions, form at the same time a system of *sufficient* conditions for the representation $(Q'.)$. The properties of the number n on which the possibility to represent those cyclotomic expressions via ψ-functions is based are thus grounded in the theory of the complex numbers formed from $(n-1)$st roots of unity, or in other words grounded in the properties of n within the domain of rationality $[e^{\frac{2\pi i}{n-1}}]$.[13] This is the reason why *Jacobi*—who could not yet use the theory of complex numbers in the domain $[e^{\frac{2\pi i}{n-1}}]$ in his 1836 lectures—was brought to the statement we have quoted before: "that it is very difficult to give a general procedure, because each prime number (n) appears here as an *individual*, for which very specific methods have to be applied."

In order to illustrate the above developments for a number of examples, let me first start from *Jacobi*'s representation of $\tilde{\omega}^{19}$ or $(\alpha, x)^{19}$ which we have given above (after the quote from *Jacobi*'s lectures). There we have

$$n = 19, \quad \gamma = 2, \quad \psi_2(\alpha) = \tilde{\omega}_1\, \tilde{\omega}_2\, \tilde{\omega}_3^{-1}, \quad m = 1, \quad l = 13, \quad \nu = 9,$$

since $2^{13} \equiv 3$ (mod. 19). Then we have

$$(1 - z^{13} + z) \sum_h r_h z^h \equiv 19 \ \big(\text{mod.}\ (z^9 + 1)\big),$$

provided that we take

$$(U.) \qquad \sum_h r_h z^h = 4 + 2z + z^2 + 5z^4 + 6z^6 + 3z^7 + 9z^{12} + 7z^{14} + 8z^{17}.$$

The numbers r thus defined are the exponents of the conjugate ψ-functions in *Jacobi*'s expression. If $-\zeta$ is a primitive 9-th root of unity, we then find

$$19 \equiv 0 \ \big(\text{mod.}\ (1 - \zeta^{13k} + \zeta^k)\big) \qquad (k = 1, 3, 5, \ldots, 17),$$

and this congruence holds for all the nine values of k, as soon as it is satisfied for $k = 1, 3, 9$. For $k = 1$, the complex number $1 - \zeta^{13k} + \zeta$, or equivalently $1 + \zeta + \zeta^4$, is indeed a complex prime factor of 19 and thus satisfies condition $(S.)$ above. Both for $k = 3$ and $k = 9$, though, we find

$$1 - \zeta^{13k} + \zeta^k = 1.$$

The exponents r defined by equation $(U.)$ therefore satisfy *all* of *Jacobi*'s conditions.

If, however, we take

$$n = 47, \quad \gamma = -2, \quad \nu = 23,$$

then, as we have seen above, it is impossible to meet all of *Jacobi*'s conditions, unless one disregards the congruence conditions $(R^0.)$. Every ψ-function which, in *Jacobi*'s parlance, has index γ^m, i.e.,

$$\psi_{\gamma^m} \quad \text{or} \quad \tilde{\omega}_1\, \tilde{\omega}_{\gamma^m}\, \tilde{\omega}_{1+\gamma^m}^{-1},$$

[13] Cf. Vol. 92 of this Journal, p. 14 ([7], p. 14 / 260–261).

is characterized completely by the number m or, alternatively, by the complex number

$$1-\zeta^{\ell}+\zeta^{m},$$

where ℓ is determined by the congruence

$$1-\gamma^{\ell}+\gamma^{m}\equiv 0 \pmod{n}.$$

For every set of six of these ψ-functions that can be reduced to each other according to *Jacobi*'s observation discussed above, the two numbers ℓ, m are, respectively

$$(\ell,m),\ (\ell-m,-m),\ (m+\nu,\ell+\nu),\ (m-\ell,-\ell-\nu),\ (-m+\nu,\ell-m+\nu),\ (-\ell,m-\ell-\nu).$$

Therefore the complex numbers

$$1-\zeta^{\ell}+\zeta^{m}, 1-\zeta^{\ell-m}+\zeta^{-m}, 1+\zeta^{m}-\zeta^{\ell}, 1-\zeta^{m-\ell}-\zeta^{-\ell}, 1+\zeta^{-m}-\zeta^{\ell-m}, 1-\zeta^{-\ell}-\zeta^{m-\ell}$$

correspond to the six functions ψ. These differ from each other, either not at all, or by simple units $\zeta^{m}, -\zeta^{\ell}, \ldots$. It is therefore the *one* complex number

$$1-\zeta^{\ell}+\zeta^{m}$$

that characterizes all the six ψ-functions that can be reduced to one another, and for the number ℓ, for instance, only one of the following six values

$$\ell,\ \ell-m,\ m+\nu,\ m-\ell,\ -m-\nu,\ -\ell$$

has to be taken. Putting now $\beta=-\zeta$ (so that, for $n=47$, β denotes a primitive 23rd root of unity), then the only eight functions ψ that have to be considered are

$$\frac{\tilde{\omega}_1^2}{\tilde{\omega}_2}, \frac{\tilde{\omega}_1\tilde{\omega}_2}{\tilde{\omega}_3}, \frac{\tilde{\omega}_1\tilde{\omega}_3}{\tilde{\omega}_4}, \frac{\tilde{\omega}_1\tilde{\omega}_4}{\tilde{\omega}_5}, \frac{\tilde{\omega}_1\tilde{\omega}_5}{\tilde{\omega}_6}, \frac{\tilde{\omega}_1\tilde{\omega}_6}{\tilde{\omega}_7}, \frac{\tilde{\omega}_1\tilde{\omega}_9}{\tilde{\omega}_{10}}, \frac{\tilde{\omega}_1\tilde{\omega}_{10}}{\tilde{\omega}_{11}},$$

or equivalently characterized by the complex numbers

$$2-\beta, 1+\beta-\beta^{-4}, 1+\beta^{-4}-\beta^{2}, 1+\beta^{2}+\beta^{9}, 1-\beta^{9}-\beta^{-3}, 1+\beta^{-3}-\beta^{-11},$$
$$1+\beta^{-8}+\beta^{10}, 1-\beta^{10}+\beta^{-6}.$$

The second and the sixth of these numbers, and also the fourth and the seventh, are conjugates of each other. Up to a simple unit, the fifth and the eighth are also conjugate, because

$$-\beta^{3}\left(1-\beta^{9}-\beta^{-3}\right)=1-\beta^{3}+\beta^{12}=1-\beta^{-2\cdot 10}+\beta^{2\cdot 6}.$$

Since each of the eight complex numbers contains the algebraic prime divisor

$$\operatorname{div}[47u+2-\beta],$$

those six complex numbers, i.e., the second, the fourth, the fifth, the sixth, the seventh, and the eighth, contain another prime divisor, apart from $\operatorname{div}[47u+2-\beta]$, which is conjugate to it. The first and the third number, though, i.e., $2-\beta$ and $1+\beta^{-4}-\beta^{2}$, contain *only* $\operatorname{div}[47u+2-\beta]$, and none of its conjugates. Since $\operatorname{div}[47u+2-\beta]$ is not absolutely equivalent to any complex number β, those two complex numbers have to contain other algebraic divisors of prime numbers *different* from 47; they cannot be divisors of the number 47, as the congruence (T'.) would have it. As for the six other complex numbers, they do actually divide the number 47, and each of them is absolutely equivalent to the product of two conjugate prime divisors

$$\operatorname{div}[47u + 2 - \beta].$$

For instance, it turns out that

$$1 + \beta - \beta^{-4} \sim \operatorname{div}[47u + 2 - \beta] \cdot \operatorname{div}[47u + 2 - \beta^{-8}],$$

since, on the one hand, the complex number on the left is divisible by each of the factors on the right, and since, on the other hand, the product of the eleven conjugate numbers

$$1 + \beta^{h^2} - \beta^{-4h^2} \qquad (h = 1, 2, 3, \ldots, 11)$$

equals 47, so that the number $1 + \beta - \beta^{-4}$ can only contain prime divisors of 47, and of those only two conjugate ones. Since the number $1+\beta-\beta^{-4}$ characterizes the ψ-function $\tilde{\omega}_1\tilde{\omega}_2\,\tilde{\omega}_3^{-1}$, the 47th power of $\tilde{\omega}$ can be written in terms of this ψ-function in the form (Q.), even though the exponents r in this representation do not fulfill the congruences (R^0.).

Jacobi seems to have conjectured that the cyclotomic expressions he wrote $(\alpha, x)^\lambda$ could *always* be expressed as products of conjugate ψ-functions. This is hardly to be expected in view of the conditions (T'.) we have found for this above: for this to be true, a number m would always have to exist such that each of the complex numbers

$$1 + \zeta^{km} - \zeta^{k\,\text{ind.}\,(1+g^m)} \qquad (k = 1, 3, \ldots, \lambda - 2),$$

where ζ is a root of the equation $\zeta^{\frac{1}{2}(\lambda-1)} + 1 = 0$, would have to be either a complex unit or a product of conjugate algebraic prime divisors of λ. However, I have not been able to determine, for any value of λ, that these conditions cannot be fulfilled. The first prime number that seems suited to check this is $\lambda = 83$.[14)]

This being said, the representation of $(\alpha, x)^\lambda$ that *Jacobi* was looking for has no theoretical value whatsoever *without* the congruence conditions (R'.), whereas *with* these congruence conditions we have seen above that it is already impossible for $\lambda = 47$. Whenever the congruence conditions (R'.) can be satisfied, the complex number $1-\zeta^\ell+\zeta^m$, which characterizes the function ψ_{γ^m}, is a complex prime factor of the prime number n, which *Jacobi* writes λ. It is this property—and not the straightforward formation and shape of the ψ-functions—that accounts for their theoretical value, both for the cyclotomic equations and for the *Abelian* equations in general. As a matter of fact, if $1 - \zeta^\ell + \zeta^m$ is replaced by any complex number

$$\sum_h q_h \zeta^h \qquad (h = 0, 1, \ldots, n - 2)$$

that satisfies the conditions (R'.)

$$\sum_h q_h \zeta^{hk} \cdot \sum_h r_h \zeta^{hk} = n \qquad (h = 0, 1, \ldots, n - 2; k = 1, 3, 5, \ldots, n - 2),$$

14) *Translator's Note.* As pointed out in the reprint [9], p. 157 and p. 500, this case is addressed in [11], pp. 72–75. Here Schwering also adds: "As Herr *Kronecker* told me in a letter, Herr *Wolfskehl* obtained the same result."

and such that the numbers r_h satisfy the congruence conditions $(R^0.)$, then we may argue as above that, according to the equation $(S.)$, we have to have

$$(S'.) \qquad \mathrm{Nm} \sum_h q_h \zeta^h = n.$$

Such a complex number $\sum_h q_h \zeta^h$ then characterizes an expression Ψ_t that is determined by the equation

$$\Psi_t = \prod_h \tilde{\omega}_{t\gamma_h}^{q_h}$$

and which has the property that $\tilde{\omega}^n$ can be written in the form $(Q.)$, once those more special functions ψ are replaced by these more general Ψ. And not only in the case where the prime number n can be written as the norm of a complex number in ζ according to the condition $(S'.)$, but also in full generality, when we replace $1 - \zeta^\ell + \zeta^m$ by *any* complex number in ζ, the above deduction yields results about the various ways of expressing the roots of *Abelian* equations. This will be elaborated in another article.

In the case of cyclotomy, as *Jacobi* stressed in his lecture courses, the ψ-functions afford a number-theoretic definition for the coefficients of the expressions that we have called $\tilde{\omega}^n$, and this definition also seems very useful for the computation of the $\tilde{\omega}^n$. If, however, in the interest of a simple number-theoretic definition, we abandon the idea of a simple actual computation, then it is preferable to write the value of the cyclotomic expression $\tilde{\omega}^n$ directly, without appealing to the ψ-functions.

III.

Denoting, as in (I.) above, by w a variable, x a primitive p-th root of unity, and by g a primitive congruence root modulo p, then, for any divisor n of $p-1$ and ω a primitive n-th root of unity, we have according to the formula $(A.)$ above, where we take $\ell = n, m = 1$:

$$\Big(\sum_r w^r x^{g^r}\Big)^n \equiv p \sum_t c_t w^t \ \big(\text{mod. } (w^n - 1)\big) \qquad (r = 0, 1, \ldots, p-2; t = 0, 1, \ldots, n-1)$$

for suitable integers $c_0, c_1, \ldots, c_{n-1}$. It is precisely the determination of these integers c that is required for cyclotomy. According as k has one of the values $1, 2, \ldots, p-1$, or the value p, one has

$$\sum_r w^r x^{kg^r} \equiv w^{n - \text{ind. } k} \sum_r w^r x^{g^r}, \quad \text{or} \quad \equiv 0 \ \big(\text{mod. } (w^n - 1)\big) \qquad (r = 0, 1, \ldots, p-2).$$

It follows that

$$\sum_k \Big(\sum_r w^r x^{kg^r}\Big)^n \equiv (p-1) \Big(\sum_r w^r x^{kg^r}\Big)^n \ \big(\text{mod. } (w^n - 1)\big)$$

$$(k = 0, 1, \ldots, p-1; r = 0, 1, \ldots, p-2).$$

The $(n+1)$-fold sum

$$\sum_k \sum_{r_1} \sum_{r_2} \cdots \sum_{r_n} w^{r_1 + r_2 + \cdots + r_n} x^{k(g^{r_1} + g^{r_2} + \cdots + g^{r_n})}$$

$$(k = 0, 1, \ldots, p-1; r_i = 0, 1, \ldots, p-2)$$

is therefore congruent modulo $(w^n - 1)$ to

$$p(p-1)\sum_t c_t w^t \qquad (t = 0, 1, \ldots, n-1).$$

If, however, we only sum over k:

$$\sum_k x^{k(g^{r_1}+g^{r_2}+\cdots+g^{r_n})} \qquad (k = 0, 1, \ldots, p-1),$$

this gives the value p, or *zero*, according as $g^{r_1} + g^{r_2} + \cdots + g^{r_n}$ is, or is not, divisible by p. Whence

$$(V.)\quad \sum_t c_t w^t \equiv \frac{1}{p-1} \sum_{r_1, r_2, \ldots, r_n} w^{r_1+r_2+\cdots+r_n} \ (\text{mod. } (w^n - 1)) \quad (t = 0, 1, \ldots, n-1),$$

where the summation on the right-hand side extends to all values $r_1, r_2, \ldots, r_n = 0, 1, \ldots, p-2$ satisfying the congruence condition

$$g^{r_1} + g^{r_2} + \cdots + g^{r_n} \equiv 0 \quad (\text{mod. } p).$$

Substituting ω^h for the variable w, the coefficients are characterized *algebraically* by the fact that one has, for all integral values of h, the equation

$$\left(x + \omega^h x^g + \omega^{2h} x^{g^2} + \cdots + \omega^{(p-2)h} x^{g^{p-2}}\right)^n = p \sum_t c_t \omega^{ht} \qquad (t = 0, 1, \ldots, n-1).$$

Each one of the coefficients c_t is defined in the most elegant way, via the equation (*V.*), in a purely *arithmetical manner as being the* $(p-1)$*-th part of the number of systems of values* $r_1, r_2, \ldots, r_n$*, which are different modulo* $(p-1)$*, and which satisfy both of the following congruences:*

$$r_1 + r_2 + \cdots + r_n \equiv t \ (\text{mod. } n), \quad g^{r_1} + g^{r_2} + \cdots + g^{r_n} \equiv 0 \quad (\text{mod. } p).$$

This same definition can be reformulated in a slightly different manner by introducing, instead of the numbers r, the numbers s defined by the congruences $g^r \equiv s \ (\text{mod. } p)$. Then one has to take into account for the numbers $s_1, s_2, \ldots, s_n$ all those values $1, 2, \ldots, p-1$ for which both of the following congruences are satisfied

$$s_1 s_2 \cdots s_n \equiv g^{t+hn}, \quad s_1 + s_2 + \cdots + s_n \equiv 0 \quad (\text{mod. } p),$$

where h denotes an arbitrary integer. Dividing now all the integers that are distinct modulo p into n groups one of which is made up by the n-th power residues, then the value of

$$(p-1)c_t \qquad (t = 0, 1, \ldots, n-1)$$

can be defined simply as *the number of systems of* n *numbers* $s_1, s_2, \ldots, s_n$ *that are different modulo* p*, whose sum is divisible by* p*, and whose product belongs to the group of* $\frac{p-1}{n}$ *numbers containing* g^t. Finally, we would like to highlight the very equation that results from replacing w with ω^h in (*V.*):

$$(V^0.)\qquad \left(\sum_r \omega^{rh} x^{g^r}\right)^n = \frac{p}{p-1} \sum_{r_1, r_2, \ldots, r_n} \omega^{h(r_1+r_2+\cdots+r_n)} \qquad (r = 0, 1, \ldots, p-2),$$

where the summation on the right-hand side is limited to only those values $r_1, r_2, \ldots, r_n = 0, 1, \ldots, p-2$ which satisfy the congruence condition

$$g^{r_1} + g^{r_2} + \cdots + g^{r_n} \equiv 0 \pmod{p}.$$

It hardly needs to be mentioned that, when counting different systems of values $r_1, r_2, \ldots, r_n$ and $s_1, s_2, \ldots, s_n$, the two systems

$$r_1 = 0, r_2 = 1, r_3 = 2, \ldots; \quad r_1 = 1, r_2 = 0, r_3 = 2, \ldots;$$

for instance, have to be accounted for as different. Denoting now by m_k the number of those numbers r that are equal to k, then the two congruences above

$$r_1 + r_2 + \cdots + r_n \equiv t \pmod{n}, \quad g^{r_1} + g^{r_2} + \cdots + g^{r_n} \equiv 0 \pmod{p}$$

transform into

$$(W.) \qquad \sum_k k m_k \equiv t \pmod{n}, \quad \sum_k k m_{\text{ind.}\,(k+1)} \equiv -n \pmod{p},$$

with the conditions

$$\sum_k m_k \leq n, \quad k = 1, 2, \ldots, p-2.$$

Indeed, that congruence $g^{r_1} + g^{r_2} + \cdots + g^{r_n} \equiv 0 \pmod{p}$ yields in the first place:

$$\sum_{k=0}^{h=p-2} m_k\, g^k \equiv 0 \quad \text{or} \quad n + \sum_{h=1}^{h=p-2} m_h\,(g^h - 1) \equiv 0 \pmod{p},$$

because $n = m_0 + m_1 + \cdots + m_{p-2}$. The latter congruence immediately yields the second congruence of (*W.*), once we put $g^h - 1 \equiv k \pmod{p}$. Thus the coefficient c_t can also be written in the form

$$\frac{n!}{p-1} \sum \frac{1}{m_0!\, m_1!\, m_2! \, \cdots \, m_{p-2}!},$$

where the summation extends to all values $m_0, m_1, \ldots m_{p-2} = 0, 1, 2, \ldots, n$ which, apart from the requirement

$$m_0 + m_1 + m_2 + \cdots + m_{p-2} = n$$

also satisfy the two congruences (*W.*). Here, as usual, one has to take $m!$ to be equal to the product $1 \cdot 2 \cdot 3 \cdots m$ as long as $m > 0$, and for $m = 0$ one has to take $m! = 1$.

If we define $m_k = \ell_{g^k}$, then ℓ_h is the number of s that are equal to h. Thus

$$c_t = \frac{n!}{p-1} \sum \frac{1}{\ell_1!\, \ell_2! \, \cdots \, \ell_{p-1}!},$$

where the summation extends to all those values $\ell_1, \ell_2, \ldots, \ell_{p-1} = 1, 2, \ldots, n$ for which

$$\ell_1 + \ell_2 + \ell_3 + \cdots + \ell_{p-1} = n, \quad \ell_1 + 2\ell_2 + 3\ell_3 + \cdots + (p-1)\ell_{p-1} \equiv 0 \pmod{p},$$

and at the same time the remainder modulo p of the number

$$2^{\ell_2} \cdot 3^{\ell_3} \cdots (p-1)^{\ell_{p-1}}$$

falls into the group of the residues of $g^t, g^{t+n}, g^{t+2n}, \dots$. For $n = 3, p = 7$, the following systems of values (where those ℓ that are not listed are equal to 0)

$$\ell_1 = 1, \ell_2 = 1, \ell_4 = 1; \ell_3 = 1, \ell_5 = 1, \ell_6 = 1;$$

$$\ell_1 = 2, \ell_5 = 1; \ell_1 = 1, \ell_3 = 2; \ell_2 = 2, \ell_3 = 1; \ell_4 = 2, \ell_6 = 1;$$
$$\ell_4 = 1, \ell_5 = 2; \ell_2 = 1, \ell_6 = 2$$

satisfy the conditions

$$\ell_1 + \ell_2 + \ell_3 + \ell_4 + \ell_5 + \ell_6 = 3, \quad \ell_1 + 2\ell_2 + 3\ell_3 + 4\ell_4 + 5\ell_5 + 6\ell_6 \equiv 0 \text{ (mod. 7)}.$$

For the first two systems of ℓ one finds $t = 0$, for the other six, $t = 2$. Since we have here $n! = p - 1$, this implies

$$c_0 = 2, \; c_1 = 0, \; c_2 = 3.$$

For an *arbitrary* prime number p of the form $6k + 1$, and $n = 3$, the number of systems of values s_1, s_2, s_3 such that $s_1 + s_2 + s_3 \equiv 0$ (mod. p) is equal to $(p-1)(p-2)$. Therefore $c_0 + c_1 + c_2 = p - 2$, and since

$$\Big(\sum_r \omega^{hr} x^{g^r}\Big)^3 = p(c_0 + c_1\omega^h + c_2\omega^{2h}) \qquad (r = 0, 1, \dots, p-2),$$

the number $\pm a$ in the equation $4p = a^2 + 27b^2$ is given by

$$3c_0 - (c_0 + c_1 + c_2) = \pm a \quad \text{or} \quad 3c_0 - (p-2) = \pm a.$$

By what was said before, $(p-1)c_0$ is the number of systems of values s_1, s_2, s_3 such that $s_1 s_2 s_3$ is a cubic residue modulo p, and since $s_1 + s_2 + s_3 \equiv 0$ (mod. p), i.e.,

$$s_1^3 + s_2^3 + s_3^3 \equiv 3 s_1 s_2 s_3 \text{ (mod. } p),$$

we obtain the following remarkable solution of the equation $4p = a^2 + 27b^2$:

One has

$$a^2 = (3c_0 - p + 2)^2,$$

where $(p-1)c_0$ *is the number of different systems of numbers*

$$s_1, \; s_2, \; s_3 \; = 1, 2, 3, \dots, p-1$$

for which the sum $s_1 + s_2 + s_3$ *is divisible by* p *and nine times the sum of their cubes* $9\,(s_1^3 + s_2^3 + s_3^3)$ *is a cubic residue modulo* p.

To mention but a few simple examples, for $p = 7$, the only systems of numbers meeting the conditions are

$$s_1, \; s_2, \; s_3 \; = 1, 2, 4; \;\; s_1, \; s_2, \; s_3 \; = 3, 5, 6,$$

i.e., twelve systems altogether, so that one gets $c_0 = 2$ and $a^2 = 1$. For $p = 13$, the only systems are

$$s_1, \; s_2, \; s_3 \; = 1, 3, 9; \; 2, 5, 6; \; 4, 10, 12; \; 7, 8, 11,$$

twenty-four of them altogether, so that again $c_0 = 2$, whereas $a^2 = (3 \cdot 2 - 11)^2$, and thus $a^2 = 25$. For $p = 19$ we find the following systems

$$s_1,\ s_2,\ s_3\ \equiv h, 3h, -4h \text{ (mod. 19)} \qquad (h = 1, 2, \ldots, 18);$$

$$s_1,\ s_2,\ s_3\ \equiv k, 7k, -8k \text{ (mod. 19)} \qquad (k = 1, 2, 2^2, 2^3, 2^4, 2^5).$$

There are $6 \cdot 18 + 6 \cdot 6$ systems altogether, which yields the value of $18c_0$. This gives $c_0 = 8$ and $a^2 = (3 \cdot 8 - 17)^2 = 49$. And indeed, $4 \cdot 19 = 49 + 27$. Let me finish by noting that similarly elegant determinations for the representations of prime numbers in the form $a^2 + b^2$ and $a^2 + 2b^2$ can be deduced from the arguments above, but also from well-known results of the cyclotomic theory.

References

[1] *P. Bachmann.* Die Lehre von der Kreistheilung und ihre Beziehungen zur Zahlentheorie. Academische Vorlesungen. Leipzig (Teubner) 1872.

[2] *C.G.J. Jacobi.* Theorie der Zahlen. Handwritten notes of lecture course taught in Königsberg in the Winter term 1836–37, Leopold Kronecker's personal copy; passed on to Kurt Hensel after Kronecker's death. Bibliothèque de l'IRMA, Strasbourg, catalogue no. L3052. Digital copy available via `https://numistral.fr/fr` (consulted 27 January 2025).

[3] *C.G.J. Jacobi.* Über die Kreistheilung und ihre Anwendung auf die Zahlentheorie. Auszug aus einem Schreiben an die Akademie der Wissenschaften zu Berlin vom 16. October 1837. Monatsber. Kgl. Akad. Wiss., October 1837, 127–136. Reproduced in J. reine angew. Math. **30** (1847), 166–364. Reprint in C.G.J. Jacobi's Gesammelte Werke, Volume 6. Karl Weierstrass, ed. Berlin (Reimer) 1891; pp. 254–274. Page numbers quoted in the form n/m refer to page n in the *Crelle* version, and page m in the Collected Papers.

[4] *C.G.J. Jacobi.* Vorlesungen über Zahlentheorie. Wintersemester 1836/37, Königsberg. Edited by F. Lemmermeyer & H. Pieper. *Algorismus*, Heft 62. Augsburg (Erwin Rauner Verlag) 2007.

[5] *L. Kronecker.* De unitatibus complexis. *Dissertatio inauguralis arithmetica*, 10 September 1845. Fully published in J. reine angew. Math. **93** (1882), 1–52. Reprint in L. Kronecker's Werke, Volume 1, Berlin (Teubner) 1895; pp. 5–73.

[6] *L. Kronecker.* Über Abelsche Gleichungen. *Monatsber. Akad. Wiss. Berlin*, December 1877, pp. 845 et seq. Reprint in [9]; pp. 65–71.

[7] *L. Kronecker.* Grundzüge einer arithmetischen Theorie der algebraischen Größen. Festschrift zu Herrn Ernst Eduard Kummers fünfzigjährigem Doctor-Jubiläum, 10 September 1881. J. reine angew. Math. **92** (1882), 1–122. Reprint in L. Kronecker's Werke, Volume 2, Berlin (Teubner) 1897; pp. 237–387.

[8] *L. Kronecker.* Zur Theorie der Abelschen Gleichungen. Bemerkungen zum vorstehenden Aufsatz des Herrn *Schwering*. J. reine angew. Math. **93** (1882), 338–364. Reprint in [9]; pp. 131–162.

[9] *L. Kronecker. Werke*, Volume 4. Kurt Hensel, ed. Leipzig (Teubner) 1929.

[10] *K. Schwering.* Zur Theorie der arithmetischen Functionen, welche von *Jacobi* $\psi(\alpha)$ genannt werden. J. reine angew. Math. **93** (1882), 334–337.

[11] *K. Schwering.* Beitrag zur Theorie gewisser komplexer Zahlen. J. reine angew. Math. **102** (1888), 56–75.

Translation by Norbert Schappacher

On "Kronecker's journal", handwritten lecture notes, and Jacobi sums

By *Norbert Schappacher* at Strasbourg

1. Borchardt, Kronecker, Weierstrass – and Jacobi

The 1882 paper by Leopold Kronecker [29], which the editors of this special issue have chosen as an illustration of the third quarter century of *Crelle's Journal*, is certainly interesting for its mathematical content, of which more below. However, what strikes us first and foremost today is its peculiar presentation. It tells us quite a bit about mathematical publishing in the 19th century in general, and about the way in which Leopold Kronecker (1823–1891) understood his role as managing editor of *Crelle's Journal* in particular. Jointly with Karl Weierstrass (1815–1897) he took over the responsibility for the Journal when Karl Wilhelm Borchardt's (1817–1880) death in June 1880 had ended the period of *Borchardt's Journal*. Indeed, Borchardt had been in charge since 1856 (Volume 53).

As Carl Gustav Jacob Jacobi (1804–1851) plays a central part in the paper we are discussing, let us mention right away that it was Weierstrass, not Kronecker, who took over from Borchardt the edition of Jacobi's *Collected Works* on behalf of the Berlin Academy. Here Weierstrass's first duty was to complete the first volume [21], which Borchardt had almost finished before his death, before turning to the following ones. This project of Jacobi's *Collected Works* was actually the second attempt to publish a collection of Jacobi's production, this time striving for completeness. In fact, three volumes of Jacobi's *Opuscula Mathematica* had appeared before. All three of them contained exclusively papers that had previously been published in *Crelle's Journal*: Volume I was edited by Jacobi himself and appeared in 1846; Volume II was completed by Peter Gustav Lejeune Dirichlet (1805–1859) after Jacobi's death and appeared in 1851. Finally, Volume III of the *Opuscula Mathematica* was published only in 1871, by Borchardt. It collects papers reconstructed from texts that Jacobi left behind at his death, and that Borchardt had already published in *Crelle's Journal* under his direction. Let us mention in passing that Weierstrass was also involved in the edition of Jakob Steiner's (1796–1863) Collected Works—see the chores awaiting him in the fall of 1880, after Borchardt's death, that Weierstrass lists in letter 95 of [2]—whereas Dirichlet's *Werke* were edited by Kronecker (and finished by Lazarus Fuchs).

Possibly also because of its long name, the *Journal für die reine und angewandte Mathematik* is often referred to as *Crelle's Journal* even today, in reference to its founder, just like the label *Journal de Liouville* may still point to the *Journal de Mathématiques Pures et Appliquées*.

In the 19th century, there was also the habit to adapt such appelations to the current editor in chief, like passing from *Crelle's Journal* to the *Borchardtsches Journal*. Changing the common name of a journal with every passage from one editor in chief to the next would gradually fall out of use around the end of the 19th century. Yet, in the 1880s, the *Journal für die reine und angewandte Mathematik* was officially referred to as *'Kronecker J.'* in the *Jahrbuch über die Fortschritte der Mathematik*, and as *Kronecker's Journal* in occasional references—see for instance [5], p. 571, footnote.

Let us mention in passing that both Borchardt and Kronecker commanded considerable personal wealth. So did Gösta Mittag-Leffler (1846–1927) when he founded the journal *Acta Mathematica* in 1882. And the same is true of Giovanni Battista Guccia (1855–1914), who as of 1884 created the *Circolo matematico di Palermo* and its journal *Rendiconti*.

An interesting example of how the managing editor could mark a scientific journal in the nineteenth century is provided by the *Archiv der Mathematik und Physik*, or in full: *Archiv der Mathematik und Physik mit besonderer Rücksicht auf die Bedürfnisse der Lehrer an höheren Unterrichtsanstalten*. It was often referred to as *Grunert's Archiv der Mathematik und Physik*, after its founder and long term editor Johann August Grunert (1797–1872) in Greifswald. Grunert's frequent contributions to his own journal are often listed as papers by the *Herausgeber*, i.e., the editor. And in general, quoting from [39], p. 435: "a somewhat familiar style prevails between the editor, the authors, and the readers, which we do not find in scientific journals today." We do not see exactly the same style in *Kronecker's Journal*. Yet it also differs noticeably from what we are used to in distinguished scientific periodicals of the present day.

2. Kronecker's Journal

Among the best-known mathematicians of the 19th century, Kronecker stands out for the fact that his active lifetime, from 1823 to 1891, spanned markedly different periods in the development of mathematics. This is particularly true for the emerging disciplines of Number Theory and Algebraic Geometry. Kronecker was attracted to mathematical research by the influence of Ernst Eduard Kummer (1810–1893), first in Kronecker's hometown Liegnitz (today Legnica, Poland), and then in Breslau (today Wrocław, Poland). He obtained his doctorate in Berlin in 1845 under the supervision of Peter Gustav Lejeune Dirichlet.

In his thesis, Kronecker studied the units of the cyclotomic field of the ν-th roots of unity, for an odd prime number ν, expressing the fundamental units of the totally real subfields of degree $\lambda \mid \nu - 1$ in terms of Gauss's *periods*. He did this at a time when Dirichlet's general unit theorem, for arbitrary algebraic number fields, was not yet known. To be sure, nobody spoke of 'number fields' back then, and Kronecker never would.

Only a shortened version of his thesis, containing but sections 1 through 16, was printed after the thesis defense. Indeed, Sections 14–16 restrict to the case where the degree λ is prime. Thirty-seven years later, in 1882, Kronecker placed the *complete* text of his thesis [26] at the beginning of the very Volume 93 of *Crelle's Journal* that also contains the paper [29] we are featuring here. Moreover, this article [29] is followed there by yet another note of Kronecker's [30], just two pages long.

Heinrich Weber in his obituary speech about Kronecker, after recalling the particular importance that cyclotomy had for Kronecker, politely remarks ([44], p. 4) that the author was right in considering his thesis worthy to be published in full in 1882. Yet, Kronecker's multiple presence in volume 93 indicates the strong mark he imprinted on the famous *Crelle's Journal* that he was now in charge of.

This impression is confirmed by looking at other *Crelle* volumes from around the same time—and it takes on another meaning: Beyond the desire to set things right or to complete the documentation of mathematics from the middle of the 19th century, we also see Kronecker effectively promoting his own mathematical agenda of the 1880s, which was based on his unified theory of what in Dedekind's parlance would be called number fields on the one hand, and function fields on the other. Indeed, when Heinrich Weber submitted the manuscript of his famous joint paper with Richard Dedekind [9] about an arithmetic treatment of Riemann surfaces to *Crelle's Journal*, Kronecker apparently decided not to look at it before publishing his own article [27] in the upcoming Volume 91 (1881), which he presented as an old write-up from 1862, together with a preface explaining its history. The paper by Dedekind and Weber was thus delayed until the subsequent 1882 Volume 92 of *Crelle's Journal*, only to be in turn preceded there by the 122-page reprint of Kronecker's momentous *Grundzüge* [28], which set out his unified arithmetic theory of numbers and polynomials in terms of his 'modular systems.'[1)] Note that Kronecker's 1882 paper [29], which we present here, also employs his language of 'modular systems' at various places, in particular in Part II, starting two paragraphs after formula (*S.*).

Looking at Kronecker's overall publication chronology—see [34], pp. 517–527—we see that he published a total of 35 papers in *Crelle's Journal*. The first one appeared in Volume 29 (1845). The last three are to be found in Volume 108 (1891). Not a single paper by Kronecker was printed in *Borchardt's Journal* between 1870 and 1880 (Volumes 73–90); during that decade Kronecker published mostly in the *Monatsberichte* of the Berlin Academy of Sciences. Then however, as soon as he had taken over the *Journal* and until his death in 1891, he placed 27 of his papers in the *Journal für die reine und angewandte Mathematik*, which greatly vary in length and nature. Only two among the altogether eighteen volumes of which Kronecker was the managing editor do not contain any contribution authored by him.

Given Kronecker's resolute personal editorial policy, the question arises how Kronecker and Weierstrass got along in their joint responsibility for the Journal as of 1880. It is all the more important to address this issue at least briefly, as an unfortunate glitch in a paper by Gert Schubring has seriously distorted the historical facts.

In his short note [40], Schubring reported on his rediscovery of letters to Weierstrass in the *Repository 92* of the *Geheimes Staatsarchiv Preußischer Kulturbesitz*, Berlin-Dahlem.[2)] Schubring misinterpreted the handwritten copy of a letter from Kronecker dated 9 November 1883 as a humiliating letter addressed to Weierstrass, and went on to hypothesize that Weierstrass had submitted a paper to the Journal, which Kronecker refused to publish without major modifications, and that Weierstrass then withdrew. This incident, Schubring suggests, factually ended the editorial collaboration between the two mathematicians—see [40], p. 426.

In reality, however, as explained in Kronecker's letter to Weierstrass dated 12 November 1883, his original letter of 9 November 1883 was actually addressed, not to Weierstrass, but to Georg Frobenius (1849–1917), who at the time was teaching in Zurich. It concerned last minute corrections to the text (and the title) of the paper [12], in line with Kronecker's editorial policy as described above. Kronecker then enclosed a handwritten copy of this letter to Frobenius

---

1) For more reflections and references on this story, cf. [37], Section 3.

2) Cf. the provisional transcriptions of those letters in [6].

with his letter to Weierstrass of 12 November, in order to keep Weierstrass duly informed of the editorial process.

As is well-known, the long-time friends Kronecker and Weierstrass eventually did fall out with one another in the mid 1880s. Their different views about the arithmetization of analysis were an important element in this rift; Weierstrass also complained increasingly about Kronecker's vanity.[3)] To what extent editorial issues with the Journal stoked their tensions seems to be hard to evaluate. At any rate, both of them signed the preface to the big Crelle Volume 100 (1887), stressing the international character of the Journal, and their exchanges about editorial issues continued. In June 1888, however, Weierstrass asked Kronecker not to be his coeditor of the Journal any more. Kronecker objected (in his letter to Weierstrass dated 30 June 1888) and together with the publisher Reimer obliged Weierstrass to at least remain on the larger editorial board.

3. Printed Journals and Handwritten Lecture Notes

The very last paper that Kronecker published in the *Journal* [32] returns to the legacy of Jacobi's lecture courses. One of them, the Number Theory course [23] that Jacobi taught at Königsberg in the Winter 1836–37, is a key issue in the article [29] that we are presenting here.

Let us put things into perspective. The present special issue of *Crelle's Journal* commemorates the bicentennial of one of those printed journals that were essential elements of the growing professionalization of mathematics in the nineteenth century. Yet, in spite of that centennial reshaping of mathematical research, the circulation of grey literature, in particular of handwritten copies of lecture notes, continued to play a key role when it came to spreading information in the mathematical community, all the way from students to the active researchers. Those handwritten volumes were based on notes taken during the lecture courses, often in shorthand, which were subsequently worked out into fair copies by a student or an assistant, with or without the supervision of the lecturer himself, as the case may be. When such a fair copy circulated, others interested in the material would copy it by hand, or pay a scribe to copy it for them.

This system of lecture notes practised in Germany is a far cry from, say, the official courses reproduced in the nineteenth century for the students of the Paris *Ecole Polytechnique*. Its remains can still be found today in many older libraries, or in archives. Whenever one finds two handwritten lecture notes of the same course, the wording will not be identical. If they are based on different notes taken during the lectures in the first place, they may show substantial variations. An extreme example of two very different sets of notes are those taken by Moritz Pasch on the one hand, and by Hermann Amandus Schwarz on the other, of Weierstrass's 1865/66 Berlin lectures on Analytic Functions.

Contrary to Weierstrass, Jacobi had been quite successful in selecting listeners who prepared notes of his lectures. In his short report about the papers left behind by Jacobi, written about half a year after Jacobi's death, Dirichlet explained[4)]:

> Since Jacobi always spoke freely when he taught his lectures, not using any written composition, his legacy contains nothing related to his courses, except a few

---

[3)] See for instance [2], pp. 291–293, 360–362, 368; or the extract from Weierstrass's letter to H.A. Schwarz of 12 August 1885 in [11], pp. 141–142.

[4)] See [10], p. 92; cf. [24], p. 518.

> short notes. There are, however, very precise lecture notes of his crucial lecture courses, which are due to several of his most distinguished listeners and which he had collected over the years. ... Based on these notes and on other ones of equal quality, Jacobi's friends will be able to reproduce his most important lecture courses ... in separate volumes.

The editors of Jacobi's Collected Works also pondered the possibility of publishing supplementary volumes with "well attested lecture notes of Jacobi's university courses" (*Supplementbände mit Mittheilungen aus wohlbeglaubigten Nachschriften von Jacobi's Universitäts-Vorlesungen*—see [21], p. VII). Indeed, most of the copies of Jacobi's lecture notes that are known today tend to show only relatively minor variations.

In spite of all these projects, it seems that only Jacobi's lectures on *Dynamics* (Königsberg, Winter 1842–43) were printed in the nineteenth century, edited by Alfred Clebsch in 1866 [20]. And they were even re-edited in 1884, as the unique supplementary volume (Volume VIII) of Jacobi's *Collected Works* that ever appeared. The printing of other lectures by Jacobi had to wait till the 1990s—see [22], [23]; cf. [18].

Kronecker's comments [29] on Schwering's paper [41] lead us to Jacobi's lecture course on Number Theory, taught at Königsberg in the Winter term 1836–37. The notes of this course have been worked out by Johann Georg Rosenhain (1816–1887). The recent commented edition by Lemmermeyer and Pieper [23] mentions five extant copies, two in Berlin, two in Strasbourg, and one in Zürich. Only the Berlin copies contain the *Additamenta* at the end; they are reproduced and translated from Latin into German in [23], pp. 274–304.

Kronecker had direct access to at least one of the extant copies of these lecture notes (without the *Additamenta*): [18] carries his *exlibris*. After his death, it passed to Kurt Hensel; after Hensel's death, during World War II, his library was sold to the Nazi *Reichsuniversität* in Strasbourg—see [38]. This is why the mathematical library of IRMA there happens to have both Kronecker's and Weierstrass's copies of these lecture notes.

Comparing the passages from Jacobi's lectures that Kronecker quotes in [29], pp. 346–347, with pages 161 and 170–172 of the pdf file of [18] shows large agreement, without being a perfect match throughout the quote. We leave open the question if Kronecker quoted a bit freely, or whether he was in fact using yet another copy of Jacobi's lectures to which he had access.

Whichever copy of the lecture notes Kronecker was using, neither Paul Bachmann nor Karl Schwering appear to have had independent access to Jacobi's Number Theory Lectures when they were writing their publications. Indeed, Bachmann thanked Kronecker for having given him access to his personal copy of Jacobi's lectures ([1], p. IV):

> On this very subject [cyclotomy] Jacobi has given lectures, of which various notebooks (*Hefte*) are still extant. Thanks to the kindness of Herr Professor Kronecker, the author was allowed to go through one of them, when the selection and ordering of the subject matters to be treated in this book had already been fixed.

His late acquaintance with this source induced Bachmann to explicitly use Jacobi's lectures only once in his book, in the context of the law of biquadratic reciprocity.

As for Schwering, who in the early 1880s was a high school teacher in the Westphalian province, he could but speculate in [41] on what Jacobi had developed in his lectures, based on Bachmann's vague allusions. Thus one of the reasons for Kronecker to publish his extensive follow-up to Schwering's short paper obviously was the desire to create a printed record of Jacobi's results on those "functions that Jacobi called $\psi(\alpha)$", as Schwering writes in the title of [41], i.e., of Jacobi's own results on what today we call *Jacobi sums*.

4. Gauss Sums and Jacobi Sums

In order to put Kronecker's paper in perspective, we shall now point to a few instances from the history of exponential sums. We set out after Lagrange's resolvent. The following sequence of remarks carries no pretense whatsoever of dressing an exhaustive timeline. The history of exponential sums illustrates in a particularly striking manner a problem that is well-known to historians of mathematics in general: the objects we are talking about in the first place are just (algebraic) complex numbers. They receive their significance, which varies considerably over time, from theoretical contexts in which they pop up. It is these contexts that tend to shape the historical narratives about Gauss and Jacobi sums, and the attention span of those who write about them.

To fix ideas, let us look at the Preliminary Note preceding the offprints of Kronecker's 1882 paper. There p and $\lambda > 3$ denote prime numbers such that $p = 2m\lambda + 1$; x a p-th root of unity, α is a λ-th root of unity, and g is a primitive root modulo p. The Gauss sums, which depend on an integer μ, are (up to sign) the expressions

$$\sum_{r=0}^{p-2} \alpha^{\mu r} x^{g^r}.$$

And the Jacobi sums $\psi_{\mu,\nu}(\alpha)$ arise from the multiplication of Gauss sums:

$$\sum_r \alpha^{\mu r} x^{g^r} \cdot \sum_r \alpha^{\nu r} x^{g^r} = \psi_{\mu,\nu}(\alpha) \sum_r \alpha^{(\mu+\nu)r} x^{g^r}.$$

Nowadays they are usually written in terms of additive and multiplicative characters of the integers modulo p, or more generally of finite fields.[5)]

4.1. Carl Friedrich Gauss (1777–1855). The final Part VII of Gauss's *Disquisitiones Arithmeticae* [13] is devoted to cyclotomy, the division of the circle into p equal parts—we change notation from Gauss's n to the letter p used in Kronecker's paper—where p is an odd prime number. For any decomposition $p - 1 = e \cdot f$, Gauss introduces what he calls the corresponding *periods*:

$$(f, \lambda) = \sum_{j=0}^{f-1} \zeta^{\lambda g^{ej}},$$

where ζ is a primitive p-th root of unity, g a primitive root modulo p, and λ is any given integer. The principal usage of these algebraic complex numbers, which Gauss presents in the very last sections §359–§366 of his *Disquisitiones*, is the proof of the explicit, step-by-step solvability of the equation $\frac{x^p-1}{x-1} = 0$. In today's field theoretic perspective, each of the periods

5) See for instance [17], Chap. 8, or [43], Chapter 6. Cf. [35] for a wealth of detailed historical references.

$(f,1),(f,g),(f,g^2),\ldots,(f,g^{e-1})$ is a primitive element of the unique subfield of $\mathbf{Q}(\zeta)$ that has degree e over the rationals.

Before these concluding sections on the cyclotomic equation, however, by way of discussing the special cases $e=2$ and $e=3$, Gauss places a number of observations that connect the periods and related exponential sums to other central themes of the *Disquisitiones Arithmeticae*. Thus in [13], §356, we encounter what is well-known today as a (quadratic) Gauss sum and the corresponding problem of determining its sign—cf. [36]; Leopold Kronecker would return to the different methods of determining this sign in a Crelle paper [31] two years before his death.

And in [13], §358, when discussing products of the periods $(f,1),(f,g),(f,g^2)$ in the case $f=\frac{1}{3}(p-1)$, which occurs for primes p of the form $3k+1$, Gauss points out a remarkable connection with the theory of quadratic forms: The unique representation $4p = M^2+27N^2$ that exists for every prime number $p \equiv 1 \pmod 3$ can be obtained explicitly from a careful analysis of the way in which the cubes are distributed in $(\mathbf{Z}/p\mathbf{Z})^*$. For the representation of $4p$, cf. the historical remarks in the footnote on p. 97 of [23]. Since Gauss's proof thereof involves solutions of cubic equations modulo a prime number, it bears some analogy with the famous 'Last Entry' of Gauss's mathematical diary.

Furthermore, later readers of the *Disquisitiones Arithmeticae*, after the reedition of this book in the first volume of Gauss's Collected Works (1863), would find at the very end of §358 a footnote that starts out in Latin, and ends in German. It states the following *Corollary* of the previous analysis:

$$(f,1)+\varepsilon(f,g)+\varepsilon^2(f,g^2)=\frac{p}{2}(M+N\sqrt{-27}),$$

where ε is a primitive third root of unity.[6)] The sum on the left is referred to today as a (cubic) Gauss sum.

The general term "Gauss sum" does not seem to have existed in the first half of the nineteenth century, and we do not know who invented it—cf. [4], footnote 16. Neither Jacobi nor Kronecker use this expression.

4.2. Augustin-Louis Cauchy (1789–1857). In Cauchy's arithmetical papers—which were published in two waves (before and after his exile) in 1829–1831 and 1839–1840, altogether more than six hundred pages—one finds Gauss sums and Jacobi sums, certain products of Gauss sums, relations between Gauss sums and Bernoulli numbers, as well as a certain problem linked to quadratic forms: to find representations of the kind $4p^\mu = x^2+ny^2$. The latter naturally reminds us of the special case where the exponent $\mu=1$, and $p=3k+1$, $n=27$, that we have encountered in §358 of Gauss's *Disquisitiones Arithmeticae*. Not surprisingly, given the central importance that Gauss attributed to the law of quadratic reciprocity in his arithmetical works, speculations about higher order reciprocity laws are lurking in the background of Cauchy's developments around Gauss sums. However, according to Jenny Boucard's in-depth analysis of Cauchy's arithmetic, the heart of the matter seems to be to establish that peculiar relation between cyclotomy and the theory of quadratic forms: to obtain representations $4p^\mu = x^2+ny^2$ from the evaluation of specific products of Gauss sums—see [4], starting from Section 4.1.3.

While Cauchy's arithmetical studies were certainly read by the number theorists of the nineteenth century, they seem to have received less and less attention as time went on.

6) For some more details, cf. [7], (4.24), (4.25), and Exercise 4.29.

4.3. C.G.J. Jacobi. For Jacobi, Gauss sums and the sums that today carry his own name, were the gate to systematically develop "those fundamental theorems of cyclotomy that are not yet known", in such a way that one can "effectively execute the operations" in every given case, thanks to "simple mechanical rules, which greatly simplify the computations." This is what he announced in Lecture XI of his 1836–37 lecture course on Number Theory—see [23], p. 83. The corresponding developments are then presented all through Lecture XXXI. Note that these lectures were taught half way between the two periods of Cauchy's arithmetic publications. There is indeed a certain amount of overlap—see [3], pp. 302–304. In particular, the connection between multiplicative relations linking Gauss or Jacobi sums on the one hand and the representation of numbers related to the given prime by certain quadratic forms appears several times, for instance in Lectures XIII, XVI, and XXI.

In the recent edition [23] of Jacobi's lecture notes, the editors have enriched the table of contents ([23], pp. xiii–xxv) with concise remarks summarizing the content, and translating it into modern notation; their transcription of the lecture notes is also accompanied by helpful footnotes.

4.4. L. Kronecker. Gauss and Jacobi sums, and the basic relations between them are introduced in the first part of Kronecker's 1882 paper [29]. In the spirit of his *Grundzüge* [28], Kronecker presents them as consequences of polynomial identities, once these are read with respect to certain moduli systems. The whole setup is sufficiently abstract to be valid not only for cyclotomic, but for arbitrary cyclic, and in fact abelian extensions.

In the second part of his paper, Kronecker quotes from the handwritten notes of Jacobi's 1836–67 lectures, in particular Lecture XVII, to point out to Herr Schwering and to the mathematical world the considerable extent to which Jacobi had already inspected multiplicative relations between Jacobi sums. After discussing the general multiplicative relations, Jacobi's peculiar investigations concerning individual small primes p are presented; the special case $p = 47$ turns out to be interesting. This second part ends with critical reflections on the potential generality of some of Jacobi's calculations.

The final third part of Kronecker's paper connects the question of those multiplicative relations, in a few special cases, with quadratic representations such as $4 \cdot 19 = 7^2 + 27 \cdot 1^2$, which reminds us of Gauss, and more generally of Cauchy. One may regret that Kronecker mentions neither Gauss nor Cauchy.

4.5. Ludwig Stickelberger (1850–1936). Stickelberger's Theorem, published in 1890, tends to be seen today as a game changer of the theory of Gauss and Jacobi sums, in that it shifts attention to annihilating the ideal class group of the cyclotomic field in question—see modern treatments such as [43], Section 6.2. Even André Weil in his overview of the history of cyclotomy switches to a somewhat anachronistic style of presentation when discussing Stickelberger's work—see [45], §14, pp. 256–258. Stickelberger's own presentation of his "generalization of cyclotomy", however, revolves around the remarkable relationship "between Algebra and Number Theory", explicitly between periods, Gauss sums and Jacobi sums, on the one hand, and quadratic forms on the other. This is stressed from the very first sentence of the introduction, and picked up again in the penultimate section §7 of [42]. Thus unlike Kronecker, Stickelberger, while taking the subject much further, consciously wrapped up the long story that starts from Gauss's *Disquisitiones Arithmeticae*, including notably Cauchy.

4.6. David Hilbert (1862–1943). As befits a report about the state of the theory of algebraic number fields, David Hilbert's famous *Zahlbericht* [16] duly mentions Kronecker's paper [29], at the end of §112—if only in a summary list of publications related to what Hilbert idiosyncratically chose to call *Lagrange root numbers* (*Lagrangesche Wurzelzahl*) in [16], §111 and §112. Hilbert also uses the term Gauss sum (*Gaußsche Summe*), but reserves it for the quadratic case.

4.7. New points of view in the twentieth century. Two new ways of looking at Gauss and Jacobi sums would increasingly mark the subject as of the 1920s and 1930s—cf. [45], §17–§19. On the one hand, they appear as constants in the functional equation of L-functions. On the other hand, they turn up as zeros of zeta- or L-functions of varieties over finite fields. The last aspect enabled Harold Davenport (1907–1969) and Helmut Hasse (1898–1979) to derive multiplicative relations between Gauss sums in a completely new way—see [8]. The new point of view inspired Hasse to delve into general reflections about Gauss sums, and multiplicative relations between them in particular. The comparison of the two editions of his textbook *Lectures on Number Theory* [15] shows the evolution of Hasse's thoughts on this subject. Hasse's suspicion that the Davenport-Hasse relations exhaust all multiplicative relations between Gauss sums turned out to be not quite true—see for instance [46] and [25].

Acknowledgement. I thank Franz Lemmermeyer for helpful remarks.

References

[1] *P. Bachmann.* Die Lehre von der Kreistheilung und ihre Beziehungen zur Zahlentheorie. Academische Vorlesungen. Leipzig (Teubner) 1872.

[2] *R. Bölling* (ed.). Briefwechsel Karl Weierstraß – Sofja Kowalewskaia. Berlin (Akademie Verlag) 1993.

[3] *J. Boucard.* « Un rapprochement curieux de l'algèbre et de la théorie des nombres » : Études sur l'utilisation des congruences en France de 1801 à 1850. Thèse de doctorat (Université Pierre et Marie Curie) 2011. `https://www.imj-prg.fr/theses/pdf/jenny_boucard.pdf` (consulted 24 June 2025).

[4] *J. Boucard.* Cyclotomie et formes quadratiques dans l'œuvre arithmétique d'Augustin-Louis Cauchy (1829–1840). Arch. Hist. Ex. Sciences **67** (2013), 349–414.

[5] *F. Caspary.* Über einen einfachen Beweis der Rosenhain'schen Fundamentalformeln. Math. Ann. **30** (1887), 571–577.

[6] *E. Confalonieri.* Beiträge zur Geschichte der mathematischen Werke von Karl Weierstrass, Parts II and III. Self-edition, online.

[7] *D.A. Cox.* Primes of the form $x^2 + ny^2$. Fermat, Class Field Theory, and Complex Multiplication. New York etc. (John Wiley & Sons) 1989.

[8] *H. Davenport & H. Hasse.* Die Nullstellen der Kongruenzzetafunktionen in gewissen zyklischen Fällen. J. reine angew. Math. **172** (1934), 151–182.

[9] *R. Dedekind & H. Weber.* Theorie der algebraischen Functionen einer Veränderlichen. J. reine angew. Math. **92** (1882), 181–290.

[10] *G.L. Dirichlet.* Nachricht über Jacobi's Wissenschaftlichen Nachlass. J. reine angew. Math. **42** (1851), 91–92.

[11] *P. Dugac.* Eléments d'analyse de Karl Weierstrass. Arch. Hist. Ex. Sciences **10** (1973), 41–176.

[12] *G. Frobenius.* Ueber die principale Transformation von Thetafunctionen mehrerer Variabeln. J. reine angew. Math. **95** (1883), 264–296.

[13] *C.F. Gauss.* Disquisitiones Arithmeticae. Lipsiae [= Leipzig] (Gerhard Fleischer) 1801.

[14] *C. Goldstein, N. Schappacher, J. Schwermer.* The Shaping of Arithmetic after C.F. Gauss's Disquisitiones Arithmeticae. Berlin etc. (Springer) 2007.

[15] *H. Hasse. Vorlesungen über Zahlentheorie.* First edition 1950. // Zweite neubearbeitete Auflage. Berlin etc. (Springer) 1964.

[16] *D. Hilbert.* Die Theorie der algebraischen Zahlkörper. Jahresber. DMV **4** (1897), 175–546.

[17] *K. Ireland & M. Rosen.* A Classical Introduction to Modern Number Theory. 2nd edition. New York (Springer) 1990.

[18] *C.G.J. Jacobi.* Theorie der Zahlen. Handwritten notes of lecture course taught in Königsberg in the Winter term 1836–37, Leopold Kronecker's personal copy; passed on to Kurt Hensel after Kronecker's death. Bibliothèque de l'IRMA, Strasbourg, catalogue no. L3052. Digital copy available via `https://numistral.fr/fr` (consulted 27 January 2025).

[19] *C.G.J. Jacobi.* Über die Kreistheilung und ihre Anwendung auf die Zahlentheorie. Auszug aus einem Schreiben an die Akademie der Wissenschaften zu Berlin vom 16. October 1837. Monatsber. Kgl. Akad. Wiss., October 1837, 127–136. Reproduced in J. reine angew. Math. **30** (1847), 166–364. Reprint in C.G.J. Jacobi's Gesammelte Werke, Volume 6. Karl Weierstrass, ed. Berlin (Reimer) 1891; pp. 254–274. Page numbers quoted in the form n/m refer to page n in the *Crelle* version, and page m in the Collected Papers.

[20] *C.G.J. Jacobi.* Vorlesungen über Dynamik. Nebst fünf hinterlassenen Abhandlungen desselben herausgegeben von A. Clebsch. Berlin (Reimer) 1866.

[21] *C.G.J. Jacobi.* C.G.J. Jacobi's Gesammelte Werke. Herausgegeben auf Veranlassung der Königlich Preussischen Akademie der Wissenschaften. Erster Band mit dem Bildnisse Jacobi's. Herausgegeben von C.W. Borchardt. Berlin (Reimer) 1881.

[22] *C.G.J. Jacobi.* Vorlesungen über Analytische Mechanik. Berlin 1847/48. Nach einer Mitschrift von Wilhelm Scheibner herausgegeben von Helmut Pulte. Deutsche Mathematiker-Vereinigung: Dokumente zur Geschichte der Mathematik, Band 8. Braunschweig/Wiesbaden (Vieweg) 1996.

[23] *C.G.J. Jacobi.* Vorlesungen über Zahlentheorie. Wintersemester 1836/37, Königsberg. Edited by F. Lemmermeyer & H. Pieper. *Algorismus*, Heft 62. Augsburg (Erwin Rauner Verlag) 2007.

[24] *L. Koenigsberger.* Carl Gustav Jacobi. Festschrift zur Feier der hundertsten Wiederkehr seines Geburtstages. Leipzig (Teubner) 1904.

[25] *T. Kohno.* The sign ambiguity of Multiplicative Relations of Gauss Sums of Order $4l$. Commentarii Mathematici Universitatis Sancti Pauli 68 (2020), 49–68.

[26] *L. Kronecker.* De unitatibus complexis. *Dissertatio inauguralis arithmetica*, 10 September 1845. Fully published in J. reine angew. Math. **93** (1882), 1–52. Reprint in L. Kronecker's Werke, Volume 1, Berlin (Teubner) 1895; pp. 5–73.

[27] *L. Kronecker.* Über die Discrimante algebraischer Funktionen einer Variabeln. J. reine angew. Math. **91** (1881), 301–334. Reprint in L. Kronecker's Werke, Volume 2, Berlin (Teubner) 1897; pp. 193–236.

[28] *L. Kronecker.* Grundzüge einer arithmetischen Theorie der algebraischen Größen. Festschrift zu Herrn Ernst Eduard Kummers fünfzigjährigem Doctor-Jubiläum, 10 September 1881. J. reine angew. Math. **92** (1882), 1–122. Reprint in L. Kronecker's Werke, Volume 2, Berlin (Teubner) 1897; pp. 237–387.

[29] *L. Kronecker.* Zur Theorie der Abelschen Gleichungen. Bemerkungen zum vorstehenden Aufsatz des Herrn *Schwering*. J. reine angew. Math. **93** (1882), 338–364. Reprint in [33]; pp. 131–162.

[30] *L. Kronecker.* Zur arithmetischen Theorie der algebraischen Formen. J. reine angew. Math. **93** (1882), 365–366.

[31] *L. Kronecker.* Summierung der Gauß'schen Reihen $\sum_{h=0}^{h=n-1} e^{\frac{2h^2\pi i}{n}}$. J. reine angew. Math. **105** (1889), 267–268.

[32] *L. Kronecker.* Über die Zeit und die Art der Entstehung der Jacobi'schen Thetaformeln, nebst einem Verzeichnis der von C.G.J. Jacobi gehaltenen Vorlesungen. J. reine angew. Math. **108** (1891), 325–334. Reprint in [34], pp. 361–370, 487–493.

[33] *L. Kronecker. Werke*, Volume 4. Kurt Hensel, ed. Leipzig (Teubner) 1929.

[34] *L. Kronecker. Werke*, Volume 5. Kurt Hensel, ed. Leipzig (Teubner) 1930.

[35] *F. Lemmermeyer.* Reciprocity Laws. From Euler to Eisenstein. Berlin – Heidelberg – New York (Springer) 2000.

[36] *S.J. Patterson.* Gauss Sums. Chapter VII.2 in [14], pp. 505–528.

[37] *N. Schappacher.* Rewriting Points. In Proc. ICM Hyderabad, India, 2010; pp. 3258–3291.

[38] *N. Schappacher.* National Pride in the Interest of Science? The Strasbourg Mathematical Library between Germany and France. *Historia Math.* 73 (2025). `https://doi.org/10.1016/j.hm.2025.09.004`

[39] *P. Schreiber.* Johann August Grunert and his *Archiv der Mathematik und Physik* as an integrative factor of everyone's mathematics in the middle of the nineteenth century. In Mathematical Europe. History, Myth, Identity; C. Goldstein, J. Gray, J. Ritter (eds.). Paris (Éditions de la Maison des sciences de l'homme) 1996; pp. 431–441.

[40] *G. Schubring.* An unknown part of Weierstrass's Nachlass. Historia Math. 25 (1998), 423–430.

[41] *K. Schwering.* Zur Theorie der arithmetischen Functionen, welche von *Jacobi* $\psi(\alpha)$ genannt werden. J. reine angew. Math. **93** (1882), 334–337.

[42] *L. Stickelberger.* Über eine Verallgemeinerung der Kreistheilung. Math. Ann. **37** (1890), 321–367.

[43] *L.C. Washington.* Introduction to Cyclotomic Fields. Graduate Texts in Mathematics 83. New York (Springer) 1982.
[44] *H. Weber.* Leopold Kronecker. Math. Ann. **43** (1893), 1–25.
[45] *A. Weil.* La cyclotomie jadis et naguère. Ens. math. **XX** (1974), 247–263.
[46] *K. Yamamoto.* On a Conjecture of Hasse Concerning Multiplicative Relations of Gaussian Sums. J. Comb. Theory **I** (1966), 476–489.

Norbert Schappacher, IRMA, Université de Strasbourg, 7 rue René Descartes, 67084 Strasbourg cedex, France
e-mail: n.schappacher@unistra.fr

1901–1925

Über die Darstellung der endlichen Gruppen durch gebrochene lineare Substitutionen.

(Von Herrn *J. Schur* in Berlin.)

Das Problem der Bestimmung aller endlichen Gruppen linearer Substitutionen bei gegebener Variabelnzahl n ($n > 1$) gehört zu den schwierigsten Problemen der Algebra und hat bis jetzt nur für die binären und ternären Substitutionsgruppen seine vollständige Lösung gefunden.*) Für den allgemeinen Fall ist nur bekannt, daß die Anzahl der in Betracht kommenden Typen von Gruppen eine endliche ist;**) dagegen fehlt noch jede Übersicht über die charakteristischen Eigenschaften dieser Gruppen.

Die Umkehrung dieses Problems bildet in einem gewissen Sinne die Aufgabe: alle Gruppen von höchstens h ganzen oder gebrochenen linearen Substitutionen zu finden, die einer gegebenen endlichen Gruppe $\mathfrak{H}$ der Ordnung h ein- oder mehrstufig isomorph sind, oder auch, wie man sagt, alle Darstellungen der Gruppe $\mathfrak{H}$ durch lineare Substitutionen zu bestimmen.

Diese Aufgabe hat nun im Gegensatz zu der zuerst genannten eine erfolgreiche Behandlung gefunden, indem die Herren *Molien****) und *Frobenius*†) gezeigt haben, daß das Problem der Bestimmung aller Darstellungen der

---

*) Vergl. *A. Wiman*, Encykl. d. math. Wiss., Bd. I, S. 522.

**) *C. Jordan*, dieses Journal, Bd. 84, S. 89.

***) Sitzungsberichte der Naturforscher-Gesellschaft zu Dorpat, 1897, S. 259.

†) „Über Gruppencharaktere", Sitzungsberichte der Berliner Akademie, 1896, S. 985; „Über die Primfaktoren der Gruppendeterminante", ebenda, S. 1343; „Über die Darstellung der endlichen Gruppen durch lineare Substitutionen", ebenda, 1897, S. 994; „Über die Darstellung der endlichen Gruppen durch lineare Substitutionen II.", ebenda, 1899, S. 482. — Diese Abhandlungen werde ich im Folgenden mit „Gruppencharaktere", „Primfaktoren", „D. I." und „D. II." zitieren.

J. reine angew. Math. **127** (1904), 20–50
DOI 10.1515/crll.1904.127.20

Journal für die reine und angewandte Mathematik

On the representation of finite groups by fractional linear substitutions

By *Herr J. Schur* at Berlin

The problem of determining all finite groups of linear substitutions for a given number of variables n $(n > 1)$ belongs to the most difficult problems in Algebra and by now it has found complete solutions only for the binary and ternary substitution groups.[1)] In the general case it is only known that the number of types of groups to be considered is finite.[2)] However, any kind of clear view on the characteristic properties of these groups is still missing.

In a certain sense, the converse of this problem is formed by the following task: to find all groups of at most h integer or rational substitutions, which are isomorphic to a given finite group $\mathfrak{H}$ of order h in one or more steps, or, as one may put it, to determine all representations of the group $\mathfrak{H}$ by linear substitutions.

This task, in contrast to the first one, has now found a successful treatment. Indeed, *Molien*[3)] and *Frobenius*[4)] showed that the problem of determining all representations of the group $\mathfrak{H}$ by linear substitutions is identical to the problem of splitting the group matrix of $\mathfrak{H}$ into partial matrices that cannot be further decomposed. The first and most essential step for solving this task is, as the investigations of *Frobenius* showed, the decomposition of the group determinant of $\mathfrak{H}$ into its prime factors, or, what is essentially the same, calculating the group characters of $\mathfrak{H}$.

In the current work I have attempted to explain how the problem of determining all representations of a given finite group by fractional linear substitutions can also be addressed using the theory of the group matrix and group characters.

If one assigns to the h different elements $A, B, \ldots$ of the group $\mathfrak{H}$ the h linear substitutions of nonzero determinant

$$\{A\} \qquad x_\nu = \frac{a_{\nu 1}y_1 + \cdots + a_{\nu,n-1}y_{n-1} + a_{\nu n}}{a_{n1}y_1 + \cdots + a_{n,n-1}y_{n-1} + a_{nn}},$$

$$\{B\} \qquad x_\nu = \frac{b_{\nu 1}y_1 + \cdots + b_{\nu,n-1}y_{n-1} + b_{\nu n}}{b_{n1}y_1 + \cdots + b_{n,n-1}y_{n-1} + b_{nn}}, \ldots \qquad (\nu = 1, 2, \ldots, n-1)$$

[1)] Cf. A. Wiman, Encykl. d. math. Wiss., Bd. I, p. 522.

[2)] C. Jordan, this journal, Bd. 84, p. 89.

[3)] Sitzungsberichte der Naturforscher-Gesellschaft zu Dorpat, 1897, p. 259.

[4)] "Über Gruppencharaktere", Sitzungsberichte der Berliner Akademie, 1896, p. 985; "Über die Primfaktoren der Gruppendeterminante", ibid., p. 1343: "Über die Darstellung der endlichen Gruppen durch lineare Substitutionen", ibid., 1897, p. 994; "Über die Darstellung der endlichen Gruppen durch lineare Substitutionen II.", ibid., 1899, p. 482. — I will cite these treatises in the following as "Gruppencharaktere", "Primfaktoren", "D. I." and "D. II.".

then these substitutions form a representation of the group if, for any two elements A, B of $\mathfrak{H}$, the equation $\{A\}\{B\} = \{AB\}$ holds. If one denotes the matrices $(a_{ik}), (b_{ik}), \ldots$ by $(A), (B), \ldots$, then the matrix $(A)(B)$ will be different from the matrix (AB) only by a constant factor, which might be equal to 1. Conversely, any system of matrices, whose determinants are all nonzero and which have the property that for any two elements A, B in the group an equality of the form $(A)(B) = r_{A,B}(AB)$ holds for constants $r_{A,B}$, corresponds to a representation of the group by fractional linear substitutions[5]. Therefore, in the following I will call every such system of matrices a representation of the group by linear substitutions,[6] namely as one that belongs to the system of numbers $r_{A,B}$. The degree n of the matrices $(A), (B), \ldots$ shall be called the *degree* of the representation. Hence, if in particular all numbers $r_{A,B}$ are equal to 1, the system of matrices $(A), (B), \ldots$ also corresponds to a representation of the group by the ordinary linear substitutions

$$x_\nu = a_{\nu 1} y_1 + a_{\nu 2} y_2 + \cdots + a_{\nu n} y_n, \qquad (\nu = 1, 2, \ldots, n)$$
$$x_\nu = b_{\nu 1} y_1 + b_{\nu 2} y_2 + \cdots + b_{\nu n} y_n, \ldots .$$

Two representations $(A), (B), \ldots$ and $(A'), (B'), \ldots$, for which the h constants $a, b, \ldots$ can be determined, such that $(A') = a(A)$, $(B') = b(B), \ldots$, have to be considered essentially the same, because they correspond to the same system of fractional linear substitutions. In the following two such representations will be called *associated.*

The notions of *equivalence* and *primitivity* are defined exactly as for representations by ordinary linear substitutions (cf. *Frobenius,* D. II.): Two representations $(A), (B), \ldots$ and $(A'), (B'), \ldots$ are *equivalent,* if there is a matrix P with nonzero determinant satisfying

$$(A') = P^{-1}(A)P, \;\; (B') = P^{-1}(B)P, \; \ldots$$

Hence, two equivalent representations belong to the same factor system $r_{A,B}$. Further, a representation is *primitive,* if there is no equivalent representation $(A'), (B'), \ldots$ of the form

$$(A') = \begin{pmatrix} (A_1), & 0 \\ 0, & (A_2) \end{pmatrix} \quad (B') = \begin{pmatrix} (B_1), & 0 \\ 0, & (B_2) \end{pmatrix}, \quad \ldots$$

where $(A_1), (B_1), \ldots$ are matrices of the same degree.

For the primitive representations of a group $\mathfrak{H}$ by fractional linear substitutions the analogous theorem as for representations by ordinary linear substitutions (*Frobenius,* Primfaktoren, § 12) is valid, which states that their degree always divides the order of $\mathfrak{H}$ (§ 4).

The investigation of representations of a group by fractional linear substitutions is tightly connected to the theory of central elements[7] of finite groups.

If $\mathfrak{G}$ is a finite group, which contains a subgroup $\mathfrak{A}$ consisting of central elements in $\mathfrak{G}$, such that $\frac{\mathfrak{G}}{\mathfrak{A}}$ is isomorphic to the given group $\mathfrak{H}$ (in one step), then let

$$\mathfrak{G} = \mathfrak{A}A' + \mathfrak{A}B' + \cdots$$

[5] Translator's note: in this work Schur considers both ordinary representations as well as those via fractional linear transformations [gebrochene lineare Substitution] — such representations are the primary consideration of this work. When he wishes to emphasise that a representation is of the former type, he writes that a representation is via "ganze linear substitutionen", which we translate as "ordinary linear substitutions". (Another possible translation would be "integral linear substitutions" which would potentially lead to confusion with matrices all of whose entries are integers.)

[6] Translator's note: presumably Schur means by *fractional* linear substitutions.

[7] Translator's note: What Schur calls *invariant elements* [invariante Elemente] are today known as *central elements.* We use the nowadays well-established term "central" throughout.

and let the cosets[8] $\mathfrak{A}A', \mathfrak{A}B', \dots$ correspond to the elements $A, B, \dots$ of the group $\mathfrak{H}$. — I will in the following denote such a group $\mathfrak{G}$ as *a group* $\mathfrak{H}$ *augmented*[9] *by the group* $\mathfrak{A}$, or sometimes I will write more briefly, "let $\mathfrak{G}$ be group $(\mathfrak{A}, \mathfrak{H})$". — If one considers an arbitrary primitive representation of $\mathfrak{G}$ by ordinary linear substitutions, then each matrix corresponding to an element J of $\mathfrak{A}$ has the form $j(E)$, where j denotes a root of unity and (E) is the unit matrix. The matrices $(A'), (B'), \dots$ corresponding to the elements $A', B', \dots$ in the sense defined above provide a (primitive) representation of the group $\mathfrak{H}$ by fractional linear substitutions.

Now, one can also find augmented groups $\mathfrak{G}$ of $\mathfrak{H}$ with the property that every primitive representation of $\mathfrak{H}$ by fractional linear substitutions (or an associated representation) is equivalent to at least one of the representations of $\mathfrak{H}$ provided by the group $\mathfrak{G}$ in the manner discussed above. I will call an augmented group of $\mathfrak{H}$ with this property a *sufficiently augmented* group, and a sufficiently augmented group of lowest possible order will be called a *representation group* of the group $\mathfrak{H}$.

Several non-isomorphic representation groups can correspond to a given group $\mathfrak{H}$. However, the subgroups $\mathfrak{A}$ (consisting of central elements) of $\mathfrak{G}$, for which $\frac{\mathfrak{G}}{\mathfrak{A}} = \mathfrak{H}$ holds, are isomorphic to each other, for all representation groups of $\mathfrak{H}$. In this way, each group $\mathfrak{H}$ corresponds to a well-defined *Abel*ian group, which I will call the *multiplier* of the group $\mathfrak{H}$.[10] A group, whose multiplier is the trivial group, will be called a *closed* group (§§ 1–3).

Thus, in order to obtain all the primitive representations of a given group $\mathfrak{H}$ by fractional linear substitutions, one must first determine a representation group $\mathfrak{G}$ of $\mathfrak{H}$, which can be considered a pure group theoretic problem. Then, using the methods given by *Frobenius*, one has to investigate the primitive representations of $\mathfrak{G}$ by (ordinary) linear substitutions. In particular, for a closed group each representation by fractional linear substitutions can be replaced by a representation by ordinary linear substitutions.

In §5, I develop some theorems, which in special cases can serve to determine the multiplier and the representation groups of a given group.

The general theory presented here shall be further pursued in a future work to appear later. There, it will be applied to several special cases of groups.

§1.

Let $\mathfrak{H}$ be a finite group of order h, whose elements are $H_0 = E, H_1, \dots, H_{h-1}$. Assume that the matrices $(H_0), (H_1), \dots, (H_{h-1})$ form a representation of $\mathfrak{H}$ associated with the factor system $r_{P,Q}$, i.e. satisfying the following h^2 equations:

$$(1.) \qquad (P)(Q) = r_{P,Q}(PQ) \qquad {\scriptstyle (P,Q=H_0,H_1,\dots,H_{h-1})}$$

If P, Q, R are three arbitrary elements of the group, then by the associative law the matrix $(P)(Q)(R)$ on the one hand is equal to

$$r_{P,Q}(PQ)(R) = r_{P,Q}r_{PQ,R}(PQR)$$

[8] Translator's note: We adopt the by now standard *cosets* in place of Schur's *complexes* [Komplexe].

[9] Translator's note: Mathematicians today woulds write *extended* in place of *augmented* [ergäntze], see the commentary for further discussion. Because of various possible interpretations of "extended", we keep Schur's terminology throughout. Note that Schur appears to only ever consider central extensions in this work.

[10] Translator's note: Today, this group is known as the *Schur multiplier.*

but on the other hand also to

$$r_{Q,R}(P)(QR) = r_{P,QR}r_{Q,R}(PQR).$$

Thus, the h^2 quantities $r_{P,Q}$ must satisfy the h^3 equations

$$\text{(A.)} \qquad r_{P,Q}r_{PQ,R} = r_{P,QR}r_{Q,R} \qquad {\scriptstyle (P,Q,R=H_0,H_1,\dots,H_{h-1})}$$

Conversely, for any system of h^2 nonzero numbers $r_{P,Q}$ satisfying the equations (A.), there is a representation of $\mathfrak{H}$ associated with the system $r_{P,Q}$. To prove this, I introduce h independent variables $x_{H_0}, x_{H_1}, \dots, x_{H_{h-1}}$ and consider the following matrix of degree h,

$$X = (r_{PQ^{-1},Q} \cdot x_{PQ^{-1}})$$

whose rows and columns are obtained, by replacing P and Q by the h elements $H_0, H_1, \dots,$ H_{h-1} in this order. Then X can clearly be written in the form

$$X = \sum (R)x_R$$

where the sum is over all elements of $\mathfrak{H}$. The expression (R) means a certain matrix of degree h, which is obtained from X by setting the variable x_R to 1, and all other variables to 0. I will show that the h matrices $(H_0), (H_1), \dots, (H_{h-1})$ obtained in this way satisfy the equalities in (1.).

To this end, let $y_{H_0}, y_{H_1}, \dots, y_{H_{h-1}}$ be another set of independent variables, and let further the h quantities $z_{H_0}, z_{H_1}, \dots, z_{H_{h-1}}$ be determined by the equations

$$z_P = \sum r_{R,S}x_R y_S, \qquad {\scriptstyle (P=H_0,H_1,\dots,H_{h-1})}$$

where the sum is over all elements R and S satisfying $RS = P$. Let X become Y and Z, if the variables x_P are replaced by y_P and z_P, respectively. Then we obtain

$$XY = (\sum_R r_{PR^{-1},R}r_{RQ^{-1},Q} \cdot x_{PR^{-1}}y_{RQ^{-1}})$$

Now, by (A.), we have

$$r_{PR^{-1},RQ^{-1}}r_{PQ^{-1},Q} = r_{PR^{-1},R}r_{RQ^{-1},Q}.$$

Hence,

$$\sum_R r_{PR^{-1},R}r_{RQ^{-1},Q} \cdot x_{PR^{-1}}y_{RQ^{-1}} = r_{PQ^{-1},Q} \sum_R r_{PR^{-1},RQ^{-1}}x_{PR^{-1}}y_{RQ^{-1}}.$$

But, the right side of this equation is equal to $r_{PQ^{-1},Q} \cdot z_{PQ^{-1}}$. Thus we obtain

$$(X)(Y) = (r_{PQ^{-1},Q} \cdot z_{PQ^{-1}}) = Z$$

or

$$\sum_{R,S}(R)(S)x_R y_S = \sum_T (T)z_T = \sum_{R,S} r_{R,S}(RS)x_R y_S.$$

Thus, in fact we have,

$$(R)(S) = r_{R,S}(RS) \tag{1'.}$$

which is what was to be proved.

Now, we can further see that the matrix (R) contains exactly one nonzero element in each of its rows, as well as in each of its columns. Hence, also the determinants $d_{H_0}, d_{H_1}, \dots, d_{H_{h-1}}$ of the matrices $(H_0), (H_1), \dots, (H_{h-1})$ all are nonzero.[11)] If in equation (1'.) we take the determinant on both sides, we further obtain:

$$r^h_{R,S} = \frac{d_R d_S}{d_{RS}}. \tag{2.}$$

We obtain the theorem:

Theorem 1. *To each system of h^2 numbers $r_{P,Q}$ there are associated representations of the group $\mathfrak{H}$, if and only if the numbers $r_{P,Q}$ satisfy the equations* (A.).

The system of equations (A.) has an infinite number of solutions.[12)] For if the numbers $r_{P,Q}$ constitute such a solution, and if $c_{H_0}, c_{H_1}, \dots, c_{H_{h-1}}$ are h arbitrary nonzero quantities, then the h^2 quantities

$$r'_{P,Q} = \frac{c_P c_Q}{c_{PQ}} r_{P,Q} \tag{3.}$$

satisfy the equations (A.). Two solutions $r_{P,Q}$ and $r'_{P,Q}$ of (A.), for which h quantities c_P can be determined such that the equalities (3.) are satisfied, will be called *associated* solutions or *associated factor systems.*[13)]

Then, two representations of $\mathfrak{H}$ with corresponding solutions $r_{P,Q}$ and $\tilde{r}_{P,Q}$ of (A.) can only be associated (cf. Introduction), if these solutions are associated with each other. If, however, the solutions of (A.) for two representations of $\mathfrak{H}$ are not associated to each other, we will say that these representations are of *different type.*

If we regard all associated solutions of the equations (A.) as combined into one class, then the number of such classes is finite. For if $d_{H_0}, d_{H_1}, \dots, d_{H_{h-1}}$ are the quantities appearing in equation (2.), $\delta_{H_0}, \delta_{H_1}, \dots, \delta_{H_{h-1}}$ denote arbitrary roots of the h equations $\delta^h_P = d_P$, and the solution $r_{P,Q}$ is replaced by its associated solution

$$s_{P,Q} = \frac{\delta_P^{-1}\delta_Q^{-1}}{\delta_{PQ}^{-1}} r_{P,Q}$$

then according to (2.) we get $s^h_{P,Q} = 1$. Hence, each of the classes contains solutions $s_{P,Q}$, for which all numbers $s_{P,Q}$ are h^{th} roots of unity. Thus, the number of different classes is therefore certainly not larger than $(h)^{h^2}$.

---

11) If r is the order of an element R in $\mathfrak{H}$, then we have, as can be shown easily

$$d_R = (-1)^{h-\frac{h}{r}} \prod_S r_{R,S}.$$

12) Translator's note: Schur refers to this collection of numbers $r_{A,B}$ as a systems of numbers [Zahlensystem], and we have adopted this convention above. Today, solutions to (A.) are called "factor systems" or "2-cocycles". (For further pointers to the modern literature, see the commentary.) From now on we will use the terminology "factor system" for solutions to (A.).

13) Translator's note: Today, rather than *associated*, we would say that the two factor systems or cocycles are *cohomologous.*

Let this number be m, and denote the m different classes of solutions by $K_0, K_1, \dots, K_{m-1}$. If $r^{(\lambda)}_{P,Q}$ and $r^{(\mu)}_{P,Q}$ are two solutions of (A.) belonging to the classes K_λ and K_μ, then the numbers $r^{(\lambda)}_{P,Q} r^{(\mu)}_{P,Q}$ will also satisfy the equations (A.) The class K_ν of this solution obviously does not depend on the actual choice of the solutions $r^{(\lambda)}_{P,Q}$ and $r^{(\mu)}_{P,Q}$ within the classes K_λ and K_μ, hence it is uniquely determined by the classes K_λ and K_μ. If we define $K_\nu = K_\lambda K_\mu$, then we also have $K_\mu K_\lambda = K_\nu$. Further, as one can easily see, each equation of the form $K_\alpha K_\beta = K_\alpha K_\gamma$ implies equality of the classes K_β and K_γ. Now, as the composition of classes defined above satisfies the associative law, the m classes $K_0, K_1, \dots, K_{m-1}$ form a well-defined *Abel*ian group $\mathfrak{M}$ of order m, in which the unit element is the class K_0, which contains the solution $r_{P,Q} = 1$. I will call this group the *multiplier* of the group $\mathfrak{H}$.[14)]

The order of the multiplier of $\mathfrak{H}$ tells us how many different types of representations of the group $\mathfrak{H}$ by fractional linear substitutions exist.

Equation (2.) shows that for any class K_λ we have the relation $K^h_\lambda = K_0$. *Therefore, the number m cannot have a prime factor that is coprime to h.*

§2.

Now, let $\mathfrak{G}$ be an arbitrary group $\mathfrak{H}$ augmented by an *Abel*ian group $\mathfrak{A}$. We let

$$\mathfrak{G} = \mathfrak{A}G_0 + \mathfrak{A}G_1 + \cdots + \mathfrak{A}G_{h-1} \qquad (G_0 = E)$$

where the coset $\mathfrak{A}G_\lambda$ shall correspond to the element H_λ of the group $\mathfrak{H}$ that is isomorphic to $\frac{\mathfrak{G}}{\mathfrak{A}}$. If the group $\mathfrak{H}$ is determined by the h^2 equations

$$(4.) \qquad H_\lambda H_\mu = H_{\varphi(\lambda,\mu)} \qquad (\lambda,\mu = 0,1,\dots,h-1)$$

the elements $G_0, G_1, \dots, G_{h-1}$ of $\mathfrak{G}$ satisfy h^2 relations of the form

$$(5.) \qquad G_\lambda G_\mu = A_{\lambda,\mu} G_{\varphi(\lambda,\mu)} \qquad (\lambda,\mu = 0,1,\dots,h-1;\, A_{0,\lambda} = A_{\lambda,0} = E)$$

where the $A_{\lambda,\mu}$ denote elements of the subgroup $\mathfrak{A}$ of $\mathfrak{G}$.

Let $A_0 = E, A_1, \dots, A_{a-1}$ be the elements, and

$$\psi^{(0)}(A), \psi^{(1)}(A), \dots, \psi^{(a-1)}(A)$$

the a characters of the *Abel*ian group $\mathfrak{A}$. It is well-known that these characters can be associated to the elements $A_0, A_1, \dots, A_{a-1}$ in such a way that, if we set $\psi^{(a)}(A) = \psi_{A_a}(A)$, for each A the equations

$$(6.) \qquad \psi_{A_\alpha}(A)\psi_{A_\beta}(A) = \psi_{A_\alpha A_\beta}(A)$$

are valid.[15)]

Now, if we are given an arbitrary primitive representation D of the group $\mathfrak{G}$ by ordinary linear substitutions, in which the element R of $\mathfrak{G}$ corresponds to the matrix (R), then for each

[14)] In the following it will become clear that this definition of the multiplier coincides with the one given in the introduction.

[15)] Cf. *Weber,* Lehrbuch der Algebra, Bd. II, Abschnitt II. — In the following, if the characters of an *Abel*ian group are written in the form $\psi_{A_a}(A)$, we always assume the equations (6.) to hold.

element A of $\mathfrak{A}$, the matrix (A) has the form $\psi^{(a)}(A) \cdot (E)$, where E denotes the unity matrix and the numbers $\psi^{(a)}(A)$ represent one of the a characters of $\mathfrak{A}$. — In this case we will say (cf. *Frobenius*, D. II.) that the representation D *affords* the character $\psi^{(a)}(A)$ of $\mathfrak{A}$. — If we let

$$\psi^{(a)}(A_{\lambda,\mu}) = r^{(a)}_{H_\lambda,H_\mu} \tag{7.}$$

then because of (5.) we obtain the equations

$$(G_\lambda)(G_\mu) = r^{(a)}_{H_\lambda,H_\mu}(G_{\varphi(\lambda,\mu)}).$$

The h matrices $(G_0), (G_1), \ldots, (G_{h-1})$ thus form a representation of the group $\mathfrak{H}$ by fractional linear substitutions belonging to the factor system $r^{(a)}_{H_\lambda,H_\mu}$. Moreover, by considering all the representations of $\mathfrak{G}$ affording the character $\psi^{(a)}(A)$ of $\mathfrak{A}$, one also gets all the representations of $\mathfrak{H}$ by fractional linear substitutions belonging to the factor system (7.). Indeed, if the matrices $(H_0), (H_1), \ldots, (H_{h-1})$ constitute such a representation, and if we set

$$(A_\beta G_\lambda) = \psi^{(a)}(A_\beta) \cdot (H_\lambda) \qquad {\scriptstyle(\beta=0,1,\ldots,a-1,\lambda=0,1,\ldots,h-1)}$$

then, as one can easily check, the ah matrices $(A_\beta G_\lambda)$ form a representation of $\mathfrak{G}$ by ordinary linear substitutions, affording the character $\psi^{(a)}(A)$.

By the theorems on the group determinant[16] each of the a characters of $\mathfrak{A}$ correspond to representations of $\mathfrak{G}$. Thus the a systems

$$r^{(\alpha)}_{H_\lambda,H_\mu} \qquad {\scriptstyle(\alpha=0,1,\ldots,a-1)}$$

yield a solutions to the equations (A.).[17] Among these a solutions, some may be associated. The question then arises, what is the number m' of different classes K_λ represented by the solutions considered here. In other words: how many different types of representations of the group $\mathfrak{H}$ result from considering the group $\mathfrak{G}$.

Let

$$\psi_{B_0}(A), \psi_{B_1}(A), \ldots, \psi_{B_{b-1}}(A)$$

be those characters of $\mathfrak{A}$, which also correspond to linear characters[18] $\chi(R)$ of $\mathfrak{G}$. Then, the elements $B_0, B_1, \ldots, B_{b-1}$ form a subgroup $\mathfrak{B}$ of $\mathfrak{A}$. Indeed, if the characters $\psi_{B_\alpha}(A)$ and $\psi_{B_\beta}(A)$ of $\mathfrak{A}$ correspond to two linear characters $\chi^{(\alpha)}(R)$ and $\chi^{(\beta)}(R)$ of $\mathfrak{G}$, then the character $\psi_{B_\alpha B_\beta}(A)$ corresponds to the linear character $\chi^{(\alpha)}(R)\chi^{(\beta)}(R)$. Now, if for two elements B and C of $\mathfrak{A}$ the two factor systems $\psi_B(A_{\lambda,\mu})$ and $\psi_C(A_{\lambda,\mu})$ represent two solutions of (A.) which are associated, then there exist h constants $\varphi(H_0), \varphi(H_1), \ldots, \varphi(H_{h-1})$, satisfying

$$\psi_B(A_{\lambda,\mu}) = \frac{\varphi(H_\lambda)\varphi(H_\mu)}{\varphi(H_\lambda H_\mu)}\psi_C(A_{\lambda,\mu})$$

---

16) *Frobenius*, Gruppencharaktere, §7.

17) This can more easily be seen as follows. If we let $A_{\lambda,\mu} = A_{H_\lambda,H_\mu}$, then equations (5.) by the associative law yield the relations

$$A_{P,Q}A_{PQ,R} = A_{P,QR}A_{Q,R}; \qquad {\scriptstyle(P,Q,R=H_0,H_1,\ldots,H_{h-1})}$$

and thus we also obtain for each character $\psi(A)$ of $\mathfrak{A}$

$$\psi(A_{P,Q})\psi(A_{PQ,R}) = \psi(A_{P,QR})\psi(A_{Q,R}).$$

18) *Frobenius*, Primfaktoren, §2.

or, equivalently,

$$\varphi(H_\lambda)\varphi(H_\mu) = \psi_{BC^{-1}}(A_{\lambda,\mu})\varphi(H_\lambda H_\mu).$$

Now if we set

$$\chi(A_\alpha G_\lambda) = \psi_{BC^{-1}}(A_\alpha)\varphi(H_\lambda)$$

then for every two elements $R = A_\alpha G_\lambda$ and $S = A_\beta G_\mu$ of $\mathfrak{G}$ the equality $\chi(R)\chi(S) = \chi(RS)$ results. Furthermore, because of

$$\varphi(H_0)\varphi(H_0) = \psi_{BC^{-1}}(A_{0,0})\varphi(H_0) = \varphi(H_0),$$

we deduce $\phi(H_0) = 1$, and obtain the equality

$$\chi(A_\alpha) = \psi_{BC^{-1}}(A_\alpha)\varphi(H_0) = \psi_{BC^{-1}}(A_\alpha).$$

Thus, the ah numbers $\chi(A_\alpha G_\lambda)$ form a linear character of $\mathfrak{G}$ corresponding to the character $\psi_{BC^{-1}}(A)$; consequently, BC^{-1} has to be an element of $\mathfrak{B}$. One can easily conclude that this condition is also sufficient for the converse.

Therefore, the number m' we are looking for is equal to the number of elements of $\mathfrak{A}$ which are not congruent mod. $\mathfrak{B}$, hence $\frac{a}{b}$.

Let

$$\mathfrak{A} = \mathfrak{B}A_0 + \mathfrak{B}A_1 + \cdots + \mathfrak{B}A_{m'-1}$$

then the m' solutions

$$\psi_{A_\alpha}(A_{\lambda,\mu}) = r^{(a)}_{H_\lambda,H_\mu} \qquad {\scriptstyle (a=(0,1,\ldots,m'-1)}$$

of the equations (A.) belong to m' different classes $K_0, K_1, \ldots, K_{m'-1}$. But now, if $\mathfrak{B}A_\alpha \cdot \mathfrak{B}A_\beta = \mathfrak{B}A_\gamma$, then obviously the solution $r^{(\gamma)}_{H_\lambda,H_\mu}$ is associated to the solution $r^{(\alpha)}_{H_\lambda,H_\mu} r^{(\beta)}_{H_\lambda,H_\mu}$, thus $K_\gamma = K_\alpha K_\beta$. Hence, the classes $K_0, K_1, \ldots, K_{m'-1}$ form a subgroup $\mathfrak{M}'$ of the multiplier $\mathfrak{M}$ of $\mathfrak{H}$, which is isomorphic to the group $\frac{\mathfrak{A}}{\mathfrak{B}}$. Therefore, m' is a divisor of the order m of $\mathfrak{M}$, too.

There is yet another meaning of the number m' for the group $\mathfrak{G}$.

Let $\mathfrak{R}$ be the commutator[19] of the group $\mathfrak{H}$, $\mathfrak{R}'$ the commutator of the group $\mathfrak{G}$, and $\mathfrak{D}$ the intersection[20] of the groups $\mathfrak{R}'$ and $\mathfrak{A}$. Let r, r' and d be the respective orders of these groups. Then because of the isomorphism between the groups $\frac{\mathfrak{G}}{\mathfrak{A}}$ and $\mathfrak{H}$, the group $\frac{\mathfrak{R}'}{\mathfrak{D}}$ is isomorphic to the group $\mathfrak{R}$, i.e. $r' = dr$.

In general, for the commutator $\mathfrak{R}$ of a group $\mathfrak{H}$ the following theorem holds:

"If $\chi(S)$ is a linear character of $\mathfrak{H}$, then for every element R of $\mathfrak{R}$ we have $\chi(R) = 1$; conversely, every element R of $\mathfrak{H}$ having the property that $\chi(R) = 1$ for each linear character $\chi(S)$ of $\mathfrak{H}$, belongs to the commutator of $\mathfrak{H}$."[21]

We will apply this theorem to our case.

The group $\mathfrak{D}$ contains all elements J of $\mathfrak{A}$ characterized by the fact that for any linear character $\chi(R)$ of $\mathfrak{G}$ we have $\chi(J) = 1$, and no other elements. But now, for each element A of $\mathfrak{A}$ we have $\chi(A) = \psi_B(A)$, where B is an element of $\mathfrak{B}$. Thus, we obtain all elements of $\mathfrak{D}$ by searching for all elements J of $\mathfrak{A}$ satisfying

$$\psi_B(J) = 1 \tag{8.}$$

---

[19] The notion of commutator of a group has been introduced by *Dedekind* (Math. Ann. Bd. 48, p. 548). See also *Frobenius*, Primfaktoren, §2.

[20] Translator's note: Throughout, Schur writes greatest common divisor [größte gemeinsame Teiler] for the intersection of two subgroups.

[21] *Frobenius*, ibid.

where B runs through all elements of $\mathfrak{B}$. All elements J satisfying the b conditions in (8.) are known to form a subgroup of $\mathfrak{A}$, which is isomorphic to $\frac{\mathfrak{A}}{\mathfrak{B}}$, which by *Weber* (ibid.) has been called the subgroup of $\mathfrak{A}$ *reciprocal* to $\mathfrak{B}$. Hence, $\mathfrak{D}$ is also isomorphic to the group $\mathfrak{M}'$, and we have

$$m' = \frac{a}{b} = d = \frac{r'}{r}.$$

This yields a theorem, which will be important in what follows:

Theorem 2. *Whenever the group $\mathfrak{G}$ is $\mathfrak{H}$ augmented by the group $\mathfrak{A}$, the intersection of the commutator of $\mathfrak{G}$ and the subgroup $\mathfrak{A}$ of $\mathfrak{G}$ is isomorphic to a subgroup of the multiplier of $\mathfrak{H}$.*

In particular, if $\mathfrak{G}$ is a sufficiently augmented group of $\mathfrak{H}$, then among the representations of $\mathfrak{H}$ by fractional linear substitutions obtained by considering representations of $\mathfrak{G}$ by ordinary linear substitutions, all the m possible types must appear. Thus, for each such group the equality $m' = d = m$ has to hold, and consequently a has to be divisible by m. The arguments given two pages earlier also yield the fact that if $d = m$ then $\mathfrak{G}$ is always a sufficiently augmented group of $\mathfrak{H}$.

§3.

The next aim is to show that for each group $\mathfrak{H}$ it is indeed possible to find sufficiently augmented groups, even such groups of order mh.

We introduce h generating elements $Q_0, Q_1, \dots, Q_{h-1}$ and demand that the h^2 elements

$$\text{(9.)} \qquad Q_\lambda Q_\mu Q^{-1}_{\varphi(\lambda,\mu)} = J_{H_\lambda, H_\mu} \qquad {\scriptstyle (\lambda,\mu=0,1,\dots,h-1)}$$

(where the index $\phi(\lambda, \mu)$ is defined by equation (4.)) commute with the h generating elements; then the elements J_{H_λ, H_μ} commute with one another, too. The h^3 equations

$$\text{(10.)} \qquad Q_\nu \cdot J_{H_\lambda, H_\mu} = J_{H_\lambda, H_\mu} \cdot Q_\nu \qquad {\scriptstyle (\lambda,\mu,\nu=0,1,\dots,h-1)}$$

define a certain infinite group $\mathfrak{K}'$; the elements J_{H_λ, H_μ} generate a certain infinite *Abel*ian group $\mathfrak{N}'$, which is contained in $\mathfrak{K}'$ as a central subgroup. By the associative law, (9.) and (10.) yield the h^3 relations for the J_{H_λ, H_μ}:

$$\text{(B.)} \qquad J_{P,Q} J_{PQ,R} = J_{P,QR} J_{Q,R} \qquad {\scriptstyle (P,Q,R=H_0,H_1,\dots,H_{h-1})}$$

Each element of $\mathfrak{K}'$ can be brought to the form JQ_λ, where J is in the group $\mathfrak{N}'$. For $\lambda \neq \mu$, one can never have an equality of the form $JQ_\lambda = J'Q_\mu$, because this would imply $H_\lambda = H_\mu$, too, as the elements H_λ substituted for the Q_λ satisfy the equations (9.) and (10.). We may also deduce from (9.) and (10.) that amongst the elements J_{H_λ, H_μ}, there exists no relation of the form

$$\prod_{\lambda,\mu} J^{l_{\lambda,\mu}}_{H_\lambda, H_\mu} = E,$$

which does not already result from the equations (B.). For if this were the case, then not every system of numbers r_{H_λ, H_μ} satisfying the relations

$$\text{(A.)} \qquad r_{P,Q} r_{PQ,R} = r_{P,QR} r_{Q,R} \qquad {\scriptstyle (P,Q,R=H_0,H_1,\dots,H_{h-1})}$$

would satisfy the equation[22)]

$$\prod_{\lambda,\mu} r_{H_\lambda,H_\mu}^{l_{\lambda,\mu}} = 1. \tag{11.}$$

But we have seen in § 1 that for an arbitrary solution $r_{P,Q}$ of the equations (A.) one can find h matrices $(H_0), (H_1), \ldots, (H_{h-1})$, which satisfy the conditions $(H_\lambda)(H_\mu) = r_{H_\lambda,H_\mu}(H_{\varphi(\lambda,\mu)})$. Moreover, the matrices (H_λ), inserted for the Q_λ, satisfy conditions (9.) and (10.). But this would imply that the numbers r_{H_λ,H_μ} satisfy the equality (11.), which leads to a contradiction.

The h^3 equations $(B.)$ thus can be considered a complete system of defining relations[23)] for the *Abel*ian group $\mathfrak{N}'$ generated by the elements J_{H_λ,H_μ}.

Let us denote the $h^2 = p$ elements J_{H_λ,H_μ} by $X_1, X_2, \ldots, X_p$ in some order, so that the equations (B.) can be written in the form

$$X_1^{a_{\lambda 1}} X_2^{a_{\lambda 2}} \ldots X_p^{a_{\lambda p}} = E. \qquad {\scriptstyle (\lambda=1,2,\ldots,n,\, n=h^3)} \tag{12.}$$

where the exponents $\alpha_{\lambda\mu}$ are equal to 1, -1, or 0.

Now, we have the following well-known theorem:[24)]

"If p pairwise commuting elements $X_1, X_2, \ldots, X_p$ satisfy n relations of the form (12.), if the rank is $p - s$, and if $e_1, e_2, \ldots, e_\rho$ are the elementary divisors of the exponent system $(\alpha_{\lambda,\mu})$ greater than 1, then the *Abel*ian group $\mathfrak{N}'$ generated by the elements $X_1, X_2, \ldots, X_p$ can be represented as the direct product of a finite group $\mathfrak{N}$, whose invariants[25)] are the numbers $e_1, e_2, \ldots, e_\rho$, and an infinite group $\mathfrak{N}''$ of rank s, in which no non-trivial element has finite order.[26)] In other words: in the group $\mathfrak{N}'$ one can find $\rho + s$ independent basis elements

$$Y_\alpha = X_1^{s_{\alpha 1}} \ldots X_p^{s_{\alpha p}}, \qquad {\scriptstyle (\alpha=1,2,\ldots,\rho)}$$

$$Z_\beta = X_1^{t_{\beta 1}} \ldots X_p^{t_{\beta p}}, \qquad {\scriptstyle (\beta=1,2,\ldots,s)}$$

where Y_α corresponds to the exponent e_α, while there are no restrictions for Z_β at all. More precisely, let

$$X_\nu = Y_1^{a_{\nu 1}} \ldots Y_\rho^{a_{\nu\rho}} \cdot Z_1^{b_{\nu 1}} \ldots Z_s^{b_{\nu s}}$$

Then, if $x_1, x_2, \ldots, x_p$ are p numbers satisfying the equations

$$x_1^{a_{\lambda 1}} x_2^{a_{\lambda 2}} \ldots x_p^{\alpha_{\lambda p}} = 1 \qquad {\scriptstyle (\lambda=1,2,\ldots,n)} \tag{13.}$$

and if we let

$$y_\alpha = x_1^{s_{\alpha 1}} \ldots x_p^{s_{\alpha p}}, \quad z_\beta = x_1^{t_{\beta 1}} \ldots x_p^{t_{\beta p}} \qquad {\scriptstyle (\alpha=1,2,\ldots,\rho,\, \beta=1,2,\ldots,s)}$$

then the ρ quantities y_α satisfy the equations $y_\alpha^{e_\alpha} = 1$ and we have

$$x_\nu = y_1^{a_{\nu 1}} \ldots y_\rho^{a_{\nu\rho}} \cdot z_1^{b_{\nu 1}} \ldots z_s^{b_{\nu s}}. \tag{14.}$$

---

22) Cf. the fact mentioned below.

23) Cf. *Dyck*, Math. Ann., Bd. 20, p. 1.

24) Cf. *Frobenius* and *Stickelberger*, this journal, Vol. 86, p. 217.

25) Translators note: Recall that any finite abelian group is isomorphic to $\mathbb{Z}/k_1\mathbb{Z} \oplus \mathbb{Z}/k_2\mathbb{Z} \oplus \cdots \oplus \mathbb{Z}/k_m\mathbb{Z}$ with $k_1|k_2|\cdots|k_m$. The k_i, are what Schur calls invariants [Invariante]. Today they are usually referred to as "invariant factors".

26) If no elementary divisor of $(\alpha_{\lambda,\mu})$ is greater than 1, then one has to set $\mathfrak{N} = E$. The group $\mathfrak{N}$ is uniquely determined; it is the largest finite group contained in $\mathfrak{N}'$. However, if neither $\mathfrak{N} = E$ nor $s = 0$ holds, the group $\mathfrak{N}''$ can be chosen in different ways.

Conversely, all solutions of the equations (13.) can be obtained from (14.) by inserting arbitrary roots of the equations

$$y_1^{e_1} = 1, \ldots, y_\rho^{e_\rho} = 1$$

for $y_1, \ldots, y_\rho$, and arbitrary (nonzero) values for $z_1, \ldots, z_s$."

We now require that the elements $X_1, X_2, \ldots, X_p$ of the group $\mathfrak{K}'$ (and thus also the elements $Q_0, Q_1, \ldots, Q_{h-1}$ generating this group) satisfy the s conditions

$$(15.) \qquad Z_\beta = X_1^{t_{\beta 1}} \ldots X_p^{t_{\beta p}} = E. \qquad {\scriptstyle (\beta=1,2,\ldots,s)}$$

Then the relations (9.), (10.) and (15.) define a finite group $\mathfrak{K}$, in which the elements J_{H_λ, H_μ} generate the *Abel*ian group $\mathfrak{N}$ of order $\overline{m} = e_1 e_2 \ldots e_\rho$. We view $\mathfrak{K}$ as an augmented group $(\mathfrak{N}, \mathfrak{H})$ of $\mathfrak{H}$.

I want to show that $s = h$, that $\mathfrak{N}$ is isomorphic to the multiplier of $\mathfrak{H}$, and that $\mathfrak{K}$ is a sufficiently augmented group of $\mathfrak{H}$.

If we set $x_\nu = r_{P,Q}$, where X_ν means the element $J_{P,Q}$, then the system of equations (13.) is equivalent to the system of equations (A.). We further let

$$y_\alpha = x_1^{s_{\alpha 1}} \ldots x_p^{s_{\alpha p}} = f_\alpha(r_{R,S}), \quad z_\beta = x_1^{t_{\beta 1}} \ldots x_p^{t_{\beta p}} = g_\beta(r_{R,S}) \text{ [27]}$$

and

$$y_1^{a_{\nu 1}} \ldots y_\rho^{a_{\nu\rho}} = F_{P,Q}(y_\alpha), \qquad z_1^{b_{\nu 1}} \ldots z_s^{b_{\nu s}} = G_{P,Q}(z_\beta)$$

so that we obtain

$$(16.) \qquad r_{P,Q} = F_{P,Q}(y_\alpha) G_{P,Q}(z_\beta).$$

Now, if $s < h$ were to hold, then for h arbitrary variables $c_{H_0}, c_{H_1}, \ldots, c_{H_{h-1}}$, the h^2 functions of these variables $r_{P,Q} = \frac{c_P c_Q}{c_{PQ}}$ also provide a solution of (A.). Thus the h expressions

$$\prod_Q \left(\frac{c_P c_Q}{c_{PQ}}\right) = c_P^h$$

could also be expressed by $s < h$ variables $z_1, z_2, \ldots, z_s$, which is not possible. But $s > h$ is also not possible. Namely, if $r_{P,Q}$ is an arbitrary solution of (A.), we let the quantities $d_{H_0}, d_{H_1}, \ldots, d_{H_{h-1}}$ have the same meaning as in §1. Then, if we add to the equations (A.) the h conditions

$$d_{H_0} = 1, d_{H_1} = 1, \ldots, d_{H_{h-1}} = 1$$

then the numbers $r_{P,Q}$ become h^{th} roots of unity, as can be deduced from (2.). But this would be impossible, if the most general solution of the system of equalities (A.) would contain $s > h$ indeterminates $z_1, z_2, \ldots, z_s$. Thus, indeed we have $s = h$.

Moreover, if we set $r_{R,S} = \frac{c_R c_S}{c_{RS}}$ the ρ expressions $f_\alpha(r_{R,S})$ are identically equal to 1. This is true, because otherwise the equation

$$\left\{ f_\alpha \left(\frac{c_R c_S}{c_{RS}} \right) \right\}^{e_\alpha} = 1$$

[27] Accordingly, one has to define $Y_\alpha = f_\alpha(J_{R,S})$, $Z_\beta = g_\beta(J_{R,S})$.

which is satisfied no matter how the quantities c_P are chosen, would yield a condition on these quantities. Hence, we always have

$$(17.)\quad \frac{c_P c_Q}{c_{PQ}} = F_{P,Q}\left\{f_\alpha\left(\frac{c_R c_S}{c_{RS}}\right)\right\} G_{P,Q}\left\{g_\beta\left(\frac{c_R c_S}{c_{RS}}\right)\right\} = G_{P,Q}\left\{g_\beta\left(\frac{c_R c_S}{c_{RS}}\right)\right\}.$$

On the other hand, for any arbitrary choice of the quantities $z_1, z_2, \ldots, z_h$ one can determine quantities $\alpha_{H_0}, \alpha_{H_1}, \ldots, \alpha_{H_{h-1}}$ such that $z_\beta = g_\beta(\frac{\alpha_R\alpha_S}{\alpha_{RS}})$ is satisfied. In order to show this it suffices to show that the h functions $\overline{g}_\beta = g_\beta(\frac{c_R c_S}{c_{RS}})$ in h variables c_P are independent of each other. But this is indeed true: for if there was an equality of the form $\phi(\overline{g}_1, \overline{g}_2, \ldots, \overline{g}_h) = 0$, then from the fact shown in §1 that any solution of (A.) can be brought into the form $s_{P,Q}\frac{c_P c_Q}{c_{PQ}}$, where the numbers $s_{P,Q}$ are h-th roots of unity, we would know that the quantities

$$z_\beta^h = g_\beta(r_{R,S}^h) = g_\beta\left(\frac{c_R^h c_S^h}{c_{RS}^h}\right)$$

would have to satisfy this equation for each solution $r_{P,Q}$. But this is impossible, because the quantities z_β must remain indeterminates.

Hence, two solutions

$$r_{P,Q} = F_{P,Q}(y_\alpha)G_{P,Q}(z_\beta)$$

and

$$r'_{P,Q} = F_{P,Q}(y'_\alpha)G_{P,Q}(z'_\beta)$$

are associated if and only if $y'_\alpha = y_\alpha$. For if the h quantities α_P can be determined such that $r'_{P,Q} = \frac{\alpha_P\alpha_Q}{\alpha_{PQ}} r_{P,Q}$, then this yields

$$y'_\alpha y_\alpha^{-1} = f_\alpha(r'_{R,S})f_\alpha(r_{R,S}^{-1}) = f_\alpha(r'_{R,S}r_{R,S}^{-1}) = f_\alpha(\frac{a_R a_S}{a_{RS}}) = 1,$$

and thus $y'_\alpha = y_\alpha$. But if these conditions are satisfied, one can determine the quantities α_P such that

$$z'_\alpha z_\beta^{-1} = g_\beta\left(\frac{a_R a_S}{a_{RS}}\right).$$

Then, because of (17.) we obtain

$$r'_{P,Q} = F_{P,Q}(y'_\alpha)G_{P,Q}(z_\beta)G_{P,Q}(z'_\beta z_\beta^{-1}) = \frac{a_P a_Q}{a_{PQ}} r_{P,Q}.$$

Therefore, each solution of (A.) is associated to one of the $\overline{m} = e_1 e_2 \ldots e_\rho$ solutions

$$(18.)\quad r_{P,Q} = F_{P,Q}(y_\alpha)$$

obtained by inserting arbitrary roots of the equations $y_1^{e_1} = 1, \ldots, y_\rho^{e_\rho} = 1$ for $y_1, \ldots, y_\rho$. Further, amongst these m solutions, no two are associated.

Hence we have $\overline{m} = m$, and the order of the group $\mathfrak{K}$ is mh.

For each of the m characters $\psi(J)$ of the group $\mathfrak{N}$, the system of numbers $\psi(J_{P,Q})$ coincides with one of the m factor systems (18.). Now, as these belong to m different classes of solutions of the equations (A.) for the augmented group $\mathfrak{K} = (\mathfrak{N}, \mathfrak{H})$ of $\mathfrak{H}$ the number m' from the preceding section[28] is equal to m. Hence $\mathfrak{K}$ is a sufficiently augmented group of $\mathfrak{H}$, as we wanted to show.

[28] Translator's note: i.e. § 2.

At the same time we can see that $\mathfrak{N}$ is isomorphic to the multiplier $\mathfrak{M}$ of $\mathfrak{H}$, and we obtain a method of determining the invariants of $\mathfrak{M}$.

By the argument at the end of the preceding section, the order of a sufficiently augmented group of $\mathfrak{H}$ cannot be smaller than mh. This fact shows us that the group $\mathfrak{K}$ constructed here describes a representation group of $\mathfrak{H}$.

The definition of the group $\mathfrak{K}$ by virtue of the equations (9.), (10.) and (15.) depends heavily on the choice of the elements $Z_1, \dots, Z_h$ within the group $\mathfrak{N}'$. Now, since these elements can be chosen in various different ways, in general there can exist several representation groups $\mathfrak{G} = (\mathfrak{A}, \mathfrak{H})$, which are not isomorphic. But all of them have order mh, and in each of them the subgroup $\mathfrak{A}$ is isomorphic to the multiplier of $\mathfrak{H}$.

But even more can be proved.

I will use the term linear character of the infinite group $\mathfrak{K}'$ for every system of nonzero quantities $\chi(A)$, which can be associated to the elements A of $\mathfrak{K}'$ in such a way that for any two elements A and B of $\mathfrak{K}'$ the equation

$$\chi(A)\chi(B) = \chi(AB) \tag{19.}$$

holds. Let $\mathfrak{T}$ be the commutator of $\mathfrak{K}'$, i.e. the subgroup of $\mathfrak{K}'$ generated by all elements of the form $ABA^{-1}B^{-1}$. Then in complete analogy with the case of finite groups, the elements of $\mathfrak{T}$ can be characterized as follows[29]: an element T of $\mathfrak{K}'$ belongs to the commutator $\mathfrak{T}$ of $\mathfrak{K}'$, if and only if $\chi(T) = 1$ holds, for every linear character $\chi(A)$ of $\mathfrak{K}'$.

Based on this criterion it is easy to determine the intersection of the groups $\mathfrak{T}$ and $\mathfrak{N}'$.

Obviously, one gets all linear characters $\chi(A)$ of $\mathfrak{K}'$ by choosing the h numbers $\chi(Q_0)$, $\chi(Q_1), \dots, \chi(Q_{h-1})$ arbitrarily and determining the remaining quantities $\chi(A)$ using the equalities (19.). Now, if an arbitrary character satisfies $\chi(Q_\lambda) = c_{H_\lambda}$, then we obtain $\chi(J_{P,Q}) = \frac{c_P c_Q}{c_{PQ}}$ and

$$\chi(Y_\alpha) = \chi\{f_\alpha(J_{R,S})\} = f_\alpha\{\chi(J_{R,S})\} = f_\alpha\left(\frac{c_R c_S}{c_{RS}}\right).$$

However, as we saw above, this expression equals 1. Therefore, any of the elements $Y_1, Y_2, \dots, Y_\rho$ and thus also the group $\mathfrak{N}$ generated by these elements is contained in $\mathfrak{T}$. In contrast to this, no element of $\mathfrak{N}''$ (except the identity element) belongs to the group $\mathfrak{T}$. For if

$$J = Z_1^{k_1} Z_2^{k_2} \dots Z_h^{k_h}$$

were an element of $\mathfrak{T}$, then for every linear character $\chi(A)$ of $\mathfrak{K}'$

$$\chi(J) = [\chi(Z_1)]^{k_1}[\chi(Z_2)]^{k_2} \dots [\chi(Z_h)]^{k_h} = 1$$

would hold. But, again if $\chi(J_{P,Q}) = \frac{c_P c_Q}{c_{PQ}}$, we have

$$\chi(Z_\beta) = \chi\{g_\beta(J_{R,S})\} = g_\beta\{\chi(J_{R,S})\} = g_\beta\left(\frac{c_R c_S}{c_{RS}}\right).$$

However, we have seen that the numbers c_P can be chosen such that the h quantities $g_\beta(\frac{c_R c_S}{c_{RS}})$ can assume arbitrarily chosen values $z_1, z_2, \dots, z_h$. Thus, as long as all exponents $k_1, k_2, \dots, k_h$ are not all equal to 0, it cannot be that $\chi(J)$ is always be equal to 1.

[29] Cf. the theorem stated on page 148. — It can easily be seen that this theorem remains true for every group generated by finitely many elements.

Consequently, the intersection of $\mathfrak{T}$ and $\mathfrak{N}'$ is nothing other than the group $\mathfrak{N}$ of order m, which is isomorphic to the multiplier $\mathfrak{M}$ of $\mathfrak{H}$. The commutator $\mathfrak{T}$ of $\mathfrak{K}'$ is thus a finite group, which should be considered as the commutator $\mathfrak{R}$ of $\mathfrak{H}$, augmented by the group $\mathfrak{N} \cong \mathfrak{M}$. Its order is mr, where r denotes the order of $\mathfrak{R}$.

Now let $\mathfrak{L}$ be an arbitrary representation group of $\mathfrak{H}$, i.e. a group of the form $\mathfrak{H}$ augmented by the group $\mathfrak{M}$, whose commutator $\mathfrak{T}'$ contains the whole group $\mathfrak{M}$. Let

$$\mathfrak{L} = \mathfrak{M}L_0 + \mathfrak{M}L_1 + \cdots + \mathfrak{M}L_{h-1},$$

where the coset $\mathfrak{M}L_\lambda$ corresponds to the element H_λ in the group $\mathfrak{H}$, which is isomorphic to $\frac{\mathfrak{L}}{\mathfrak{M}}$. Then

$$L_\lambda L_\mu L^{-1}_{\varphi(\lambda,\mu)} = M_{H_\lambda, H_\mu}$$

is an element of $\mathfrak{M}$, hence a central element of $\mathfrak{L}$. The assumption that the commutator $\mathfrak{T}'$ of $\mathfrak{L}$ contains all elements of $\mathfrak{M}$ immediately implies that the elements $L_0, L_1, \ldots, L_{h-1}$ generate the whole group $\mathfrak{L}$.[30] But the elements $L_0, L_1, \ldots, L_{h-1}$, inserted for the Q_λ, satisfy the conditions (9.) and (10.), which define the infinite group $\mathfrak{K}'$. This implies[31] that the group $\mathfrak{K}'$ is isomorphic to the group $\mathfrak{L}$ in several steps,[32] where in particular to every element of $\mathfrak{K}'$ there is only one corresponding element of $\mathfrak{L}$. Hence, also the commutator $\mathfrak{T}$ of $\mathfrak{K}'$ in the same way is isomorphic to the commutator $\mathfrak{T}'$ of $\mathfrak{L}$ in several steps. But as the order of $\mathfrak{T}'$ is equal to the order mr of $\mathfrak{T}$, the isomorphism between $\mathfrak{T}$ and $\mathfrak{T}'$ is in one step. This means that the commutator of an arbitrary representation group is isomorphic to the well-defined group $\mathfrak{T}$.

We obtain the theorem:

Theorem 3. *The commutator groups of any two representation groups of a given group are isomorphic to each other.*

In particularly, if $\mathfrak{R} = \mathfrak{H}$, thus $r = h$, then any representation group of $\mathfrak{H}$ coincides with its commutator group. Hence, Theorem 3 yields:

Theorem 4. *If the commutator of a group $\mathfrak{H}$ contains all elements of $\mathfrak{H}$, then $\mathfrak{H}$ has only one representation group.*[33]

§4.

The considerations up to this point focussed mainly on the investigation of all factor systems $r_{P,Q}$, which may correspond to representations of the group $\mathfrak{H}$ by fractional linear substitutions. The other question, namely "how can the different primitive representations of the group $\mathfrak{H}$ corresponding to the same factor system $r_{P,Q}$ be determined?" can be solved without introducing new tools, based on the theorems of *Frobenius*.

As in § 2, let

$$\mathfrak{G} = \mathfrak{A}G_0 + \mathfrak{A}G_1 + \cdots + \mathfrak{A}G_{h-1} \qquad (G_0 = E)$$

---

[30] This follows from the fact that every commutator element of $\mathfrak{L}$ has the form $L_\lambda L_\mu L_\lambda^{-1} L_\mu^{-1}$.

[31] Cf. *Dyck*, ibid. and *Burnside*, Theory of Groups of finite Order, §182.

[32] Translator's note: Schur's language that groups $\mathfrak{K}'$ and $\mathfrak{L}$ are isomorphic "in several steps" [mehrstufig isomorph] is potentially confusing to the contemporary reader. In modern language, he means $\mathfrak{L}$ is a quotient of $\mathfrak{K}'$. (In the literature at the time, isomorph seems closer to what we would today call homomorphic image.)

[33] For the case of a group $\mathfrak{H}$, whose commutator does not contain all elements of $\mathfrak{H}$, I will determine the exact number of non-isomorphic representation groups in a subsequent work.

be an arbitrary group of $\mathfrak{H}$ augmented by the *Abel*ian group $\mathfrak{A}$; assume that $A_\alpha, A_{\lambda,\mu}, \psi^{(\alpha)}(A)$ and $r^{(\alpha)}_{H_\lambda,H_\mu} = \psi^{(\alpha)}(A_{\lambda,\mu})$ have the same meaning as in § 2. As we have seen, investigating all primitive representations of the group $\mathfrak{H}$ by fractional linear substitutions belonging to the factor system $r^{(\alpha)}_{H_\lambda,H_\mu}$ is identical to determining the various primitive representations of the group $\mathfrak{G}$ by ordinary linear substitutions corresponding to the character $\psi^{(\alpha)}(A)$ of $\mathfrak{A}$, or, what amounts to the same thing, determining those representations corresponding to certain group matrices belonging to G.[34)]

For these we now have the following theorems:[35)]

"If the ah quantities $x_{A_\beta G_\lambda} (\beta = 0, 1, \ldots, a-1; \lambda = 0, 1, \ldots, h-1)$ denote independent variables, then all primitive matrices belonging to the group $\mathfrak{G}$ and corresponding to the character $\psi^{(\alpha)}(A)$ are obtained by decomposing the matrix

$$Y^{(\alpha)} = \left(\sum_\beta \psi^{(a)}(A_\beta) x_{A_\beta G_\lambda G_\mu^{-1}}\right) \qquad (\lambda,\mu=0,1,\ldots,h-1)$$

of degree h in the group into primitive submatrices. If $\mathbf{\Phi}_1, \mathbf{\Phi}_2, \ldots, \mathbf{\Phi}_{l_a}$ are the l_a different prime factors of the determinant $Q^{(\alpha)}$ of $Y^{(\alpha)}$, and if f_λ is the degree of the prime factor $\mathbf{\Phi}_\lambda$, then

$$Q^{(\alpha)} = \mathbf{\Phi}_1^{f_1} \mathbf{\Phi}_2^{f_2} \ldots \mathbf{\Phi}_{l_\alpha}^{f_{l_\alpha}}, \tag{20.}$$

and thus

$$f_1^2 + f_2^2 + \cdots + f_{l_\alpha}^2 = h. \tag{21.}$$

If further $\chi^{(1)}(S), \chi^{(2)}(S), \ldots, \chi^{(l_\alpha)}(S)$ are the characters of $\mathfrak{G}$ corresponding to the l_α prime factors $\mathbf{\Phi}_\lambda$, and if $S = A_\beta G_\lambda$, then

$$f_1 \chi^{(1)}(S) + f_2 \chi^{(2)}(S) + \cdots + f_{l_\alpha} \chi^{(l_\alpha)}(S) = h \psi^{(\alpha)}(A_\beta) \varepsilon_\lambda \tag{22.}$$

where ε_λ equals 1 or 0, if $\lambda = 0$ or $\lambda > 0$, respectively."

Now, if for any element R of $\mathfrak{G}$ we let

$$y_R^{(\alpha)} = \sum_\beta \psi^{(\alpha)}(A_\beta) x_{A_\beta R}, \tag{23.}$$

then $Y^{(\alpha)}$ can be written in the form

$$Y^{(\alpha)} = (y^{(\alpha)}_{G_\lambda G_\mu^{-1}}).$$

Furthermore, as can be seen easily, for each element A of $\mathfrak{A}$, we have

$$y_{AR}^{(\alpha)} = \psi^{(\alpha)}(A^{-1}) y_R^{(\alpha)}. \tag{24.}$$

---

34) *Frobenius*, D. I.
Translator's note: As the reader will immediately see, Schur is referring to Frobenius' theory of the group determinant.

35) *Frobenius*, Über Relationen zwischen den Charakteren einer Gruppe und denen ihrer Untergruppen, §§2 and 3. Sitzungsberichte 1898, p. 501.

If we denote the element H_μ^{-1} of $\mathfrak{H}$ by $H_{\mu'}$, we obtain

$$G_\mu G_{\mu'} = A_{\mu,\mu'} G_0 = A_{\mu,\mu'},$$

and thus $G_\mu^{-1} = A_{\mu,\mu'}^{-1} G_{\mu'}$. Hence, if we put $H_\lambda H_\mu^{-1} = H_\rho$, we have

$$G_\lambda G_\mu^{-1} = A_{\mu,\mu'}^{-1} A_{\lambda,\mu'} G_\rho.$$

Now, denoting the expression $y_{G_\lambda}^{(\alpha)}$ by z_{H_λ}, from (24.) we obtain

$$y_{G_\lambda G_\mu{}^{-1}}^{(\alpha)} = \psi^{(\alpha)}(A_{\mu,\mu'})\psi^{(\alpha)}(A_{\lambda,\mu'}^{-1}) z_{H_\lambda H_\mu^{-1}} = \frac{r_{H_\mu,H_\mu^{-1}}^{(\alpha)}}{r_{H_\lambda,H_\mu^{-1}}^{(\alpha)}} z_{H_\lambda H_\mu^{-1}}.$$

Then, as the system of numbers $r_{H_\lambda,H_\mu}^{(\alpha)}$ satisfy the system of equations (A.) and (since $G_0 = E$) we have $r_{H_\lambda,E}^{(\alpha)} = 1$, we obtain

$$r_{H_\lambda H_\mu^{-1},H_\mu}^{(\alpha)} r_{H_\lambda,H_\mu^{-1}}^{(\alpha)} = r_{H_\lambda H_\mu^{-1},E}^{(\alpha)} r_{H_\mu,H_\mu^{-1}}^{(\alpha)} = r_{H_\mu,H_\mu^{-1}}^{(\alpha)}.$$

Hence, the matrix $Y^{(\alpha)}$ can be written in the form

$$Y^{(\alpha)} = (r_{PQ^{-1},Q}^{(\alpha)} z_{PQ^{-1}}). \qquad (P,Q=H_0,H_1,\ldots,H_{h-1}) \tag{25.}$$

Therefore, we see that $Y^{(\alpha)}$ can be viewed as the matrix of degree h associated with the factor system $r_{P,Q}^{(\alpha)}$ in §1.

Under the assumption that $\mathfrak{G} = \mathfrak{K}$ is a representation group of $\mathfrak{H}$, then every system of numbers $r_{P,Q}$ satisying the equations (A.) is associated to one of the $a = m$ number systems $r_{P,Q}^{(\alpha)}$. Hence, the cited theorem yields:

Theorem 5. *If* $r_{P,Q}$ *is an arbitrary solution of the equations* (A.), *if*

$$x_{H_0}, x_{H_1}, \ldots, x_{H_{h-1}}$$

are independent variables, and if we set

$$(r_{PQ^{-1},Q} x_{PQ^{-1}}) = \sum (R) x_R, \qquad (R=H_0,H_1,\ldots,H_{h-1})$$

then every primitive representation of the group $\mathfrak{H}$ *belonging to the number system* $r_{P,Q}$ *is equivalent to one of the primitive representations appearing in a decomposition of the representation formed by the matrices* (R).

My next aim will be to show how the number l_α of prime factors of $Q^{(\alpha)}$ or, what amounts to the same, the number of different primitive representations of $\mathfrak{G}$ corresponding to the character $\psi^{(\alpha)}(A)$ of $\mathfrak{A}$, can be determined more exactly. The method applied here to determine l_α is modeled on the method used by *Frobenius*[36] to determine the number of different prime factors of the group determinant.

For each of the l_α characters $\chi^{(\rho)}(S)$ of $\mathfrak{G}$ corresponding to the character $\psi^{(\alpha)}(A)$ we have

$$\chi^{(\rho)}(A_\beta G_\lambda) = \psi^{(\alpha)}(A_\beta)\chi^{(\rho)}(G_\lambda).$$

[36] Primfaktoren, §7.

Then, since[37)]

$$ah = \sum_{S} \chi^{(\rho)}(S^{-1})\chi^{(\rho)}(S) = \sum_{\beta,\lambda} \chi^{(\rho)}(A_\beta^{-1}G_\lambda^{-1})\chi^{(\rho)}(A_\beta G_\lambda)$$
$$= \sum_{\beta,\lambda} \psi^{(\alpha)}(A_\beta^{-1}A_\beta)\chi^{(\rho)}(G_\lambda^{-1})\chi^{(\rho)}(G_\lambda) = a\sum_{\lambda} \chi^{(\rho)}(G_\lambda^{-1})\chi^{(\rho)}(G_\lambda)$$

we obtain

$$\sum_{\lambda} \chi^{\rho}(G_\lambda^{-1})\chi^{\rho}(G_\lambda) = h. \tag{26.}$$

Furthermore, the formula[38)]

$$\sum_{S} \chi(G_\lambda^{-1}S^{-1}G_\lambda S) = \frac{ah}{f}\chi(G_\lambda^{-1})\chi(G_\lambda)$$

which is valid for any character of degree f, yields the equation

$$\sum_{\mu=0}^{h-1} \chi^{(\rho)}(G_\lambda^{-1}G_\mu^{-1}G_\lambda G_\mu) = \frac{h}{f_\rho}\chi^{(\rho)}(G_\lambda^{-1})\chi^{(\rho)}(G_\lambda).$$

Thus, from (26.) we have

$$\sum_{\lambda,\mu} \frac{f_\rho}{h}\chi^{(\rho)}(G_\lambda^{-1}G_\mu^{-1}G_\lambda G_\mu) = h$$

and

$$\sum_{\rho=1}^{l_\alpha}\sum_{\lambda,\mu} \frac{f_\rho}{h}\chi^{(\rho)}(G_\lambda^{-1}G_\mu^{-1}G_\lambda G_\mu) = hl_\alpha$$

or equivalently

$$hl_\alpha = \sum_{\lambda,\mu}\sum_{\rho=1}^{l_\alpha} \frac{f_\rho}{h}\chi^{(\rho)}(G_\lambda^{-1}G_\mu^{-1}G_\lambda G_\mu).$$

Now, let

$$G_\lambda^{-1}G_\mu^{-1}G_\lambda G_\mu = F_{\lambda,\mu}G_\nu$$

where $F_{\lambda,\mu}$ denotes an element of $\mathfrak{A}$. Then, by (22.), we have

$$\sum_{\rho=1}^{l_\alpha} \frac{f_\rho}{h}\chi^{(\rho)}(G_\lambda^{-1}G_\mu^{-1}G_\lambda G_\mu) = \psi^{(\alpha)}(F_{\lambda,\mu})\varepsilon_\nu.$$

Hence, we obtain

$$hl_\alpha = \sum_{\lambda=1}^{h-1}\sum_{\mu} \psi^{(\alpha)}(F_{\lambda,\mu}), \tag{27.}$$

---

37) *Frobenius*, Gruppencharaktere, §5, formula (10.).

38) ibid., formula (5.).

where μ ranges over those numbers $0, 1, \ldots, h-1$ for which $G_\lambda^{-1} G_\mu^{-1} G_\lambda G_\mu$ is an element of $\mathfrak{A}$, or, what amounts to the same, for which H_μ commutes with H_λ. From (27.), l_α can be determined as follows.

Let the h elements $H_0, H_1, \ldots, H_{h-1}$ of $\mathfrak{H}$ be split into k classes of conjugate elements denoted by $(0), (1), \ldots, (k-1)$; let the class (ρ) contain exactly h_ρ elements, and let (0) denote the principle class (the element E). It is obvious that two elements G_λ and G_μ can only be conjugate in $\mathfrak{G}$, if H_λ and H_μ are conjugate in $\mathfrak{H}$. Moreover, to simplify our considerations at a later point, let us right here establish the fact that the complete residue system $G_0, G_1, \ldots, G_{h-1}$ of $\mathfrak{G}$ mod. $\mathfrak{A}$ can be chosen in such a way that G_λ and G_μ are conjugate in $\mathfrak{G}$, whenever H_λ and H_μ belong to a conjugacy class (ρ). Namely, if $H_{\alpha_1}, H_{\alpha_2}, \ldots$ are the elements of $\mathfrak{H}$ conjugate to H_1, then each of the cosets $\mathfrak{A}G_{\alpha_1}, \mathfrak{A}G_{\alpha_2}, \ldots$ contains at least one element which is conjugate to G_1; then we denote these elements by $G_{\alpha_1}, G_{\alpha_2}, \ldots$. The same shall be done, for example, if H_2 is an element which is not conjugate to H_1, with respect to H_2, and so on. Now, if e.g. $H_0, H_1, \ldots, H_{k-1}$ belong to the k different classes

$$(0), (1), \ldots, (k-1)$$

then, in $\mathfrak{G}$, among the elements $G_0, G_1, \ldots, G_{h-1}$ there are exactly $h_0 = 1$ conjugate to G_0, exactly h_1 conjugate to G_1, and so on.

Now, if $H_{\mu_1}, H_{\mu_2}, \ldots$ are those elements of $\mathfrak{H}$ which commute with the element H_λ of class (ρ), then we have

$$G_\lambda G_{\mu_\sigma} = F_{\lambda,\mu_\sigma} G_{\mu_\sigma} G_\lambda. \qquad (\sigma=1,2,\ldots,\tfrac{h}{h_\rho})$$

Assume that among the $\frac{h}{h_\rho}$ elements F_{λ,μ_σ} there are exactly α_λ different ones. Then, obviously these α_λ elements form a subgroup $\mathfrak{A}_\lambda$ of $\mathfrak{A}$; and it is also easy to see that α_λ has to be a divisor of $\frac{h}{h_\rho}$. Moreover, one can easily show that the group $\mathfrak{A}_\lambda$ does not change, if G_λ is replaced by a conjugate element G_{λ_1}. This way one obtains k subgroups

$$\mathfrak{A}_0 (= E), \mathfrak{A}_1, \ldots, \mathfrak{A}_{k-1}$$

of $\mathfrak{A}$, which correspond to the k classes $(0), (1), \ldots, (k-1)$ of $\mathfrak{H}$, respectively.[39] Furthermore, those elements B of $\mathfrak{A}$ for which we have $\psi^{(\alpha)}(B) = 1$ form a subgroup $\mathfrak{B}_\alpha$ of $\mathfrak{A}$. Let $\mathfrak{D}_{\alpha\rho}$ be the intersection of the groups $\mathfrak{A}_\rho$ and $\mathfrak{B}_\alpha$, and $d_{\alpha\rho}$ the order of $\mathfrak{D}_{\alpha\rho}$. Then, in (27.), if H_λ belongs to the class (ρ), we obtain

$$\sum_\mu \psi^{(\alpha)}(F_{\lambda,\mu}) = \frac{h}{a_\rho h_\rho} \sum_J \psi^{(\alpha)}(J) = \frac{h d_{\alpha\rho}}{h_\rho a_\rho} \sum_{J'} \psi^{(\alpha)}(J')$$

where J ranges over all elements of $\mathfrak{A}_\rho$, and J' over a complete residue system of $\mathfrak{A}_\rho$ mod. $\mathfrak{D}_{\alpha\rho}$. But now, the $\frac{\alpha_\rho}{d_{\alpha\rho}}$ numbers $\psi^{(\alpha)}(J')$ form a character of the group $\frac{\mathfrak{A}_\rho}{\mathfrak{D}_{\alpha\rho}}$, and since $\psi^{(\alpha)}(J')$ can only be 1, if J' belongs to the group $\mathfrak{D}_{\alpha\rho}$, by a well-known theorem we obtain

$$\sum \psi^{(\alpha)}(J') = 0,$$

---

[39] For example, if $\mathfrak{A} = \mathfrak{A}_\rho + \mathfrak{A}_\rho B_2 + \cdots + \mathfrak{A}_\rho B_{\frac{\alpha}{\alpha_\rho}}$, the class (ρ) of $\mathfrak{H}$ corresponds to $\frac{\alpha}{\alpha_\rho}$ different classes in $\mathfrak{G}$, each with $\alpha_\rho h_\rho$ conjugate elements represented by the elements $G_\rho, B_2 G_\rho, \ldots, B_{\frac{\alpha}{\alpha_\rho}} G_\rho$.

except in the case that $d_{\alpha\rho} = \alpha_\rho$, i.e. if $\mathfrak{A}_\rho$ is contained in $\mathfrak{B}_\alpha$. Thus we can put

$$\sum_\mu \psi^{(\alpha)}(F_{\lambda,\mu}) = \frac{h}{h_\rho}\left[\frac{d_{\alpha\rho}}{a_\rho}\right] \tag{28.}$$

where $[x]$, as usual, denotes the greatest integer $\leq x$. As the right hand side of equation (28.) only depends on the class (ρ), which contains the element H_λ, from (27.) we obtain

$$l_\alpha = \sum_{\rho=0}^{k-1}\left[\frac{d_{\alpha\rho}}{a_\rho}\right]$$

i.e. l_α equals the number of classes (ρ) for which $\mathfrak{A}_\rho$ is contained in $\mathfrak{B}_\alpha$.

This result may also be rephrased as follows:

Theorem 6. *If $r_{P,Q}$ is an arbitrary solution of the equations* (A.)*, then among the k classes of conjugate elements in $\mathfrak{H}$ there are $l \geq 1$*[40] *classes with the property that whenever an element P belongs to one of these l classes and Q is an arbitrary element commuting with P, then $r_{P,Q} = r_{Q,P}$. At the same time, l is also the number of different (pairwise non-equivalent)*[41] *primitive representations of $\mathfrak{H}$ corresponding to the factor system $r_{P,Q}$.*

The degrees of a group's characters are, as was proved by *Frobenius* (Primfaktoren, §12), divisors of the group's order. In an exactly analogous way one can prove:

Theorem 7. *If the group $\mathfrak{G}$ of order ah has a subgroup of order a consisting of central elements in $\mathfrak{G}$, then the degree of each character is a divisor not only of ah, but also of h.*

This immediately implies:

Theorem 7a. *The degree of any primitive representation of a group $\mathfrak{H}$ by (ordinary or fractional) linear substitutions is a divisor of the order of $\mathfrak{H}$.*

Here is my proof of Theorem 7:

If we demand for the ah variables x_S, where S ranges over all elements of $\mathfrak{G}$, that for any two elements S and T the equality $x_{ST} = x_{TS}$ holds, then for any two conjugate elements S and S' of the group we have $x_S = x_{S'}$. However, if k' is the number of classes of conjugate elements in $\mathfrak{G}$, and $S, S', S'', \dots$ are k' elements of $\mathfrak{G}$, such that no two of them are conjugate, then the quantities $x_S, x_{S'}, x_{S''}, \dots$ shall be considered independent variables. Then for two arbitrary elements S and T we also have

$$y_{ST}^{(\alpha)} = \sum_\beta \psi^{(\alpha)}(A_\beta)x_{A_\beta ST} = \sum_\beta \psi^{(\alpha)}(A_\beta)x_{A_\beta TS} = y_{TS}^{(\alpha)}.$$

Hence, if H_λ and H_μ are two conjugate elements of $\mathfrak{H}$, and thus by our assumption G_λ and G_μ are conjugate elements of $\mathfrak{G}$, we also obtain $y_{G_\lambda}^{(\alpha)} = y_{G_\mu}^{(\alpha)}$ or $z_{H_\lambda} = z_{H_\mu}$. Now, if $H_\lambda, H_{\lambda_1}, \dots$ are the h_ρ elements of the class (ρ) in $\mathfrak{H}$, denote all the quantities $z_{H_\lambda}, z_{H_{\lambda_1}}, \dots$, which are equal to each other, by z_ρ. Then, if (ρ) is a class, for which $\mathfrak{A}_\rho$ is not contained in $\mathfrak{B}_\alpha$, one can find elements A in $\mathfrak{A}_\rho$, such that $\psi^{(\alpha)}(A) \neq 1$; thus, since G_λ and AG_λ are conjugate elements of $\mathfrak{G}$, by (24.) we have

$$z_\rho = y_{G_\lambda}^{(\alpha)} = y_{AG_\lambda}^{(\alpha)} = \psi^{(\alpha)}(A^{-1})y_{G_\lambda}^{(\alpha)} = 0.$$

---

[40] The principle class (0) always has this property.

[41] However, if the commutator of $\mathfrak{H}$ is not equal to $\mathfrak{H}$, then among these l primitive representations there may be several ones associated to each other.

If, however, (ρ) is one of the l_α classes $(\rho_0) = (0), (\rho_1), \ldots$, for which $\mathfrak{A}_\rho$ is contained in $\mathfrak{B}_\alpha$, then

$$z_\rho = y^{(\alpha)}_{G_\lambda} = a_\rho \sum_B \psi^{(\alpha)}(B) x_{BG_\lambda}, \qquad {\scriptstyle (\rho=\rho_0,\rho_1,\ldots)}$$

(where B ranges over a complete residue system of $\mathfrak{A}$ mod. $\mathfrak{A}_\rho$) is nonzero, because among the $\frac{\alpha}{\alpha_\rho}$ elements BG_λ no two are conjugate, and hence the variables x_{BG_λ} are all different from each other. At the same time we see that the l_α quantities $z_{\rho_0}, z_{\rho_1}, \ldots$, which are non-zero, can be considered to be independent variables. Indeed, these quantities, viewed as functions of the x_S, have no variables in common.

But the prime factors of $Q^{(\alpha)}$ are prime factors of the group determinant of $\mathfrak{G}$, too. Hence, by (*Frobenius*, Primfaktoren, §6), if one puts

$$\xi_\lambda = \frac{1}{f_\lambda} \sum_S \chi^{(\lambda)}(S) x_S$$

then from the assumptions on the variables x_S we obtain

$$\mathbf{\Phi}_\lambda = \xi_\lambda^{f_\lambda}.$$

However, if $\chi^{(\lambda)}(G_\mu)$ is denoted by $\chi^{(\lambda)}_\rho$, whenever H_μ belongs to the class (ρ) in $\mathfrak{H}$, we have

$$\begin{aligned} \xi_\lambda &= \frac{1}{f_\lambda} \sum_{\beta,\mu} \psi^{(\alpha)}(A_\beta) \chi^{(\lambda)}(G_\mu) x_{A_\beta G_\mu} \\ &= \frac{1}{f_\lambda} \sum_\mu \chi^{(\lambda)}(G_\mu) y^{(\alpha)}_{G_\mu} = \frac{1}{f_\lambda} \sum_{\rho=0}^{k-1} h_\rho \chi^{(\lambda)}_\rho z_\rho. \end{aligned}$$

Consider the equation

$$Q^{(\alpha)} = \prod_{\lambda=0}^{l_\alpha} \xi_\lambda^{f_\lambda^2}.$$

Note that (cf. formula (25.)) $Q^{(\alpha)}$ is a polynomial[42)] in the l_α variables $z_{\rho_0}, z_{\rho_1}, \ldots$ whose coefficients are algebraic integers and that $\frac{h_0 \chi^{(\lambda)}_0}{f_\lambda} = 1$. It follows by a well-known theorem that the l_α quantities

$$\frac{h_\rho \chi^{(\lambda)}_\rho}{f_\lambda} \qquad {\scriptstyle (\rho=\rho_0,\rho_1,\ldots)}$$

are algebraic integers. In fact this is the case for all $\rho = 0, 1, \ldots, k-1$ because for the other $k - l_\alpha$ classes we have we have $\chi^{(\lambda)}_\rho = 0$. However, if we denote by $\chi^{(\lambda)}_{\rho'}$ the complex conjugate of $\chi^{(\lambda)}_\rho$, then from (26.) we deduce that

$$\frac{h}{f_\lambda} = \sum_\rho \frac{h_\rho \chi^{(\lambda)}_\rho}{f_\lambda} \chi^{(\lambda)}_{\rho'}$$

42) Translator's note: Schur writes entire rational function [ganze rationale Funktion], which here means polynomial.

is an integral algebraic and also rational number. This gives the theorem. [43) 44)]

§5.

The determination of the multiplier and representation group of a given group $\mathfrak{H}$ following the methods outlined in §§ 1 and 3 leads to very awkward calculations. In the following we derive several theorems, which allow one to bypass these calculations in many cases.

Each representation group $\mathfrak{K}$ of $\mathfrak{H}$ is characterized by the following properties:

1. $\mathfrak{K}$ contains a subgroup $\mathfrak{M}$ consisting of central elements, and $\frac{\mathfrak{K}}{\mathfrak{M}}$ is isomorphic to $\mathfrak{H}$;
2. the commutator of $\mathfrak{K}$ contains all elements of $\mathfrak{M}$;
3. there exists no group $\mathfrak{G}$ satisfying conditions 1. and 2., whose order is greater than that of $\mathfrak{K}$.

The following theorems serve to derive an upper bound on the order m of the multiplier of $\mathfrak{H}$. This is often enough to determine a representation group of $\mathfrak{H}$. For if we know, for example, that m is not greater than a certain integer $\overline{m}$, and if we know of a group $\mathfrak{K}$ of order $\overline{m}h$, which satisfies properties 1. and 2., then we conclude, on the basis of Theorem 2, that $\overline{m} = m$, and that $\mathfrak{K}$ is a representation group of $\mathfrak{H}$.

I begin by making the following observation.

Property 2. implies that for each linear character $\chi(R)$ of $\mathfrak{K}$, we must have $\chi(J) = 1$ as soon as J is an element of $\mathfrak{M}$.

Now let $\mathfrak{K} = \mathfrak{M}Q_0 + \mathfrak{M}Q_1 + \cdots + \mathfrak{M}Q_{h-1}$, where $\mathfrak{M}Q_\lambda$ corresponds to the element H_λ under the isomorphism between $\frac{\mathfrak{K}}{\mathfrak{M}}$ and $\mathfrak{H}$. Let $M_0, M_1, \ldots, M_{m-1}$ be the elements and

$$\psi^{(\mu)}(J) = \psi_{M_\mu}(J) \qquad {\scriptstyle (\mu=0,1,\ldots,m-1)}$$

the m characters of the *Abel*ian group $\mathfrak{M}$.

---

43) Theorem 7 can be proved more simply in the following way, as *Frobenius* was kind enough to point out to me.

Let $(0), (1), \ldots$ be the k' conjugacy classes of elements into which the $ah = g$ elements of the group $\mathfrak{G}$ decompose, with the αth class containing g_α elements, and let (α') denote the class inverse to (α). Let $e_{\alpha,\beta} = 0$ when $\beta \neq \alpha'$, and $e_{\alpha\alpha'} = 1$. Furthermore, let (*Frobenius*, Gruppencharaktere, §4) $\frac{gp_{\alpha,\beta}}{g_\alpha}$ be the number of solutions of the equation $SR = RSAB$, where A is a specified element of the αth conjugacy class, B runs over the g_β elements of the βth conjugacy class traversed, and R and S independently run over all elements of $\mathfrak{G}$. Now (ibid. § 3, (12)), if f denotes the degree of a character of $\mathfrak{G}$, then $|f^2 p_{\alpha,\beta'} - gg_\alpha e_{\alpha\beta'}| = 0$, and hence also $|\frac{g}{g_\alpha} p_{\alpha,\beta'} - \frac{g^2}{f^2} e_{\alpha\beta'}| = 0$. Hence the k' roots $(\frac{g}{f})^2$ of this equation are integers.

If $q_{\alpha\beta}$ denotes the number of solutions of the equation $QP = PQAB$, where P and Q run over the h elements $G_0, G_1, \ldots, G_{h-1}$, then we have $\frac{gp_{\alpha,\beta}}{g_\alpha} = \alpha^2 q_{\alpha\beta}$. Indeed if P, Q is a solution, then one obtains α^2 different solutions of $SR = RSAB$, by letting R take the a values $PA_0, PA_1, \ldots, PA_{\alpha-1}$, and S independently the α elements $QA_0, QA_1, \ldots, QA_{a-1}$. It follows that

$$\left| q_{\alpha,\beta'} - \frac{g^2}{\alpha^2 f^2} e_{\alpha,\beta'} \right| = 0,$$

and thus the k' roots $(\frac{g}{\alpha f})^2 = \left(\frac{h}{f}\right)^2$ of this equation are integers.

44) Translator's note: It is quite hard to follow the argument in this footnote. There is a standard proof (due to Frobenius) that character degrees divide the order of the group, by showing that $\sum RSR^{-1}S^{-1}$ acts as the scalar $(|G|/\chi(1))^2$ in any irreducible representation with character χ. This scalar is both rational and algebraic, and hence is an integer. This is easily adapted to give the strengthening of Theorem 7 above, which looks like what Schur is explaining in the second paragraph.

If one considers any primitive representation of degree f of $\mathfrak{K}$ by ordinary linear transformations corresponding to the character $\psi(J) = \psi_M(J)$ of $\mathfrak{M}$, then if (J) is the matrix corresponding to J, we have $(J) = \psi_M(J) \cdot E$. Now the determinants of the representing matrices provide a linear character of $\mathfrak{K}$. It follows that for each J we must have

$$\psi(J)^f = \psi_{M^f}(J) = 1,$$

and thus $M^f = E$. *Hence f is divisible by the order of the elements of M.* If, for example, the character $\psi(J)$ of $\mathfrak{M}$ corresponds to l characters of the degrees $f_1, f_2, \ldots, f_l$, then according to formula (21.):

$$f_1^2 + f_2^2 + \cdots + f_l^2 = h.$$

It follows that h must be divisible by the square of the order of the element M. As this holds for each element M of $\mathfrak{M}$ we deduce:

Theorem 8. *The square of the largest invariant*[45] *(and hence of any invariant) of the multiplier of $\mathfrak{H}$ divides the order h of $\mathfrak{H}$.*

Thus if only the first power of a prime power p divides h, then m is not divisible by p. *In particular, all groups of square-free order are closed.*

This theorem is a special case of a much more general theorem.

If the elements $H_0, H_1, \ldots, H_{s-1}$ build a subgroup $\mathfrak{S}$ of $\mathfrak{H}$, then the elements contained in the s cosets $\mathfrak{M}Q_0, \mathfrak{M}Q_1, \ldots, \mathfrak{M}Q_{s-1}$ build a subgroup $\mathfrak{S}'$ of order ms of $\mathfrak{K}$, which can be seen as a group $\mathfrak{S}$ augmented by the group $\mathfrak{M}$. Let $\mathfrak{R}'$ be the commutator of $\mathfrak{S}'$ and $\mathfrak{D}$ the intersection of $\mathfrak{M}$ and $\mathfrak{R}'$. Further, let

$$\mathfrak{K} = \mathfrak{S}'A_0 + \mathfrak{S}'A_1 + \cdots + \mathfrak{S}'A_{n-1}$$

so that each element P of $\mathfrak{K}$ can be written uniquely as $S'_\alpha A_\nu$, where S'_α denotes an element of $\mathfrak{S}'$. In this case set $S'_\alpha = S'_P$ and denote the coset

$$\mathfrak{R}' \prod_{\nu=0}^{n-1} S'_{A_\nu P}$$

by $\mathfrak{R}'_P$. Then for every two elements P and Q of $\mathfrak{K}$ we have the relationship $\mathfrak{R}'_P\mathfrak{R}'_Q = \mathfrak{R}'_{PQ}$, as shown in my work "Neuer Beweis eines Satzes über endliche Gruppen"[46]. Furthermore, for each element P of $\mathfrak{K}$ belonging to the commutator of $\mathfrak{K}$ we have $\mathfrak{R}'_P = \mathfrak{R}'$. Now if $P = J$ is an element from $\mathfrak{S}'$ that is contained in $\mathfrak{M}$, then for each ν, $A_\nu J = JA_\nu$, so $S'_{A_\nu J} = J$, from which follows $\mathfrak{R}'_J = \mathfrak{R}'J^n$. But J is also an element of the commutator of $\mathfrak{K}$, and so $\mathfrak{R}'_J = \mathfrak{R}'$. We obtain, for each element J, the equation

$$J^n\mathfrak{R}' = \mathfrak{R}'.$$

In other words, the nth power of each element of $\mathfrak{M}$ lies in $\mathfrak{R}'$, and thus also in $\mathfrak{D}$.

In particular, if the order j of the element J is coprime to n, then $J^j = E$ and J^n are both elements of $\mathfrak{D}$, and hence so is J. Taking into account that $\mathfrak{D}$ is isomorphic to a subgroup of the multiplier of $\mathfrak{S}$ by Theorem 2, we obtain the theorem:

[45] Translator's note: See footnote 25.

[46] Sitzungsberichte der Berliner Academie, 1902, S. 1013

Theorem 9. *Suppose that $\mathfrak{H}$ contains a subgroup $\mathfrak{S}$ of order s and index n and denote by $\mathfrak{M}^{(n)}$ the subgroup of the multiplier $\mathfrak{M}$ of $\mathfrak{H}$ consisting of all elements of order coprime to n, then $\mathfrak{M}^{(n)}$ is isomorphic to a subgroup of the multiplier of $\mathfrak{S}$.*

In the special case when s' is a divisor of h which is coprime to n, then the elements of $\mathfrak{M}$ whose order is divisible by s' build a subgroup of $\mathfrak{M}$ contained in $\mathfrak{M}^{(n)}$. From Theorem 9 we deduce in particular:

Theorem 10. *If p^α is the highest power of a prime p dividing h, and $\mathfrak{S}$ is a subgroup of $\mathfrak{H}$ whose order is divisible by p^α, then the order m of the multiplier of $\mathfrak{H}$ is not divisible by any higher power of p than that of the order m' of the multiplier of $\mathfrak{S}$.*

In order to examine which powers of p divide m, one can first examine the multipliers of the subgroups of $\mathfrak{H}$ whose order is divisible by p^α. In the special case where m' is coprime to p for such a group, then this is also the case for m.

I would like to remark that if s and n are coprime, the groups $\mathfrak{D}$ and $\mathfrak{M}^{(n)}$ considered above must agree. Indeed, $\mathfrak{D}$ is a subgroup of the multipliers of $\mathfrak{S}$, and thus by Theorem 8 the order of each element of $\mathfrak{D}$ is divides s. Thus $\mathfrak{D}$ is in $\mathfrak{M}^{(n)}$ by our assumptions. At the same time $\mathfrak{M}^{(n)}$ is also a subgroup from $\mathfrak{D}$. It follows that indeed $\mathfrak{D} = \mathfrak{M}^{(n)}$.

From Theorem 10 one deduces almost immediately:

Theorem 11. *If a group $\mathfrak{H}$ has the property that all of its subgroups of prime power order are cyclic, then $\mathfrak{H}$ is closed.*

By Theorem 10 it is enough to show that every cyclic group $\mathfrak{C}$ is closed. This is easy to see. Indeed if $\mathfrak{S} = (\mathfrak{A}, \mathfrak{C})$ is an augmented group of $\mathfrak{C}$ by an arbitrary *Abel*ian group $\mathfrak{A}$, then $\mathfrak{S}$ is abelian, because $\mathfrak{A}$ consists of central elements in $\mathfrak{S}$ and $\frac{\mathfrak{S}}{\mathfrak{A}}$ is cyclic. Thus the commutator of $\mathfrak{S}$ is E. This implies that the multiplier of $\mathfrak{C}$ is indeed equal to E.

Translation by Volker Diekert, Ulrich Hertrampf, and Geordie Williamson.

The translators would like to thank Bob Howlett for many detailed comments which improved the translation significantly.

Schur and the birth of projective representations

By *Geordie Williamson* at Sydney

Abstract. We provide a commentary on Schur's paper [19] on the representations of a group via fractional linear translations, which was published in Crelle in 1904. We give some historical context for Schur's work, and discuss the results and methods of the paper in detail. Our aim is to make this beautiful paper more accessible to the modern reader.

1. Introduction

Schur's paper [19], see also the translation [48], on the representations of groups via fractional linear transformations is remarkable in several aspects. It emerged only a few years after Frobenius, Molien, Burnside and Schur began studying the representation theory of finite groups. The concepts of central extension and (Schur) multiplier which it introduces have had a profound impact on group theory. One sees one of the first theorems proved via group presentations, and the first glimmer of cohomological methods. Both ideas would go on to deeply influence 20th century algebra and topology. It is also interesting to study a paper written at the point where the group concept was clearly articulated and understood, but before abstract axiomatic mathematics had become mainstream.

We attempt to provide a motivated historical introduction to Schur's work. We also include a modern discussion of all results of the paper, closely following Schur's line of argument. We warn the reader that the author is not a historian of mathematics, and is far from an expert on this period of early 20th century algebra. On the other hand, even as an amateur the reading of Schur's and other papers of the time is a meaningful and rewarding activity. Our goal is to help the modern mathematician read and achieve a greater appreciation of Schur's work.

This paper is structured as follows: we first discuss the evolution of the group concept through the works of Lagrange, Galois, Cayley, von Dyck and Frobenius. We then discuss the emergence of the theory of group representations in the work of Frobenius and Molien. We also discuss the theory of associative algebras (then called hypercomplex number systems) and the comparatively small impact this had on Frobenius and Schur. We discuss his results in detail, as a companion to the English translation of the paper [50], and point out some historical curiosities along the way. We do not discuss Schur's life; for this the reader is referred to [30,40,41].

The author was supported by Australian Research Council Laureate Fellowship FL230100256.

 | HTTPS://DOI.ORG/10.1515/9783111260440-009

Acknowledgement. J. Avigad, J. Baine, B. Howlett, D. Huybrechts, J-P. Serre (relayed via C. Goldstein and D. Huybrechts) and A. Sherman all provided invaluable feedback on a preliminary version.

2. Historical context

Studying the history of the development of a subject has a fractal nature: the more one examines it, the more complexity and interconnectedness it reveals. Our aim here is to provide some context for Schur's paper, whet the reader's appetite, and to point out the excellent existing literature. Our principal sources when preparing this section were the books of Curtis [30] and Wussing [52] and the articles of Hawkins [33–35].

2.1. The concept of a group. The group concept is surely one of the most useful in mathematics. The three axioms (associativity, unit, and existence of inverses) are close to universally accepted, and it is difficult to anticipate the incredible richness that emerges from these three axioms.[1) 2)] One can view the axiomatization of groups as one of the first successes of the formal approach to mathematics, advocated by Hilbert and others. However it took great effort (and most of the 19th century) for this idea to crystallize.

In [52], Wussing undertakes a detailed study of the emergence of the group concept. His conclusion is that the widespread notion that the group concept arose via abstraction of the notion of a permutation group is simplistic. In fact, group theory had three distinct and equally important historical roots in the 19th century: the theory of algebraic equations, number theory and geometry.

For most of the 19th century, the modern group concept was not explicitly articulated, which makes the tracing of its historical roots all the more difficult. For example, in Galois' work, the concept of a "group" is fluid, becoming more and more precise as his work evolves [52, p. 112]. Throughout the 19th century, it seems that most authors had in mind the object that Burnside defines in his first English-language textbook on group theory [1, Chapter 2], which was published in 1897:

> Let $A, B, C \ldots$ denote a set of operations, which can be performed on the same object, or set of objects. Suppose this set of operations has the following characteristics...

Burnside then goes on to what nowadays appears a rather convoluted discussion of associativity and the existences of inverses. The key point is that a group is a concrete collection of operations, and the modern idea of "abstract group" is entirely absent.

As an example of the impact of a missing definition, the classification theorem for finite abelian groups is implicit in Gauß's Disquisitiones Arithmeticae (see [26, p. 184]) and explicit in Kronecker's 1870 work [14], although without reference to the group concept. In 1879, Frobenius and Stickelberger reprove the same theorem [6], this time with the group concept

---

1) Mathematicians were already aware of this richness at the time of Schur's work. As Hawkins [34, §4] points out, Pierpont [48] wrote in 1904 that "the group concept, hardly noticeable at the beginning of the century, has at its close become one of the fundamental and most fruitful notions in the whole range of our science."

2) Other mathematical concepts like "space" and "function" are arguably more universal, but there is no widely agreed upon single set of axioms describing these concepts [28, 32].

in mind. It is clear from their introduction that they are beginning to be aware of the group concept as a unifying framework.[3)]

2.2. Emergence of the axioms of a group. In a remarkable paper in 1854, Cayley [2] implicitly defines what we would today call an abstract group. He describes the multiplication table and classifies groups of prime order and of order up to 6. Cayley's point of view seems to have been ignored for over two decades, before obtaining general currency. Cayley returned to this theme in 1878 [3], where he articulated his famous dictum: "A group is defined by means of the laws of combination of its symbols."

A turning point in the development of the group concept is provided by Dyck's papers [4, 5]. He defines what we would today call a finitely-presented group. As is apparent when reading the introductions to these papers, Dyck is motivated by an attempt to unify the various appearances of the group concept in number theory, algebra and geometry. Dyck's paper plays a crucial role in Schur's work: in order to construct the Schur covering group, Schur first considers an infinite finitely-generated group and then constructs the covering group as a quotient of this group, which he argues is finite. We thus encounter a compelling early example where infinite groups are used crucially in a question concerning only finite groups.

By the 1880s it is clear that the group concept has reached something close to its modern conceptual form. However, interestingly the axioms used by mathematicians in the 1880s and 1890s still differed from the modern axioms. Here is Frobenius' definition from his famous Crelle paper of 1884 on the Sylow theorems [7]:

(i) Each two elements A and B determine a *unique* third element, that will be denoted by AB.

(ii) From the equation $AC = BC$ or $CA = CB$ follows $A = B$.

(iii) For the operation, which out of A and B produces AB, the associative rule holds $(AB)C = A(BC)$. (However, the commutative law $AB = BA$ does not necessarily hold.)

(iv) The number of elements is finite.

Frobenius then goes on to *prove* the existence and unicity of the unit element, and establish the existence of inverses.

Remark 2.1. It is interesting to trace the emergence of the inverse axiom. All early sources studying finite groups prove the existence of inverses, rather than asserting their existence. Klein states that in his investigations with Lie in the 1870s "Lie found it necessary to require that for each A the group should contain its inverse A^{-1}" [13, Kapitel 8]. Dyck's work alluded to above is notable in that the existence of an inverse is axiomatized (see [4, p. 74]). However, even around 1904 at the time of Schur's article [19] it was still not standard to include the existence of the inverse as part of the definition of a group (see e.g. [12]). In [19], Schur checks the cancellative law (i.e. axiom (ii) above), rather than demonstrate inverses.

The modern mathematician is now used to axioms, and usually doesn't think twice about the axioms of a group. However, it is interesting to reflect on what might be lost in our modern

[3)] In [19], Schur uses the classification of *finitely-generated* abelian groups in a crucial way. Amusingly, he cites [6] which only proves the theorem for *finite* abelian groups.

approach. In this direction, the following opinion of Klein [13] from a lecture delivered around 1915 is thought-provoking. After introducing the modern group axioms, he writes:

> ...there is a complete loss of the appeal to the imagination. On the other hand, the logical skeleton is carefully prepared. ... This abstract formulation is excellent for the working out of proofs but it does not help one find new ideas and methods. Rather, it represents the conclusion of an advance. It therefore seems to ease the learning process, for it helps one prove known theorems in a flawless and simple manner. On the other hand, it makes matters far more difficult for the mind of the student, for he confronts something closed, does not know how one arrives at these definitions, and can imagine absolutely nothing. In general, the disadvantage of the method is that it fails to encourage thought. All one must beware of is that one does not violate the four commandments [i.e. the group axioms].[4)]

2.3. Representations and the character table. Following a detailed analysis of the 19th century literature, Hawkins [34, §5] traces the roots of character theory and representation theory to at least four places: 1) the structure theory of finite-dimensional algebras (Molien, Cartan); 2) the theory of Lie groups and algebras (Cartan, Burnside); 3) Klein's form problem (Maschke); 4) number theory and group determinants (Dedekind, Frobenius). For Schur's paper, it is this last motivation that exerts the strongest influence, and we discuss this first. For a detailed historical treatment the reader is referred to the papers of Hawkins [33, 35] and for a more mathematically driven discussion to [30, 46].

The notion of a character of an *abelian* group emerged naturally at the beginning of the 19th century in Gauß's work on quadratic forms, and on Dirichlet's work on primes in arithmetic progressions. Dedekind played a substantial role in advancing it as a unifying concept [36, §2].

Around 1880, motivated by number-theoretic questions, Dedekind began studying the *group determinant*, which he defined as follows. For a finite group G, consider commuting variables x_g indexed by elements of g, and consider the matrix

$$(x_{gh^{-1}})_{g,h\in G} \tag{2.1}$$

for some fixed ordering $g_1, g_2, g_3, \ldots$ of its elements. (Thus, this matrix is, up to reordering of the columns, simply an encoding of the multiplication table of G.) The group determinant is its determinant:

$$\Phi_G = \det((x_{gh^{-1}})_{g,h\in G}) \in \mathbb{C}[x_g \mid g \in G].$$

Dedekind was able to prove that if G is abelian then Φ_G factors into linear factors, given by the characters of G:

$$\Phi_G = \prod_{\chi\in\hat{G}} (\sum_{g\in G} \chi(g)x_g).$$

In two letters to Frobenius in 1896, Dedekind defined the group determinant, provided the factorization for several small non-abelian groups, and stated some conjectures.

Motivated by Dedekind's questions, Frobenius was able to completely settle the matter in two papers [8, 9] which appeared at the end of 1896.[5)] He proved that for any finite group

---

[4)] Translated by Wussing [52, p. 228].

[5)] Frobenius' struggle towards these theorems in the second half of 1896 is vividly recounted, based on letters exchanged between Dedekind and Frobenius, in [33, 35].

one has a factorization

$$\Phi_G = \prod_{\chi \in \hat{G}} (\Phi_\chi)^{d_\chi} \tag{2.2}$$

indexed by the characters $\hat{G}$ of G, where Φ_χ is an irreducible polynomial of degree d_χ. He also proved that if one sets $x_g = x_{g'}$ whenever g and g' are conjugate in G, then each irreducible polynomial may be factored into linear factors

$$\Phi_\chi = \left(\sum_{C \in cl(G)} \frac{|C|}{d_\chi} \chi(C) x_C \right)^{d_\chi}, \tag{2.3}$$

where $cl(G)$ denotes the set of conjugacy classes in G, $\chi(C)$ denotes the value of χ on such a conjugacy class, and $x_C = x_g$ for any $g \in C$.

Here we are being somewhat ahistorical: we assume the reader knows what a character is, and use them to factor the group determinant. In Frobenius' work in 1896 the theory proceeded the other way around—the characters of a finite group G are defined so as to enable the factorization (2.3). Thus the group determinant is central to early treatments. The following year Frobenius established [10] that his characters were indeed the traces of elements of g in the irreducible representations of G, recovering the modern definition.

As Frobenius' student, Schur was deeply familiar with the group determinant, and the connection between the characters and the irreducible representations of G. The group determinant plays an important role in [19]. Frobenius' arguments are very complicated, and in [20] Schur achieved a major simplification in 1902 which became a standard reference. It is this paper that introduces Schur's lemma [20, Satz I].

2.4. Hypercomplex number systems. The modern reader should have no trouble translating both Dedekind's question and Frobenius' answer into modern language. Indeed, if $\mathbb{C}G$ denotes the group algebra, then (2.1) is simply the matrix of the left action of a "general element"

$$\gamma = \sum_{g \in G} x_g g$$

in the regular representation. Basic representation theory provides the decomposition

$$\mathbb{C}[G] = \bigoplus_{\chi \in \hat{G}} V_\chi^{\oplus d_\chi}$$

and (2.2) is immediate.[6] Moreover, for any conjugacy class C in G the element $\sum_{g \in C} g$ is central, and hence acts as a scalar c_χ in any irreducible representation, by Schur's lemma. Taking traces yields the relation

$$c_\chi = \frac{|C|}{d_\chi} \chi(C),$$

which implies Frobenius' linear factorization (2.3).

6) With the exception of the fact that each Φ_χ is irreducible, which follows with a little more thought. For a modern proof see [46].

It is the absence of the group algebra that is most striking to the modern reader of Frobenius and Schur. At the time, associative algebras were known as hypercomplex number systems, and there was considerable interest in these structures from several mathematicians [35, §1]. It is noteworthy that in his letters, Dedekind repeatedly asked Frobenius whether his theory can be interpreted in terms of hypercomplex number systems [33].

Remarkably, at almost the same time as Frobenius was proving his theorems on the factorization of the group determinant, Molien established many of the same theorems using the theory of hypercomplex number systems developed in his thesis. Molien appears to have been deeply influenced by Killing's work on Lie algebras, as well as Klein's work. He realized that finite groups provide a rich class of hypercomplex number systems via their group algebras. He was able to use a result from his thesis [15] to deduce that group algebras were isomorphic to a direct product of matrix rings, and deduce the fundamental theorems of group representation theory [16, 17]: that the number of irreducible representations agrees with the number of conjugacy classes, the fundamental orthogonality relations for characters, and that the character degrees divide the order of the group [34, §4].

Throughout the early development of the subject, Frobenius was aware of the theory of hypercomplex number systems and this had some impact on his work [35, § 6]. In a letter to Dedekind in 1896 he wrote:

> Your ideas involve thoughts about which I presently have a very obscure notion. The theory of groups is concerned with only one operation which obeys certain rules. In this respect, it reminds one of tightrope-walking, where the feet have only one dimension at their disposal. Many times I have thought about how, to the multiplication of elements, one could adjoin another operation (addition).[7)]

When Frobenius became aware of Molien's work he was quick to acknowledge its importance (see [12]). The group algebra and module point of view seems never to have been appealing to Schur [42].[8)] We owe its prominence today largely to the work of E. Noether [53].

3. Discussion of the paper

We now turn to Schur's paper in detail. We discuss its contents, using modern notation. Our aim is to help the reader follow what Schur is doing as efficiently as possible.

3.1. On the introduction to Schur's paper. Throughout the second half of the 19th century the classification of permutation groups on n letters for small values of n was a well-known problem. It appears that the corresponding problem for matrix groups (i.e. the determination of finite subgroups of $GL_n(\mathbb{C})$ for small values of n) received much less attention. Famously, Burnside wrote in his introduction to [1]:

> It may then be asked why, in a book which professes to leave all applications on one side, a considerable space is devoted to substitution groups; while other particular modes of representation, such as groups of linear transformations, are not even referred to. My answer to this question is that while, in the present state of our knowledge, many results in the pure theory are arrived at most readily

7) See [35, § 6], translation due to Hawkins.

8) See [41, § 5] for an interesting anecdote.

> by dealing with properties of substitution groups, it would be difficult to find a result that could be most directly obtained by the consideration of groups of linear transformations.[9)]

In contrast, Klein emphasized the importance of this problem in the 1880s and 1890s [34, §4]. A well-known result of Jordan stated that this number is finite.

Schur opens his paper with a discussion of the question of the finite subgroups of $GL_n(\mathbb{C})$, and its apparent difficulty. He points out that representation theory furnishes a sort of converse to this problem—namely, given a finite group G, its representations provide all ways in which G can be realized as a matrix group. He notes that the theory of Frobenius and Molien provides a solution to this problem. (Interestingly, solving this problem seems to have been a primary motivation for Molien [34, §4], whereas for Frobenius this consequence appears as a happy accident.) Schur then goes on to address the problem studied in this work, namely the classification of representations of a fixed finite group via fractional linear transformations. (Thus, rather than classifying realizations inside $GL_n(\mathbb{C})$, Schur aims at classifying realizations inside the projective general linear group $PGL_n(\mathbb{C})$.)

He then observes that a representation by fractional linear transformations of a finite group G is the same thing as an assignment $x \mapsto A_x$ of matrices to each element of G so that for all $x, y \in G$ the relation $A_x A_y = r_{x,y} A_{xy}$ holds, for some complex numbers $r_{x,y}$. In the body of the paper the numbers $r_{x,y}$ play a central role. At this point Schur simply points out that representations correspond to such systems of numbers, and that if all $r_{x,y} = 1$ then one recovers the theory of representations via linear transformations. He introduces crucial basic notions like degree [Grad], equivalence [Äquivalenz] and irreducibility [which Schur calls Primitivität, following Frobenius], exactly as in the case of "ordinary" representations.

He also points out the crucial role played by central extensions. Namely, consider a short exact sequence

$$A \hookrightarrow G \twoheadrightarrow H$$

where A is a central subgroup of G. By Schur's lemma, A acts by scalars on any irreducible representation of G, and hence furnishes an irreducible projective representation of H. This tight connection between central extensions and projective representations is a major theme of the paper.

Schur then summarizes the main results of the paper:

(i) For any finite group H, there exists a central extension as above such that all projective representations of H lift to ordinary representations of G. Schur calls such a group sufficiently augmented [hinreichend ergänzt]. Such a group of minimal order is called a representation group [Darstellungsgruppe]. Nowadays, such groups are called *Schur covering groups*.

(ii) In general, for each H there are several non-isomorphic covering groups G, however they all have the same order and the kernel of the surjection $G \twoheadrightarrow H$ provides a finite abelian group, which depends only on H. Schur terms this group the multiplier [Multiplikator]. Nowadays, this is called the *Schur multiplier* of H and may be identified with the second homology group $\mathrm{H}_2(H, \mathbb{Z})$.

[9)] Burnside writes in the preface to the second edition (published in 1911) that "... the theory of groups of linear substitutions has been the subject of numerous and important investigations by several writers; and the reason given in the original preface for omitting any account of it no longer holds good. In fact, it is now more true to say that for further advances in the abstract theory one must look largely to the representation of a group as a group of linear substitutions."

Schur points the reader to a number of results at the end of the paper which often allow the efficient determination of the multiplier of a finite group. He promises further theory and computations for specific groups in future works, which include [21,22].

Remark 3.1. The reader may notice that several of Schur's early papers (including this one) are signed "J. Schur". An explanation is given in [43, pp. i-li] quoting Alfred Brauer:[10] "In the first of these papers Schur's first name had been abbreviated wrongly and Schur considered it proper not to change this for the subsequent papers in the journal. But in fact he always attached great importance to being quoted as I. Schur. Nevertheless, even today his first name is sometimes wrongly abbreviated."

3.2. Interlude concerning group cohomology. In the following discussion, it will be useful to have the modern terminology of group cohomology at our disposal. Given a discrete group G its cohomology $\mathrm{H}^*(G, \mathbb{C}^*)$ is defined to be the cohomology of the complex

$$C^n(G) = \{f \colon G^n \to \mathbb{C}^*\}$$

(where $C^0(G)$ is understood to be maps $f : \{1\} \to \mathbb{C}^*$) and differential

$$d^n f(g_1, \dots, g_{n+1}) = \prod_{i=0}^{n+1} f(g_1, \dots, g_i g_{i+1}, \dots, g_{n+1})^{(-1)^i}$$

where, by convention, the 0th and final terms are $f(g_2, \dots, g_{n+1})$ and $f(g_1, \dots, g_n)^{(-1)^{n+1}}$.

One checks easily that d^0 is trivial, and that a 1-cocycle is a map $f : G \to \mathbb{C}^*$ satisfying

$$1 = f(g_1) f(g_1 g_2)^{-1} f(g_2) \quad \text{for all } g_1, g_2 \in G$$

or in other words

$$f(g_1 g_2) = f(g_1) f(g_2).$$

Thus $\mathrm{H}^0(G, \mathbb{C}^*) = \mathbb{C}^*$ and $\mathrm{H}^1(G, \mathbb{C}^*) = G^\vee$ is the character group of G. Now, 2-cocycles are maps satisfying

$$1 = f(g_2, g_3) f(g_1 g_2, g_3)^{-1} f(g_1, g_2 g_3) f(g_1, g_2)^{-1} \quad \text{for all } g_1, g_2, g_3 \in G,$$

or in other words,

$$f(g_1, g_2) f(g_1 g_2, g_3) = f(g_2, g_3) f(g_1, g_2 g_3) \quad \text{for all } g_1, g_2, g_3 \in G.$$

Such a two cocycle is cohomologous to zero if it is of the form $\frac{f(g_1)f(g_2)}{f(g_1 g_2)}$ for some $f : G \to \mathbb{C}^*$. In a similar way one defines homology groups $\mathrm{H}_*(G, \mathbb{Z})$ and one has

$$\mathrm{H}^i(G, \mathbb{C}^*) = \mathrm{H}_i(G, \mathbb{Z})^\vee$$

where $(-)^\vee$ denotes the dual abelian group.

Remark 3.2. Group cohomology is a big subject with connections to many areas of mathematics. For detailed treatments, see [31, Chapter 15] and [24,25,27]. For an inspiring history of the development of group cohomology see [47] (see also Remark 3.6 below).

[10] Alfred Brauer served as Schur's assistant for many years in Berlin, and is the brother of the famous group theorist Richard Brauer.

3.3. Schur's Section 1. Schur begins with an assignment $x \mapsto A_x$ of matrices to the elements $x \in H$, "belonging to the number system $r_{x,y}$", i.e.

$$A_x A_y = r_{x,y} A_{xy} \quad \text{for all } x, y \in H. \tag{3.1}$$

By expanding the associative law, he deduces that $r_{x,y}$ is a 2-cocycle.

He then proceeds to the first theorem of the paper, which states that there exist projective representations associated to any 2-cocycle. He proves this by producing explicit $|H| \times |H|$-matrices, satisfying (3.1). Let us give an alternative explanation using the group algebra $\mathbb{C}H = \bigoplus_{z \in H} \mathbb{C}z$ which is close to Schur's argument. Consider the operator of left multiplication by x, which in the basis of group elements is given by

$$L_x \colon z \mapsto xz.$$

(When written in a basis, we get a permutation matrix of size $|H|$.) We can twist this operator using our scalars $r_{x,y}$:

$$L'_x \colon z \mapsto r_{x,z} xz.$$

(When written in a basis, we get a matrix with one non-zero entry per row and column, with entries determined by our 2-cocycle $r_{x,z}$.) Now we easily calculate, using the 2-cocycle condition

$$L'_x L'_y(z) = L'_x(r_{y,z} \cdot yz) = r_{x,yz} r_{y,z} \cdot xyz = r_{x,y} r_{xy,z} \cdot xyz = r_{x,y} \cdot L'_{xy}(z).$$

Or in other words, the operators L'_x satisfy the relations

$$L'_x L'_y = r_{x,y} L'_{xy} \quad \text{for all } x, y \in H. \tag{3.2}$$

These are Schur's matrices satisfying (3.1).

Remark 3.3. One can consider an algebra $\mathbb{C}^r H$ with generators corresponding to the elements of the group, and multiplication determined by $g \cdot h = r_{g,h} gh$. This object is usually called the *twisted group algebra*, and representations of $\mathbb{C}^r H$ are exactly projective representations of H associated to the 2-cocycle r. For more on twisted group algebras and their relations to projective representations, see [29, 39, 44, 45].

Schur then points out that if we replace our matrices A_x above by $A'_x = c_x A_x$ for some scalars $c_x \in \mathbb{C}^\times$ then our 2-cocycle above is transformed into a new 2-cocycle

$$r'_{x,y} = \frac{c_x c_y}{c_{xy}} r_{x,y}.$$

Schur calls such 2-cocycles associated [assoziiert]. We saw in the previous section that this is precisely the condition for $r_{x,y}$ and $r'_{x,y}$ be cohomologous, i.e. represent the same element in $\mathrm{H}^2(H, \mathbb{C}^*)$.

He then proceeds to argue that $\mathrm{H}^2(H, \mathbb{C}^*)$ is finite. To this end, let d_x denote the determinant of L'_x. Taking determinants on both sides of (3.2) yields

$$r_{x,y}^{|H|} = \frac{d_x d_y}{d_{xy}}.$$

Thus, if we replace $r_{x,y}$ by the cohomologous 2-cocycle

$$s_{x,y} = \frac{\delta_x^{-1}\delta_y^{-1}}{\delta_{xy}^{-1}} r_{x,y} \quad \text{for all } x, y \in H$$

where δ_x are fixed $|H|$-roots of d_x, then

$$s_{x,y}^{|H|} = 1 \quad \text{for all } x, y \in H.$$

In other words, any cocycle is cohomologous to one in which all values are $|H|^{th}$ roots of unity. Finiteness of $\mathrm{H}^2(H, \mathbb{C}^*)$ follows. It also follows that $\mathrm{H}^2(H, \mathbb{C}^*)$ is of exponent dividing $|H|$.

Having already defined the (Schur) multiplier in the introduction, Schur then redefines the multiplier to be the group of 2-cocycles modulo the equivalence relation of being associated, i.e. $\mathrm{H}^2(H, \mathbb{C}^*)$. Nowadays, authors typically reserve the term multiplier for the group $\mathrm{H}_2(H, \mathbb{C}^*)$ dual to $\mathrm{H}^2(H, \mathbb{C}^*)$. This agrees with Schur's definition in the introduction (see e.g. [49, § 11.4]). Schur does not appear to distinguish a finite abelian group from its dual, possibly because he is only concerned with these groups up to isomorphism.

Remark 3.4. Interestingly, it is only at the end of this section that Schur shows that cycles up to equivalence form an abelian group. Rather than demonstrate inverses, he shows the "cancellative property": that an identity $K_\alpha K_\beta = K_\alpha K_\gamma$ implies that $K_\beta = K_\gamma$. Thus he seems to have the axioms stated by Frobenius (discussed in §2.2) in mind.

3.4. Schur's Section 2. In this section, Schur studies central extensions and the projective representations that they give rise to. Consider a central extension

$$A \hookrightarrow G \overset{p}{\twoheadrightarrow} H. \tag{3.3}$$

Schur chooses a set-theoretic section of p

$$s\colon H \to G\colon x \mapsto s_x$$

(i.e. $p \circ s = id$). Because (3.3) is exact, we can find unique $a_{x,y} \in A$ such that

$$s_x s_y = a_{x,y} s_{xy} \quad \text{for all } x, y \in H. \tag{3.4}$$

Schur observes that in any irreducible representation D of G on a complex vector space, any element $a \in A$ necessarily acts by a scalar $\psi(a) \in \mathbb{C}^*$, and we obtain in this way a character ψ of A. (Indeed, A is central and hence $\rho(A)$ is scalar by Schur's lemma.) In both Schur's and modern language, we say that D *affords* the central character ψ. On the other hand, for any character ψ of A there exists a representation of G affording ψ. (Schur deduces this result from Frobenius [11, §7]. The modern reader can convince themselves of this fact by observing that all characters occur when the regular representation of G is decomposed according to characters of A.) In any such representation, we obtain a 2-cocycle via

$$r_{x,y} = \psi(a_{x,y}).$$

As Schur points out in a footnote, one can expand (3.4) to deduce directly that $\psi(a_{x,y})$ yields a 2-cocycle. In particular, the 2-cocycle does not depend on the irreducible representation D affording ψ.

Thus, we obtain a homomorphism

$$m\colon A^\vee \to \mathrm{H}^2(H, \mathbb{C}^*),$$

where $A^\vee$ denotes the character group of A. (Recall that $\mathrm{H}^2(H, \mathbb{C}^*)$ is the finite abelian group dual to the Schur multiplier.)

Lemma 3.1. The kernel of m consists of those characters of A which extend to characters of G.

Proof. The image of ψ in $\mathrm{H}^2(H, \mathbb{C}^*)$ is trivial if and only if there exists $\varphi : H \to \mathbb{C}^*$ such that

$$\psi(r_{x,y}) = \frac{\varphi(x)\varphi(y)}{\varphi(xy)},$$

which is easily seen to be equivalent to the fact that

$$\psi\colon G \to \mathbb{C}^*\colon as_x \mapsto \psi(a)\varphi(x) \quad \text{for } a \in A,\, x \in H$$

is a one-dimensional character of G.[11)] □

Now, one-dimensional characters of G are the same thing as characters of $G/[G,G]$, and thus a character of A extends to a character of G if and only if it it factors over $A \to G/[G,G]$ if and only if it is trivial on $A \cap [G,G]$. Hence we obtain an injection

$$m'\colon (A \cap [G,G]))^\vee \hookrightarrow \mathrm{H}^2(H, \mathbb{C}^*),$$

which is Schur's Theorem 2. Schur also points out that if $A \cap [G,G]$ is of cardinality equal to that of $\mathrm{H}^2(H, \mathbb{C}^*)$, then m' must be an isomorphism. He is able to deduce that in this case every projective representation lifts to G. Thus, in this case G is a sufficiently augmented [hinreichend ergänzte] group.

Remark 3.5. A few remarks of a historical nature concerning this section:

(i) The fact that one-dimensional characters are precisely those that factor over $G/[G,G]$ plays an important role, and appeared early in the theory, as such characters account for the linear factors of the group determinant. (This was conjectured to hold in Dedekind's second letter to Frobenius.)

(ii) The modern distinction between an abelian group and its dual is not made by Schur. He points out in [19, § 1(6)] that one can always choose an isomorphism between an abelian group A and group of characters, and uses it when needed. This lack of distinction between A and $A^\vee$ makes some of the proofs (particularly of Schur's version of Lemma 3.1 above, which occurs just before his proof of Satz II) difficult to read.

11) One-dimensional characters are often called "linear characters" in the literature. However this terminology can lead to confusion, and we avoid it here.

3.5. An overview of Schur's Section 3. From a modern point of view this is probably the most surprising section of the paper. We first give a high-level overview before discussing the details.

Given a finite group H, Schur's goal is to establish the existence of a "Schur cover". In other words a central extension

$$A \hookrightarrow G \overset{p}{\twoheadrightarrow} H,$$

such that (a) every projective representation of H lifts to an ordinary representation of G and (b) the order of G is minimal with this property.

Why is it surprising that Schur is able to construct G? For one, there were limited means at the time of constructing new groups from old. Secondly, Schur was aware that in general G is not unique, and this ambiguity must arise somehow in the construction of G.

Schur proceeds as follows. He considers a free group F on generators corresponding to the elements of H, and considers the corresponding presentation of H

$$R \hookrightarrow F \twoheadrightarrow H.$$

Note that R is finitely generated. He then forces each element of R to be central, by quotienting by the commutator subgroup $[F, R] \subset F$. The result is an extension

$$N' \hookrightarrow K' \twoheadrightarrow H$$

with N' finitely generated and central (in particular abelian). Schur then *chooses* a free subgroup $Z \subset N'$ such that N'/Z is isomorphic to the torsion subgroup of N'. The resulting quotient

$$N \hookrightarrow K \twoheadrightarrow H$$

is the sought-after Schur cover of H. Thus the two challenges discussed above are addressed as follows:

(i) Schur uses *infinite* groups and presentations to construct the *finite* group G. He relies crucially on the influential treatment of von Dyck [4,5] from 1882, which first introduced the concept of a finitely-presented group.

(ii) Given a finitely-generated abelian group A the torsion subgroup is canonical but in general there are many possible free subgroups $A' \subset A$ such that A/A' is isomorphic to the torsion subgroup. The non-uniqueness of the Schur cover arises from different possible choices of such an A'.

Remark 3.6. Hopf's paper [37] from 1942 is widely regarded as ushering in the study of group cohomology. For any group H, Hopf defines its second homology to be

$$N' = R \cap [F, F]/[F, R] = N' \cap [K', K'],$$

which—as we will see below—is exactly Schur's construction! This formula for the second homology is often called "Hopf's formula" (e.g. [49, § 11.4.15], [27, II,§ 5]). As Mac Lane [47, §2] points out, "Hopf, while a student in Berlin, had been an assistant to Schur, but his 1942 paper does not mention the connection with the multiplicator."

3.6. Nuts and Bolts of Schur's Section 3. We now explain how Schur goes about proving that the construction outlined in the previous section furnishes the Schur cover of H. Schur's proof is beautiful, and has some advantages over modern treatments (e.g. [38, §23]), and so we give some detail.

Consider the central extension

$$N' \hookrightarrow K' \twoheadrightarrow H$$

considered in the previous section. All three groups are finitely generated and N' is central. Let

$$N'_{\text{tors}} \hookrightarrow N' \twoheadrightarrow N'_{\text{free}}$$

denote the torsion subgroup and free quotient of N'. The crucial statements are:

$$N'_{\text{free}} \text{ is of rank } |H|, \tag{3.5}$$

$$(N'_{\text{tors}})^\vee \cong \mathrm{H}^2(H, \mathbb{C}^*), \tag{3.6}$$

$$[K', K'] \cap N' = N'_{\text{tors}}. \tag{3.7}$$

(Note that (3.7) implies the non-trivial fact that the commutator subgroup $[K', K']$ is finite.)

Indeed, if we know (3.6), then any choice of free subgroup $Z \subset N'$ such that N'/Z is isomorphic to the torsion subgroup will provide (by quotienting) a central extension

$$N \hookrightarrow K \twoheadrightarrow H$$

such that $N \subset [K, K]$ and $|N| = |\,\mathrm{H}^2(H, \mathbb{C}^*)|$ which then must be a Schur covering group, by facts established in § 2.[12)]

Instead of working with the abelian group N' in question directly, Schur works with "arbitrary assignments of values satisfying the relations". (For an example, see the statement of the classification of finitely-generated abelian groups, part way into § 3.) This makes parts very difficult to follow, however this difficulty is only a question of language. For the modern reader we can interpret what Schur is saying as follows: We begin with a finitely-generated abelian group M and consider group homomorphisms

$$X_M = \mathrm{Hom}(M, \mathbb{C}^*).$$

This is an algebraic group in a natural way (see e.g. [51, § 3.2]). Schur's arguments are easily interpreted as algebro-geometric statements about groups of the form X_M. It will be important below to note that elements of M naturally provide functions on X_M. We will use the language of algebraic groups in what follows, as it makes Schur's arguments rather transparent.

Recall that F denotes a free group on generators corresponding to H. Let us write q_h for $h \in H$ for these generators. Then Schur's group K' is generated by q_h for $h \in H$ subject only to the condition that the $|H|^2$ elements

$$j_{x,y} := q_x q_y q_{xy}^{-1} \quad \text{for } x, y \in H, \tag{3.8}$$

which generate the kernel of the canonical map $K' \to H$, are central. The associativity relation implies that the elements $j_{g,h}$ satisfy the 2-cocycle condition

$$j_{x,y} j_{xy,z} = j_{x,yz} j_{y,z} \quad \text{for all } x, y, z \in H. \tag{3.9}$$

12) Actually, Schur proceeds slightly differently. He establishes (3.5) and (3.6), and then uses the natural isomorphism in (3.6) to conclude that K is a Schur covering group, before moving on to (3.7).

Schur argues that (3.9) is in fact a presentation of N'. It follows that $X_{N'}$ is nothing other than the group of 2-cocycles of H. Let us define

$$Y_H = \{r \colon H \to \mathbb{C}^*\}$$

to be all maps of sets from H to $\mathbb{C}^*$, which is an algebraic group in a natural way (i.e. $Y_H \cong (\mathbb{C}^*)^H$). We have a natural map

$$d : Y_H \to X_{N'}$$

that associates to each 1-cocycle its coboundary:

$$j_{x,y} = \frac{r(x)r(y)}{r(xy)}.$$

As we saw above, $X_{K'}$ is the set of all 2-cocycles, and $d(X_H)$ consists of all coboundaries. Thus

$$X_{N'}/d(Y_H) = \mathrm{H}^2(H, \mathbb{C}^*).$$

Finally, Schur argues that Y_H and $X_{N'}$ have the same dimension $|H|$.

In summary, we have a long exact sequence, in which the first two terms both have dimension $|H|$: [13)]

$$Y_H \xrightarrow{d} X_{N'} \longrightarrow \mathrm{H}^2(H, \mathbb{C}^*) \longrightarrow \{1\}. \tag{3.10}$$

Dualizing (3.10) (i.e. passing back to finitely-generated abelian groups) yields a short exact sequence of finitely-generated abelian groups

$$0 \to \mathrm{H}_2(H, \mathbb{C}^*) \to N' \to \mathbb{Z}^H.$$

Thus $N'/\,\mathrm{H}_2(H, \mathbb{C}^*)$ embeds inside the free module $\mathbb{Z}^H$, from which it follows that $\mathrm{H}_2(H, \mathbb{C}^*)$ is isomorphic to the torsion subgroup of N'. It also follows that N' and $\mathbb{Z}^H$ have the same rank $|H|$. This establishes (3.5) and (3.6).

To establish (3.7), Schur notes that $g \in K'$ belongs to $[K', K']$ if and only if $\chi(g) = 1$ for any one-dimensional character of K'. By construction, one-dimensional characters of K' are simply arbitrary assignments of values to each q_h. In other words, we have a canonical identification

$$Y_H = \text{one-dimensional characters of } K'.$$

Moreover, given such a one-dimensional character $\chi : q_h \mapsto c_h$, its value on $j_{x,y}$ is, by (3.8),

$$\chi(j_{x,y}) \mapsto \frac{c_x c_y}{c_{xy}}.$$

In other words, the map from one-dimensional characters of K' to characters of N' is nothing other than the differential d above. Thus (3.7) follows by following a sequence of beautiful tautologies!

Schur then points out that although the central extension

$$N \hookrightarrow K \twoheadrightarrow H$$

[13)] Of course, the kernel of the first map in (3.10) consists of the one-dimensional characters of H, also known as $\mathrm{H}^1(H, \mathbb{C}^*)$.

depends on choices, the induced central extension

$$N \hookrightarrow [K,K] \twoheadrightarrow [H,H]$$

does not. He concludes that the commutator subgroups of any two Schur covering groups agree, and that the Schur cover is unique up to isomorphism if H is perfect (i.e. $H = [H,H]$).

Remark 3.7. A well-known result of Schur states that if the centre of an infinite group G is of finite index, then the derived subgroup $[G,G]$ is finite. Moreover, it is common to cite [19], although this statement is not explicit in the paper. Most modern proofs use the transfer map (see e.g. [49, § 10.1.4] or [38, Satz 2.3]). However, one may derive a different proof from this section. Consider the central extension (with G potentially infinite)

$$Z(G) \hookrightarrow G \twoheadrightarrow H = G/Z(G)$$

and consider the subgroup G' of G generated by a fixed choice $g_0, g_1, \ldots, g_{h-1}$ of preimages of elements of $G/Z(G)$. As Schur points out in this section, every commutator in G is a commutator of these elements, and hence $[G,G] \subset G'$. Now, G' is a homomorphic image of K' discussed above, where we have already seen that there the commutator subgroup is finite. It is clear (in particular from the line that cites von Dyck and Burnside) that Schur is aware of the possibility of reasoning along these lines.

3.7. Schur's Section 4. In this section, Schur studies the question of determining all irrreducible projective representations associated to a fixed 2-cocycle $r_{x,y}$. If $A \hookrightarrow G \twoheadrightarrow H$ denotes a Schur cover, then Schur first points out that irreducible projective representations of H corresponding to $r_{x,y}$ can be obtained via the group determinant of G, using results of Frobenius. (In modern language, one may see this easily as a consequence of the fact that the twisted group algebra $\mathbb{C}^r H$ of H can be realized as a direct summand inside the group algebra of G. This also explains the connection which Schur makes with the matrices considered in Schur's § 1.)

Schur then goes on to count the number of irreducible projective representations corresponding to r in terms of the conjugacy classes of H satisfying the condition that $r_{x,y} = r_{y,x}$ whenever x belongs to the centralizer of y. The argument here is somewhat involved; modern readers can convince themselves of this theorem by computing the centre of the twisted group algebra (see [29, §1, Remark 4]). Schur finishes this section by proving that the dimension of any irreducible projective representation divides the order of $|H|$. The argument is similar to Frobenius' classic argument for ordinary representations, which is still used today.

3.8. Schur's Section 5. In this section, Schur obtains several bounds on the cardinality m of the multiplier of H. Indeed, as Schur points out, one often has a candidate for the Schur multiplier, and knowing good bounds can sometimes allow one to deduce equality. The techniques of this section are standard, and a modern reader should have little difficulty in following the original paper. It is perhaps worth noting that the proof of Theorem 10 exploits the transfer map, which was introduced by Schur in [18], was rediscovered by Artin in 1929 (see [23, pp. 159–164]) and went on to play an important role in number theory and topology.

References

Archival sources

[1] *W. Burnside*, Theory of groups of finite order, Cambridge University Press. xvi + 388 pp. (1897).
[2] *A. Cayley*, On the theory of groups, as depending on the symbolic equation $\theta^n = 1$, The London, Edinburgh, and Dublin Philosophical Magazine and Journal of Science, 7:42, 1854.
[3] *A. Cayley* Desiderata and Suggestions: No. 1. The Theory of Groups, American Journal of Mathematics, 1878, Vol. 1, No. 1, pp. 50–52.
[4] *W. v. Dyck*, Gruppentheoretische Studien, Math. Ann. 20 (1882), 1–44.
[5] *W. v. Dyck*, Gruppentheoretische Studien. II. Ueber die Zusammensetzung einer Gruppe discreter Operationen, über ihre Primitivität und Transitivität. Math. Ann. 22 (1883), 70–108.
[6] *G. Frobenius, L. Stickelberger*, Ueber Gruppen von vertauschbaren Elementen, J. reine angew. Math. 86 (1879), 217–262.
[7] *G. Frobenius*, Neuer Beweis des Sylowschen Satzes, J. reine angew. Math. 100 (1887), 179.
[8] *G. Frobenius*, Ueber Gruppencharaktere, Berl. Ber. 1896, 985–1021 (1896).
[9] *G. Frobenius*, Ueber die Primfactoren der Gruppendeterminante, Berl. Ber. 1896, 1343–1382 (1896).
[10] *G. Frobenius*, Ueber die Darstellung der endlichen Gruppen durch lineare Substitutionen, Berl. Ber. 1897, 994–1015 (1897).
[11] *G. Frobenius*, Ueber die Darstellung der endlichen Gruppen durch lineare Substitutionen. II, Berl. Ber. 1899, 482–500 (1899).
[12] *G. Frobenius*, Theorie der hyperkomplexen Größen I. II., Berl. Ber. 1903, 504–537, 634–645 (1903).
[13] *F. Klein*, Vorlesungen über die Entwicklung der Mathematik im 19. Jahrhundert, Vol. 1. Ed. R. Courant and O. Neugebauer, Berlin 1926.
[14] *L. Kronecker*, Auseinandersetzung einiger Eigenschaften der Klassenanzahl idealer complexer Zahlen, MB. Akad. Berlin 1870. WA: [420], pp. 271–282.
[15] *T. Molien*, Ueber Systeme höherer complexer Zahlen, Math. Ann. XLI, 83–156 (1893).
[16] *T. Molien*, Eine Bemerkung zur Theorie der homogenen Substitutionsgruppen, S'ber. Naturforscher-Ges. Univ. Dorpat, 11, (1897) 259–74.
[17] *T. Molien*, Ueber die Anzahl der Variabeln einer irreductibelen Substitutionsgruppe, S'ber. Naturforscher-Ges. Univ. Dorpat, 11, (1897) 277–88.
[18] *I. Schur*, Neuer Beweis eines Satzes über endliche Gruppen, Berl. Ber. 1902, 1013–1019 (1902).
[19] *I. Schur*, Über die Darstellung der endlichen Gruppen durch gebrochen lineare Substitutionen, J. reine angew. Math. 127 (1904), 20–50.
[20] *I. Schur*, Neue Begründung der Theorie der Gruppencharaktere, Berl. Ber. 1904, 406–432 (1904).
[21] *I. Schur*, Über die Darstellung der symmetrischen und der alternierenden Gruppe durch gebrochene lineare Substitutionen, J. reine angew. Math. 139, 155–250 (1911).
[22] *I. Schur*, Untersuchungen über die Darstellung der endlichen Gruppen durch gebrochene lineare Substitutionen, J. für Math. 132, 85–137 (1907).

Published material

[23] *E. Artin*, The collected papers of Emil Artin. Edited by Serge Lang and John T. Tate, Addison-Wesley Publishing Co., Inc., Reading, Mass.-London, 1965. xvi+560 pp.
[24] *D. Benson*, Representations and cohomology. I. Basic representation theory of finite groups and associative algebras, Cambridge Stud. Adv. Math. 30, Cambridge University Press, Cambridge, 1991. xii+224 pp.
[25] *D. Benson*, Representations and cohomology. II. Cohomology of groups and modules, Cambridge Stud. Adv. Math. 31, Cambridge University Press, Cambridge, 1991. x+278 pp.
[26] *N. Bourbaki*, Algebra I. Chapters 1–3. Translated from the French. Reprint of the 1989 English translation, Elem. Math. (Berlin) Springer-Verlag, Berlin, 1998. xxiv+709 pp.
[27] *K. Brown*, Cohomology of groups, Grad. Texts in Math. 87, Springer-Verlag, New York-Berlin, 1982. x+306 pp
[28] *P. Cartier*, La folle journée, de Grothendieck à Connes et Kontsevich: évolution des notions d'espace et de symétrie, Les relations entre les mathématiques et la physique théorique, 23–42. Institut des Hautes Études Scientifiques, Bures-sur-Yvette, 1998.
[29] *S. B. Conlon*, Twisted group algebras and their representations, J. Austral. Math. Soc. 4 (1964), 152–173.
[30] *C. Curtis*, Pioneers of representation theory: Frobenius, Burnside, Schur, and Brauer, Hist. Math. 15, American Mathematical Society, Providence, RI; London Mathematical Society, London, 1999. xvi+287 pp.

[31] *M. Hall*, The theory of groups, Reprinting of the 1968 edition Chelsea Publishing Co., New York, 1976. xiii+434 pp.
[32] *T. Hawkins*, Lebesgue's theory of integration. Its origins and development, University of Wisconsin Press, Madison, Wis.-London, 1970. xv+227 pp.
[33] *T. Hawkins*, The Origins of the Theory of Group Characters, Archive for History of Exact Sciences, Vol. 7, No. 2 (3.III.1971), pp. 142–170.
[34] *T. Hawkins*, Hypercomplex numbers, Lie groups, and the creation of group representation theory, Arch. History Exact Sci. 8 (1972), no. 4, 243–287.
[35] *T. Hawkins*, New light on Frobenius' creation of the theory of group characters, Arch. History Exact Sci. 12 (1974), 217–243.
[36] *T. Hawkins*, The Origins of the Theory of Group Characters, History of analysis (Proc. Conf., Rice Univ., Houston, Tex., 1977) Rice Univ. Stud. 64 (1978), no. 2–3, 57–71.
[37] *H. Hopf*, Fundamentalgruppe und zweite Bettische Gruppe, Comment. Math. Helv. 14 (1942), 257–309.
[38] *B. Huppert*, Endliche Gruppen I, Die Grundlehren der mathematischen Wissenschaften, Band 134, Springer-Verlag, Berlin-New York, 1967. xii+793 pp.
[39] *M. Isaacs*, Character theory of finite groups, Pure Appl. Math., No. 69, Academic Press [Harcourt Brace Jovanovich, Publishers], New York-London, 1976. xii+303 pp.
[40] *A. Joseph, A. Melnikov, R. Rentschler (eds).*, Studies in memory of Issai Schur, Progr. Math. 210, Birkhäuser Boston, Inc., Boston, MA, 2003. clxxxvii+365 pp.
[41] *W. Ledermann*, Issai Schur and his school in Berlin, Bull. London Math. Soc. 15 (1983), no. 2, 97–106.
[42] *W. Ledermann*, Schur's influence on my career, Studies in memory of Issai Schur (Chevaleret/Rehovot, 2000), xxxvii–xxxix. Progr. Math. 210, Birkhäuser Boston, Inc., Boston, MA, 2003.
[43] *W. Ledermann, P. Neumann*, The life of Issai Schur through letters and other documents, Studies in memory of Issai Schur (Chevaleret/Rehovot, 2000), xlv–xc. Progr. Math., 210 Birkhäuser Boston, Inc., Boston, MA, 2003
[44] *G. Karpilovsky*, Projective representations of finite groups, Monogr. Textbooks Pure Appl. Math. 94, Marcel Dekker, Inc., New York, 1985. xiii+644 pp.
[45] *G. Karpilovsky*, The Schur multiplier, London Math. Soc. Monogr. (N.S.) 2, The Clarendon Press, Oxford University Press, New York, 1987. x+302 pp.
[46] *T. Lam*, Representations of finite groups: a hundred years I, Notices Amer. Math. Soc. 45 (1998), no. 3, 361–372.
[47] *S. Mac Lane*, Origins of the cohomology of groups, Enseign. Math. (2) 24 (1978), no. 1–2, 1–29.
[48] *J. Pierpont*, The history of mathematics in the nineteenth century, Bull. Amer. Math. Soc. 11 (1904), no. 3, 136–159.
[49] *D. Robinson*, A course in the theory of groups, Second edition Grad. Texts in Math. 80, Springer-Verlag, New York, 1996. xviii+499 pp.
[50] *J. Schur*, On the Representation of Finite Groups by Fractional Linear Substitutions, (English Translation by Volker Diekert, Ulrich Hertrampf and Geordie Williamson), this volume.)
[51] *T. Springer*, Linear algebraic groups, Mod. Birkhäuser Class. Birkhäuser Boston, Inc., Boston, MA, 2009. xvi+334 pp.
[52] *H. Wussing*, The genesis of the abstract group concept, A contribution to the history of the origin of abstract group theory, (Translated from the German by Abe Shenitzer and Hardy Grant). MIT Press, Cambridge, MA, 1984. 331 pp.
[53] *B. L. van der Waerden*, A history of algebra, From al-Khwarizmī to Emmy Noether Springer-Verlag, Berlin, 1985. xi+271 pp.

Geordie Williamson, Sydney Mathematical Research Institute, University of Sydney, NSW, 2006, Australia
e-mail: g.williamson@sydney.edu.au

1926–1950

Beweis eines Hauptsatzes in der Theorie der Algebren.

Von *R. Brauer* in Königsberg, *H. Hasse* in Marburg und *E. Noether* in Göttingen [1]).

Endlich ist es unseren vereinten Bemühungen gelungen, die Richtigkeit des folgenden Satzes zu beweisen, der für die Strukturtheorie der Algebren über algebraischen Zahlkörpern sowie auch darüber hinaus von grundlegender Bedeutung ist:

Hauptsatz. *Jede normale Divisionsalgebra über einem algebraischen Zahlkörper ist zyklisch (oder, wie man auch sagt, vom Dicksonschen Typus).*

Es ist uns eine besondere Freude, dieses Ergebnis, als einen im wesentlichen der p-adischen Methode zu dankenden Erfolg, Herrn Kurt Hensel, dem Begründer dieser Methode, zu seinem 70. Geburtstag vorzulegen.

Unser Beweis besteht in drei Reduktionen, von denen jeder von uns eine beigesteuert hat [2]).

1. Die erste Reduktion gab H. Hasse auf Grund der von ihm kürzlich entwickelten Theorie der zyklischen Algebren über algebraischen Zahlkörpern [3]).

Reduktion 1. *Der Hauptsatz ist bewiesen, wenn gezeigt ist:*

I. *Jede überall zerfallende Algebra über Ω ist $\sim \Omega$.*

Zur Abkürzung der Ausdrucksweise soll dabei hier wie im folgenden „Algebra A über Ω" stets „normale einfache Algebra A über dem algebraischen Zahlkörper Ω" („einfaches hyperkomplexes System mit dem Zentrum Ω") bedeuten. Ferner bezeichnet $A \sim \Omega$, daß A eine vollständige Matrixalgebra in Ω ist, also die Zugehörigkeit von A zu der speziellen durch Ω selbst als Divisionsalgebra über Ω bestimmten Klasse im Sinne der Ähnlichkeit (Gleichheit der zugehörigen Divisionsalgebren über Ω) [H, 5]. Schließlich ist unter einer „überall zerfallenden" Algebra über Ω eine solche verstanden, bei der für jede Primstelle $\mathfrak{p}$ von Ω die $\mathfrak{p}$-adische Erweiterung (d. h. die durch Erweiterung des Koeffizientenkörpers Ω auf den zugehörigen $\mathfrak{p}$-adischen Körper $\Omega_\mathfrak{p}$ entstehende Algebra) $A_\mathfrak{p} \sim \Omega_\mathfrak{p}$ ist.

Beweis. Sei D eine Divisionsalgebra über Ω. Ist dann Z ein zyklischer Körper über Ω derart, daß für jede Primstelle $\mathfrak{p}$ von Ω der $\mathfrak{p}$-Grad von Z ein Multiplum des $\mathfrak{p}$-Index von D ist [vgl. d. Anm. zu H, 17 Bb], so ist für die zugehörigen Primteiler $\mathfrak{P}$ in Z jeweils der $\mathfrak{P}$-adische Körper $\mathsf{Z}_\mathfrak{P}$ Zerfällungskörper von $D_\mathfrak{p}$ [H, 18.1], und daraus ergibt sich, daß die durch Erweiterung des Koeffizientenkörpers Ω von D auf Z ent-

[1]) Die Abfassung dieser Note übernahm H. Hasse.

[2]) Diese werden in der Reihenfolge ihrer Entstehung wiedergegeben, die der systematischen Reihenfolge entgegengesetzt ist.

[3]) H. Hasse, Theorie der zyklischen Algebren über einem algebraischen Zahlkörper, Gött. Nachr. 1931. — Eine ausführliche Darstellung der Beweise, in der insbesondere auch die von E. Noether in einer Vorlesung entwickelte, dort wie hier grundlegende Theorie der Zerfällungskörper und verschränkten Produkte entwickelt wird, erscheint demnächst in den Trans. Amer. Math. Soc. (Theory of cyclic algebras over an algebraic number field). Die letztere Arbeit wird im folgenden mit H zitiert. H, 1—6 machen — bis auf den Unterschied in der Sprache — die erstere Note aus.

J. reine angew. Math. **167** (1932), 399–404
DOI 10.1515/crll.1932.167.399

Journal für die reine und angewandte Mathematik

Proof of a main theorem in the theory of algebras

By *R. Brauer* at Königsberg, *H. Hasse* at Marburg and *E. Noether* at Göttingen[1)]

Finally, our combined efforts have succeeded in proving the correctness of the following theorem, which is of fundamental importance for the structure theory of algebras over algebraic number fields and also beyond that.

Main Theorem. *Every normal division algebra over an algebraic number field is cyclic (or, as one also says, of Dickson's type).*

It is a particular pleasure for us to present this result, as an achievement in the spirit of the p-adic method, to Mr. Kurt Hensel, the founder of this method, on the occasion of his 70th birthday.

Our proof consists of three reductions, each of which was contributed by one of us[2)].

1. The first reduction was given by H. Hasse on the basis of his recently developed theory of cyclic algebras over algebraic number fields[3)].

Reduction 1. *The main theorem is proved, if it is shown*:

I. *Each algebra over* Ω *which splits everywhere is* $\sim \Omega$.

For the sake of simplicity, "algebra A over Ω" here and in the following will always mean "normal simple algebra A over the algebraic number field Ω" ("simple hypercomplex system with centre Ω"). Furthermore $A \sim \Omega$ denotes that A is a complete matrix algebra in Ω, therefore that A belongs to the special class determined by Ω itself as a division algebra over Ω in the sense of similarity (equality of the corresponding division algebras over Ω) [H, 5]. Finally, an algebra over Ω which "splits everywhere" is understood to be an algebra for which, given[0)] any prime $\mathfrak{p}$ of Ω, the $\mathfrak{p}$-extension $A_\mathfrak{p}$ (that is, the algebra obtained by extending the coefficient field Ω to the corresponding $\mathfrak{p}$-adic field $\Omega_\mathfrak{p}$) satisfies $A_\mathfrak{p} \sim \Omega_\mathfrak{p}$.

Proof Let D be a division algebra over Ω. Then if Z is a cyclic field over Ω such that for every prime $\mathfrak{p}$ of Ω the $\mathfrak{p}$-degree of Z is a multiple of the $\mathfrak{p}$-index of D [cf. the note to H, 17 Bb], given any corresponding prime divisor $\mathfrak{P}$ in Z, the $\mathfrak{P}$-field $Z_\mathfrak{P}$ is a spltting field of $D_\mathfrak{p}$ [H, 18.1], and therefore it follows that the algebra D_Z over Z obtained by extension of

1) H. Hasse took over the composition of this note.

2) These are reproduced in the order in which they were unfolded, which is opposite to the systematic order.

3) H. Hasse, Theorie der zyklischen Algebren über einem algebraischen Zahlkörper, Gött. Nachr. 1931. – A detailed exposition of the proofs, in which, in particular, also the fundamental theory of splitting fields and crossed products, as developed in a lecture by E. Noether, is developed, will appear soon in the Trans. Amer. Math. Soc. (Theory of cyclic algebras over an algebraic number field). The latter paper is referred to as H in the following. H, 1 - 6, account for the former note, up to the difference in language.

0) *Translator's note*: A prime (or place) $\mathfrak{p}$ of Ω is a class of equivalent valuations of Ω. In the original these are called "Primstelle" (prime spot). The non-archimedean equivalence classes are called finite primes and the archimedean ones infinite primes. In the case of a finite prime the letter $\mathfrak{p}$ also stands for the prime ideal in the ring of integers of Ω.

the coefficient field Ω of D to Z splits everywhere. If I is already certain (for each Ω), then $D_Z \sim Z$ follows. This means that D has the cyclic splitting field Z, that is, D is cyclically representable. But then, D also possesses such a cyclic splitting field whose degree is equal to the degree of D itself, that is, D is cyclic [H, 6, Satz 6]. Therefore, under the assumption I, the Main Theorem is established.

2. R. Brauer gave the decisive impetus for the second reduction by informing H. Hasse by letter how the related question regarding the exact value of the exponent of a division algebra (which, incidentally, is also solved by the Main Theorem – see below) can be traced back via Sylow's group theorem to the case of a solvable splitting field. This idea could then also be used for a corresponding further reduction of the question of cyclicity.

Reduction 2. *Assertion* I *is proven, if it is shown*:

II. *Each algebra over* Ω *which splits everywhere with a solvable splitting field over* Ω *is* $\sim \Omega$.

Proof Let A be an algebra over Ω which splits everywhere. If then K is a Galois splitting field for A, p is any prime number, and Σ is one of the Sylow fields of K over Ω in respect of p (that is, the field of invariants under a Sylow group related to p of the Galois group), then A_Σ is an algebra over Σ which also splits everywhere and which has the solvable splitting field K. If II is already known (for every Ω) then $A_\Sigma \sim \Sigma$ follows. This means that A has the splitting field Σ with a degree prime to p. Thus the index of A, being a divisor of this degree, is prime to p. Since this holds for every prime number p, it is one, i.e. $A \sim \Omega$. Therefore, under the assumption II, assertion I is established.[4)]

3. The third reduction, which then, as H. Hasse – in possession of the first two reductions – recognised, leads to the final proof, was given by E. Noether, prompted by a message from H. Hasse; in this message the Main Theorem was proven in the special case of an abelian splitting field, namely by means of the general reduction I and a formulaic reduction of the corresponding factor system, the kernel of which E. Noether carved out as the third reduction. Incidentally, independently of E. Noether, R. Brauer had also considered this third reduction.

[4)] The idea of reduction to solvable splitting fields via Sylow's group theorem was already applied by R. Brauer, namely to show that each prime divisor of the index also occurs in the exponent (Über den Zusammenhang von arithmetischen und invariantentheoretischen Eigenschaften von Gruppen linearer Substitutionen, Berl. Akad.-Ber. 1926). Recently A. A. Albert developed simple proofs, independent of representation theory, for this idea, as well as generally for a series of general theorems of R. Brauer's and E. Noether's theory (1. On direct products, cyclic division algebras, and pure Riemann matrices; 2. On direct products; both in Trans. Amer. Math. Soc. **33** (1931); for the reduction in question, see notably Theorem 23 in 2.)

Addendum during printing. Furthermore A. A. Albert, having a mail notification from H. Hasse that he proved the Main Theorem for abelian algebras (see later in the text), deduced directly out of it, independent of us, the following facts:

a) the Main Theorem for degrees 2^s

b) Theorem 1 following below (exponent = index)

c) aside from the basic idea of Reduction 2 also the subsequent one of Reduction 3, naturally without the relation to Reduction 1, and accordingly with the result: for division algebras D of prime power degree p^s over Ω there exists an extension field Ω' of Ω of degree prime to p such that $D_{\Omega'}$ is cyclic.

All three results are, of course, superseded by our now given proof of the Main Theorem. However, they show that A.A. Albert has an independent share in the proof of the Main Theorem.

Finally, A. A. Albert (having regard to our proof of the Main Theorem) has also remarked that our central Theorem I follows in a few lines from the Theorems 13, 10, 9 of a work of his currently in print (Bull. Amer. Math. Soc. **87** (1931)). The proof of these Theorems is essentially based on the same conclusions as our Reductions 2 and 3.

Reduction 3. *Assertion* II *is proven, if it is shown*:

III. *Each algebra over* Ω *which splits everywhere with a cyclic splitting field of prime degree over* Ω *is* $\sim \Omega$.

Proof. Let A be an algebra which splits everywhere with a solvable splitting field K over Ω. Then let $K = \Lambda_0 > \Lambda_1 > \cdots > \Lambda_r = \Omega$ be a chain of fields between K and Ω such that Λ_i over Λ_{i+1} is always cyclic of prime degree. Then first of all A_{Λ_1} is an algebra over Λ_1 which also splits everywhere and which has the cyclic splitting field $K = \Lambda_0$ of prime degree. If III is already known (for every Ω), it also follows $A_{\Lambda_1} \sim \Lambda_1$. This means that Λ_1 is a splitting field for A. Now, beginning with Λ_1 instead of Λ_0, one analogously proves $A_{\Lambda_2} \sim \Lambda_2$, etc., until finally $A_{\Lambda_r} \sim \Lambda_r$, thus, $A \sim \Omega$. Therefore, under the assumption III, assertion II is established.

4. But now, according to the results of H. Hasse [H, 24], III holds. Therefore, the correctness of the Main Theorem follows from the Reductions 3, 2, 1 backwards.

E. Noether also points out the following:

The correctness of III – more generally for cyclic splitting fields of arbitrary degree – is based on Hasse's norm theorem[5]. For Reduction 3, however, only the special case of prime degree, i.e. the original Hilbert-Furtwängler norm theorem[6] is used. Therefore Reduction 3 yields a new simple proof of Hasse's norm theorem. The proof would go like this:

Let Z be a cyclic field over Ω and α a number in Ω, for which the norm residue symbol $\left(\frac{\alpha, Z}{\mathfrak{p}}\right) = 1$ for every prime of Ω. Then each cyclic algebra $A = (\alpha, Z)$ splits everywhere [H, 17.7]. According to reduction 3, using only the Hilbert-Furtwängler norm theorem, it follows $A \sim \Omega$. But then α is the norm of an element in Z [H, 15.4].

Relating to the norm theorem we further remark that Assertion II has to be viewed as its correct generalisation to higher (even non-abelian) cases whereas the word-for-word generalisation, as H. Hasse has shown[7], is not generally valid.

Conclusions (H. Hasse)

5. The obvious conclusion to be drawn from the Main Theorem is that the theory of cyclic algebras over algebraic number fields as developed by H. Hasse has now become a general structure and invariant theory for normal simple algebras (in particular, normal division algebras) over algebraic number fields. Especially, it is now generally proven:

Proposition 1. *The exponent of a normal simple algebra over an algebraic number field is equal to its index.*

Moreover, Proposition 1 gives rise to the remarkable fact:

Proposition 2. *The ground ideal of a proper normal division algebra over* Ω *is divisible at least by one prime of* Ω *(and in fact by at least two).*

Here the ground ideal is defined, for instance, as the (reduced) norm of the (reduced) different[8]. In addition, an infinite prime is included in the ground ideal if and only if the

---

[5] Cited in H, 3.11; proven in : H. Hasse, Beweis eines Satzes und Widerlegung einer Vermutung über das allgemeine Normrestsymbol, Gött. Nachr. 1931.

[6] See H. Hasse, Bericht über neuere Untersuchungen und Probleme in der Theorie der algebraischen Zahlkörper II, Jahresber. der D.M.V., Erg.-Bd. **6** (1930), Â§ 8, as well as the literature cited there.

[7] See footnote 5.

[8] In the special case of a rational quaternion algebra, this relates to the notion "Grundzahl" as introduced by H. Brandt (Idealtheorie in Quaternionenalgebren, Math. Ann. 99 (1928)). Since this is not a number but an ideal, "Grundideal" had to be used; do not confuse this with Dedekind's labelling Grundideal and Grundzahl for different and discriminant, which proves, for the very reason given, useless for relative fields. – With regard to the definition of (red.) different see the following footnote. The (red.) norm of an ideal is also defined "prime-wise" by E. Noether, namely, – following the fact, that for each prime every ideal is a principal ideal – simply by forming the

corresponding algebra is reduced to the quaternion algebra (and not the real or complex number field).

Proof. According to the results of H. Hasse[9] a prime $\mathfrak{p}$ of Ω divides the ground ideal if and only if the $\mathfrak{p}$-index is different from 1. Now, if all $\mathfrak{p}$-indices were equal to 1, the algebra would split everywhere, and, using assertion I, would not be a proper division algebra. (The fact that a $\mathfrak{p}$-index alone cannot be different from 1 results from the fact that the product theorem (reciprocity law) holds for the norm residue symbols whose orders are the $\mathfrak{p}$-indices [H, 3.8]).

6. Furthermore, assertion I allows for determining (arithmetic characterisation) all splitting fields K relevant to an algebra A over Ω. In exact generalisation of the facts already established by H. Hasse in the special case of A cyclic [H, 6, Satz 2] one obtains:

Proposition 3. *Given an algebra A over Ω, an algebraic number field K over Ω is a splitting field for A if and only if for all prime divisor $\mathfrak{P}_i$ in K of all the primes $\mathfrak{p}$ of Ω the $\mathfrak{P}_i$-degree $n_{\mathfrak{P}_i}$ of K is a multiple of the $\mathfrak{p}$-index $m_\mathfrak{p}$ of A.*

Thereby the $\mathfrak{P}_i$-degree of K is the degree of the $\mathfrak{P}_i$-adic field $K_{\mathfrak{P}_i}$ over the $\mathfrak{p}$-adic field $\Omega_\mathfrak{p}$ attached to $\mathfrak{P}_i$, i.e. if

$$\mathfrak{p} = \prod_i \mathfrak{P}_i^{e_{\mathfrak{P}_i}}, \quad N_{K/\Omega}(\mathfrak{P}_i) = \mathfrak{p}^{f_{\mathfrak{P}_i}}$$

is the decomposition of $\mathfrak{p}$ in K, the product of the degree $f_{\mathfrak{P}_i}$ and the ramification index $e_{\mathfrak{P}_i}$ of $\mathfrak{P}_i$ with respect to $\mathfrak{p}$, $n_{\mathfrak{P}_i} = f_{\mathfrak{P}_i} e_{\mathfrak{P}_i}$ [H, 4]. And the $\mathfrak{p}$-index of A is the index $m_\mathfrak{p}$ of the $\mathfrak{p}$-adic extension $A_\mathfrak{p}$ [H, 6].

Proof. According to assertion I, the fact that K is a splitting field for A, i.e. $A_K \sim K$, is equivalent to the fact that $A_{\mathfrak{P}_i} \sim K_{\mathfrak{P}_i}$ for each $\mathfrak{P}_i$.

In the case of a finite prime $\mathfrak{p}$, let now $A_\mathfrak{p} \sim (\pi, W_\mathfrak{p})$ be the arithmetically distinguished representation of $A_\mathfrak{p}$ [H, 16; in addition loc. cit.[10]], thus $W_\mathfrak{p}$ is the unramified field of degree $m_\mathfrak{p}$ over $\Omega_\mathfrak{p}$, and π a number in $\Omega_\mathfrak{p}$ which is exactly divisible by p^l. Furthermore, let $\mathsf{K}_{\mathfrak{P}_i}$ the compositum and $\Delta_{\mathfrak{P}_i}$ the intersection of $W_\mathfrak{p}$ and $K_{\mathfrak{P}_i}$. Since the largest unramified subfield contained in $K_{\mathfrak{P}_i}$ is of degree $f_{\mathfrak{P}_i}$ over $\Omega_\mathfrak{p}$, and since there is only one unramified field over $\Omega_\mathfrak{p}$ for each degree, $\Delta_{\mathfrak{P}_i}$ has degree $d_{\mathfrak{P}_i} = (m_\mathfrak{p}, f_{\mathfrak{P}_i})$ over $\Omega_\mathfrak{p}$. Therefore, $\mathsf{K}_{\mathfrak{P}_i}$ is of degree $m_\mathfrak{p}/d_{\mathfrak{P}_i}$ over $K_{\mathfrak{P}_i}$, and thereby unramified.

Now $A_{K_{\mathfrak{P}_i}} = (A_\mathfrak{p})_{K_{\mathfrak{P}_i}} \sim (\pi, \mathsf{K}_{\mathfrak{P}_i})$ [H, 15.5]. Therefore, $A_{K_{\mathfrak{P}_i}} \sim K_{\mathfrak{P}_i}$ is equivalent to the fact that π is a norm from $\mathsf{K}_{\mathfrak{P}_i}$ with respect to $K_{\mathfrak{P}_i}$. Taking into account that $\mathsf{K}_{\mathfrak{P}_i}$ is unramified over $K_{\mathfrak{P}_i}$, this is the case if and only if the ramification index $e_{\mathfrak{P}_i}$ of π in $\mathfrak{P}_i$ is a multiple of the degree $m_\mathfrak{p}/d_{\mathfrak{P}_i}$ of $\mathsf{K}_{\mathfrak{P}_i}$ over $K_{\mathfrak{P}_i}$, or, equivalently, because $(m_\mathfrak{p}/d_{\mathfrak{P}_i}, f_{\mathfrak{P}_i}/d_{\mathfrak{P}_i}) = 1$, if $(f_{\mathfrak{P}_i}/d_{\mathfrak{P}_i})e_{\mathfrak{P}_i} = (n_{\mathfrak{P}_i}/d_{\mathfrak{P}_i})$ is a multiple of $(m_\mathfrak{p}/d_{\mathfrak{P}_i})$ that is, if $n_{\mathfrak{P}_i}$ is a multiple of $m_\mathfrak{p}$, as claimed.

In the case of an infinite prime $\mathfrak{p}$, and we are not in the trivial case $m_\mathfrak{p} = 1$, then $m_\mathfrak{p} = 2$ and $A_\mathfrak{p}$ is the quaternion algebra over the real field $\Omega_\mathfrak{p}$. Then the fact $A_{K_{\mathfrak{P}_i}} = (A_\mathfrak{p})_{K_{\mathfrak{P}_i}} \sim K_{\mathfrak{P}_i}$ is equivalent to the fact that $K_{\mathfrak{P}_i}$ is the complex field, this means $n_{\mathfrak{P}_i} = f_{\mathfrak{P}_i} = 2 = m_\mathfrak{p}$ (in this case, $e_{\mathfrak{P}_i}$ does not occur), which again yields the claim ($n_{\mathfrak{P}_i}$ as well as $m_\mathfrak{p}$ can only take the values 1 or 2).

(red.) norms of the principal ideal generators for the respective primes.

[9] H. Hasse, Über $\mathfrak{p}$-adische Schiefkörper und ihre Bedeutung für die Arithmetik hyperkomplexer Zahlsysteme, Math. Ann. **104** (1931); see there Satz 42, 59.

[10] H. Hasse, Über $\mathfrak{p}$-adische Schiefkörper und ihre Bedeutung für die Arithmetik hyperkomplexer Zahlsysteme, Math. Ann. **104** (1931), Satz 38.

7. One arrives at significant results if one turns around the question about all splitting fields K of a fixed algebra A, as answered in Proposition 3, that is to say, given a fixed algebraic field over Ω, one considers the entirety/aggregate of all algebras over Ω which split over K. These algebras A form a subgroup $\mathfrak{K}$, determined by K, of the R. Brauer group $\mathfrak{A}$ of all algebras (more precisely, of all classes of similar algebras) over Ω[11]. If one assumes K to be Galois over Ω, one arrives at theorems which are to be viewed as generalisations of the Main Theorems of class field theory (theory of relatively abelian number fields) to general relatively Galois number fields.

Proposition 4. (Decomposition Proposition). *The relative degree of a prime divisor in K of a prime ideal in Ω which does not occur in the relative discriminant of K is equal to the smallest exponent for which $A_{\mathfrak{p}}^f \sim \Omega_{\mathfrak{p}}$ holds for all algebras A in the group $\mathfrak{K}$ attached to K.*

Proof. Since f is the $\mathfrak{p}$-degree of K (being the same for all prime divisors of $\mathfrak{p}$ in K), while on the other hand the $\mathfrak{p}$-index of A is equal to the index, that is, the index of $A_{\mathfrak{p}}$, then, by Theorem 3, f is certainly a multiple of that smallest exponent, and it suffices for the proof to show that there are in the group $\mathfrak{K}$ algebras A with exact $\mathfrak{p}$-index f.

Using the Frobenius Density Theorem there exists another prime ideal $\mathfrak{p}'$ in Ω which does not occur in the relative discriminant of K and whose prime divisors in K have the relative degree f. Then let Z be a cyclic field over Ω whose $\mathfrak{p}$-degree and $\mathfrak{p}'$-degree both take the value f. Furthermore, let α be a number in Ω, for which the norm residue symbols $\left(\frac{\alpha,Z}{\mathfrak{p}}\right)$ and $\left(\frac{\alpha,Z}{\mathfrak{p}'}\right)$ take the reciprocal value of the order f, whereas for all other divisors $\mathfrak{q}$ of the conductor of Z $\left(\frac{\alpha,Z}{\mathfrak{q}}\right) = 1$ holds. According to the generalised theorem of arithmetic progressions α can be chosen in such a way that, aside from possibly $\mathfrak{p}, \mathfrak{p}'$ and the divisors $\mathfrak{q}$, it only contains one single prime ideal $\mathfrak{r}$ of Ω exactly to the first power. For this, by the product theorem of the norm residue symbol (reciprocity law), $\left(\frac{\alpha,Z}{\mathfrak{r}}\right) = 1$ [H, 3.8 - 10]. Each algebra $(\alpha, Z) = A$ then has $\mathfrak{p}$-index and $\mathfrak{p}'$-index f, whereas for all other primes of Ω it has the index 1 [H, 17.7]. By Theorem 3, it follows from this, because of the assumption on $\mathfrak{p}$ and the choice of $\mathfrak{p}'$, that K is a splitting field of A. Thereby, algebras A of $\mathfrak{p}$-index f are found in the group $\mathfrak{K}$.

Proposition 5. (Uniqueness and Order Proposition). *If $K \subset K'$, then $\mathfrak{K} \subset \mathfrak{K}'$, and conversely.*

In particular, the assignment of the group $\mathfrak{K}$ of algebras to the Galois fields K is reversibly unique.

Proof. a) From $K \subset K'$ trivially follows $\mathfrak{K} \subset \mathfrak{K}'$; for every algebra A split by K is a fortiori an algebra split by K'.

b) Conversely, let $\mathfrak{K} \subset \mathfrak{K}'$. Then, according to the Decomposition Theorem, for each prime $\mathfrak{p}$ which does not divide the relative discriminant $f_{\mathfrak{p}}|f'_{\mathfrak{p}}$. In particular, $f_{\mathfrak{p}} = 1$ for almost all $\mathfrak{p}$ with $f'_{\mathfrak{p}} = 1$. The outcome of this is, according to the usual analytic reasoning[12], $K \subset K'$.

8. Finally, it should be noted that the Main Theorem makes a significant contribution to the question dealt with by I. Schur[13] concerning the number fields in which absolutely irreducible representations of a finite group are possible.

Proposition 6. *The absolutely irreducible representations of a finite group $\mathfrak{G}$ are all realisable over cyclotomic fields; in particular, always over the field of n^h-th roots of unity,*

[11] R. Brauer, Über Systeme hyperkomplexer Grössen, Jahresber. d. D. M.-V. **38**, (1929), S. 47/48, see also H, 13.1.

[12] See, for example, H. Hasse [loc. cit. footnote 6], Â§ 25, III.

[13] I. Schur, Arithmetische Untersuchungen über endliche Gruppen linearer Substitutionen. Berl. Akad.-Ber. 1906.

where n is the order of $\mathfrak{G}$, and h is sufficiently large.

Proof. Passing over to the rational group ring G of $\mathfrak{G}$, the absolutely irreducible representations Γ_i of $\mathfrak{G}$ become the absolutely irreducible representations of the simple components G_i of the semi-simple algebra G, and their centres are always the fields Ω_i of the corresponding characters[14)], hence in any case, due to the finiteness of $\mathfrak{G}$, are cyclotomic fields, and in fact certainly subfields of the field of n-th roots of unity.

According to Proposition 3, a cyclic field Z_i over Ω_i is a splitting field for G_i if for each prime $\mathfrak{p}$ of Ω_i its $\mathfrak{p}$-degree $n_{i\mathfrak{p}}$ is a multiple of the $\mathfrak{p}$-index $m_{i\mathfrak{p}}$ of G_i. Now $m_{i\mathfrak{p}}$ is different from 1 only for the prime divisors of the ground ideal of G_i with respect to Ω_i, thus certainly only for the prime divisors of the absolute discriminant of G; since this occurs in n^n – where n^n is the discriminant of a (non-maximal) order in G[15)] – therefore at most for the prime divisors of n. In order to make $n_{i\mathfrak{p}}$ for these $\mathfrak{p}$ a multiple of $m_{i\mathfrak{p}}$, it suffices to prescribe that the $\mathfrak{p}$ in question are ramified in Z_i to an index divisible by $m_{i\mathfrak{p}}$. But then, as one easily sees, the field of n^h-th roots of unity for sufficiently high h accomplishes this.

In the final sentence of the work as cited, I. Schur asserts that in all hitherto known cases one already gets by with the field of n-th roots of unity. Whether this is always true, and whether the methods developed here suffice to decide this question, remains subject to further investigations.

Translation by Joachim Schwermer

[14)] See: a) R. Brauer and E. Noether, Über minimale Zerfällunskörper irreduzibler Darstellungen, Berl. Akad.-Ber. 1927; §1 b) R. Brauer, Über System hyperkomplexer Zahlen, Math. Zeitschr. **30** (1929); Satz 3. c) E. Noether, Hyperkomplexe Grössen und Darstellungstheorie, Math. Zeitschr. **30** (1929); §§21, 24, 26.

[15)] See E. Noether [loc. cit. footnote 11 c)], §26.

Class field theory – Cyclic algebras, Brauer groups, and a local-global principle

By *Joachim Schwermer* at Bonn and Vienna

1. Introduction

The title "Proof of a main theorem in the theory of algebras" of the paper by R. Brauer, H. Hasse and E. Noether [15], published in 1932, places its content in the theory of linear associative algebras[1)], but basically it presents an important step, based on non-commutative methods, towards a "hypercomplex approach" to class field theory. This article aims to provide a possible basis for extending the theory, as developed by T. Takagi, H. Hasse and E. Artin, among others, to non-abelian algebraic number fields, i.e. those with a non-abelian Galois group. We attempt to outline the various lines of research leading to the questions dealt with in this note, to describe its mathematical results and the methods used, and to position the findings in their historical context.

2. Historical background – Hypercomplex number systems and linear algebras

2.1. Linear associative algebras. The paper "On hypercomplex numbers", published in 1907, by J.H.M. Wedderburn had a significant influence on the development of modern algebra. He "aimed to set the theory of hypercomplex numbers on a rational basis" [75, p. 77], that is, to develop the theory over an arbitrary field. Wedderburn was able to merge two different methodological strands of previous investigations.

On the one hand, there was the more abstract approach, beginning, among others, with B. Pierce's memoir "Linear associative algebra", read before the National Academy of Sciences in 1870 and posthumously published in 1881 [65], [66], followed by the works of L.E. Dickson. He already gave a tentative but useful general definition of an associative algebra over an arbitrary field in 1903 [29].

1) Following the generally accepted concept from 1933 (see [29] or [28]), an algebra over a field K is a module A of finite rank over K equipped with a bilinear multiplication $(x, y) \mapsto xy$. If the multiplication also satisfies the associative law (i.e. $(xy)z = x(yz)$ for all $x, y, z \in A$) and has a unit element, then it is called an associative algebra with 1, or a K-algebra for short. This terminology corresponds to current usage. In the past, the term 'system of hypercomplex numbers' was also commonly used in this context, but it originally referred only to algebras of finite rank over the complex or real numbers. The extension of complex numbers to quaternions in 1843 by W.R. Hamilton and beyond was one of the origins of this research area.

On the other hand, mainly G. Scheffers [68], Th. Molien [58], E. Cartan [16], [17] and G. Frobenius [39], by investigating hypercomplex number systems in their original form (i.e. associative algebras with unit element over the real or complex field), advanced the understanding of essential structural elements of the theory of associative algebras. For example, in modern terms, the concept of the radical of an algebra had been devised, and the seminal notion of a semi-simple resp. simple associative algebra had been formulated.[2)]

Moreover, S. Lie and Scheffers emphasised in their book "Vorlesungen über continuierliche Gruppen mit geometrischen und anderen Anwendungen" [57] the close connection between hypercomplex number systems and Lie groups. Given such a system, its group of units forms a Lie group. It is this transformation group point of view which facilitated the transfer of group-theoretical concepts to the theory of associative algebras over an arbitrary field. The analogue of the notion of a complex, as introduced by Frobenius in his work on finite groups [38], was effectively used in the context of associative algebras by Wedderburn. In fact, he developed a calculus of complexes (see [75, p. 80]) which, in particular, he used to coin the decisive general concept of an invariant sub-complex in the algebra in question, or, in current terms, of a two-sided ideal in an associative algebra of finite dimension over an arbitrary field. Following Wedderburn [75, p. 81], "an algebra which has no invariant sub-complex is said to be simple". However, in his paper "Sur les groupes bilinéaires et les systèmes de nombres complexes", published in 1898, E. Cartan had already introduced this notion (see [16, no. 69]) within the realm of hypercomplex number systems. Further, many of the results in [75] are generalisations of the analogues in [16] asserted in the case of associative algebras over the field of complex numbers. The following achievements in [75] are of fundamental significance:

> Theorem 22. - *Any simple algebra can be expressed as the direct product of a primitive algebra and a simple matric algebra.*

In other words, given a simple finite-dimensional algebra R over a field K, there exists a unique positive number r and a primitive algebra D over K, uniquely determined up to isomorphism, such that $R \cong D \times M_r(K) \cong M_r(D)$.[3)]

> Theorem 23. - *The direct product A of a primitive algebra B and a quadratic matric algebra C is simple; and any element which is commutative with every other element of A is an element of B.*

In Wedderburn's terminology, a primitive algebra is a division algebra. By definition, a non-zero associative algebra, defined over a field F, is a division algebra if and only if it is an associative ring with unit element in which every non-zero element is invertible and that is finite-dimensional over its centre. Note that the centre is a field containing F.

2.2. Cyclically generated algebras. At the turn of the 20th century, the field of real (resp. complex numbers) and the 4-dimensional non-commutative algebra $\mathbb{H}$ of real quaternions as constructed by W.R. Hamilton in 1843, were the only known examples of primitive algebras. In 1905 Wedderburn showed that every primitive algebra with finite centre is a field. However, given any infinite field K different from $\mathbb{R}$, the quest to describe the primitive algebras over K demanded an answer. In discussing such algebras, without any loss of generality, one may

---

2) Th. Hawkins [51] details the historical significance of Molien's work.

3) Note that the direct product of two K-algebras, as formally defined e.g. in [28, I, 2], coincides with their tensor product.

assume that the centre coincides with the field of definition. "A primitive algebra will be called normal when reduced in this manner", as Wedderburn points out ([76, p. 162]).[4)]

In a short note [30], delivered on April 14, 1906, during the American Mathematical Society meeting held in Evanston, Dickson indicated the construction of a new type of linear associative algebra. In his 1914 publication "Linear associative algebras and abelian equations" [31], he finally gave details, starting from an abelian equation over K of degree n whose automorphism group is cyclic. However, there and also in his later book "Algebras and their Arithmetics", published in 1923, he avoided any use of a Galois-theoretic point of view or group theoretical methods. At any rate, in the case $n = 3$, Dickson provided new examples of primitive algebras. Given an extension field L of K of degree n with cyclic Galois group G, let A be an algebra with L-basis $1, u, u^2, \ldots, u^{n-1}$ and with the relations $ux = \sigma(x)u$ for every $x \in L$ where σ denotes a generating element of G, and $u^n = a \neq 0$, $a \in K$. Then A, also denoted by $A = (L/K, \sigma, a)$, is a K-algebra of dimension n^2 which is normal and simple.

Wedderburn [76] took a closer look at Dickson's construction for general n and clarified its structural underpinnings. Thereby, given such an algebra, to be called an algebra of Dickson's type, he stated an arithmetic condition under which A is a primitive algebra. This condition reads as follows: if a^n is the least power of a which is the norm of an element in L, then A is a primitive algebra.

2.3. Classes of similarity. Let R and S be normal simple K-algebras. By Wedderburn's structure theorem, there exist unique positive integers r, s and primitive algebras D, D' over K, uniquely determined up to isomorphism, such that $R \cong M_r(D)$ and $S \cong M_s(D')$. By definition, R and S are similar, denoted by $R \sim S$, if the primitive K-algebras D and D' are isomorphic. The notion of similarity defines an equivalence relation on the set of normal simple K-algebras. In each equivalence class, say $[R]$, there is a normal primitive K-algebra D, unique up to isomorphism. Therefore, in particular, $R \sim K$ means that R is a matrix algebra over K. If R is similar to a cyclically generated algebra, one says that R is cyclically representable. Given an equivalence class $[R] = [D]$, the index of R, or, more generally, of the class $[R]$, is defined to be the degree of the primitive K-algebra D, formally $\mathrm{ind}R := \dim_K D^{1/2}$. The exponent is the least positive integer e such that the direct product R^e is similar to K, that is, R^e is a matrix algebra over K. The exponent of R divides the index of R.

Given two normal simple K-algebras R and S, their direct product is again a normal simple K-algebra. The similarity classes of normal simple K-algebras, endowed with the composition $([R], [S]) \mapsto [R \times_K S]$, form an abelian group $Br(K)$, called the *Brauer group of K*. The identity element is the class $[K]$, and the class $[R^{op}]$ of the opposite algebra of R is the inverse of $[R]$. Given a field extension L/K, the extension $R_L := R \times_K L$ is a normal simple L-algebra. Since $R \sim S$ implies $R_L \sim S_L$, and $(R \times S)_L = R_L \times S_L$, the assignment $[R] \mapsto [R \times_K L]$ defines a group homomorphism $r_{L/K} \colon Br(K) \longrightarrow Br(L)$. One says that L splits R, or L is a splitting field for R if $[R] \in \ker r_{L/K}$, that is, $R \times_K L \cong M_n(L)$ for some $n \in \mathbb{N}$. One writes $Br(L/K)\colon = \ker r_{L/K}$ for the kernel of the restriction $r_{L/K}$, to be called the relative Brauer group.

2.4. Existence of splitting fields. Using Wedderburn's theorem, given a normal primitive K-algebra $R = M_r(D)$, an extension field L of K splits R if and only if L splits the

4) In modern terms, a normal primitive algebra is a central division algebra. Generally, given any finite-dimensional K-algebra $R \neq 0$, the image of the injective ring homomorphism $i : K \longrightarrow R$ is contained in the centre $Z(R)$ of R. Upon identification of K with $i(K)$, we say that R is a central K-algebra if $K = Z(R)$.

division K-algebra D. If the degree of a splitting field L of R is finite, it is a multiple of the index of R. Moreover, following Brauer [13], there exist splitting fields of degree equal to the index of R. More precisely, regarding the existence of splitting fields for D, one has

Theorem 2.1. *Let R be a normal primitive K-algebra, Then there exists a Galois extension L/K of finite degree such that L is a splitting field for R. Further, R is similar to some normal primitive K-algebra S which contains L as a strictly maximal subfield, that is, the centraliser of L in S is L itself, formally, $Z_S(L) = L$. Consequently, the Brauer group $Br(K) = \bigcup Br(L/K)$ is the union over all relative Brauer groups, where L ranges over all Galois extension fields L of K of finite degree.*

2.5. Crossed products and factor systems. In her lecture series, entitled "Algebra der hyperkomplexen Grössen", in 1929 in Göttingen,[5] Noether unveiled her theory of crossed-product algebras and factor systems, starting off with Brauer's preceding theory of factor systems [12]. In fact this theory had its origin in questions regarding absolutely irreducible representations of groups, as previously dealt with by Speiser, I. Schur and Brauer (see [60]). Noether's central result is the isomorphism of the relative Brauer group $Br(L/K)$ associated with a given Galois extension L/K of finite degree with a group formed by equivalence classes of factor systems attached to the Galois group $G(L/K)$. We now define the notions alluded to in technical terms.

Let L/K be a finite Galois extension of degree $n = [L : K]$, with Galois group $G = G(L/K)$, and let $f : G \times G \longrightarrow L^\times$ be a map. To simplify matters, we may write $f_{\sigma,\tau} := f(\sigma, \tau)$ respectively $f^\rho_{\sigma,\tau} := \rho(f(\sigma, \tau))$ for all $\rho, \sigma, \tau \in G$. Let S be a (left) K-vector space of dimension n, endowed with a (left) L-basis $\{u_\sigma\}_{\sigma \in G}$. We define an algebra structure on $S = \bigoplus_{\sigma \in G} Lu_\sigma$ by the formulas:

$$u_\sigma \cdot x = \sigma(x) \cdot u_\sigma, \quad u_\sigma u_\tau = f_{\sigma,\tau} u_{\sigma\tau} \quad \text{for all } x \in L,\ \sigma, \tau \in G. \tag{2.1}$$

One checks that this multiplication on S, combined with the given addition, satisfies the distributive law. Note that K is contained in the centre of S. With regard to the question under which conditions the multiplication is associative, we have to analyse this using the monomials $\lambda_\sigma u_\sigma$, $\sigma \in G$, $\lambda_\sigma \in L$. In fact, this question can be reduced to the monomials $u_\sigma, \sigma \in G$. The K-algebra S will be associative if and only if $u_\rho(u_\sigma u_\tau) = (u_\rho u_\sigma)u_\tau$, for all $\rho, \sigma, \tau \in G$, or, equivalently, if and only if,

$$f^\rho_{\sigma,\tau} \cdot f_{\rho,\sigma\tau} = f_{\rho,\sigma} \cdot f_{\rho\sigma,\tau} \quad \text{for all} \quad \rho, \sigma, \tau \in G. \tag{2.2}$$

A map $f : G \times G \longrightarrow L^\times$ satisfying condition (2.2) is called a factor system from G to $L^\times$. Given such a factor system $f : G \times G \longrightarrow L^\times$, then the algebra S, to be denoted by $S_{L/K,f}$, is an (associative) normal simple K-algebra. The centraliser Z of L in $S_{L/K,f}$ is L itself, that is, L is a strictly maximal subfield. Moreover, L is a splitting field of S. We call the K-algebra $S_{L/K,f}$ a crossed-product algebra of L with G.[6]

---

[5] Posthumously published (based on notes taken by M. Deuring) as [61].

[6] Hasse reviews (with proofs) [47, Sections 8–12] the construction of crossed-product algebras and their properties. His treatment also contains a discussion of the results of R. Brauer and E. Noether regarding splitting fields and general normal primitive algebras as crossed-products.

Replacing the L-basis $\{u_\sigma\}_{\sigma\in G}$ by an L-basis $\{u'_\sigma\}_{\sigma\in G}$ where $u'_\sigma = \delta_\sigma u_\sigma$ for any set $\{\delta_\sigma \in L^\times \mid \sigma \in G\}$, the analogues of formulas (2.1), in particular, yield

$$u'_\sigma u'_\tau = g_{\sigma,\tau} u'_{\sigma\tau} \quad \text{with } g_{\sigma,\tau} = f_{\sigma,\tau} \cdot \delta_\sigma \sigma(\delta_\tau)\delta_{\sigma\tau}^{-1} \text{ for all } \sigma, \tau \in G. \tag{2.3}$$

It turns out that the assignment $(\sigma,\tau) \mapsto \delta_\sigma\sigma(\delta_\tau)\delta_{\sigma\tau}^{-1}$ defines a factor system $\hat{\delta} : G\times G \longrightarrow L^\times$ from G to $L^\times$. We call any factor system of this form a principal factor system. The collection of factor systems from G to $L^\times$ forms an abelian group under pointwise multiplication. We call two factor systems f and g equivalent if there exists a principal factor system $\hat{\delta}$ such that $g = f \cdot \hat{\delta}$. Consequently, given two equivalent factor systems f and g from G to $L^\times$, the associated crossed products $S_{L/K,f}$ and $S_{L/K,g}$ are isomorphic as K-algebras.

Theorem 2.2. *Let L/K be a Galois extension of finite degree with Galois group G. There exists an isomorphism between the group of equivalence classes of factor systems from G to $L^\times$ to the relative Brauer group $Br(L/K)$, assigning the class $[S_{L/K,f}]$ of the crossed product $S_{L/K,f}$ to the class $[f]$ represented by the factor system f from G to $L^\times$.*[7)]

Note that in each equivalence class of factor systems one finds a normalised factor system f which satisfies $f_{\sigma,id} = f_{id,\sigma} = 1$ for all $\sigma \in G$.

2.6. Cyclic algebras. A normal simple K-algebra R of degree n is said to be cyclic if it contains a cyclic subfield L of maximal degree n, that is, the Galois group $G = G(L/K)$ is a cyclic group of order n. Then R is cyclically generated: given a generating element σ of G, R is of the form $R = L1 \oplus Lu \oplus Lu^2 \oplus \cdots \oplus Lu^{n-1}$ with the relations $ux = \sigma(x)u$ for every $x \in L$, and $u^n = a \neq 0$, $a \in K$. The map $f_a : G \times G \longrightarrow L^\times$, defined by the assignment

$$(\sigma^i, \sigma^j) \mapsto \begin{cases} 1 & \text{if } i + j < n, \\ a & \text{if } i + j \geq n, \end{cases}$$

is the corresponding factor system from $G = \langle\sigma\rangle$ to $L^\times$. In this case, the factor system is determined by a single element $a \in K^\times$. Therefore, we may write $R = (L/K, \sigma, a)$. In this notation, given elements $a, b \in L^\times$, one has

$$(L/K, \sigma, a) \times (L/K, \sigma, b) \sim (L/K, \sigma, ab) \tag{2.4}$$

and $(L/K, \sigma, a) \sim (L/K, \sigma, b)$ if and only if $ba^{-1} \in n_{L/K}(L^\times)$. Thus, the unit element in the group of equivalence classes of factor systems is given by $n_{L/K}(L^\times)$, the image of $L^\times$ under the norm map. Therefore, we have

$$Br(L/K) \cong K^\times / n_{L/K}(L^\times) \tag{2.5}$$

as an isomorphism of groups in the case of a cyclic Galois extension L/K.

7) In current terms, the group of equivalence classes of factor systems from G to $L^\times$ is the second cohomology group of the Galois group G with coefficients in $L^\times$, to be denoted $H^2(G, L^\times)$, see Section 7.

3. Historical background – From the Small to the Large

In his thesis work, completed in early 1921, and later on at the end of that year in his Habilitationsschrift, H. Hasse illustrated the importance of K. Hensel's theory of p-adic numbers [52] for classical questions in number theory, in particular, the arithmetic theory of quadratic forms over algebraic number fields. Hasse's approach was based on the valuation-theoretic point of view, introduced by J. Kürschak in 1912. He gave the notion of a p-adic number a firm foundation, in analogy with the completion $\mathbb{R}$ of the field of rational numbers with respect to the absolute value. More generally, given a number field k, analogous to the usual real or complex absolute values on k, one introduced absolute values associated with prime ideals in the ring of integers of k. These give rise to the infinite number of non-archimedean completions k_v of k with respect to the topology induced on k by (an equivalence class of) such an absolute value. The fields k_v are parametrised by the "finite" places of k, one for each prime ideal. The set V_k of places of k is completed by the finite number of archimedean (or, infinite) places, given by embeddings of k into $\mathbb{R}$ or $\mathbb{C}$ (up to complex conjugation in the latter case) and the respective absolute values on k. At that time, instead of place, the notion "Primstelle" was used, embracing both of the two types of absolute values possible.

In the period 1920–1924, Hasse developed the idea of the p-adic inference from the "Small" to the "Large", as he later referred to it [6]. One of Hasse's results was that quadratic forms are equivalent over $\mathbb{Q}$ if and only if they are equivalent over all completions $\mathbb{Q}_p$, where (p) is a prime ideal in $\mathbb{Z}$ or $p = \infty$ the infinite place of $\mathbb{Q}$. The essence of this work and its generalisation to the case of quadratic forms over algebraic number fields lies in the distinction between the notions "local" versus "global" and their mutual influence. This methodological approach would become the effective principle of Hasse's research plans in the years to come. In simple words, it says that a Diophantine problem has a solution over k if and only if it has a solution over all completions k_v where v ranges over all places of k.[8)]

4. Beyond class field theory

Artin's general reciprocity law [10], in its abstract form, is the heart of the theory of abelian extensions ℓ of algebraic number fields k. It classifies such extensions by certain subgroups $n_{\ell/k}(C_\ell)$ of a group C_k, a modification of the class group of k. Then the result states that there exists a canonical isomorphism $\mathrm{Gal}(\ell/k) \xrightarrow{\sim} C_k/n_{\ell/k}(C_\ell)$, thereby giving a description of the Galois group in terms of the arithmetic of the base field k. As shown by Hasse, this general reciprocity law implies the explicit reciprocity laws obtained previously by C.-F. Gauss, G. Eisenstein, D. Hilbert and T. Takagi.[9)]

On the one hand, these results entailed a certain completion of the theory of abelian field extension within algebraic number theory. On the other hand, a certain stagnancy was perceived in the investigation of arbitrary Galois extensions. Therefore, in the second half of the 1920s, a certain interest unfolded towards a generalisation in the non-commutative theory, that is, the arithmetic and algebra of hypercomplex number systems. Mainly following ideas of Noether, presented in her lectures and seminars in Göttingen, it was thought that the theory

[8)] The beginnings of this principle, especially in the work of H. Minkowski, are discussed in detail in [69]. For a general view on the concept, in particular, its analytic aspects, see [26].

[9)] For background information, we refer to Hasse's *History of class field theory* [50], written 1965, and K. Conrad's more recent overview of the historical development of the theory [27].

of non-commutative algebras could play a significant role in dealing with questions regarding abelian extensions of number fields. Results of Speiser, H. Brandt and E. Artin, among others, supported this impression. Dickson's book [32], now available in the considerably extended version in German, entitled "Algebren und ihre Zahlentheorie" and published in 1927, provided basic material and attracted attention. [10)]

Using the theory of algebras, one seeks to obtain, only relying on structural properties of the underlying algebra, invariant and simple formulations for arithmetic results regarding cyclic field extensions already proved in another framework. Subject to a new proof of this formulation a transfer to the general case of a Galois field extension should be possible. As phrased later on by Noether in her ICM talk in 1932, entitled "Hyperkomplexe Systeme in ihren Beziehungen zur kommutative Algebra und Zahlentheorie" [62],

> "...it should be noted that the main difficulty lies in obtaining the formulation for general Galois fields, for which there was no starting point at all without the hypercomplex method; [...], the underlying transition to the non-commutative is obtained by the simultaneous consideration of fields and groups by means of the "crossed product" and its multiplication constants, the "factor systems".

5. Hasse's research program

Being aware of this approach to class field theory via hypercomplex arithmetic, Hasse combined it with his idea of the p-adic inference from the "Small" to the "Large". He had in mind to go beyond the innovative work of Brauer and Noether forming the theory of hypercomplex number systems in terms of representation theory, and to leave behind the classification of equivalence classes of normal simple K-algebras encoded in the Brauer group. Continuing his previous close contact with Brauer and Noether, Hasse intended to understand completely the internal structure of division K-algebras.

In quick succession, in 1930 and 1931, Hasse presented the results of his investigations and his research plans in various letters, talks and corresponding publications.[11)] These creative efforts culminated in the note of Brauer, Hasse and Noether, the joint work with A.A. Albert, and the publications [47], [48].

5.1. Local case. As a first step, Hasse studied the structure of division algebras defined over a local field given as the $\mathfrak{p}$-adic field k_v attached to a non-archimedean place v of a number field k. We denote by $\mathcal{O}_v$ the valuation ring in k_v. The associated residue class field has finite cardinality, say q. This approach is based on an extension of Hensel's arithmetic of k by passing over to hypercomplex number systems. Hasse [44] gives a precise insight into the algebraic and arithmetic structure of these algebras. The manuscript was completed on 26 February, 1930.

More precisely, given a normal division algebra D over k, the corresponding k_v-algebra $D_v := D \otimes k_v$ is a simple normal k_v-algebra. The structure of these $\mathfrak{p}$-adic division algebras is much simpler than the one in the global case D/k, and it differs essentially from the case where D_v is an extension field ℓ of the local field k_v. Firstly, D_v contains a unique maximal $\mathcal{O}_v$-order, to be denoted Δ_v; we denote by $\mathfrak{p}_v$ the unique maximal ideal of Δ_v. Moreover, the

---

10) It was reviewed by Hasse [43].

11) These range from his work [44], completed on February 26, 1930, a letter to Artin, over a colloquium talk in Göttingen in January 1931, to the paper [47], submitted in May, 1931.

absolute value v on k can be uniquely extended (up to equivalence) to one on D_v. Secondly, the ramification index $e(D_v/k_v)$ and the inertial degree $f(D_v/k_v)$ coincide, and both are equal to the degree m_v of D_v.

Next, on the one hand, D_v contains a unique (up to conjugacy) unramified extension field $k_v(\omega)$ of k_v of degree m_v as a maximal subfield, where ω is a primitive $(q^{m_v}-1)$-th root of unity. The extension $k_v(\omega)/k_v$ is a Galois extension with cyclic Galois group, generated by the Frobenius automorphism σ_v, defined by the assignment $\omega \mapsto \omega^q$. Such a field is called an inertia field of D_v.

On the other hand, given any prime element π of the order Δ_v, the field extension $k_v(\pi)/k_v$ is a totally ramified extension of degree m_v, and $k_v(\pi)$ is a maximal subfield of D_v. The two elements ω and π do not commute if $m_v > 1$, and we have $D_v = k_v(\omega, \pi)$.

In a normalising process we may adapt the choice of π to the previous selection of ω and the specification of any given prime element $a \in \mathcal{O}_v$. Indeed, there exists a prime element $\pi_{D_v} \in \Delta_v$ such that

$$\text{(5.1)} \qquad \pi_{D_v}\omega\pi_{D_v}^{-1} = \omega^{q^{r_v}}, \quad \pi_{D_v}^{m_v} = a$$

where r_v is a positive integer, prime to m_v, and $1 \leq r_v \leq m_v$. Therefore D_v is an algebra of Dickson's type, i.e. $D_v \sim (k_v(\omega)/k_v, \sigma_v, a)$. Clearly the prime element a is not an invariant of D_v whereas the division algebra D_v uniquely determines the integer r_v; it is independent of the choices of ω or a. It follows that D_v is a cyclic division algebra.

By taking intrinsic properties of cyclic algebras (as in (2.4)) into account, one derives that only the value $\mathrm{ord}_v(a) \bmod m_v$ matters in the presentation $D_v \sim (k_v(\omega)/k_v, \sigma_v, a)$. Furthermore, the residue class of the ratio $\mathrm{ord}_v(a)/m_v$ modulo 1 only depends on the equivalence class $[D_v]$, thus, we may call this value the invariant of D_v, to be denoted $\mathrm{inv}_{k_v}[D_v]$. Consequently there is a well-defined map $\mathrm{inv}_{k_v} : Br(k_v) \longrightarrow \mathbb{Q}/\mathbb{Z}, \quad [D_v] \mapsto \mathrm{inv}_{k_v}[D_v]$, from the Brauer group of the local field k_v to the additive group $\mathbb{Q}/\mathbb{Z}$. In fact, the map $\mathrm{inv}_{k_v} : Br(k_v) \longrightarrow \mathbb{Q}/\mathbb{Z}$ is a bijection, and, since $\mathrm{inv}_{k_v}[D_v \times D_v'] = \mathrm{inv}_{k_v}[D_v] + \mathrm{inv}_{k_v}[D_v']$, it is an isomorphism of groups.

Given an archimedean place $v \in V_k$, the structure of the Brauer group $Br(k_v)$ is very simple. If $k_v \cong \mathbb{C}$, then $Br(\mathbb{C})$ is trivial; it consists of the class $[\mathbb{C}]$. If $k_v \cong \mathbb{R}$, then $Br(\mathbb{R}) \cong \mathbb{Z}/2\mathbb{Z}$; with the non-trivial element of order two represented by the $\mathbb{R}$-algebra $\mathbb{H}$ of Hamilton quaternions. Accordingly, we have $\mathrm{inv}_{\mathbb{C}}[\mathbb{C}] = 0$, $\mathrm{inv}_{\mathbb{R}}[\mathbb{R}] = 0$, and $\mathrm{inv}_{\mathbb{R}}[\mathbb{H}] = 1/2$.

5.2. Norm residue symbol. Hasse's approach to these structural results is strongly interwoven with the theory of norm residue symbols, that is, local class field theory. On the one hand, its properties, partly previously proved by global methods and tied to Artin's general reciprocity law, could be reformulated in terms of cyclic crossed product algebras. On the other hand, the final structural result discussed in Section 5.1 enables one to come up with a new proper definition of the norm residue symbol in the local case. Namely, given a cyclic field extension ℓ_v/k_v of degree n and a generating element τ of its Galois group, we consider the invariant $\mathrm{inv}[S] \equiv r/n \bmod 1$ of the equivalence class of the cyclic algebra $(\ell_v/k_v, \tau, a)$, $a \in k_v^\times$. We define the norm residue symbol by $(a, \ell_v/k_v) := \tau^{-r}$. Using the known properties of cyclic algebras, this gives rise to a surjective homomorphism $(-, \ell_v/k_v) : k_v^\times \longrightarrow \mathrm{Gal}(\ell_v/k_v)$, a variant of the norm residue symbol, defined by local data. The kernel of $(-, \ell_v/k_v)$ is equal to the norm group $n_{\ell_v/k_v}(\ell_v/k_v) \subset k_v^\times$. In other words, the cyclic algebra $(\ell_v/k_v, \tau, a)$ splits over ℓ_v if and only if a is in the image of the norm map n_{ℓ_v/k_v}.

5.3. Norm theorem. Hasse shared his results concerning the structure of local algebras at various times with Brauer, Artin and Noether, among others. Noether edited the submission of the relevant manuscript for publication in Mathematische Annalen on 18 June 1930. Artin wrote to Hasse on 27 November of the same year [2]:

> "I also thank you for sending me the corrections of your beautiful and interesting work on hypercomplex arithmetic. Using it, everything turns out to be very simple. What pertains to the proposition on skew fields, I believe, is that each skew field (finite) over a number field is going to be cyclic...Would it be possible for you to prove this with your method, or do you not believe this, or do you even have a counterexample."

Artin saw very clearly which methodological path Hasse had taken in his investigations into the structural theory of division algebras over number fields k and encouraged him in this. In fact, Hasse had already taken a further step in the note [45] of 7 March 1930 by generalising Furtwängler's norm theorem to the following case: If ℓ/k is a cyclic field extension, then an element $a \in k^\times$ is the norm of an element $\alpha \in \ell^\times$ if and only if a is a norm locally everywhere.[12)] This is another manifestation of Hasse's principle of passage from the "Small" to the "Large". At the same time, Hasse disproved the analogous statement in the case of a field extension that is not cyclic by giving a counterexample.

5.4. The year 1931. Without doubt, Artin's statement concerning the structure of division algebras had long been a crucial conjecture that guided Hasse's work in the arithmetic and algebra of hypercomplex number systems. In fact, besides extensive correspondence, he engaged in a wealth of activities in which he expounded his ideas on the proof of the conjecture.

In January 1931, Hasse held a colloquium in Göttingen entitled "Über Schiefkörper". The 17-page handwritten version of his talk [3] provides a detailed insight into his previous investigations and his further, not yet fully formed ideas. Shortly afterwards, on 26–28 February, 1931, Hasse organised a conference in Marburg, devoted specifically to questions on the structure and arithmetic of algebras.[13)] In addition to Hasse, the speakers included Brauer, Noether, M. Deuring, G. Köthe, and H. Brandt. The manuscript for Hasse's lecture "Dicksonsche Schiefkörper von Primzahlgrad"[4] contained not only a comprehensive introduction, but also ideas for a meaningful concept of invariants to characterise cyclic algebras. He started from his results in the $\mathfrak{p}$-adic case in order to indicate a certain direction for the treatment of the general case in question. This lecture required a great familiarity not only with the technical details in the theory of algebras, but also with details of class field theory. In particular, there were open questions about norm residue symbols and the validity of the norm theorem.

5.5. A digression —The correspondence with Adrian A. Albert. One of the participants at the Marburg meeting was Ralph Archibald, a post-graduate from the University of Chicago, visiting Göttingen in the winter of 1930. "That is where my knowledge of American literature comes from", Noether had told Hasse in December 1930. But Hasse himself, no

12) More precisely, given a Galois extension ℓ/k, then all extension fields ℓ_w of k_v where w ranges over all places lying over $v \in V_k$ are isomorphic over k_v. Then a is a norm locally if and only if it is the norm from one of the extensions ℓ_w, and thus from all of them.

13) Noether referred to this meeting as the "Schiefkongress" in her correspondence, for example, the letter to Hasse, 8 February, 1931, [7].

doubt inspired by his reading of Dickson's book "Algebren und ihre Zahlentheorie", was also very interested in the work of the members of the circle around Dickson and Wedderburn.

On the one hand, on 6 February 1931, Adrian A. Albert answered an enquiry from Hasse concerning some results in his dissertation that Noether had previously pointed out to Hasse. The latter had enclosed his publication on division algebras over $\mathfrak{p}$-adic fields with his letter. Albert's reply led to further questions, in particular, concerning division algebras of degree 4. This marked the beginning of an extended correspondence [1]. The first five letters, the last of which Albert wrote to Hasse on 30 June, 1931, form something of a preliminary exchange between the two mathematicians. The final five letters in 1931, particularly those written on 6 November, 26 November, and 9 December, proved crucial to the delicate collaboration that resulted between Albert and Hasse on what they (and others) considered the principal problem in the theory of normal simple algebras over number fields.[14)]

On the other hand, Hasse had long endeavoured to make the results of German mathematicians in the theory of linear associative algebras known to the American mathematical community. It was important to him to demonstrate the deep number-theoretical insights and methods on which many algebraic results were based. Hasse's American publication "Theory of cyclic algebras over an algebraic number field" [47] was written precisely for this reason, submitted on 29 May 1931 to the editors of the Transactions of the AMS. Using the concept of the norm residue symbol, he develops his theory of invariants for the algebras in question. At the same time, in the second section, he deals in detail with the theory of crossed products as developed by Noether in her lectures in the winter term 1929–30 in Göttingen.

6. Structure and results of the paper by Brauer, Hasse and Noether

6.1. The formal structure and the reduction steps. It is Hasse who took over the composition of this note. It begins with a short introduction in which the main theorem is formulated, followed by a proof. This first part is divided into three reductions, whose assertions are labelled I–III, and arranged in the chronological order of their evolvement. Hasse is responsible for the idea underlying step I, Brauer for step II, and Noether for step III. A remark by Noether concludes the proof. The second part of the note, entitled "Conclusions", identifies Hasse as the author of this part.

There are several reasons for this unusual structural organisation of the manuscript. First, according to the correspondence between Hasse, Noether and Brauer, Hasse was only able to add the components contained in reductions 2 and 3 to his argumentation as a conclusion within the period 8–10 November 1931. Secondly, this work was to appear in the volume of Crelle's Journal that was to be presented on the occasion of Hensel's 80th birthday on 29 December 1931. Therefore, a certain haste is noticeable in the presentation of the results in the manuscript, which is also characterised by an abundance of different footnotes. Eventually it was submitted on 11 November, and volume 167 of Crelle's Journal, dedicated to Hensel, was published on 6 January 1932.

Within a short time later, Hasse took up the subject again and completed his treatment of the determination of the Brauer group attached to an algebraic number field, supplemented by a derivation of the reciprocity law from hypercomplex arithmetic. He dedicated this work to Noether and submitted it on 23 March 1932 for publication to Mathematische Annalen.

---

14) In [35], the authors examine the correspondence between Hasse and Albert in 1931, and their final joint publication highlighting Albert's share in dealing with the structural problem.

6.2. The result. Right at the beginning of the paper, as its focal point, it is stated:

Main Theorem. *Every normal division algebra over an algebraic number field is cyclic (or, as one also says, of Dickson's type).*

Its proof is a consequence of Assertion I, given in reduction 1, which reads as

I. *Each algebra over Ω which splits everywhere is $\sim \Omega$.*

Clearly this statement is a manifestation of Hasse's principle of the transfer from the "Small" to the "Large".

Without repeating in detail the lines of argumentation that lead to the proof of this result, we would like to point out three very different characteristic elements that are important for understanding the work, its creation and the interplay of the people involved.

First, unfortunately, the deduction from Assertion I of the Main Theorem contains an unproven statement about the existence of algebraic number fields with given local properties. In the notation of the paper in question, it is the existence of the field Z. This gap could only be closed in full generality much later (see the next section).

Secondly, Assertion I is the centrepiece of the present work. Hasse had been working all summer to find a proof. In his initial approach, he used the theory of associated factor systems, combined with a consideration of Diophantine equations derived from it. However, as this approach proved to be impractical in the general case, he turned to Brauer on 27 July 1931 explaining his line of thought and asking some specific questions in order to make progress.

At this point, neither Brauer nor Artin and Noether, who had also been informed, were able to provide any help. Only when Hasse was able to treat the case of a division algebra D/k which admits a splitting field ℓ with abelian Galois group $\mathrm{Gal}(\ell/k)$ did the arguments of Brauer and Artin given in the reduction steps 2 and 3 come into play. Indeed, with regard to Assertion II, the reduction of the general case via Sylow's theorems to the case of an algebra D which admits a solvable splitting field is already laid out in Brauer's work [13].

Concerning Assertion III, added to the previous steps is the inductive argument presented by Noether in the appendix to her letter of 8 November 1931 to Hasse, which makes it possible to generalise the now familiar case of an abelian algebra to the case of algebras which admit a solvable splitting field. In the decisive induction step, only the Hilbert-Furtwängler norm theorem for extensions of prime degree plays a role; the generalisation to cyclic fields by Hasse is not required.

In subsequent letters to Hasse as the author of the first version of the note, Noether suggested some improvements and additions to the text. She insisted that it be clearly formulated that Hasse had combined the last steps in the proof of the Main Theorem. However, a substantial simplification of the reduction steps, as suggested by Noether, only appeared in Hasse's next publication [48, Sect. 6.12].

Thirdly, in the final part of the note, entitled "Conclusions", Hasse stated that the structure and invariant theory of cyclic algebras developed by him is also valid in the general case of normal simple algebras defined over algebraic number fields. Propositions 1–3 contain the relevant results. In particular, in Proposition 3, an arithmetic characterisation of all the splitting fields of a given normal simple algebra over Ω is given. By asking the reverse question, given an algebraic extension L/Ω, Propositions 4 and 5 assert results concerning the elements of the relative Brauer group $Br(L/\Omega)$, that is, those similarity classes of algebras which split over L. Finally, Hasse answered one of Schur's questions in the affirmative, namely, the absolutely irreducible representations of a finite group of order n are always possible in a cyclotomic field of the n^hth roots of unity with sufficiently large h.

6.3. Existence of algebraic number fields with prescribed local properties. Given a normal division algebra over k, the first step in the proof of the main theorem starting from assertion I requires the existence of a cyclic field extension ℓ/k with prescribed local properties. More precisely, one needs a cyclic extension field ℓ of k such that for every place $v \in V_k$ the degree of ℓ_w is a multiple of the index of D_v. In such a case, the completion ℓ_w with respect to some place w above v is a splitting field for D_v. For a possible proof of this result Hasse referred (via [cf. the note to H. 17Bb]) to his paper [47], at that time a still unpublished manuscript. However, there Hasse referred the reader to the forthcoming joint paper with Albert, published in 1932, where he announced that a complete proof will be published elsewhere [9, p. 723]. One year later, recent work of Hasse's student W. Grunwald seemingly resolved the issue. Yet, in 1935, Hasse was still not perfectly satisfied with the situation when, in a letter to Ph. Furtwängler, he offered a contribution to the envisaged W. Wirtinger Festband of the Monatshefte für Mathematik. He intended "to summarise earlier investigations of mine on the existence of algebraic number fields with prescribed local properties, as they are needed in the theory of algebras, and ... to bring in a readable form and to simplify the proof of the main existence theorem basic for the theory of algebras which a student of mine (Grunwald) did not present quite to my satisfaction." (see [35, Section 6]). In the end, Hasse, jointly with E. Witt, submitted a paper related to algebraic function fields. However, years later, in 1948 in a seminar on class field theory led by Artin at Princeton University, the graduate student S. Wang provided a counterexample to Grunwald's result. Indeed, the latter had overlooked in his work that his result would give an affirmative response to the general question:

> Let k be an algebraic number field. Given $a \in k^\times$ which is an n-th power in k_v for all but finitely many finite places of k, is a an n-th power in k?

Yet, in this generality, this is incorrect. Wang published his counter-example [73], and shortly afterwards he provided a correct statement of the existence theorem in his 1950 publication [74]. At the same time, Hasse [49] (see also [59]) again took up the theme, rectified Grunwald's formulation and gave a proof. Both authors indicated the range of counter-examples. In current terms, they are bound together in the set of *special cases*, whose members are denoted by (k, n, S) where k is an algebraic number field, $n \in \mathbb{N}$, and $S \subset V_k$ a finite set of places of K. All three items are subject to certain conditions, interwoven with one another, see, for example, [74]. The fact that in the case of the field $\mathbb{Q}$ the powers $2^r, r \geq 3$, require special attention and the formulation of the result an adjustment, was already shown by E. Trost in 1934 [71], but remained unnoticed at that time.

6.4. A remarkable footnote: Adrian A. Albert and H. Hasse. The correspondence between Hasse and Albert, alluded to in Section 5, and the immediately resulting discussion of the approach to the various steps in the proof of the Main Theorem were reflected in the footnote "Addendum during printing". It presents results found independently by Albert that overlap with the steps in Reduction 1 and 2. At the same time, it emphasised that the essential ingredient of the proof of the Main Theorem, namely the transfer from the "small" to the "large", i.e. Assertion I, was not available to Albert. In other, more precise words, Assertion I is not found in Albert's letters or publications.

As is evidenced by Albert's reply on 26 November 1931, Hasse informed his colleague of the Main Theorem and its proof. From whatever proof Hasse made available to Albert, the latter immediately recognised the difficulties behind the question of the existence of algebraic numbers with given local properties (see above). Nevertheless, Albert expressed his appreciation for the result obtained. He frankly admitted, "In all my work on division algebras, the principal difficulty has been to somehow find a cyclic splitting field. This your p-adic method accomplishes" (see [1]).

However, from Albert's perspective, parts of the proof of the Main Theorem overlapped with results independently obtained and published by him. He therefore also claimed a certain share of the main result of the note of Brauer, Hasse, and Noether for himself. The footnote in question is a consequence of this request, among others. In addition, following a suggestion by Hasse, Albert shortly afterwards wrote the 5-page paper "A determination of all normal division algebras over an algebraic number field" [9]. It contains, as Albert pointed out, "a historical sketch of the proof, my short proof, and a slight revision (to make it more suitable for American readers) of your proof".[15)]

7. Further developments

7.1. Structure of the Brauer group of an algebraic number field. Let A be a central simple algebra over an algebraic number field k. Given a place $v \in V_k$, there exist a unique positive integer r_v and a (unique up to isomorphism) central division algebra D_v over the completion k_v (associated to the absolute value given by v) such that $A_v \cong M_{r_v}(D_v)$. The local index $m_v := \text{index}[A_v]$ of A_v at $v \in V_k$ is defined to be the degree of D_v. The k-algebra A splits at $v \in V_k$ if and only if $m_v = 1$. For almost all $v \in V_k$, one has $m_v = 1$. This gives rise to a well-defined homomorphism $Br(k) \longrightarrow \oplus_{v \in V_k} Br(k_v)$. The Main Theorem in the note by Brauer, Hasse and Noether [15] asserted that this homomorphism is injective. The subsequent task was to characterise its image.

This could be done by using the local invariants $\text{inv}_{k_v}[A_v]$, $v \in V_k$, attached to a given class $[A]$ in $Br(k)$, as defined in 5.3. It was carried out in detail by Hasse in a manuscript, composed almost at the same time as the joint paper with Brauer and Noether and submitted for publication on 23 March 1932 to Mathematische Annalen. To begin with, Hasse reviewed the basic facts in class field theory and the theory of Brauer groups used in [15], and recalled the proof of the Main Theorem. He then turned his attention to a detailed investigation of the relations between the local invariants, see [48, Sect. 6]. The result he obtained was that the relation

$$\sum_{v \in V_k} \text{inv}_{k_v}[A_v] \equiv 0 \bmod 1$$

holds, and this is the only one. Taking into account the properties of the local invariants, the final result is, in modern terms, that the sequence

$$1 \longrightarrow Br(k) \longrightarrow \bigoplus_{v \in V_k} Br(k_v) \xrightarrow{\text{inv}} \mathbb{Q}/\mathbb{Z} \longrightarrow 0 \tag{7.1}$$

(where $\text{inv} := \sum \text{inv}_{k_v}$) is exact as a sequence of groups. Explicitly, given any set $\{x_v\}_{v \in V_k}$ of reduced fractions such that $x_v = 0/1$ for almost all $v \in V_k$, $\sum_{v \in V_k} x_v \equiv 0 \bmod \mathbb{Z}$, and subject to the conditions: $x_v = 0/1$ for v a complex place, $x_v = 0/1$ or $1/2$ if v is a real place, then there exists a unique class $[A] \in Br(k)$ with this set of local invariants.

15) The circumstances of this manuscript, submitted for publication on 29 January 1932, and the accompanying discussion by letter, are discussed in detail in [35].

The transfer between the "Small" and the "Large", nowadays called the Local-Global principle, as encoded in the exactness of the sequence of Brauer groups is of seminal importance as a guiding principle in various mathematical areas. It can be seen as a strategic approach to link a given problem for an algebraic number field k with the totality of analogous problems for the associated completions k_v, $v \in V_k$.

7.2. Class field theory – revisited. After reading the manuscript of [44], in which Hasse clarified the structure of the local Brauer groups, Noether suggested on 12 April 1931 in a postcard that he should use the local invariants and the sum relation they satisfy for a "hypercomplex creation of the reciprocity law". In fact, using Chevalley's investigations [21] regarding the definition of the norm residue symbol, Hasse showed that the theory of central simple algebras over number fields made it possible to construct local class field theory, "die Klassenkörpertheorie im Kleinen", in a satisfactory way. This was followed by the transition to global class field theory, "der Klassenkörpertheorie im Grossen", in the form of the isomorphism theorem and Artin's reciprocity law. However, whether this successful use of the theory of central simple algebras, that is, non-commutative methods, to justify the essential statements of class field theory could also prove helpful in dealing with the decomposition laws in arbitrary fields remained an open question.

7.3. Claude Chevalley's notion of "éléments idéaux". In the academic year 1931/32, the young Claude Chevalley arrived in Hamburg as a Rockefeller fellow where he came into close contact with Artin. In particular, he attended a course by Artin on class field theory. His presentation and the numerous new proofs he gave of some of the results within the theory had a great influence on Chevalley's mathematical interests and his investigations for his thesis. In this work he intended to give a new treatment of global as well as local class field theory.[16) In the former case, his exposition went far beyond Takagi's work in its simplicity. In the latter case, Chevalley was able to leave behind the previous approach via the global theory. He developed an independent class field theory of abelian field extensions of local fields.[17)

Since March 1934, Hasse, being one of the editors of a new edition of the volume "Algebra und Zahlentheorie" of the *Encyklopädie der Mathematischen Wissenschaften mit Einschluss ihrer Anwendungen*, published since 1906 by B. G. Teubner, had approached Chevalley for a contribution on class field theory. In a letter to Hasse, dated 20 June, 1935, Chevalley outlined some ideas regarding a new approach to class field theory. The letter, published and its mathematical content analysed in [34, Chapter I], fell into two parts, the first one dealing with the local theory, the second one covering the global theory. Here, in order to characterise the abelian extensions, finite or infinite, of an algebraic number field k, Chevalley replaced the classical ideal-theoretic language by a new concept, namely that of "*éléments idéaux (e.i.)*".

In modern terms, by definition, an "*élément idéal*" is a family $x = (x_v)_{v \in V_k}$ of elements $x_v \in k_v$ such that $x_v \neq 0$ for all $v \in V_k$ and x_v is a unit in the ring $\mathcal{O}_v$ of integers in k_v for almost all $v \in V_k$. Endowed with componentwise multiplication, the set of "*éléments idéaux (e.i.)*" forms a group, called the idèle group and to be denoted J_k. The true value of this new concept was then revealed in the two publications [23], [24] in which the approach to the theory proposed in the letter was refined. In the latter one, the notion of "*élément idéal*", now being

---

16) The thesis [22], finally published in November 1933, relied in parts on already earlier communicated notes, among them [20]

17) We refer to [34, Chapter I] for a discussion of the components in Chevalley's thesis, being essentially different from previous points of view

called "*idèle*", and the related notion of the idèle class group played decisive roles. Via the diagonal map $k^\times \longrightarrow J_k$, $a \mapsto (a, \ldots, a, \ldots)$, the group of units $k^\times$ may be identified with a subgroup of the group J_k of idèles; its elements are called "*principal idèles*". The idèle class group is, by definition, the quotient group $C_k := J_k/k^\times$, depending only on the given number field k.

The group J_k carries the structure of a locally compact group. A base for the open sets in J_k is given as $\prod_{v \in V_k} U_v$ where each U_v is an open set in $k_v^\times$ and $U_v = \mathcal{O}_v^\times$ for almost all $v \in V_k$. The group $k^\times$, given the subspace topology, is a discrete, and thus closed, subgroup of J_k. Therefore, the idèle class group C_k is a locally compact Hausdorff group.

As Hasse also recognised in his reply to the letter, Chevalley's new approach to class field theory by passing over to the notion of idèles clearly brings out the overarching local-global principle. Chevalley's letter to Hasse and his subsequent works on class field theory greatly influenced not only the development of number theory but also neighbouring fields. The concept of an idèle, assigned to a number field, more generally, a global field, supplemented by the later additive version, known as adèle, has become indispensable in contemporary research.

7.4. Cohomological methods. Two decades after the publication of the note [15] of Brauer, Hasse, and Noether, G. Hochschild and T. Nakayama [53], [54] introduced cohomological methods into the treatment of class field theory. They played the role of a calculus that enabled a clear and unified structure of that theory in Chevalley's formulation. The seeds for this methodological transformation were already sown in the realm of the note, especially in the use of the concept of factor systems introduced by Brauer. In the language of the cohomology theory of groups, the description given in Theorem 2.2 of the relative Brauer group of a finite Galois extension ℓ/k with Galois group $G_{\ell/k}$ is that given by the second cohomology group of $G_{\ell/k}$ with coefficients in $\ell^\times$. In 1951–52, Artin and J. Tate elaborated the new ideas in their seminar into a comprehensive treatment of class field theory [11]. It is the notion of class formation which clarifies the cohomological aspects of class field theory, both in the local and global cases. This led to a significant simplification of this area of research.

Given a finite Galois extension ℓ/k with Galois group $G_{\ell/k}$, the idèle group J_ℓ can be endowed in a canonical way with a $G_{\ell/k}$-module structure; the submodule $J_\ell^{G_{\ell/k}}$ of elements fixed by $G_{\ell/k}$ is equal to J_k. Since $J_k \cap \ell^\times = k^\times$ holds, the assignment $xk^\times \mapsto x\ell^\times$ defines an injective homomorphism $C_k \longrightarrow C_\ell$. Hence, C_k may be viewed as a subgroup of the idèle class group C_ℓ. The idèle class group C_ℓ inherits a canonical $G_{\ell/k}$-module structure, and

$$[\mathcal{C}]: \quad 1 \to \ell^\times \to J_\ell \to C_\ell \to 1$$

is an exact sequence of $G_{\ell/k}$-modules. Passing over to the corresponding exact sequence, the fact $C_\ell^{G_{\ell/k}} = C_k$ follows because $H^1(G_{\ell/k}, \ell^\times) = \{0\}$.

The determination of the cohomology groups $H^q(G_{\ell/k}, J_\ell)$ of the idèle group can be reduced to local calculations. More precisely, we have an isomorphism

$$H^q(G_{\ell/k}, J_\ell) \cong \bigoplus_{v \in V_k} H^q(G_{\ell_w/k_v}, \ell_w^\times)$$

where w is a fixed chosen place over v. This localisation of the group $H^q(G_{\ell/k}, J_\ell)$ leads to investigations of those cohomology groups $H^q(G_{\ell_w/k_v}, \ell_w^\times)$ that are essential in local class field theory. This transition is compatible with the cohomological mappings inflation, restriction and corestriction that occur when the field extensions in question are changed. As a consequence of the analogous local result, $H^1(G_{\ell/k}, J_\ell) \cong \{0\} \cong H^3(G_{\ell/k}, J_\ell)$ holds.

Given a place $v \in V_k$, using the results of local class field theory, there is an isomorphism,

$$\mathrm{inv}_{\ell_w/k_v} : H^2(G_{\ell_w/k_v}, \ell_w^\times) \xrightarrow{\sim} \frac{1}{[\ell_w : k_v]}\mathbb{Z}/\mathbb{Z} \subset \frac{1}{[\ell : k]}\mathbb{Z}/\mathbb{Z}.$$

Note that choosing a different place w' over $v \in V_k$ does not change this invariant map because there is a canonical isomorphism $H^q(G_{\ell_w/k_v}, \ell_w^\times) \cong H^q(G_{\ell_{w'}/k_v}, \ell_{w'}^\times)$. The local isomorphisms $\mathrm{inv}_{\ell_w/k_v}$, $v \in V_k$, can be combined to form a canonical homomorphism

$$\mathrm{inv}_{\ell/k} : H^2(G_{\ell/k}, J_\ell) \longrightarrow \frac{1}{[\ell : k]}\mathbb{Z}/\mathbb{Z}.$$

If the extension ℓ/k is cyclic, then $H^3(G_{\ell/k}, \ell^\times) = H^1(G_{\ell/k}, \ell^\times) = \{0\}$. The exactness of the cohomology sequence attached to the sequence $[\mathcal{C}]$, combined with $H^1(G_{\ell/k}, C_\ell) = \{0\}$, asserts that the homomorphism $H^2(G_{\ell/k}, \ell^\times) \longrightarrow H^2(G_{\ell/k}, J_\ell)$ is injective, and that its cokernel has order $|H^2(G_{\ell/k}, C_\ell)| = [\ell : k]$. Moreover, using the local theory of the norm residue symbol, one has that $\mathrm{inv}_{\ell/k}(y) = 0$ for $y \in H^2(G_{\ell/k}, \ell^\times)$. Finally, one shows that the sequence

$$0 \longrightarrow H^2(G_{\ell/k}, \ell^\times) \longrightarrow H^2(G_{\ell/k}, J_\ell) \longrightarrow \frac{1}{[\ell : k]}\mathbb{Z}/\mathbb{Z} \longrightarrow 0$$

is exact. Just as the validity of Hasse's theorem is limited to cyclic field extensions, the last assertion is not correct in the general case of a Galois extension ℓ/k. Nevertheless, cyclic field extensions which are contained in a field obtained by adjunction of roots of unity (cyclic cyclotomic field extensions, for short) play a crucial role in global class field theory.

Indeed, given a Galois extension ℓ'/k containing the Galois extension ℓ/k, there is an exact sequence $0 \longrightarrow H^2(G_{\ell/k}, \ell^\times) \xrightarrow{\mathrm{Inf}} H^2(G_{\ell'/k}, \ell'^\times) \xrightarrow{\mathrm{Res}} H^2(G_{\ell'/\ell}, \ell'^\times)$, and, considering the cohomological interpretation of the relative Brauer group, the diagram

$$\begin{array}{ccc} Br(\ell/k) & \xrightarrow{\cong} & H^2(G_{\ell/k}, \ell^\times) \\ \downarrow & & \downarrow \\ Br(\ell'/k) & \xrightarrow{\cong} & H^2(G_{\ell'/k}, \ell'^\times) \end{array}$$

is commutative. Therefore, passing to the limit over ℓ, one obtains

$$Br(k) \xrightarrow{\sim} H^2(G_{k_s/k}, k_s^\times),$$

where k_s denotes a separable closure of k. Both sides can be replaced by the limit over cyclic cyclotomic field extensions ℓ/k, therefore a fortiori by the limit over all cyclic extensions ℓ/k. Next, for any given $n \in \mathbb{N}$, there exists a cyclic extension k'/k with $n \mid [k' : k]$, thus, the groups $\frac{1}{[\ell:k]}\mathbb{Z}/\mathbb{Z}$ where ℓ/k ranges over the cyclic extensions generate the group $\mathbb{Q}/\mathbb{Z}$. Passing to the limit, the sequence

$$0 \longrightarrow H^2(G_{k_s/k}, k_s^\times) \longrightarrow H^2(G_{k_s/k}, J_{k_s}) \longrightarrow \mathbb{Q}/\mathbb{Z} \longrightarrow 0$$

is exact, and $H^2(G_{k_s/k}, J_{k_s})$ is the union of the groups $\oplus_{v \in V_k} H^2(G_{\ell_w/k_v}, \ell_w^\times)$. This short exact sequence is the cohomological form of the sequence (7.1) of Brauer groups.

7.5. Galois cohomology and algebraic groups. The transfer from the "Small" to the "Large", as it appeared in the work of Minkowski and Hasse in the theory of quadratic forms and in the treatment of Brauer groups of algebraic number fields k in the present note, can be formulated uniformly in the framework of Galois cohomology as follows: the natural mapping

$$H^1(G_{k_s/k}, G) \longrightarrow \coprod_{v \in V_k} \lim_{\ell/k} H^1(G_{\ell_w/k_v}, G)$$

is injective when G is a special orthogonal group or the general linear projective group PGL_n. This led to the question for which other algebraic groups this statement is correct. It was answered positively by G. Harder [41], [42], M. Kneser [55], and V.I. Chernoussov [18] in the case of a connected absolutely simple group G. In the general case, the map has finite kernel.

References

Archival sources

[1] *A.A. Albert,* Letters to H. Hasse, Niedersächsische Staats- und Universitätsbibliothek, Göttingen, Handschriftenabteilung [=NSUB], Cod. Ms. H. Hasse 1:25.

[2] *E. Artin,* Letter to H. Hasse, Niedersächsische Staats- und Universitätsbibliothek, Göttingen, Handschriftenabteilung [=NSUB], Cod. Ms. H. Hasse 1:59, nr. 35.

[3] *H. Hasse,* Über Schiefkörper, Lecture in the Math. Gesellschaft Göttingen, January 1931, Niedersächsische Staats- und Universitätsbibliothek, Göttingen, Handschriftenabteilung [=NSUB], Cod. Ms. H. Hasse, 13:16.

[4] *H. Hasse,* No title given, Lecture given at the "Schiefkörper Congress", February 26, 1931, Niedersächsische Staats- und Universitätsbibliothek, Göttingen, Handschriftenabteilung [=NSUB], Cod. Ms. H. Hasse, 13:17.

[5] *H. Hasse,* Darstellungstheorie (Nach E. Noether), Lecture Notes, March, 1931, Niedersächsische Staats- und Universitätsbibliothek, Göttingen, Handschriftenabteilung [=NSUB], Cod. Ms. H. Hasse, 17:18.

[6] *H. Hasse,* Letter to H. Weyl, 15 December, 1931, ETH-Bibliothek Zürich, Archiv, HS 91:591 [publ. in [69]]

[7] *E. Noether,* Letters to H. Hasse, Niedersächsische Staats- und Universitätsbibliothek, Göttingen, Handschriftenabteilung [=NSUB], Cod. Ms. H. Hasse 1:1203A.

Published material

[8] *A.A. Albert,* A determination of all normal division algebras in sixteen units, Trans. Amer. Math. Soc. **31** (1929), 253–260.

[9] *A.A. Albert* and *H. Hasse,* A determination of all normal division algebras over an algebraic number field, Trans. Amer. Math. Soc. **34** (1932), 722–726.

[10] *E. Artin,* Beweis des allgemeinen Reziprozitätsgesetzes, Abh. Math. Semin. Univ. Hambg. **5** (1927), 353–363.

[11] *E. Artin* and *J. Tate,* Class Field Theory, Benjamin, New York-Amsterdam, 1967.

[12] *R. Brauer,* Untersuchungen über die arithmetischen Eigenschaften von Gruppen linearer Substitutionen, Math. Z. **28** (1928), 677–696.

[13] *R. Brauer,* Über Systeme hyperkomplexer Zahlen, Math. Z. **30** (1929), 79–107.

[14] *R. Brauer* and *E. Noether,* Über minimale Zerfällungskörper irreduzibler Darstellungen, Sitzungsber. Preuss. Akad, Wiss. Berlin, **32** (1927), 221–228.

[15] *R. Brauer, H. Hasse,* and *E. Noether,* Beweis eines Hauptsatzes in der Theorie der Algebren, J. reine angew. Math., **167** (1932), 399–404.

[16] *É. Cartan,* Sur les groupes bilinéaires et les systèmes de nombres complexes, Ann. Fac. Sci. Toulouse Math., Serie 1, **12** (1898), 1–64.

[17] *É. Cartan,* Nombres complexes, d'après l'article allemand de E. Study, Encyclopédie des Sciences Mathématiques (édition francaise), I, 5, 329-468. Paris, Gauthier-Villars. Oeuvres II, 1, 107-246.

[18] *V.I. Chernousov,* On the Hasse principle for groups of type E_8, Dokl. Akad. Nauk. SSSR **306** (1989), 1059–1063; English translation in Soviet Math. Dokl. **39** (1989), 592–596.

[19] *C. Chevalley,* Sur la théorie de restes normiques, C. R. Acad. Sci. Paris, **191** (1930), 426–428.

[20] *C. Chevalley,* Sur la structure de la théorie du corps de classes, C. R. Acad. Sci. Paris, **194** (1932), 766–769.

[21] *C. Chevalley*, La théorie du symbole de restes normiques, J. reine angew. Math., **169** (1933), 140–157.
[22] *C. Chevalley*, Sur la théorie du corps de classes dans les corps finis et les corps locaux, J. Fac. Sci. Univ. Tokyo (1933), I-2, 365–475.
[23] *C. Chevalley*, Généralisation de la théorie du corps de classes pour les extensions infinies, J. Math. Pures Appl. **15** (1936), 359–371.
[24] *C. Chevalley*, La théorie du corps de classes, Annals of Math. **41** (1940), 394–418.
[25] *C. Chevalley* and *J. Herbrand*, Nouvelle démonstration du théoreme d'existence en théorie du corps de classes, C. R. Acad. Sci. Paris, **193** (1931), 814–815.
[26] *R. Chorlay*, "Local-Global": the first twenty years, Arch. Hist. Exact Sci. **65** (2011), 1–66.
[27] *K. Conrad*, History of class field theory, https://kconrad.math.uconn.edu/blurbs/gradnumthy/cfthistory.pdf
[28] *M. Deuring*, Algebren, Ergeb. Math. Grenzgebiete, Bd. 4, Verlag J. Springer, Berlin, 1935.
[29] *L.E. Dickson*, Definition of a linear assocative algebra by independent postulates, Trans. Amer. Math. Soc. **4** (1903), 21–26.
[30] *L.E. Dickson*, Linear algebras in which division is always uniquely possible, Bull. Amer. Math. Soc. **12** (1906), 441–442.
[31] *L.E. Dickson*, Linear associative algebras and abelian equations, Trans. Amer. Math. Soc. **15** (1914), 31–46.
[32] *L.E. Dickson*, Algebras and their Arithmetics, University of Chicago Press, Chicago, 1923.
[33] *L.E. Dickson*, Algebren und ihre Zahlentheorie, Orell Füssli Verlag, Zürich & Leipzig, 1927.
[34] *D. Dumbaugh* and *J. Schwermer*, Emil Artin and Beyond - Class Field Theory and L-functions, Heritage of European Mathematics, Volume: 9; EMS Publishing House, Zürich, 2015.
[35] *D. Fenster* and *J. Schwermer*, A delicate collaboration: Adrian Albert and Helmut Hasse and the Principal Theorem in division algebras in the early 1930s, Arch. Hist. Exact Sci. **59** (2005), 349–379.
[36] *D. Fenster* and *J. Schwermer*, Beyond class field theory: Helmut Hasse's arithmetic in the theory of algebras in early 1931, Arch. Hist. Exact Sci. **61** (2007), 425–456.
[37] *G. Frobenius*, Gesammelte Abhandlungen, ed. by J-P. Serre, volumes I - III, Springer, Berlin, 1968.
[38] *G. Frobenius*, Über endliche Grupppen, Sitzungsberichte Kgl. Preuss. Akad. Wiss. Berlin, (1895), 163–194. [Ges. Abh. vol. II, 632–663].
[39] *G. Frobenius*, Theorie der hypercomplexen Systeme, Sitzungsberichte Kgl. Preuss. Akad. Wiss. Berlin, (1903), 504–537. [Ges. Abh., vol. III, 284–329]
[40] *W. Grunwald*, Ein allgemeines Existenztheorem für algebraische Zahlkörper, J. reine angew. Math., **169** (1933), .103–107.
[41] *G. Harder*, Über die Galoiskohomologie halbeinfacher Matrizengruppen.I. Math. Z. **90** (1965), 404–428; II. Math. Z. **92**, (1966) 396–415.
[42] *G. Harder*, Bericht über neuere Resultate der Galoiskohomologie halbeinfacher Gruppen. Jahresbericht der DMV **70** (1968) 182–216.
[43] *H. Hasse*, Review of "L.E. Dickson, Algebren und ihre Zahlentheorie", Jahresber. Dtsch. Math.-Ver., **37** (1928), 90–97.
[44] *H. Hasse*, Über p-adische Schiefkörper und ihre Bedeutung für die Arithmetik hyperkomplexer Zahlsysteme, Math.Ann. **104** (1931), 495–536.
[45] *H. Hasse*, Beweis eines Satzes und Widerlegung einer Vermutung über das allgemeine Normenrestsymbol, Nachr. Ges. Wiss. Göttingen, Math.-Phys. Kl., 1931, 64–69.
[46] *H. Hasse*, Theorie der zyklischen Algebren über einem algebraischen Zahlkörper, , Nachr. Ges. Wiss. Göttingen, Math.-Phys. Kl., 1931, 70–79.
[47] *H. Hasse*, Theory of cyclic algebras over an algebraic number field, Trans. Amer. Math. Soc. **34** (1932), 171–214, 727–730.
[48] *H. Hasse*, Die Struktur der R. Brauerschen Algebrenklassengruppe über einem algebraischen Zahlkörper, Math.Ann. **107** (1933), 731–760.
[49] *H. Hasse*, Zum Existenzsatz von Grunwald in der Klassenkörpertheorie, J. reine angew. Math., **188** (1950), 40–64.
[50] *H. Hasse*, History of class field theory. In *Algebraic Number Theory*, ed. by J.W.S. Cassels and A. Fröhlich, Academic Press. London, 1967, 266–279.
[51] *Th. Hawkins*, Hyperomplex numbers, Lie groups, and the creation of group representations, Arch. Hist. Exact Sci. **8** (1972), 243–287.
[52] *K. Hensel*, Neue Grundlagen der Arithmetik. J. reine angew. Math. **127** (1904), 51–84.
[53] *G. Hochschild*, Local class field theory Annals of Math. **51**, (1950), 331–347.
[54] *G. Hochschild* and *T. Nakayama*, Cohomology in class field theory Annals of Math. **55**, (1952), 348–366.
[55] *M. Kneser*, Lectures on Galois Cohomology of Classical Groups, Tata Institute of Fundamental Research, Bombay 1969.

[56] *F. Lemmermeyer* and *P. Roquette*, Helmut Hasse und Emmy Noether. Die Korrespondenz 1925–1935, Universitätsverlag, Göttingen, 2006.
[57] *S. Lie* and *G. Scheffer*, Vorlesungen über continuierliche Gruppen mit geometrischen und anderen Anwendungen, Leipzig, B.G. Teubner, 1893.
[58] *Th. Molien*, Ueber Systeme höherer complexer Zahlen, Math. Ann. **41** (1893), 83–156.
[59] *P. Morton*, A correction to Hasse's version of the Grunwald-Hasse-Wang theorem, J. reine angew. Math. **659** (2011), 169–174.
[60] *E. Noether*, Hyperkomplexe Grössen und Darstellungstheorie, Math. Z. **30**, (1929), 641–692.
[61] *E. Noether*, Algebra der hypercomplexen Grössen, In: Ges. Abhandl., Ed. N. Jacobson, Springer, Berlin-Heidelberg-New York, 1983, pp. 711–763.
[62] *E. Noether*, Hyperkomplexe Systeme in ihren Beziehungen zur Algebra und Zahlentheorie, Verhandl. Int. Math. Kongress Zürich 1932, Bd. 1, Orell-Füssli, Zürich-Leipzig, 1932, 189-194.
[63] *E. Noether*, Gesammelte Abhandlungen, Ed. N. Jacobson, Springer, Berlin-Heidelberg-New York, 1983.
[64] *K. Parshall*, Joseph H.M. Wedderburn and the structure theory of algebras, Arch. Hist. Exact Sci. **32** (1985), 223–349.
[65] *B. Peirce*, Linear associative algebra, Amer. J. Maths. **4** (1881), 97–229.
[66] *H.M. Pycior*, Benjamin Peirce's *Linear associative algebra*, Isis **70** (1979), 537–551.
[67] *P. Roquette*, The Brauer-Hasse-Noether Theorem in Historical Perspective, Springer, Berlin-Heidelberg, 2005.
[68] *G. Scheffers*, Zurückführung complexer Zahlensysteme auf typische Formen, Math. Ann. **39** (1891), 294–340.
[69] *J. Schwermer*, Minkowski, Hensel and Hasse - On the beginnings of the local-global principle, In: Episodes in the history of modern algebra (1850-1950), ed. J. J. Gray and K. Parshall, American Mathematical Society, Rhode Island, 2006, 153–177.
[70] *E. Study*, Theorie der gemeinen und höheren complexen Grössen, In: Encyclopädie der math. Wissenschaften, vol IA4, ed. W.F. Meyer, Vieweg Teubner, Wiesbaden, 1898, 147–183.
[71] *E. Trost*, Zur Theorie der Potenzreste, Nieuw Archief voor Wiskunde **18** (1934), 58-61.
[72] *P. Ullrich*, The genesis of Hensel's p-adic numbers, In: Charlemagne and his heritage: 1200 years of civilisation and science in Europe, vol. 2, Mathematical Arts, ed. P. Butzer et al., Brepol Publishers, Turnhout (1998), 163–178.
[73] *Sh. Wang*, A counter-example to Grunwald's theorem, Annals of Math. **49** (1948), 1008–1009.
[74] *Sh. Wang*, On Grunwald's theorem, Annals of Math. **51** (1950), 471–484.
[75] *J.H.M. Wedderburn*, On hypercomplex numbers, Proc. London Math. Soc. **s2-6**, (1908), 77–118.
[76] *J.H.M. Wedderburn*, A type of primitive algebras, Trans. Amer. Math. Soc. **15** (1914), 162–166.

Joachim Schwermer, Faculty of Mathematics, University of Vienna, Oskar-Morgenstern-Platz 1, A-1090 Vienna, Austria and Max-Planck-Institute for Mathematics, Vivatsgasse 7, D-53111 Bonn, Germany.
e-mail: Joachim.Schwermer@univie.ac.at

1951–1975

J. reine angew. Math. **212** (1963), 7–25
DOI 10.1515/crll.1963.212.7

Journal für die reine und angewandte Mathematik

Notes on elliptic curves. I.

By *B. J. Birch* at Cambridge (England) and *H. P. F. Swinnerton-Dyer* at Cambridge (England)

Siegel [10], [11] has shown that the density of rational points on a quadric surface can be expressed in terms of the densities of p-adic points; which for almost all primes p depends directly on the number of solutions of the corresponding equation in the finite field with p elements. More recently, Siegel's work has been fruitfully extended and simplified by Tamagawa (see [14]) and independently by Kneser.

It is natural to hope that something similar will happen for the elliptic curve

$$\Gamma\colon y^2 = x^3 - Ax - B \tag{1}$$

where A, B are rational. In particular, one hopes that if for most p the curve (1) has unusually many points in the finite field with p elements, then it will have a lot of rational points.

Several years ago we carried out by hand some calculations which tended to support these hopes. Since then, we have used the Cambridge University electronic computer EDSAC 2 for extensive calculations on elliptic curves. As a result, we have been able to put our conjectures into an exact form and support them with a good deal of numerical evidence. However, it has become clear to us that we are unlikely to be able to prove these conjectures.

We therefore feel it best to publish the material we have obtained, in a series of short notes. Much of the information in these notes is the result of machine computation; however the theoretical basis of these computations is not always trivial.

We wish to express our indebtedness to Prof H. Davenport and Dr J. W. S. Cassels for their advice and encouragement, and to Dr M. V. Wilkes for permission to use EDSAC 2.

1. It is well known that any elliptic curve with a rational point can be written in the form (1) or the equivalent homogeneous form

$$\Gamma\colon y^2z = x^3 - Axz^2 - Bz^3.$$

Here $4A^3 - 27B^2 \neq 0$, since otherwise the curve would have a double point and so be rational; and without loss of generality we may suppose that A and B are integers.

We write $\mathfrak{o}$ for the point $(0, 1, 0)$; then the points of Γ form an abelian group with $\mathfrak{o}$ as unit, and the rational points form a subgroup $\mathcal{A}$ which is finitely generated. It is easy to find the elements of $\mathcal{A}$ of finite order (see Lutz [7]); thus the most interesting thing to find out about Γ is the value of g, the number of independent generators of $\mathcal{A}$ of infinite order. This note is concerned with the practical computation of g, though some other results arise as by-products.

Let g_2 be the number of independent generators of $\mathcal{A}$ of finite even order; then $g_2 = 0, 1$ or 2 according as $x^3 - Ax - B = 0$ has 0, 1 or 3 rational roots. The group $\mathcal{A}/2\mathcal{A}$ has order

2^{g+g_2}, and to find g we have only to exhibit $\mathcal{A}/2\mathcal{A}$. Following Cassels [3], II, we say that there is a 2-covering U of Γ if there is a curve D defined over the rationals and a commutative triangle

$$\begin{array}{ccc} \Gamma & \longrightarrow & \Gamma \\ \updownarrow & \nearrow & \\ D & & \end{array}$$

with associated generic points

$$\begin{array}{ccc} \boldsymbol{x}_1 & \longrightarrow & \boldsymbol{x} = 2\boldsymbol{x}_1 \\ \updownarrow & \nearrow & \\ \boldsymbol{X} & & \end{array}$$

where the map $\boldsymbol{X} \to \boldsymbol{x}$ is over the rationals, and $\boldsymbol{X} \leftrightarrow \boldsymbol{x}_1$ is over the complex numbers. Another curve D' with generic point $\boldsymbol{X}'$ gives the same 2-covering if and only if there is a birational mapping $\boldsymbol{X} \leftrightarrow \boldsymbol{X}'$ over the rationals and a point $\mathfrak{d}$ on Γ with $2\mathfrak{d} = \mathfrak{o}$ such that the diagram

$$\begin{array}{ccc} \boldsymbol{X} & \longleftrightarrow & \boldsymbol{x}_1 \\ \updownarrow & & \updownarrow \\ \boldsymbol{X}' & \longleftrightarrow & \boldsymbol{x}_1' = \boldsymbol{x}_1 + \mathfrak{d} \end{array}$$

is commutative. There is a natural law of composition of 2-coverings, inherited from the law of composition of the curves D regarded as homogeneous spaces (Weil [12], [13]); under this law the 2-coverings form an abelian group and every element other than the identity has order 2. We shall be interested in two of its subgroups; G, the group of those 2-coverings U for which D has a point in each p-adic field (including the reals), and G', the group of coverings for which D actually has a rational point.

Weil has shown that G is finite, and we shall write its order as 2^k. Moreover, G' is naturally isomorphic to $\mathcal{A}/2\mathcal{A}$ and so has order $2^{\mathrm{g}+\mathrm{g}_2}$; since G can be computed, this gives a useful bound for g. In most numerical cases, we find that $k = \mathrm{g} + \mathrm{g}_2$. When $k > \mathrm{g} + \mathrm{g}_2$, the elements of G/G' are principal homogeneous spaces and in fact are just the elements of order 2 in the Tate-Šafarevič group.

It is possible to find the elements of G by the classical process of descent; and for hand calculation this is probably the easiest way. However for any given curve Γ one needs to know the structure of the appropriate algebraic number field, and it is not convenient to investigate this by means of an automatic computer. We have therefore used a different procedure. To justify it, we first find a canonical form for the 2-coverings of Γ.

Lemma 1. *If D is a curve corresponding to a 2-covering U in G (so that D has a point in every p-adic field) then D has a positive rational divisor of degree* 2.

The mapping $D \times D \to \Gamma$ given by $\boldsymbol{X}_1 \times \boldsymbol{X}_2 \to \boldsymbol{x} = \boldsymbol{X}_1 + \boldsymbol{X}_2$ induces a mapping of positive divisors of degree 2 on D to points of Γ. The mapping is defined over the rationals, and divisors map onto the same point if and only if they are linearly equivalent. Hence the divisors that map onto $\mathfrak{o}$ can be parametrised over the rationals by a curve C of genus 0. If $(\boldsymbol{X}_1, \boldsymbol{X}_2)$ is any such divisor, then the fields of definition of $\boldsymbol{X}_1$ and $\boldsymbol{X}_2$ are the same; thus C

has points in each p-adic field, since D has such points. It now follows that C has a rational point; and the divisor on D corresponding to this point satisfies the conditions of the lemma.

This is a purely ad hoc proof and cannot be extended to n-coverings, although the corresponding theorem is still true (see Cassels [3], IV, Theorem 1. 3). The condition that D has points in every p-adic field is in fact necessary, as Cassels has subsequently shown (see [3], V); an earlier example constructed by Lang and Tate [6] of a 2-covering with no rational divisor of degree 2 involves a more complicated ground field.

Lemma 2. *If U is a 2-covering in G, then we can take D in the form*

$$y^2 = g(x) \equiv ax^4 + bx^3 + cx^2 + dx + e \tag{2}$$

where a, b, c, d, e are rational.

This is essentially well-known after Lemma 1; compare Chevalley [4], p. 24. We denote by Q the rational numbers. Let $\mathfrak{a}$ be the divisor given by Lemma 1; by the Riemann-Roch theorem there is a non-constant function x with $(x) > -\mathfrak{a}$, and a further function v linearly independent of 1, x, x^2 with $(v) > -2\mathfrak{a}$. Now the functions whose divisor $> -4\mathfrak{a}$ form a linear space of dimension 8 which contains the 9 functions

$$1, x, x^2, x^3, x^4, v, xv, x^2v, v^2;$$

hence there is a linear dependence relation between them. This must actually involve v^2, for otherwise v would be in the field $Q(x)$, and since $\mathfrak{a}$ is the actual divisor of poles of x – the function field of D contains no functions with only one simple pole – this would make v linearly dependent on 1, x, x^2. Suppose then that the relation is

$$v^2 - 2v(\lambda x^2 + \mu x + \nu) = \alpha x^4 + \beta x^3 + \gamma x^2 + \delta x + \varepsilon.$$

If we now write $y = v - \lambda x^2 - \mu x - \nu$, which is not a constant, we obtain an equation of the form (2). Since the function field of D is quadratic over $Q(x)$, it must be precisely $Q(x, y)$; this proves the lemma.

Conversely, any curve (2) gives rise to a 2-covering of its Jacobian by mapping the point $(\xi, 0)$ onto $\mathfrak{o}$; here ξ is any root of $g(x) = 0$, and the covering is the same whichever root we take. Two curves D and D* of the form (2) give the same covering if and only if there is a birational mapping $D \leftrightarrow D^*$ defined over Q in which the points with $y = 0$ correspond to those with $y^* = 0$. Such a mapping must be bilinear between x and x^*; thus it exists if and only if there are rational numbers $\alpha, \beta, \gamma, \delta, \mu$ such that

$$g(x) = \mu^2(\gamma x + \delta)^4 g^* \left(\frac{\alpha x + \beta}{\gamma x + \delta}\right). \tag{3}$$

In this case we shall say that the two quartics $g(x)$ and $g^*(x^*)$ are *equivalent;* this is slightly weaker than the usual equivalence of binary quartic forms under rational linear transformation.

The 2-coverings produced by the classical method of descent are all of the form (2); to prove that such a descent gives all the members of G (as opposed to G') one may use the exact sequence of Lang and Tate [6], or else proceed explicitly, using Lemma 4 of Cassels [3], II.

2. In order to find the elements of G for a given Γ, we have to find a representative of each equivalence class of quartic forms $g(x)$ corresponding to Γ. To do this, we define a reduced quartic form in such a way that each equivalence class contains at least one reduced form and that the reduced forms corresponding to any given Γ are finite in number and reasonably easy to compute. Ideally, we would like just one reduced form in each equivalence class, but this does not seem to be practicable.

We write the invariants of $g(x)$ as

$$(4) \qquad I = 12ae - 3bd + c^2, \quad J = 72ace + 9bcd - 27ad^2 - 27eb^2 - 2c^3.$$

Then (2) is a 2-covering of

$$(5) \qquad y^2 = x^3 - 27Ix - 27J;$$

and conversely any 2-covering of (5) has invariants $I\lambda^4$, $J\lambda^6$ for some rational λ.

Replacing x and y by suitable multiples, we can bring any curve (1) into the form (5) where I, J are integers and there is no integer $m > 1$ with $m^4 \mid I$, $m^6 \mid J$. Henceforth we shall assume this done. Similarly, by absorbing denominators into y, we may assume in Lemma 2 that a, b, c, d, e are integers. Thus the invariants of $g(x)$ will be $n^4 I$, $n^6 J$ for some integer n. Our next object is to show that we need only consider small values of n.

Lemma 3. *Let $g(x) = ax^4 + bx^3 + cx^2 + dx + e$, where a, b, c, d, e are rational integers, have invariants I, J. Suppose that p is a prime other than 2 or 3, that $p^4 \mid I$ and $p^6 \mid J$, and that $y^2 = g(x)$ is p-adically soluble. Then there is a $g^*(x)$ equivalent to $g(x)$ which has integral coefficients and whose invariants are $p^{-4}I, p^{-6}J$.*

The proof is straightforward, though somewhat tedious because of the number of cases to be considered.

Case 1: $p \nmid (a, b, c, d, e)$. Since $p \geqq 5$ there is an m such that $p \nmid g(m)$. Replacing y by yx^{-2} and x by $m + x^{-1}$, we can assume that $p \nmid a$. Since $I \equiv J \equiv 0 \bmod p$, $g(x) \equiv 0$ has a triple root. Taking this to be zero, we have $p \mid (c, d, e)$. There are now two possibilities:

Case 1.1: $p \mid b$. Now we find successively $p^2 \mid e$ since $p^2 \mid I$; $p^2 \mid d$ since $p^3 \mid J$; $p^2 \mid c$ since $8c^3 \equiv 6cI - J \equiv 0 \bmod p^4$; $p^3 \mid e$ since $p^3 \mid I$; $p^3 \mid d$ since $p^5 \mid J$; $p^4 \mid e$ since $p^4 \mid I$. Now writing px for x and $p^2 y$ for y we get the curve

$$D^*\colon y^2 = ax^4 + bx^{-1}x^3 + cp^{-2}x^2 + dp^{-3}x + ep^{-4},$$

equivalent to D, with integer coefficients and invariants $p^{-4}I, p^{-6}J$.

Case 1.2: $p \nmid b$. By adding a suitable multiple of p to x we may further ensure that $p^4 \mid c$. Now we find $p^2 \mid e$ since $p^2 \mid J$; $p^2 \mid d$ since $p^2 \mid I$; $p^4 \mid e$ since $p^4 \mid J$; $p^4 \mid d$ since $p^4 \mid I$; $p^6 \mid e$ since $p^6 \mid J$. Now writing $p^2 x$ for x and $p^3 y$ for y we get the curve

$$D^*\colon y^2 = ap^2x^4 + bx^3 + cp^{-2}x^2 + dp^{-4}x + ep^{-6},$$

which does what we want.

Case 2: $p \mid (a, b, c, d, e)$. If $p^2 \mid (a, b, c, d, e)$ we can write py for y, which has the effect of dividing all the coefficients by p^2. If not, we may assume that $p^2 \nmid a$ and $p^2 \mid (c, d, e)$ by an argument similar to that of Case 1. Again there are two possibilities:

Case 2. 1: $p^2 \mid b$. We have $p^3 \mid e$ since $p^4 \mid I$; and $p^3 \mid d$ since $p^6 \mid J$. But now $p^4 \mid e$; for otherwise the power of p that divides $\mathrm{g}(x)$ would be odd whatever the value of x, and so $y^2 = \mathrm{g(x)}$ would not be soluble in p-adic numbers. Now we can use the reduction in Case 1. 1.

Case 2. 2: $p^2 \nmid b$. Now we find $p^3 \mid e$ since $p^5 \mid J$; $p^3 \mid d$ since $p^4 \mid I$; $p^4 \mid e$ since $p^6 \mid J$. Thus we can again use the reduction in Case 1. 1.

This completes the proof of the lemma.

Lemma 4. *With the notation of Lemma* 3, *suppose that either* $3^5 \mid I$, $3^9 \mid J$ *or* $3^4 \parallel I$, $3^6 \parallel J$, $3^{15} \mid (4I^3 - J^2)$. *Then if* $y^2 = g(x)$ *is* 3*-adically soluble, there is a* $g^*(x)$ *equivalent to* $g(x)$ *with integer coefficients and invariants* $3^{-4}I$, $3^{-6}J$.

Lemma 5. *With the notation of Lemma* 3, *suppose that* $2^6 \mid I$, $2^9 \mid J$, $2^{10} \mid (8I + J)$. *Then if* $y^2 = g(x)$ *is* 2*-adically soluble, there is a* $g^*(x)$ *equivalent to* $g(x)$ *with integer coefficients and invariants* $2^{-4}I$, $2^{-6}J$.

The proofs of these lemmas are similar to that of Lemma 3, but far more tedious: we have therefore omitted them. Lemma 4 is best possible in the sense that if the conditions on I, J are not satisfied no such reduction is possible. For if $g^*(x)$ is such that $3 \mid c^*$ then $3 \mid I^*$ and $3^3 \mid J^*$; if $3 \nmid c^*$ then $3 \nmid I^*, 3 \nmid J^*, 3^3 \mid (4I^{*3} - J^{*2})$.

It follows from these three lemmas that in looking for 2-coverings of a given curve we can confine ourselves to curves of the form (2) with integral a, b, c, d, e and one of two possible pairs of values of I, J. We have still available the possibility of making unimodular integral bilinear transformations on x; that is, of taking

$$g'(x) = (\gamma x + \delta)^4 g\left(\frac{\alpha x + \beta}{\gamma x + \delta}\right)$$

where $\alpha, \beta, \gamma, \delta$ are integers with $\alpha\delta - \beta\gamma = 1$. Hermite has given a general theory of the reduction of binary forms of arbitrary degree under such transformations; and for quartic forms this has been worked out in detail by Julia [5]. The basic idea is to find a positive definite quadratic form which is a covariant of the original form under real linear transformations, and to say that the original form is reduced if this quadratic form is.

Julia's results are stated in a geometric form, and in one of the three cases that have to be considered his definition leads to inequalities very inconvenient to use; we have therefore developed the theory independently, and have chosen our definition of reduction purely for computational convenience. If the discriminant $\Delta = 4I^3 - J^2$ is positive, our inequalities like Julia's ensure that except on the boundaries of the reduced domain there is only one reduced form in each class of forms equivalent under integral unimodular transformation; accordingly, the size of the region of search for a, b, c, and so the amount of computation needed, is approximately proportional to $|\Delta|^{\frac{1}{2}}$. If Δ is negative, our definition fails to ensure this, and the amount of computation involved is approximately proportional to $|\varphi^2 - I|^{\frac{3}{2}}$, where φ is defined by (10) below: accordingly, we lose very little unless Δ is small and negative. Since the equivalence we are really interested in is not integral unimodular equivalence, but the equivalence given by (3) above, some reprocessing to remove equivalent forms would be necessary anyway.

If $g(x)$ has a rational root, it gives rise to the trivial 2-covering defined by the map $X \to 2X$ of Γ onto itself. We may therefore assume that this does not happen, and in particular that $a \neq 0$. Now we can write $g(x)$ as a product of real quadratic factors

$$(6) \qquad g(x) = a(x^2 + px + q)(x^2 + p'x + q').$$

If $g(x)$ has four real roots this can be done in more than one way: we choose that one for which the roots of either factor separate those of the other. We shall also assume that

$$(7) \qquad 4q - p^2 \geqq 4q' - p'^2,$$

and, writing $-x$ for x if necessary, that

$$(8) \qquad p' \geqq p.$$

It follows immediately from (6) that

$$(9) \qquad p + p' = \frac{b}{a},\ pp' = \frac{2c - \varphi}{3a},\ q + q' = \frac{c + \varphi}{3a},\ pq' + p'q = \frac{d}{a},\ qq' = \frac{e}{a},$$

where

$$(10) \qquad \varphi^3 - 3I\varphi + J = 0.$$

Write

$$(11) \qquad H = 8ac - 3b^2,$$

$$(12) \qquad \lambda^2 = 4q - p^2,\ \lambda'^2 = 4q' - p'^2;$$

then λ^2, λ'^2 (which need not be positive) are the roots of

$$(13) \qquad 3a^2\Lambda^2 - (H + 2a\varphi)\,\Lambda + \left(4I - \varphi^2\right) = 0.$$

Since p, p' are real, it follows from (9) that

$$(14) \qquad H \leqq 4a\varphi.$$

We shall now have to split cases according to the number of real roots of $g(x)$. Our basic idea, however, is the same in all three cases. We find a quadratic $h(x)$ which is a covariant of $g(x)$ under real linear transformations. By a suitable integral unimodular transformation we could ensure that $h(x)$ was reduced in the classical sense. This in particular would imply that the square of the leading coefficient of $h(x)$ was at most one third of its discriminant. We shall in fact merely require the latter condition, which still leaves us free to replace x by $x + n$ for any integer n. (Such a transformation does not alter the validity of (7) or (8)). We now choose n, as is obviously possible, so that

$$(15) \qquad 2\,|\,a\,| \geqq b \geqq -2\,|\,a\,|.$$

In splitting cases, we note that if $4I^3 > J^2$ then $g(x)$ has no or four real roots, while if $4I^3 < J^2$ it has just two.

Case 1: *No real roots.* We may now assume that $a > 0$, for if $a < 0$ the equation $y^2 = g(x)$ has no real solutions. The cubic (10) has three real roots; let them be $\varphi_1 > \varphi_2 > \varphi_3$.

Since (14) is the condition that the factorisation (6) is real, which happens for just one of the values of φ, we have $\varphi = \varphi_1$ and

$$(16) \qquad 4a\varphi_1 \geqq H \geqq 4a\varphi_2.$$

Since $x^2 + px + q$, $x^2 + p'x + q'$ have no real factors, λ^2 and λ'^2 in (12) are positive and we may take λ, λ' real and positive. The positive definite quadratic form

$$h'(x) = \lambda^{-1}(x^2 + px + q) + \lambda'^{-1}(x^2 + p'x + q')$$

is a covariant of $g(x)$ under real transformations; for it is just the sum of the real quadratic factors each normalised to have discriminant 1. Write for convenience

$$K = \lambda^2\lambda'^2 a^{-2} = \frac{1}{3}\left(4I - \varphi_1^2\right);$$

then $K > 0$ and the discriminant of $h(x)$ is $2 + 2K^{-\frac{1}{2}}\varphi_1$. The condition that the square of the leading coefficient of $h(x)$ should be at least one third of the discriminant gives, after some manipulation,

$$(17) \qquad H \leqq 2K + 2K^{\frac{1}{2}}\varphi_1 - 6K^{\frac{1}{2}}a - 2a\varphi_1;$$

combining this with (16) we find

$$(18) \qquad 0 < a \leqq \frac{K + K^{\frac{1}{2}}\varphi_1}{3K^{\frac{1}{2}} + \varphi_1 + 2\varphi_2},$$

since the denominator in this last expression is positive. We note that (17) is stronger than the first half of (16) if and only if $3a > K^{\frac{1}{2}}$.

Case 2: *Four real roots.* The cubic (10) again has three real roots φ_1, φ_2, φ_3; we shall number them so that $\varphi_1 > \varphi_2 > \varphi_3$ if $a > 0$, but $\varphi_1 < \varphi_2 < \varphi_3$ if $a < 0$. By considering the values of $(p' - p)^2 = \frac{4a\varphi - H}{3a^2}$, we see that the factorisation (6) that we have chosen corresponds to $\varphi = \varphi_2$; moreover since all three factorisations are real we have

$$(19) \qquad H \leqq 4a\varphi_3.$$

Suppose the roots of $g(x)$ are $\alpha_1 > \alpha_2 > \alpha_3 > \alpha_4$, so that α_1, α_3 are the roots of $x^2 + px + q$; and write

$$\Delta^2 = (\alpha_1 - \alpha_2)(\alpha_1 - \alpha_3)(\alpha_1 - \alpha_4)(\alpha_2 - \alpha_3)(\alpha_2 - \alpha_4)(\alpha_3 - \alpha_4).$$

Then the quadratic form

$$h(x) = \frac{1}{2}\Delta^{-1}\left[(x - \alpha_1)^2\frac{(\alpha_2 - \alpha_3)(\alpha_3 - \alpha_4)}{\alpha_1 - \alpha_3} + \text{three similar terms}\right]$$

is a sum of squares with positive coefficients, and is clearly a covariant of $g(x)$ under real transformations, for these do not alter the cyclic order of the roots. We may write $\Delta h(x)$ as

$$(p' - p)x^2 + 2(q' - q)x + (pq' - p'q)$$

with discriminant $\frac{4(I-\varphi_2^2)}{3a^2}$. The condition that the square of the leading coefficient is at most one third of the discriminant gives, after some manipulation,

$$H \geqq 4a\varphi_2 - \frac{4}{3}(I - \varphi_2^2); \tag{20}$$

and this and (19) give

$$0 < a(\varphi_2 - \varphi_3) \leqq \frac{1}{3}(I - \varphi_2^2). \tag{21}$$

Case 3: *Two real roots.* Now the cubic (10) has just one real root φ; and $x^2 + px + q$ has complex roots, while $x^2 + p'x + q'$ has real ones. The positive definite quadratic form

$$h(x) = \lambda^{-1}(x^2 + px + q)$$

is a covariant of $g(x)$ under real transformations, and it has discriminant 1; the condition on the leading coefficient is therefore $\lambda^2 \geqq 3$. Since we know that $\lambda'^2 < 0$, the left-hand side of (13) must be non-positive for $\Lambda = 3$; this implies

$$H \geqq 9a^2 - 2a\varphi + \frac{1}{3}(4I - \varphi^2). \tag{22}$$

Combining this with (14), we find also

$$\left(a - \frac{1}{3}\varphi\right)^2 \leqq \frac{4}{27}(\varphi^2 - I). \tag{23}$$

We can now define a reduced quartic form, the definition being designed to make Theorem 1 below valid. We shall say that $g(x) = ax^4 + bx^3 + cx^2 + dx + e$ is *reduced* if

(i) a, b, c, d, e are integers with $a \neq 0$;

(ii) there is no prime $p > 3$ such that $p^4 \mid I$, $p^6 \mid J$;

(iii) it is not true that $3^5 \mid I$, $3^9 \mid J$, nor that $3^4 \parallel I$, $3^6 \parallel J$, $3^{15} \mid (4I^3 - J^2)$, nor that $2^6 \mid I$, $2^9 \mid J$, $2^{10} \mid (8I + J)$;

(iv) if $g(x)$ has no real roots then $a > 0$ and (15), (17) hold;

(v) if $g(x)$ has two real roots then (15), (22) hold;

(vi) if $g(x)$ has four real roots then (15), (20) hold.

Theorem 1. *If Γ is an elliptic curve and U is a non-trivial 2-covering of Γ in G, then U can be represented by a curve $y^2 = g(x)$ where $g(x)$ is reduced. The reduced $g(x)$ which give 2-coverings of a given Γ are finite in number and computable.*

The first sentence of this Theorem merely summarises the results we have obtained in §§1 and 2. The proof of the second sentence is implicit in the computational procedure which we give in §3.

3. In the previous two sections we have given the theory underlying our method of calculating the locally soluble 2-coverings of an elliptic curve; in this section we give an account of the actual programs involved. The main calculation falls naturally into three stages; there is then a fourth stage to find which of these 2-coverings actually have rational points. (There is no reliable way known of determining whether a 2-covering has a rational point: all we can do is to say that a curve on which we have not been able to find a rational point probably has not got one.) The stages are

(i) find the permissible values of I, J and list the reduced forms corresponding to them;

(ii) reject those forms $g(x)$ which are equivalent to a form previously obtained, or which yield the trivial 2-covering;

(iii) reject the forms $g(x)$ for which $y^2 = g(x)$ is insoluble in some p-adic field;

(iv) find out whether $y^2 = g(x)$ has a rational point.

In discussing these stages we give the algebraic background to the calculation and also note such special difficulties as arise from doing the calculations automatically. (These are matters which the hand computer would deal with by common sense without even consciously noticing that they could give trouble: but they are more serious when they have to be predicted and guarded against in advance.) At the end we discuss the possibilities of error.

We start with a curve Γ given by (1). We take Γ in its simplest form, so that A, B are integers and there is no prime p such that $p^4 \mid A$, $p^6 \mid B$. The invariants of Γ are $3A$, $27B$. From the definition of a reduced form, the only permissible values of the invariants of $g(x)$ are $3A$, $27B$ or $48A$, $1728B$, with the proviso that we must divide I by 3^4 and J by 3^6 if $3^3 \,\|\, A$, $3^3 \,\|\, B$, $3^{12} \mid (4A^3 - 27B^2)$; we may ignore the second pair of values if $2^2 \mid A$, $2^3 \mid B$, $2^4 \mid (2A + B)$.

For a given pair I, J we find all the reduced forms. If $4I^3 < J^2$ we have only to consider case 3; if $4I^3 > J^2$ we must consider cases 1 and 2, and in case 2 we must distinguish according to the intended sign of a, since this controls the choice of φ_1, φ_2, φ_3. We first find the range of values of a, from (18), (21) or (23). For each value of a, the range of values of b is given by (15); and the range of values of H is given by (16) and (17), or (19) and (20), or (14) and (22). For given a and b, we use the bounds for H to give bounds for c. (Otherwise, most trial values of H would give non-integral values for c, which would waste time.) For fixed a, b, c we find λ^2, λ'^2 from (7) and (13); and p, p' from (8) and the first two equations (9). (Special care is needed if $H = 4a\varphi_i$ since one of the quadratics has equal roots; as the square root of a small error is much larger, we are liable to lose accuracy if we solve the quadratics in the obvious way.) Then we find q, q' from (12) and finally d, e from the last two equations (9). It only remains to check that the resulting values of d, e are integers.

In practice, difficulties arise because we are trying to calculate integers as a combination of irrational numbers, and small round-off errors creep in. First, this may mean that an inequality may appear to be false, when in fact it holds with exact equality; to cope with this, we extend any ranges for which this may happen by a number big enough to overcome any accumulation of round-off error. This may introduce a few more forms, but they do not cost much time to dispose of. Second, we cannot look at the resulting values of d and e and see if they are integers; we therefore see if they are near integers, and if so round them off to integral values and verify that these satisfy (4).

This stage of the calculation takes up most of the computer time needed for the whole job, since it involves a search over a large region. It is worth taking trouble to make the program, and particularly its inner loop, as fast as possible. Since in cases 2 and 3, if $g(x)$ is a reduced form then $-g(x)$ is reduced and has invariants $I, -J$, it saves time to deal with the curves $y^2 = x^3 - Ax \mp B$ together; but there is no advantage to be gained by dealing simultaneously with a whole range of values of A, B. We can also save time by using congruence conditions to reject some trial values of a, b, c. We know that $3 \mid I$ implies $3 \mid c$, by (4). Moreover we can use the details of the proof of Lemma 5 to restrict further the case when $2^4 \mid I$, $2^6 \mid J$, by rejecting forms $g(x)$ which are equivalent to a reduced form with smaller I, J, or with the same I, J and smaller $|a|$, or for which $y^2 = g(x)$ and $y^2 = -g(x)$ are both 2-adically insoluble. In fact, when $2^4 \mid I$, $2^6 \mid J$ we require that $4 \mid b$, $2 \mid c$, $4 \nmid a$; if a is odd then $2^4 \nmid e$, $2^4 \nmid (a+b+c+d+e)$, while if $2 \,\|\, a$ then $4 \mid c$, $2^3 \nmid e$, $2^3 \nmid (a+b+c+d+e)$.

We now describe stage (ii), in which we reject forms $g(x)$ which are equivalent to a form previously obtained. The precautions we took in stage (i) excluded any form which was equivalent to a form with integer coeffients and smaller I, J; thus we need only consider equivalences between forms with the same I, J. We can therefore write (3) in the simpler form

$$g(x) = \frac{(\gamma x + \delta)^4}{(\alpha\delta - \beta\gamma)^2} g^* \left(\frac{\alpha x + \beta}{\gamma x + \delta} \right). \tag{24}$$

Let x_1, x_2, x_3, x_4 be the roots of $g(x) = 0$, where we choose the notation so that

$$\begin{aligned} x_1 + x_2 &= -p, & x_1 x_2 &= q, \\ x_3 + x_4 &= -p', & x_3 x_4 &= q'; \end{aligned} \tag{25}$$

after (12) we have

$$x_1 - x_2 = i\lambda, \qquad x_3 - x_4 = i\lambda' . \tag{26}$$

It is also convenient to write

$$\vartheta = \frac{(x_1 - x_3)(x_2 - x_4)}{(x_1 - x_4)(x_2 - x_3)} = \frac{\varphi - a\lambda\lambda'}{\varphi + a\lambda\lambda'}. \tag{27}$$

Suppose that $g^*(x^*)$ has the same invariants as $g(x)$; then for (24) it is necessary and sufficient that

$$x_j^* = \tfrac{\alpha x_j + \beta}{\gamma x_j + \delta} \qquad (j = 1, 2, 3, 4)$$

should be the zeros of g^* in some order. We can find complex numbers $\alpha, \beta, \gamma, \delta$ to satisfy this if and only if the cross-ratios ϑ, ϑ^* are equal. (We define starred letters by the analogues of (25)–(27) and (9); it follows that equalities still hold in starred form, though inequalities may not.) In general there are just four orderings of the roots of g^* which give $\vartheta^* = \vartheta$; there are eight if $J = 0$, and twelve if $I = 0$. We have to find the corresponding transformations and see if any of them can be written in rational form.

We are certainly only interested in a transformation if it is real. The condition for this is that real x_j^* should correspond to real x_j, and complex conjugate x_j^* to complex conjugate x_j. The transformation that takes x_j into x_j^*, given that $\vartheta = \vartheta^*$, is

$$\frac{x_3^* - x_1^*}{x_3^* - x_2^*} \frac{x - x_1}{x - x_2} = \frac{x_3 - x_1}{x_3 - x_2} \frac{x^* - x_1^*}{x^* - x_2^*}.$$

This involves complex numbers, which are not very convenient for machine calculation; but we found that any attempt to put it into real form caused more trouble than it saved. From this we find the ratios $\alpha : \beta : \gamma : \delta$, which should be real; in practice they will have small imaginary parts because of round-off error, but we ignore these.

For the actual program it is necessary to know which orderings of the roots of g^* have to be considered. We have

$$\vartheta + \vartheta^{-1} = \frac{\varphi^2 + 2I}{\varphi^2 - I};$$

thus $\varphi = \varphi^*$ except perhaps when $I = 0$ or $J = 0$. Special arguments using the reality of the transformation establish it in these cases also. If $\varphi = 0$ then $J = 0$; either $I > 0$ and the x_j are all real, when we have to consider all eight corresponding orderings of the x_j^*; or $I < 0$ and only x_3, x_4 are real, when we can pair these to the two real x_j^* and x_1, x_2 to the two complex x_j^* in any manner, giving four possible orderings. If $\varphi \neq 0$ then (27) gives

$$a\lambda\lambda' = a^*\lambda^*\lambda'^*$$

which rejects four of the eight orderings that give the correct value of φ^*. The four that are left are those that take (x_1, x_2, x_3, x_4) respectively into $(x_1^*, x_2^*, x_3^*, x_4^*)$, $(x_2^*, x_1^*, x_4^*, x_3^*)$, $(x_3^*, x_4^*, x_1^*, x_2^*)$ and $(x_4^*, x_3^*, x_2^*, x_1^*)$. All of these are available if $4I^3 > J^2$; only the first two if $4I^3 < J^2$.

Given a real transformation we have still to see if it is rational. We can always multiply $\alpha, \beta, \gamma, \delta$ by a common factor; thus we may take them to be integers with highest common factor one. We can write

$$\begin{pmatrix} \alpha & \beta \\ \gamma & \delta \end{pmatrix} = U \begin{pmatrix} \alpha\delta - \beta\gamma & 0 \\ 0 & 1 \end{pmatrix} V$$

for some integral unimodular matrices U, V; it follows that we must have

$$(\alpha\delta - \beta\gamma)^2 \mid (4I^3 - J^2).$$

There are therefore only finitely many values possible for $\alpha\delta - \beta\gamma$ under the assumptions above. For each such value, a given transformation to be examined can be written so that it has the correct value of $\alpha\delta - \beta\gamma$. In principle we see if the values of $\alpha, \beta, \gamma, \delta$ thus obtained are integers. In practice, we see if they are near integers; if so we round them off to integers and see whether the resulting transformation does in fact take $g(x)$ into $g^*(x^*)$. If it does, the forms are equivalent. If it does not do so for any of the transformations and values of $\alpha\delta - \beta\gamma$ allowed, then the forms are not equivalent.

We have now rejected those forms $g(x)$ which give equivalent curves $y^2 = g(x)$. We must also reject those $g(x)$ which give the trivial 2-covering: they are just those for which $g(x) = 0$ has a rational root, and since the denominator of such a root must be a factor of a they are easily found. This concludes the description of stage (ii).

It is desirable to do stage (ii) at the same time as stage (i); for we then only have to find the roots of the relevant equations once. At the end of stage (ii) we output a list of inequivalent reduced forms to be processed through the remaining stages.

In stage (iii) we examine the local solubility of $y^2 = g(x)$ for each reduced form given by stages (i) and (ii). This equation is certainly soluble in the reals, for we have already ensured that $g(x)$ cannot be negative definite; it remains to consider p-adic fields for finite p. We use the ideas of Hensel's lemma, noting that if we write

$$g_1(x) = x^4 g(x^{-1}) = a + bx + cx^2 + dx^3 + ex^4$$

then for $y^2 = g(x)$ to be soluble in p-adic numbers it is necessary and sufficient that at least one of $y^2 = g(x)$ and $y^2 = g_1(x)$ should be soluble in p-adic integers. For given p, x_0 we define λ, μ, α by

$$p^{\lambda} \,||\, g(x_0), \quad p^{\mu} \,||\, g'(x_0), \quad p^{\alpha} \,||\, (4I^3 - J^2).$$

and see if we can make $g(x)$ a p-adic square with x near x_0. Since $4I^3 - J^2$ is the discriminant of $g(x)$, we have

$$\min\,[\lambda, \mu] \leqq \alpha. \tag{28}$$

Lemma 6. *Let p be an odd prime. We can solve $y^2 = g(x)$ in p-adic numbers with $p^{\nu} \,|\, (x - x_0)$ if $g(x_0)$ is a p-adic square or if $\lambda - \mu \geqq \nu > \mu$; we may be able to do so if $\lambda \geqq 2\nu$, $\mu \geqq \nu$; we definitely cannot in any other case.*

We write $x = x_0 + \xi$ and use the identity

$$g(x) = g(x_0) + \xi g'(x_0) + [\text{multiple of } \xi^2]. \tag{29}$$

If $\nu > \mu$ this takes all values congruent to $g(x_0)$ mod $p^{\mu+\nu}$ and no others; if $\nu \leqq \mu$ this is congruent to $g(x_0)$ mod $p^{2\nu}$, but may not take all such values. The lemma now follows at once.

If $\alpha = 0$, the polynomial $g(x)$ cannot be formally congruent to a constant mod p; and so there is a solution x_0 of $y^2 \equiv g(x)$ mod p. By the lemma, this certainly leads to a p-adic solution; thus we are only concerned with primes for which $\alpha > 0$ (and of course 2, which is treated below).

A similar but more complicated argument gives, for $p = 2$,

Lemma 7. *We can solve $y^2 = g(x)$ in 2-adic numbers with $2^{\nu} \,|\, (x - x_0)$ if $g\,(x_0)$ is a 2-adic square, or if $\lambda - \mu \geqq \nu > \mu$, or if $\nu > \mu$ and $\lambda = \mu + \nu - 1$ is even, or if $\nu > \mu$ and $\lambda = \mu + \nu - 2$ is even and $2^{-\lambda} g\,(x_0) \equiv 1$ mod 4; we may be able to do so if $\mu \geqq \nu$ and $\lambda \geqq 2\nu$, or if $\mu \geqq \nu$ and $\lambda = 2\nu - 2$ and $2^{-\lambda} g(x_0) \equiv 1$ mod 4; we definitely cannot in any other case.*

To see if $y^2 = g(x)$ is soluble in p-adic integers, for the finitely many p we need consider, we proceed as follows. We have a list of residue classes to examine – initially consisting of the classes $x_0 \equiv 0, 1, \ldots, p - 1$ mod p. We take a class x_0 mod p^{ν} from the list and apply Lemma 6 (or Lemma 7 if $p = 2$). If it certainly gives rise to a solution we have no more to do. If it certainly cannot, we delete this class from the list. If Lemma 6 gives us no definite information we replace the class by the corresponding p classes mod $p^{\nu+1}$. Eventually we must either find that the equation is soluble or exhaust our list (in which case it is insoluble); for after (28) we never have to split classes with $\nu > \alpha$. If $y^2 = g(x)$ is insoluble in p-adic integers we go on to consider $y^2 = g_1(x)$, starting now with the one class 0 mod p; if this too does not give a solution we may reject the form $g(x)$. The program for this stage presents no hidden difficulties: all the calculations are done on integers of moderate size and are therefore exact.

This completes the description of the calculations needed to find the elements of G – that is, the locally soluble 2-coverings. There is no algorithm known which will certainly decide which of them admit rational points: in practice one proves that a curve has a rational point by exhibiting it, and that it has no rational points by a descent argument. All that we can do in a computer is to search for small rational points. This seems to be fairly effective. We write our curve in the quasi-homogeneous form

$$y^2 = ax^4 + bx^3z + cx^2z^2 + dxz^3 + ez^4$$

and search for solutions with small $|x|+|z|$. There is a preliminary search with $|x|, |z| \leqq 5$; this is very fast and has therefore been put before stage (iii) – for if a curve has a rational point we need not enquire whether it is everywhere locally soluble. Nearly all the curves found in stage (i) either have a small rational point or are rejected in stage (iii). For those which do not, we make a further search with $|x|+|z|$ running up to 100; this takes only a few seconds for each curve tested.

A curve which is left undecided after this second search is unlikely to have rational points, but there are a few exceptions: for instance, the covering $y^2 = (x^2+z^2)(7x^2+xz-7z^2) = F(x,z)$, say, of the curve $y^2 = x^3+197x$ has a rational point since $F(56249, 28882)$ is a square, and we have been unable to find a non-trivial rational point for the curve $y^2 = x^3 - 309$, though it presumably has one, since it has a single locally soluble 2-covering and also a single locally soluble 3-covering. We believe that we have found rational points in all other cases where they exist. In a good many cases we have verified this by a further descent, done by hand: but we have done these calculations in an unsystematic way, and with a fixed idea of what answer they should give; and this may affect their reliability. In every case the 2-coverings on which we have found rational points form a set whose order is a power of 2: but this is not a useful check since it is obviously also a property of the set of 2-coverings which admit a small rational point.

It remains to discuss the question of error. One can take precautions against machine errors either by running all the calculations twice or by checks included in the program. Moreover, machines – unlike human beings – are so designed that the errors they make are usually too gross to be overlooked. We are satisfied that there are in our results no undetected errors of this sort. On the other hand, in translating an elaborate scheme of calculation into a machine program one is bound to make mistakes. Most of these are found before the program is used for production runs; they show up because the program grinds to a halt or produces ridiculous results. But a program which is believed to work may still contain logical errors which only have an effect in rare circumstances: and indeed most computers have anomalies which cause them occasionally not to behave in the way that their specifications suggest. In fact, our program for stage (ii) was imperfect in that a very few equivalences were missed by the machine.

For these reasons we believe that results obtained from a computer should not be automatically trusted. In some cases they can be checked because they have properties which were not essentially used in the course of the calculation and which would be unlikely to survive if an error had been made. (For example, if a table of a smooth function has been calculated without the use of interpolation, it can be checked by differencing.) But if checks of this sort are not available, results should not be fully trusted until they have been independently reproduced by a different programmer using a different machine. We do not think this sets an unreasonable standard, now that computers are becoming so widely available; and we are satisfied that lower standards have already led to a number of untrue results being published and believed.

In our case some such checks are available. A comparison with previous tables has not revealed any errors; for the details see §4. Moreover nothing in our method of calculating G suggests that its order should be a power of 2 – indeed we have found G only as a set, not as a group. This is a good check against error except perhaps when the order of G is 1 or 2.

4. In this section we describe our results. We have applied the methods described in §§2, 3 to the curves

$$\Gamma\colon y^2 = x^3 - Ax - B \tag{30}$$

for all integers A, B with $0 < |A| \leqq 20$, $0 < |B| \leqq 30$; to the curve

$$y^2 = x^3 - D \tag{31}$$

for all integers D with $|D| \leqq 400$; and to the curve

$$y^2 = x^3 - Dx \tag{32}$$

for all integers D with $|D| \leqq 200$. We have also dealt with a few higher ranges of A, B and a few larger values of D. The calculations needed for each curve take very little time; as we had to put a great deal of effort into producing a working program, it would have been uneconomical not to do a lot of cases.

For any particular curve, the final output from the computer was a list of the everywhere locally soluble 2-coverings; for those on which a rational point had been found, its coordinates were also printed. Write 2^t for the number of elements of order 2 of the Tate-Šafarevič group. Then the number of 2-coverings (including the original curve itself) is 2^k, where $k = g + g_2 + t$; and the number of these on which there are rational points is 2^{k-t}. Since it is easy to find g_2, we now know $g + t$. If we have failed to find rational points on a 2-covering that actually possesses them, this will lead to too high an estimate for t and too low for g. From the rational points on the 2-coverings, one may also find generators of infinite order of the group of rational points on the original curve; but the numbers involved are very large and we have not thought this calculation worth while.

Since g_2 is easy to find, we have saved space by tabulating only the values of g and of t. Table 1 gives values for the curve (30). In most cases we find that all the 2-coverings produced by the computer have rational points; thus $t = 0$ and g is determined. For these cases the entry in the table is a single number, the value of g. If there are 2-coverings for which we have not found rational points, then we know the value of $g + t$, say m, and we have a lower bound n for g (which will normally be its true value). In this case the entry in the table is n^m. In every such case $m - n$ is even, as is suggested by the theorem of Cassels [3], II, III that t is even whenever the Tate-Šafarevič group is finite. (In all such cases in the range of Table 1 we actually have $n = 0$, $m = 2$; this is simply because the table does not go very far.) The four cases in Table 1 in which Γ has genus zero have been marked R.

For the curves (31) and (32) a table of this sort is inconvenient to print, and in Tables 2, 3 and 4 we have listed our results the other way round. Table 2 refers to equation (31), Table 3 to (32) with $D > 0$ and Table 4 to (32) with $D < 0$. In each of these we simply list the curves (values of D) that correspond to an assigned pair of presumed values of g and t. As in Table 1, our machine calculations certainly give the correct values of $g + t$, but only lower bounds for g.

Billing [1] has given results for the curve (30) for $0 < |A|, |B| \leqq 3$. Cassels [2] and Selmer [9] have given results for (31) for $|D| \leqq 100$; and Billing [1] has given results for (32) for $|D| \leqq 50$. (There is a known error in Billing's table at $D = 33$.) All these agree with the values we have obtained.

Tables 2, 3 and 4 cover two types of curve for which a complex multiplication exists; and for these a further check on the values of g and t is practicable. For the curve (31) to which Table 2 refers, it is easy to do a $\sqrt{-3}$ descent by hand, writing a pair of 3-isogenous equations as $x^3 = y^2 + D$ and $x^3 = y^2 - 27D$ and factorising the right-hand sides (c. f. [8], [9], [3]). In this way we have verified the results with $t > 0$ and $|D| \leqq 200$. For the curve (32), to which Tables 3 and 4 refer, we know that for any N the curves with $D = N$ and $D = -4N$

Table 1

B ↓ / ← A	−20	−19	−18	−17	−16	−15	−14	−13	−12	−11	−10	−9	−8	−7	−6	−5	−4	−3	−2	−1
30	2	1	1	1	.	.	1	2	.	1	1	1	1	0^2	1	.	2	.	1	1
29	2	1	1	2	.	1	.	.	1	1	1	2	1	1	2	1	.	.	1	1
28	1	.	2	1	1	.	0^2	1	1	1	.	.	.	1	1	.	.	.	.	.
27	1	.	.	.	1	.	1	.	2	1	2	.	.	1	1	1	.	2	.	1
26	.	.	.	1	2	1	1	.	.	2	1	1	1	.	.	1	.	1	.	2
25	1	.	.	0^2	1	.	.	1	0^2	1	.	2	0^2	.	.	.	1	.	.	1
24	.	1	1	.	1	1	1	.	.	2	1	1	.	.	2	1	1	.	1	1
23	2	1	1	1	1	1	.	1	1	1	2	.	1	1	1	1	1	.	1	1
22	1	.	1	0^2	.	1	.	.	.	0^2	.	1	1	1	.	.	1	.	.	.
21	.	2	.	1	1	1	1	1	1	**3**	2	.	1	1	1	1	.	.	1	1
20	1	1	.	.	.	.	2	.	.	1	1	.	1	.	1	2	1	1	.	.
19	1	2	1	.	.	1	1	.	1	.	1	1	1	.	2	0^2	.	1	1	1
18	1	1	1	1	0^2	1	1	1	.	.	1	.	2	2	1	.	1	1	1	.
17	1	.	1	2	.	2	.	1	1	1	.	1	1	.	1	2	.	1	1	.
16	.	2	.	1	1	.	1	1	1	1	1	1	.	1	1	0^2	1	.	.	1
15	.	.	1	1	2	1	.	1	.	.	.	1	2	.	.	.	2	.	.	1
14	1	1	1	2	1	1	2	.	1	1	.	1	1	1	.	1	2	.	1	1
13	.	1	.	1	2	2	1	1	.	1	1	.	1	1	.	.	1	1	0^2	.
12	2	1	.	1	.	2	.	.	1	1	1	1	.	1	.	.	1	1	.	.
11	.	1	.	1	2	.	2	1	.	2	1	.	.	1	1	.	.	.	1	1
10	1	1	1	.	.	1	.	1	1	1	2	1	1	1	1	.	1	1	1	.
9	2	.	.	2	1	.	.	1	1	.	.	1	1	1	1	1	1	1	1	2
8	.	.	2	1	1	1	.	.	.	2	1	1	2	1	.	.	.	.	2	1
7	.	1	.	1	1	1	1	.	1	.	1	.	.	1	.	1	1	1	.	1
6	1	1	.	1	1	1	2	.	.	.	1	1	.	1	1	1	1	1	.	1
5	2	1	1	.	1	.	1	2	.	1	1	1	1	1	1	1	.	1	.	1
4	0^2	2	.	0^2	2	1	.	1	1	.	1	1	.	1	1	1	1	.	.	1
3	.	1	1	1	.	2	.	.	1	1	1	1	.	.	1	1	1	1	1	.
2	1	.	1	1	1	1	.	1	.	1	2	.	1	1	.	1	.	.	1	.
1	1	.	1	1	2	.	0^2	.	1	.	1	2	.	1	.	.	2	.	.	1

B ↓ / A →	1	2	3	4	5	6	7	8	9	10	11	12	13	14	15	16	17	18	19	20
30	1	.	.	1	.	.	2	.	1	1	1	1	.	2	.	1	1	1	.	.
29	1	1	1	1	.	1	1	.	.	.	2	1	1	2	.	1	.	.	2	1
28	1	.	.	.	1	.	0^2	1	.	1	.	.	.	1	1	.	1	.	0^2	1
27	1	.	1	1	.	.	1	.	1	1	.	.	1	1	1	1	.	1	.	1
26	1	.	1	1	1	.	1	.	.	2	.	1	.	.	1	.	.	2	2	.
25	0^2	1	.	1	.	1	1	0^2	1	1	.	1	.	.	1	1	2	1	1	.
24	2	.	.	1	.	1	.	1	2	1	1	.	1	.	.	.	1	1	.	2
23	1	1	.	1	.	1	.	1	1	1	.	1	.	.	.	1	1	1	.	1
22	.	1	.	1	1	.	.	1	0^2	.	.	.	.	1	.	0^2	2	1	1	1
21	.	1	.	.	1	2	1	1	.	1	1	1	1	1	1	.	1	1	**3**	2
20	2	2	.	.	.	.	1	1	1	1	.	.	.	1	.	2	1	.	.	1
19	.	1	1	1	2	.	.	1	1	1	2	1	.	2	1	.	.	1	.	.
18	1	.	.	1	1	1	.	1	1	1	.	.	1	.	1	.	.	1	.	2
17	1	2	1	1	1	1	.	1	1	1	.	.	2	.	1	2	1	.	1	2
16	1	.	1	.	1	.	1	1	1	1	2	R	.	1	.	.	.	1	1	1
15	1	1	1	.	1	1	2	.	.	1	1	1	.	.	.	1	2	.	1	2
14	1	1	1	1	.	1	.	.	.	1	.	.	1	1	1	1	1	1	2	1
13	1	1	.	.	1	.	.	.	1	0^2	1	1	.	.	2	0^2	1	2	1	.
12	1	2	1	1	.	1	.	.	1	1	1	1	.	1	1	1	1	1	1	1
11	1	.	1	1	1	1	1	.	.	1	1	.	2	2	1	1	1	1	1	.
10	1	.	1	1	.	.	.	.	1	1	1	2	0^2	.	1	.	1	1	.	1
9	1	.	1	1	.	.	1	.	1	.	1	.	1	2	1	2	.	.	1	2
8	1	1	.	.	2	1	.	.	.	2	1	1	2	.	1	1	1	1	1	1
7	.	1	.	.	.	.	.	1	1	.	1	1	1	1	.	.	1	2	1	2
6	.	.	.	1	.	.	.	2	1	.	1	1	1	2	1	2	1	.	.	.
5	1	2	.	1	.	1	2	.	.	1	2	1	.	.	1	1	.	1	2	1
4	.	.	1	.	.	1	.	1	1	.	1	.	.	2	.	.	1	1	1	.
3	.	1	1	.	1	1	.	1	1	1	.	1	2	.	1	.	2	1	1	1
2	1	.	R	1	.	.	1	1	.	1	1	1	1	.	1	.	.	1	1	.
1	.	.	1	.	1	1	.	.	1	.	1	1	1	.	1	1	1	1	.	1

← *A*

B ↓	−20	−19	−18	−17	−16	−15	−14	−13	−12	−11	−10	−9	−8	−7	−6	−5	−4	−3	−2	−1
−1	2	1	1	1	1	1	**3**	1	2	1	2	1	1	2	1	1	1	1	1	1
−2	1	1	.	.	1	.	1	1	.	0^2	1	.	.	.	1	.	.	1	.	.
−3	1	2	1	2	1	1	.	.	2	.	1	.	1	1	1	1	.	.	1	1
−4	2	2	2	2	1	1	1	1	1	2	1	1	1	1	2	1	1	1	2	1
−5	1	2	1	1	.	1	1	2	.	.	1	.	.	.	1	1	.	1	.	.
−6	1	.	1	1	1	2	1	1	.	1	.	1	.	.	.	.	1	.	1	1
−7	1	.	1	1	2	.	.	1	.	1	2	.	1	.	.	2	.	.	1	1
−8	.	1	1	1	1	.	1	.	.	1	1	.	.	2	1	.	.	1	.	1
−9	1	1	1	2	1	2	1	2	1	1	1	1	1	2	1	2	2	1	2	2
−10	1	.	1	.	.	.	1	.	1	.	1	1	1	.	.	1	1	.	.	.
−11	1	.	.	.	1	1	0^2	1	1	1	1	1	1	.	1	.	1	1	1	2
−12	1	1	.	1	1	.	1	.	1	.	1	1	1	1	1	.	.	1	1	.
−13	1	.	.	.	1	1	1	1	.	2	1	1	2	.	.	.	.	1	2	1
−14	1	.	.	.	1	.	1	.	2	.	1	1	1	2	1	.	2	.	.	2
−15	1	1	1	1	1	.	1	2	1	1	1	1	1	1	1	1	1	1	1	1
−16	1	**3**	2	1	1	1	2	1	1	2	1	1	1	2	2	2	1	1	2	1
−17	.	1	1	0^2	.	1	1	.	2	0^2	1	.	.	1	.	.	1	.	.	.
−18	1	2	.	1	2	1	0^2	1	1	1	.	.	2	1	1	1	1	1	.	1
−19	.	1	1	1	1	.	1	.	.	1	1	1	.	1	2	2	1	.	1	1
−20	.	.	2	.	1	2	1	.	.	1	1	.	.	.	.	0^2	2	1	.	.
−21	.	1	.	.	2	.	1	1	.	2	2	1	.	.	1	1	1	1	1	.
−22	1	1	.	2	.	.	1	1	1	1	1	.	1	1	1	1	1	2	1	.
−23	1	.	2	1	.	.	1	2	1	0^2	1	1	.	2	2	.		2	.	1
−24	.	2	1	.	1	.	2	.	.	1	1	1	.	1	1	1	1	1	1	1
−25	2	1	1	2	2	1	1	2	1	2	1	2	**3**	1	1	1	2	1	1	1
−26	.	1	.	1	2	1	.	1	.	1	.	1	1	1	1	.	.	.	1	2
−27	.	1	.	1	2	.	1	.	.	.	2	1	1	2	.	1	1	.	.	2
−28	1	.	1	1	.	2	1	1	.	1	1	1	1	1	.	.	.	2	2	.
−29	1	2	1	1	1	.	.	.	2	.	1	1	0^2	1	2	1	1	1	1	.
−30	2	.	1	.	.	1	2	2	.	1	.	1	1	1	.	1	2	1	.	.

A →

B ↓	1	2	3	4	5	6	7	8	9	10	11	12	13	14	15	16	17	18	19	20
−1	1	.	1	2	1	1	1	2	2	2	1	1	1	2	2	**3**	1	2	2	1
−2	.	1	R	.	1	.	.	2	.	1	1	.	1	.	1	1	.	1	.	1
−3	1	1	1	.	1	.	2	.	1	.	1	1	.	1	1	.	.	.	.	1
−4	2	1	1	1	1	1	1	1	2	2	2	1	**3**	1	1	2	1	1	2	2
−5	.	2	.	.	1	1	.	.	.	.	1	1	.	1	2	1	2	.	1	1
−6	.	1	1	1	1	1	.	1	1	2	.	.	1	.	.	1	1	.	2	1
−7	1	.	1	1	1	1	1	1	1	2	1	1	1	1	1	.	1	.	1	.
−8	.	2	.	.	1	1	.	1	1	.	.	.	.	1	2	.	1	2	.	.
−9	2	1	1	2	1	1	1	1	1	1	1	1	1	2	1	2	2	2	1	2
−10	2	1	.	1	1	1	2	.	.	1	1	1	.	2	.	1	1	.	.	.
−11	.	.	1	.	2	1	.	1	1	.	2	.	.	1	1	1	1	.	.	2
−12	1	1	1	1	1	1	.	1	1	1	.	1	.	2	2	1	2	1		.
−13	.	1	.	1	2	1	1	1	.	2	.	1	2	.	1	.	1	1	2	.
−14	.	.	1	1	1	1	.	.	1	.	.	1	1	1	1	.	1	1	.	.
−15	2	2	.	.	.	.	2	1	1	.	2	1	.	.	1	1	.	.	1	.
−16	2	1	1	2	2	2	1	1	2	2	2	R	2	2	1	1	1	1	2	1
−17	.	1	.	.	.	.	.	2	1	.	1	2	.	2	.	.	1	1	1	2
−18	.	1	1	1	2	1	.	1	.	2	1	2	1	.	1	1	.	1	1	1
−19	1	1	1	2	1	.	1	2	1	1	2	.	1	1	1	2	.	1	2	.
−20	.	1	.	.	2	2	1	.	1	.	.	1	.	.	1	.	2	.	1	2
−21	1	1	.	1	2	2	.	.	1	1	1	.	2	1	1	1	1	2	2	.
−22	1	2	1	1	.	.	2	1	1	1	1	.	1	2	.	1	.	1	1	2
−23	0^2	.	1	.	1	1	.	2	.	.	1	.	1	1	1	1	1	1	.	1
−24	1	1	.	1	1	1	.	1	1	1	1	.	2	2	.	1	1	1	1	1
−25	3	2	1	2	1	2	1	1	2	2	1	2	1	1	1	2	2	1	1	1
−26	.	1	.	1	.	1	.	.	2	1	1	1	1	1	1	.	.	1	.	1
−27	.	.	2	.	1	1	.	1	.	1	.	2	2	1	1	.	1	.	1	1
−28	1	1	.	1	1	.	2	2	1	2	.	.	2	1	1	1	1	1	1	1
−29	.	1	1	.	1	1	.	1	1	.	2	.	.	2	1	.	1	1	1	1
−30	2	1	.	1	1	1	2	.	.	.	2	1	1	1	1	1	2	.	.	1

are 2-isogenous. Hence they must have the same value of g, but they need not have the same value of t. In the few cases in which both of them have a non-zero value of t, we have verified this by doing a second descent by hand.

There are also a few cases of 2-isogeny between two curves in Table 1; but in each case both isogenous curves have $t = 0$. For reference, we give the formulae involved. Suppose that $(c, 0)$ is a rational 2-division point of (30); then there is a 2-isogeny between (30) and

$$\Gamma' \colon Y^2 = X^3 - (15c^2 - 4A)X - (22c^3 - 8Ac),$$

the map $\Gamma' \to \Gamma$ being given by

$$x = c + \frac{Y^2}{4(X+2c)^2}, \; y = \frac{Y(X^2 + 4cX + 7c^2 - 4A)}{8(X+2c)^2}.$$

The pairs in the range of Table 1 are given by $c = 1, -1 \leqq A \leqq 6$.

Table 2.
Curves $y^2 = x^3 - D$ for which g and t take given values.

a) Curves with $g = t = 0$.

1	3	5	6	8	9	10	12	14	16	17	24	27
31	32	33	34	36	37	41	42	46	52	62	68	69
70	73	77	78	80	82	86	88	90	92	96	97	98
99	103	108	111	114	117	119	125	132	133	134	136	144
145	154	156	158	160	161	162	168	169	176	177	178	181
183	189	190	194	196	197	205	206	208	213	217	220	221
224	226	227	230	232	240	241	247	248	250	255	257	258
260	263	266	269	275	276	278	280	283	285	296	302	304
305	306	312	313	314	315	319	321	330	333	335	338	340
341	349	351	352	358	360	361	374	376	377	378	380	385
386	388	392	394	396	399	400	1005	1008	1009			
–1	–4	–6	–7	–13	–14	–16	–20	–21	–23	–25	–27	–29
–32	–34	–42	–45	–49	–51	–53	–59	–60	–70	–75	–78	–81
–84	–85	–86	–87	–88	–90	–93	–95	–96	–104	–109	–114	–115
–116	–124	–125	–135	–137	–140	–144	–153	–157	–158	–159	–160	–162
–165	–167	–173	–175	–176	–178	–180	–181	–187	–193	–194	–200	–201
–203	–209	–216	–228	–230	–237	–239	–240	–242	–243	–244	–245	–250
–253	–258	–259	–261	–266	–267	–270	–273	–279	–284	–292	–300	–301
–302	–304	–305	–306	–308	–309	–311	–312	–317	–324	–330	–338	–339
–341	–345	–348	–358	–361	–363	–367	–372	–374	–375	–376	–378	–387
–394	–397	–400	–1008									

b) Curves with $g = 1, t = 0$.

2	4	7	13	15	18	19	20	21	22	23	25	28
29	30	35	38	40	43	44	45	48	49	50	51	54
55	56	57	58	59	60	63	65	66	71	72	74	75
79	81	84	85	87	91	93	94	95	100	101	102	107
110	112	115	120	123	124	126	127	129	130	131	135	137
138	140	143	146	148	150	151	153	157	159	163	164	165
166	167	171	172	173	175	179	180	182	184	187	188	193
195	199	201	202	203	204	209	210	215	216	218	223	225
228	229	231	234	235	237	238	239	242	243	245	246	251
252	254	259	261	264	265	267	268	270	271	272	273	274
279	281	282	284	287	288	290	292	295	297	300	301	303
308	309*	310	311	316	317	318	323	324	325	326	328	331
332	336	337	339	342	343	344	345	347	348	353	354	359
363	367	369	372	373	375	379	381	382	383	387	389	390
395	398	1001	1007	1010								
–2	–3	–5	–8	–9	–10	–11	–12	–18	–19	–22	–26	–28
–30	–31	–33	–35	–36	–38	–39	–40	–41	–44	–46	–47	–48
–50	–52	–54	–55	–56	–58	–61	–62	–66	–67	–68	–69	–71
–72	–74	–76	–77	–80	–82	–83	–91	–92	–94	–97	–98	–99
–100	–102	–103	–105	–107	–108	–110	–111	–112	–117	–118	–119	–120
–121	–126	–127	–130	–132	–133	–134	–136	–138	–139	–143	–146	–147
–149	–150	–152	–154	–155	–156	–163	–166	–169	–170	–172	–177	–179
–182	–183	–184	–185	–188	–189	–190	–191	–196	–199	–202	–205	–206
–207	–208	–210	–211	–212	–213	–215	–218	–219	–220	–221	–224	–226
–227	–234	–235	–236	–238	–241	–246	–247	–249	–251	–254	–255	–257
–262	–263	–264	–271	–272	–274	–275	–276	–277	–278	–280	–282	–283
–285	–287	–288	–289	–290	–291	–293	–296	–298	–299	–307	–310	–313
–314	–315	–318	–319	–321	–323	–325	–326	–327	–328	–329	–332	–333
–334	–335	–336	–340	–342	–343	–344	–349	–351	–352	–354	–355	–356
–357	–364	–365	–368	–370	–371	–379	–380	–382	–385	–386	–390	–391
–393	–395	–396	–398	–399	–1002	–1003	–1005	–1007	–1010			

c) Curves with $g = 2, t = 0$.

11	26	39	47	53	61	67	76	83	89	104	106	109
116	118	121	139	147	152	155	170	186	191	200	207	211
212	214	219	222	233	236	244	249	262	277	286	289	291
293	294	298	299	327	329	334	350	355	356	364	366	368
370	371	391	393	397	1003							

–15	–17	–24	–37	–43	–57	–63	–65	–73	–79	–89	–101	–106
–122	–129	–131	–142	–145	–148	–151	–161	–164	–168	–171	–186	–195
–197	–198	–204	–217	–222	–223	–225	–229	–232	–233	–248	–252	–260
–265	–268	–269	–281	–294	–295	–297	–303	–322	–331	–337	–347	–350
–353	–360	–366	–369	–373	–377	–381	–388	–389	–392	–1001	–1004	–1006
–1009												

d) Curves with $g = 3, t = 0$.

174	307	362		
–113	–141	–316	–346	–359

e) Curves with $g = 0, t = 2$.

105	113	122	141	142	149	185	198	253	322	357	365	1006
–123	–214	–231	–286	–383								

f) Curves with $g = 1, t = 2$.

346	
–174	–362

Table 3.
Curves $y^2 = x^3 - Dx$ with positive D for which g and t take given values.

a) Curves with $g = t = 0$.

1	3	4	8	9	11	13	18	19	24	27	28	29
33	35	40	43	44	51	59	61	63	67	68	75	83
88	91	92	93	98	100	104	107	108	109	115	120	121
123	125	126	129	131	139	152	153	157	163	164	168	172
173	177	179	180	187	189	195	198	200	228	388	468	548
644	708	772										

b) Curves with $g = 1, t = 0$.

2	5	6	7	10	12	14	15	20	21	22	23	25
26	30	31	34	36	37	38	39	41	42	45	46	47
49	50	52	53	54	55	57	58	60	62	66	69	70
71	72	73	74	76	78	79	84	85	86	87	89	94
95	99	101	102	103	105	106	110	111	114	116	118	119
122	124	127	130	133	134	135	137	138	140	142	143	146
147	148	149	150	151	154	158	159	165	166	167	169	170
171	174	175	178	181	182	183	185	186	188	190	191	194
196	197	199										

c) Curves with $g = 2, t = 0$.

17 56 65 77 90 97 117 132 136 141 145 155 156
161 184

d) Curves with $g = 3, t = 0$.

82

e) Curves with $g = 0, t = 2$.

113 193 248 328 568 584 776

Table 4.
Curves $y^2 = x^3 - Dx$ with negative D for which g and t take given values.

a) Curves wit $g = t = 0$.

–1 –2 –4 –6 –7 –10 –11 –12 –22 –23 –25 –26 –27
–30 –36 –38 –42 –43 –44 –45 –50 –52 –54 –58 –59 –70
–71 –72 –74 –75 –76 –78 –86 –87 –91 –102 –103 –106 –107
–108 –110 –116 –118 –119 –122 –123 –130 –132 –134 –135 –139 –140
–147 –151 –166 –167 –169 –170 –172 –174 –182 –186 –187 –190 –199

b) Curves witl $g = 1, t = 0$.

–3 –5 –8 –9 –13 –15 –18 –19 –20 –21 –24 –28 –29
–31 –35 –37 –40 –47 –49 –51 –53 –56 –60 –61 –67 –69
–77 –79 –83 –84 –85 –88 –90 –92 –93 –95 –98 –100 –101
–104 –109 –111 –115 –120 –121 –124 –125 –126 –127 –131 –133 –141
–143 –148 –149 –152 –153 –156 –157 –159 –163 –164 –165 –168 –171
–173 –175 –179 –180 –181 –184 –188 –189 –191 –195 –196 –197 –198
–200

c) Curves with $g = 2, t = 0$.

–14 –33 –34 –39 –46 –55 –63 –65 –66 –68 –73 –89 –94
–99 –105 –113 –114 –129 –138 –145 –150 –154 –155 –158 –178 –183
–185

e) Curves with $g = 0, t = 2$.

–17 –41 –57 –62 –82 –97 –117 –137 –142 –146 –161 –177 –193
–194 –392 –452 –792

f) Curves with $g = 1, t = 2$.

–136 –584 –712 –776

References

[1] *G. Billing,* Beiträge zur arithmetischen Theorie ebener kubischer Kurven. Nova Acta Reg. Soc. Scient. Upsaliensis. (4) **11**, 1 (1938), 1–165.

[2] *J. W. S. Cassels,* The rational solutions of the diophantine equation $Y^2 = X^3 - D$. Acta Math. (Stockh.) **82** (1950), 243–273.

[3] *J. W. S. Cassels,* Arithmetic on curves of genus 1. I. On a conjecture of Selmer. J. reine angew. Math. **202** (1959), 52–99. II. A general result. J. reine angew. Math. **203** (1960), 174–208. III. The Tate-Safarevic and Selmer groups. Proc. London Math. Soc. (3) **12** (1962), 259–296. IV. Proof of the Hauptvermutung. J. reine angew. Math. **211** (1962), 95–112. V. Two counter-examples. J. London Math. Soc., in press.

[4] *C. Chevalley,* Algebraic functions of one variable. Amer. Math. Soc. Surveys No. VI, New York 1951.

[5] *G. Julia,* Etude sur les formes binaires non quadratiques. Mem. Acad. Sci. l'Inst. France **55** (1917), 1–293.

[6] *S. Lang* and *J. Tate,* Principal homogeneous spaces over abelian varieties. Amer. J. Math. **80** (1958), 659–684.

[7] *E. Lutz,* Sur l'équation dans les cor $y^2 = x^3 - Ax - B$ ps p-adiques. J. reine angew. Math. **177** (1937), 238–247.

[8] *V. D. Podsypanin,* On the indeterminate equation $x^3 = y^2 + Az^6$. Mat. Sbornik N. S. **24** (1959), 391–403. (In Russian.)

[9] *E. S. Selmer,* The rational solutions of the Diophantine equation $\eta^2 = \xi^3 - D$ for $| D | \leqq 10$. Math. Scand. **4** (1956), 281–286.

[10] *C. L. Siegel,* Über die analytische Theorie der quadratischen Formen. I, II, III. Ann. Math. **86** (1935), 527 bis 606; **37** (1936), 230–263; **38** (1937), 212–291.

[11] *C. L. Siegel,* Lectures on the analytical theory of quadratic forms. Mimeo notes taken by Morgan Ward, Princeton 1935.

[12] *A. Weil,* On algebraic groups of transformations. Amer. J. Math. **77** (1955), 355–391.

[13] *A. Weil,* On algebraic groups and homogeneous spaces. Amer. J. Math. **77** (1955), 493–512.

[14] *A. Weil,* Adeles and algebraic groups. Mimeo notes taken by M. Demazure and T. Ono, Princeton 1961.

J. reine angew. Math. **218** (1965), 79–108
DOI 10.1515/crll.1965.218.79

Journal für die reine und angewandte Mathematik

Notes on elliptic curves. II

By *B. J. Birch* at Manchester and *H. P. F. Swinnerton-Dyer* at Cambridge

To Gina

1. Introduction

During the past six years, we have made extensive calculations on cubic curves of the form

$$\Gamma\colon y^2z = x^3 - Axz^2 - Bz^3 \tag{1.1}$$

where A, B are rational integers. In a previous paper [2], we have shown how $\mathcal{A}$, the group of rational points on (1.1), can conveniently be found on an electronic computer; and we have given extensive tables. In this and subsequent papers, we put forward some conjectures connecting it with other more accessible quantities, together with the numerical evidence on which they are based.

We begin this paper with a history of our investigations. This may make the rest of the paper more understandable; and it enables us to acknowledge the very considerable help we have obtained from other workers in this field.

After the work of Siegel [19] on quadratic forms, it is natural to look at the product

$$\prod N_p/p, \tag{1.2}$$

where N_p is the number of rational points on the curve defined by (1.1) over the finite field of p elements. Write for convenience

$$f(P) = \prod N_p/p \text{ over all } p \leqq P.$$

In the autumn of 1958 we calculated, for a number of curves (1.1), the behaviour of $f(P)$ as P increased. It turned out that the rate of increase of $f(P)$ was fairly closely correlated with the number of generators of infinite order of $\mathcal{A}$; and there was some tendency for $f(P)$ to be large when the generators of $\mathcal{A}$ (or more precisely the corresponding values of x, y, z) were small. We were in fact able to predict the number of generators of $\mathcal{A}$ for specific curves Γ, with fairly consistent success, by examining the values of $f(P)$.

Since the behaviour of $f(P)$ as $P \to \infty$ is at least formally linked to the behaviour of $\zeta_\Gamma(s)$ near $s = 1$ (see §2 below), we were led to make the two linked conjectures:

(A) *If g is the number of generators of $\mathcal{A}$ of infinite order, then $f(P) \sim C(\log P)^g$ as $P \to \infty$, and $\zeta_\Gamma(s) \sim C'(s-1)^{g-1}$ as $s \to 1$, for some constants C, C' depending on Γ.*

At this point we ran into difficulties. The value of $f(P)$ oscillates vigorously as P increases, and there seems little hope of being able to find the constant C by this method with an error of less than say 10 %. (It takes $O(p)$ operations to calculate N_p, so we cannot let P become very big without spending an unreasonable amount of time.) Moreover, the terms of the product (1.2) corresponding to primes p dividing $6\left(4A^3 - 27B^2\right)$ are almost certainly the wrong ones; and it was not then known what the right ones were. Thus, even if we had been able to find the value of C for a particular curve Γ, there would have been little prospect of identifying it; this difficulty has since been removed by the work of Tamagawa [23]; see §7 and later on in this introduction. We hope to return to these direct calculations in a subsequent note.

For those curves Γ which admit complex multiplication an alternative approach is possible; for in this case it is known that $\zeta_\Gamma(s)$ can be expressed in terms of the Riemann zeta function $\zeta(s)$ and Hecke L-series. In particular, for the curve

$$\Gamma = \Gamma_D \colon y^2 z = x^3 - Dxz^2 \tag{1.3}$$

(to which we confine ourselves for the rest of this paper) we have

$$\zeta_\Gamma(s) = \frac{\zeta(s)\,\zeta(s-1)}{L_D(s)} \tag{1.4}$$

where $L_D(s)$ is the Hecke L-series defined by (2.7) below. Formally, $L_D(1) = \Pi(N_p/p)^{-1}$. In the summer of 1960 we found an approximation to $L_D(1)$ as a multiple of a reasonably rapidly convergent series. In this way we were able with fair confidence to determine whether $L_D(1)$ vanished or not; and obtained a good deal of evidence for the conjecture

$$L_D(1) = 0 \textit{ if and only if } g > 0, \tag{1.5}$$

which is a weaker form of conjecture (A) above. However, when $g = 0$ we were still not able to calculate $L_D(1)$ accurately enough to be able to interpret it.

At this point Davenport came to our rescue, by suggesting that it should be possible to express $L_D(1)$ in finite form in terms of division values of the Weierstrass $\wp$-function defined by the equation

$$\wp'^2 = 4\wp^3 - 4\wp.$$

The details of this transformation are given in §3 below. Exact (algebraic) evaluation of the formulae given there in a few simple cases – together with a remark of Kneser that 'the answer should be an integer' – suggested that

$$L_D(1) = \begin{cases} D^{-\frac{1}{4}} \cdot \omega \cdot \sigma(D) \text{ for } D > 0, \\ (-4D)^{-\frac{1}{4}} \cdot \omega \cdot \sigma(D) \text{ for } D < 0, \end{cases} \tag{1.6}$$

where ω is the real period of $\wp(u)$ and $\sigma(D)$ is in general a small positive integer. We prove in §4, by means of class field theory, that $\sigma(D)$ is in general a rational integer; but we can prove nothing about its sign. Using *EDSAC* II, we calculated $\sigma(D)$ from these formulae in a large number of cases, obtaining the results given in Table 1.

To compare these results with our conjecture (1.5) we needed to be able to find the corresponding values of g. For large D the methods of our previous paper proved too slow and we had to use a method tailored to our special curve (1.3). This is described in §5: we do not regard it as novel, but have to describe it at length in the absence of any satisfactory reference. All the results we have obtained are compatible with (1.5); but there are many curves for which we were not able to find the value of g.

The rest of this paper is concerned with the interpretation of the integer $\sigma(D)$ when it does not vanish. We hoped from the beginning that this would turn out to be essentially the order of the Tate-Šafarevič group of Γ_D (for a concise popular exposition, see for instance Cassels [6]), and this hope was supported by the fact that $\sigma(D)$ was always a power of 2 times a square. (Cassels [5] has shown that if the Tate-Šafarevič group is finite, its order is a square.) But we found it difficult to get the details right; the trouble is that if p $\mid 2D$ then N_p/p is not usually the natural factor to take in (1.2), and it was not obvious what one should replace it by. Eventually, much helped by prodding from Cassels, we realized that we should replace the product (1.2) by $\tau(D)$, the Tamagawa number of the curve Γ_D (see [23]). This too is an infinite product, and differs from (1.2) only in finitely many terms; the details are described in §6. We conjecture:

(B) *If* $g = 0$ *then the order of the Tate-Šafarevič group of* Γ_D *is* $\eta^2(D)/\tau(D)$, *where* $\eta(D)$ *is the number of rational points on* Γ_D.

We can almost interpret this in terms analogous to Tamagawa's form of Siegel's theorem. For there is a natural group structure on the set of points lying on everywhere locally soluble coverings, and its Tamagawa number is $\tau(D)$ times the order of the Tate-Šafarevič group – in other words, $\eta^2(D)$ according to our conjecture. That the factor involved should be the square of the number of rational points rather than the number of rational points seems a little odd; subsequently, this has been partly explained by Cassels [7].

The tangible evidence for our conjecture (B) is not great, for unfortunately little is known about the Tate-Šafarevič group – indeed there is not even any case in which it has been proved to be finite. We have never computed more than its 2-component: in every numerical case this at any rate has the order suggested by the conjecture. The labour involved in finding any other component appears to us prohibitive. We note that in our tables $\tau(D)$ is always a perfect square, as it must be in order to be consistent with Cassels' work. But at Stockholm we were informed that Šafarevič possessed results which, in view of the values for $\tau(D)$ given in Table 2, seemed inconsistent with our conjectures. Recently, Cassels [7] has given additional evidence which helps to confirm our conjecture; and the analogy with the work of Ono [18] is very striking.

Certain cases of the function field analogue of (A) are known to be true [20]. In particular, Mumford has shown that two elliptic curves over a finite field are isogenous if and only if they have the same number of rational points (in the classical case, and in older language, this must have been known to Weber; see §113 of his Algebra [21]; see also Manin [26]); he deduced the analogue of (1.5) for an elliptic curve with coefficients in a finite field k, considered over a function field of genus 1 over k. Tate [27] has given conjectures concerning the zeta function of a variety over a finite field which generalise this analogue, and which he has confirmed in many cases.

In conclusion, we must apologise for our rather considerable delay in publication; we hope that no harm has been done, since fairly detailed accounts of our results have been publicised – in particular, Cassels was kind enough to describe our work in his Stockholm talk [6], and one of the authors has given an account in a lecture delivered in Pasadena [3]. The informed reader will recognise that this introduction is already somewhat dated.

2. The zeta-function of Γ

Suppose that Γ is defined over the rationals by (1.1), and is a complete non-singular curve of genus 1. For almost all p, the curve defined by (1.1) over the finite field of p elements retains these properties; and if it contains N_p points defined over this finite field, its zeta-function is

$$\zeta_{\Gamma,p}(s) = \frac{1+(N_p-p-1)\,p^{-s}+p^{1-2s}}{(1-p^{-s})\,(1-p^{1-s})}. \tag{2.1}$$

The global zeta-function of Γ is therefore

$$\zeta_\Gamma(s) = \prod_p \zeta_{\Gamma,p}(s) = \frac{\zeta(s)\,\zeta(s-1)}{L_\Gamma(s)}, \tag{2.2}$$

where $\zeta(s)$ is the usual Riemann zeta-function,

$$L_\Gamma(s) = \prod_p \left\{1+(N_p-p-1)\,p^{-s}+p^{1-2s}\right\}^{-1} \tag{2.3}$$

and finitely many factors in the product may not be the most natural ones. Formally we have

$$L_\Gamma(1) = \prod (N_p/p)^{-1};$$

but for general Γ the product (2.3) is only known to converge in $R(s) > \frac{3}{2}$, and it is not known whether it can be continued analytically outside this region.

If Γ (possibly defined over an algebraic number field) admits complex multiplication, Deuring [9] has shown that $L_\Gamma(s)$ can be analytically continued over the whole plane and satisfies a functional equation. By this means he also found a natural form for the missing factors in (2.3), namely that which makes the functional equation as simple as possible.

For the special curve

$$\Gamma_D\colon y^2 z = x^3 - Dxz^2 \tag{2.4}$$

of this paper, it is simplest to use the known explicit formulae for N_p. For convenience we start by recalling some properties of the quartic residue symbol. Suppose that ν is an odd Gaussian prime and μ any Gaussian integer prime to ν. Then we define the quartic residue symbol $\left(\frac{\mu}{\nu}\right)_4$ to be that power of i which is congruent to

$$\mu^{(N\nu-1)/4} \bmod \nu,$$

where N stands for the norm. This is multiplicative in μ, and we extend it to be multiplicative in ν by the condition

$$\left(\frac{\mu}{\nu_1\nu_2}\right)_4 = \left(\frac{\mu}{\nu_1}\right)_4 \left(\frac{\mu}{\nu_2}\right)_4.$$

We also write $(\mu/\nu)_4 = 0$ when μ, ν are not coprime. The principal properties we shall need are that $(\mu/\nu)_4 = 1$ for μ, ν coprime and real, and the law of reciprocity (see for instance [12]). This latter says that if μ, ν are odd and congruent to 1 mod $(2+2i)$ – which can always be achieved by multiplication by a unit – then

$$\left(\frac{\mu}{\nu}\right)_4 = \pm\left(\frac{\nu}{\mu}\right)_4$$

where the sign is negative if both μ, and ν are congruent to $3+2i$ mod 4 and positive otherwise.

Return now to the curve (2.4), and assume for convenience that $p \nmid 2D$. For $p \equiv 3$ (mod 4) we have simply $N_p = p+1$. If $p \equiv 1$ (mod 4) we can factorize it in Gaussian integers

$$p = \pi\bar{\pi} \text{ with } \pi \equiv 1 \,(\mathrm{mod}\ 2+2i)\,;$$

then (see [8], [22]) we have

$$N_p = p + 1 - \bar{\pi}\left(\frac{D}{\pi}\right)_4 - \pi\left(\frac{D}{\bar{\pi}}\right)_4. \tag{2.5}$$

Putting these results together we obtain

(2.6)

$$\begin{aligned} L_D(s) &= \prod_{p\equiv 3(4)} \left(1+p^{1-2s}\right)^{-1} \prod_{p\equiv 1(4)} \left\{1 - \bar{\pi}\left(\tfrac{D}{\pi}\right)_4 p^{-s} - \pi\left(\tfrac{D}{\bar{\pi}}\right)_4 p^{-s} + p^{1-2s}\right\}^{-1} \\ &= \prod_{\pi\equiv 1(2+2i)} \left\{1 - \left(\tfrac{D}{\pi}\right)_4 \tfrac{\bar{\pi}}{(N\pi)^s}\right\}^{-1}, \end{aligned}$$

where the product is taken over all relevant primes, real or complex, in the Gaussian field $Q(i)$. Note that $L_D(s)$ differs rather trivially from $L_\Gamma(s)$ in that it does not contain factors corresponding to the primes which divide $2D$. Hence finally

$$L_D(s) = \sum_{\sigma\equiv 1(2+2i)} \left(\frac{D}{\sigma}\right)_4 \frac{\bar{\sigma}}{(N\sigma)^s}, \tag{2.7}$$

which is an orthodox Hecke L-series with Größencharakter (see [15]).

Over the rational field, the curve Γ_D is 2-isogenous to

$$\Gamma_{-4D}\colon y^2 z = x^3 + 4Dxz^2,$$

the isogeny arising from the rational 2-division point (0, 0, 1) on Γ. It follows that the two curves have the same value of g, the number of generators of infinite order of $\mathcal{A}$; and since $-4 = (1+i)^4$ they have the same N_p, zeta-functions and L-series. However they need not have the same Tate-Šafarevič group; for examples see [2], §4.

Even though Γ_D is defined over the rational field Q, from certain points of view its natural field of definition is the Gaussian field $Q(i)$. This is the least field over which the complex multiplication on Γ_D is defined; and it is the field of definition of the quartic residue symbol. Over $Q(i)$, Γ_D is birationally equivalent to Γ_{-4D} instead of merely isogenous. Suppose that $\mathcal{A}_i$ is the group of points on Γ_D defined over $Q(i)$, where we still assume D to be rational. In virtue of the complex multiplication on Γ_D, we can regard $\mathcal{A}_i$ as a $Z(i)$-module; and it then has as many generators of infinite order as does $\mathcal{A}$ regarded as a Z-module. We prove no relations between the various Tate-Šafarevič groups involved, though some are implicit in our conjectures; clearly the Tate-Šafarevič groups of Γ_D and Γ_{-4D} can only differ in their 2-components.

It remains to consider the zeta-function of Γ over $Q(i)$; and here it costs nothing to take D to be a Gaussian integer. If $p \equiv 3$ (mod 4) is a rational prime then

$$N_p = p^2 + 1 + p\left(\frac{D}{p}\right)_4 + p\left(\frac{\bar{D}}{p}\right)_4;$$

if $\pi \equiv 1 \pmod{2+2i}$ is a complex prime then

$$N_\pi = \pi\bar{\pi} + 1 - \bar{\pi}\left(\frac{D}{\pi}\right)_4 - \pi\left(\frac{\bar{D}}{\bar{\pi}}\right)_4.$$

Arguing in much the same way as before, we find

$$\zeta_{\Gamma,Q(i)}(s) = \frac{\zeta_{Q(i)}(s)\,\zeta_{Q(i)}(s-1)}{L_D(s)\,L_{\bar{D}}(s)}$$

where $\zeta_{Q(i)}(s)$ is the zeta-function of the field $Q(i)$ and $L_D(s)$ is still defined by (2.7). Thus in particular, when D is a rational integer $\zeta_{\Gamma,Q(i)}(s)$ can easily be written in terms of $\zeta_\Gamma(s)$ and of well-known functions not involving Γ.

3. Formulae for $L_D(1)$

We have now to derive from (2.7) an expression for $L_D(s)$ which can be analytically continued as far as $s = 1$ and is such that $L_D(1)$ can be written in finite form. Since the extra generality costs us nothing, we shall assume only that D is a Gaussian integer; however, the important case is when D is rational. We may also assume that D is fourthpower free; for multiplying D by a fourth power only changes finitely many factors in (2.6), and changes those in an obvious way.

We write $D = EF$, where $E \equiv 1 \bmod (2+2i)$ and F is the product of a power of $1+i$ and a unit; and we define Δ to be the product of the distinct primes dividing E, normalized so that $\Delta \equiv 1 \bmod (2+2i)$. If D is real, so are E, F and Δ. For convenience we also write $\varepsilon = (-1/E)_4$, so that $\varepsilon = +1$ if $E \equiv 1 \bmod 4$ and $\varepsilon = -1$ if $E \equiv 3+2i \bmod 4$. Remembering that $\sigma \equiv 1 \bmod (2+2i)$ where σ is the variable of summation in (2.7), and using the law of quartic reciprocity, we have

$$\left(\frac{D}{\sigma}\right)_4 = \left(\frac{E}{\sigma}\right)_4\left(\frac{F}{\sigma}\right)_4 = \left(\frac{\sigma}{E}\right)_4\left(\frac{\varepsilon F}{\sigma}\right)_4. \tag{3.1}$$

Here the first factor on the right only depends on the residue class of σ modulo Δ, and the second factor on the class of σ modulo 16. Let K be such that $(2+2i)\,|K$, that $K|16$ and that $(\varepsilon F/\sigma)_4$ depends only on the residue class of σ modulo K. (We could take $K = 16$ always, but a smaller value of K reduces the complexity of some of the later formulae and so also the time taken in computation. We may for instance take $K = 8$ if F is real, and $K = 4$ if $F = \pm 1$.)

Now let B be a set of representatives for the residue classes modulo Δ, and let C be a set of representatives for those residue classes mod K which are congruent to 1 mod $(2+2i)$. We can write

$$\sigma = K\Delta\mu + K\beta + \Delta\gamma$$

where μ is a Gaussian integer, $\beta \in B$ and $\gamma \in C$; and we can replace any sum over σ by a triple sum over μ, β and Γ. For convenience we write $\varrho = K\beta + \Delta\gamma$. It follows that we can rewrite (2.7) in the form

$$L_D(s) = \sum_{\beta,\gamma}\left(\frac{D}{\varrho}\right)_4 \sum_\mu \frac{\bar{K}\bar{\Delta}\bar{\mu} + \bar{\varrho}}{\{N(K\Delta\mu + \varrho)\}^s} \tag{3.2}$$

in which provided $E \neq 1$ we have

$$\left(\frac{D}{\varrho}\right)_4 = \left(\frac{K}{E}\right)_4 \left(\frac{\varepsilon F}{\Delta}\right)_4 \left(\frac{\beta}{E}\right)_4 \left(\frac{\varepsilon F}{\gamma}\right)_4. \tag{3.3}$$

We have next to find a more convenient expression for the inner sum in (3.2). For α not a Gaussian integer, we write

$$\psi(\alpha, s) = \frac{\bar{\alpha}}{|\alpha|^{2s}} + \sum_{\mu \neq 0} \left\{ \frac{\bar{\alpha} + \bar{\mu}}{|\alpha + \mu|^{2s}} - \frac{\bar{\mu}}{|\mu|^{2s}} \left(1 - \frac{s\alpha}{\mu} + \frac{\bar{\alpha}(1-s)}{\bar{\mu}} \right) \right\}. \tag{3.4}$$

Since the expression in curly brackets is $O\left(\mu^{-2s-1}\right)$, this defines an analytic function of s in $R(s) > \frac{1}{2}$; and since the series is uniformly convergent near $s = 1$ we have

$$\psi(\alpha, 1) = \frac{1}{\alpha} + \sum_{\mu \neq 0} \left\{ \frac{1}{\mu + \alpha} - \frac{1}{\mu} + \frac{\alpha}{\mu^2} \right\} = \xi(\alpha)$$

where ξ is the Weierstrass zeta-function with periods 1, i. On the other hand, if $R(s)$ is large we can rearrange the terms in (3.4) to give

$$\begin{aligned} \sum_{\mu} \frac{\bar{\alpha}+\bar{\mu}}{|\alpha+\mu|^{2s}} &= \psi(\alpha, s) + \bar{\alpha}(1-s) \sum_{\mu \neq 0} \frac{1}{(N\mu)^s} \\ &= \psi(\alpha, s) + 4\bar{\alpha}(1-s)\, \zeta_{Q(i)}(s); \end{aligned}$$

for the other sums vanish on combining the terms arising from $\pm\mu$, $\pm i\mu$. This provides an analytic continuation of the left hand side. Substituting into (3.2) we obtain

$$L_D(s) = \frac{\bar{K}\bar{\Delta}}{\{N(K\Delta)\}^s} \left\{ \sum_{\beta,\gamma} \left(\frac{D}{\varrho}\right)_4 \psi\left(\frac{\varrho}{K\Delta}, s\right) + 4(1-s)\, \zeta_{Q(i)}(s) \sum_{\beta,\gamma} \frac{\bar{\varrho}}{\bar{K}\bar{\Delta}} \left(\frac{D}{\varrho}\right)_4 \right\}.$$

In particular, since $(s-1)\,\zeta_{Q(i)}(s) \to \frac{1}{4}\pi$ as $s \to 1$, we have

$$L_D(1) = \frac{1}{K\Delta} \sum \left(\frac{D}{\varrho}\right)_4 \xi\left(\frac{\varrho}{K\Delta}\right) - \frac{\pi}{N(K\Delta)} \sum \bar{\varrho} \left(\frac{D}{\varrho}\right)_4. \tag{3.5}$$

Since this expression does not depend on the particular choice of representatives ϱ of the residue classes mod $K\Delta$, we see that

$$\xi(u+1) = \xi(u) + \pi, \ \xi(u+i) = \xi(u) - \pi i.$$

In particular, since ξ is odd it follows that

$$\xi\left(\frac{1}{2}\right) = \frac{1}{2}\pi, \ \xi\left(\frac{1}{2} i\right) = -\frac{1}{2}\pi i. \tag{3.6}$$

Provided that $E \neq 1$, we can express the right hand side of (3.5) in terms of the Weierstrass $\wp$-function. In order to simplify the algebra later, it is convenient to make a change of period. Throughout this paper, we shall therefore denote by $\wp(u)$ the Weierstrass $\wp$-function with periods ω, $i\omega$, where

$$\omega = 2^{\frac{1}{2}} \pi e^{-\frac{\pi}{6}} \prod \left(1 - e^{-2\pi n}\right)^2 = 2.6220575\ldots$$

has been chosen so that $\wp(u)$ satisfies the equation

$$\wp'^2 = 4\wp^3 - 4\wp. \tag{3.7}$$

The Weierstrass $\wp$-function with periods $1, i$ is therefore $\omega^2 \wp(\omega u)$; and (see [25]) we can write the addition formula for $\xi(u)$ in the form

$$\xi(u+v) = \xi(u) + \xi(v) + \omega \cdot \frac{\wp'(\omega u) - \wp'(\omega v)}{2\{\wp(\omega u) - \wp(\omega v)\}}. \tag{3.8}$$

In this we write $u = \gamma/K$, $v = \beta/k$, giving an expression for $\xi(\varrho/K\Delta)$ in terms of functions with simpler arguments. If we substitute the resulting values in (3.5) and add to this the equation derived by writing $-\beta$ for β throughout, many of the terms cancel. In view of the shape of (3.3), we split cases on the value of $\varepsilon = (-1/E)_4$.

If $\varepsilon = +1$, then $(D/\varrho)_4$ is not affected by replacing β by $-\beta$ and so we obtain

$$L_D(1) = \frac{1}{K\Delta} \sum_{\beta,\gamma} \left(\frac{D}{\varrho}\right)_4 \left\{\xi\left(\frac{\gamma}{K}\right) - \frac{\pi\bar{\gamma}}{\bar{K}}\right\} + \frac{\omega}{2K\Delta} \sum_{\beta,\gamma} \left(\frac{D}{\varrho}\right)_4 \frac{\wp'(\gamma\omega/K)}{\wp(\gamma\omega/K) - \wp(\beta\omega/\Delta)}.$$

In the first term on the right, the sum over β is effectively $\sum (\beta/E)_4 = 0$, since E is not a fourth power; hence this term vanishes and we have

$$L_D(1) = \frac{\omega}{2K\Delta} \sum_{\beta,\gamma} \left(\frac{D}{\varrho}\right)_4 \frac{\wp'(\gamma\omega/K)}{\wp(\gamma\omega/K) - \wp(\beta\omega/\Delta)}. \tag{3.9}$$

If $\varepsilon = -1$, replacing β by $-\beta$ changes the sign of $(D/\varrho)_4$ and we therefore obtain

$$\begin{aligned} L_D(i) = {} & \frac{1}{K\Delta} \sum_{\beta,\gamma} \left(\frac{D}{\varrho}\right)_4 \left\{\xi\left(\frac{\beta}{\Delta}\right) - \frac{\pi\bar{\beta}}{\bar{\Delta}}\right\} \\ & - \frac{\omega}{2k\Delta} \sum_{\beta,\gamma} \left(\frac{D}{\varrho}\right)_4 \frac{\wp'(\beta\omega/\Delta)}{\wp(\gamma\omega/K) - \wp(\beta\omega/\Delta)}. \end{aligned} \tag{3.10}$$

(Here the terms for which $\beta|\Delta$ are defined to be zero.) If ϵF is not an exact fourth power, then the sum over Γ in the first term on the right is effectively $\sum (\varepsilon F/\gamma)_4 = 0$; hence this term vanishes. If ϵF is an exact fourth power, the first term on the right of (3.10) is

$$\frac{\bar{K}}{8\Delta} \sum_{\beta} \left(\frac{\beta k}{E}\right)_4 \left\{\xi\left(\frac{\beta}{\Delta}\right) - \frac{\pi\bar{\beta}}{\bar{\Delta}}\right\} \tag{3.11}$$

since there are $\frac{1}{8} K\bar{K}$ values of γ to sum over. Writing $i\beta$ for β and adding, we see that (3.11) vanishes if $(i/E)_4 = -i$. If $(i/E)_4 = i$ we can evaluate (3.11) by writing $(1+i)\beta$ for β and using

$$\xi(u(1+i)) = (1-i)\,\xi(u) + \frac{\omega(1+i)}{4} \frac{\wp'(\omega u)}{\wp(\omega u)}, \tag{3.12}$$

which is a special case of (3.8). We find that (3.11) is equal to

$$\frac{\bar{K}\omega}{16\Delta} \left\{1 - i + 2i\left(\frac{1+i}{E}\right)_4\right\}^{-1} \sum_{\beta} \left(\frac{\beta K(1+i)}{E}\right)_4 \frac{\wp'(\beta\omega/\Delta)}{\wp(\beta\omega/\Delta)}.$$

In the excluded cases when $E = 1$, we can evaluate $L_D(1)$ directly from (3.5); for the values of $\xi(u)$ needed can be obtained from (3.6) and (3.12).

For explicit numerical calculation on an electronic computer, it is expedient to carry out the summation over γ algebraically. We assume that D is a rational integer, so that we have $\varepsilon = +1$, and that D is not divisible by a fourth power nor, since $L_{4D}(1) = L_{-D}(1)$, by 4. There are therefore four cases to consider. In each of them we write for convenience $\wp$ for $\wp(\beta\omega/\Delta)$. The sums concerned are taken over all $\beta \in B$; but it is enough to consider the β prime to Δ since the terms with β not prime to Δ each vanish.

Case 1: $D \equiv 1 \bmod 4$. Now we have $E = D$, $F = 1$. If we choose $K = 4$ we obtain for $D \neq 1$,

$$L_D(1) = \frac{\omega}{\Delta} \sum \left(\frac{\beta}{D}\right)_4 \frac{\wp}{\wp^2 - 2\wp - 1}. \tag{3.13}$$

Alternatively, we may take $K = 2 + 2i$ and obtain, still for $D \neq 1$,

$$L_D(1) = \frac{i\omega}{2\Delta} \sum \left(\frac{\beta(2+2i)}{D}\right)_4 \frac{1}{\wp - i}. \tag{3.14}$$

For the excluded value $D = 1$ we find $L_1(1) = \frac{1}{4}\omega$.

Case 2: $D \equiv 3 \bmod 4$. Now we have $E = -D$, $F = -1$, $K = 4$ and obtain for $D \neq -1$

$$L_D(1) = \frac{\omega}{\Delta\sqrt{2}} \sum \left(\frac{\beta}{-D}\right)_4 \frac{\wp + 1}{\wp^2 - 2\wp - 1}. \tag{3.15}$$

For the excluded case $D = -1$ we find $L_{-1}(1) = \frac{1}{4}\omega\sqrt{2}$.

Case 3: $D \equiv 2 \bmod 8$. Now we have $E = \frac{1}{2}D$, $F = 2$, $K = 8$ and obtain for $D \neq 2$

$$L_D(1) = \frac{\omega\sqrt[4]{8}}{\Delta} \sum \left(\frac{\beta}{E}\right)_4 \frac{2\wp(\wp^4 - 1)(\wp^2 + 2\wp - 1)}{(\wp^2 - 2\wp - 1)^4 - 32\wp^2(\wp^2 - 1)^2}. \tag{3.16}$$

For the excluded case $D = 2$ we find $L_2(1) = 0$.

Case 4: $D \equiv 6 \bmod 8$. Now we have $E = -\frac{1}{2}D$, $F = 2$, $K = 8$ and obtain for $D \neq -2$

$$L_D(1) = \frac{\omega\sqrt[4]{2}}{\Delta} \sum \left(\frac{\beta}{E}\right)_4 \frac{(\wp - 1)\left\{2(\wp^2+1)^3 + (\wp^2 + 2\wp - 1)^3\right\}}{(\wp^2 - 2\wp - 1)^4 - 32\wp^2(\wp^2 - 1)^2}. \tag{3.17}$$

For the excluded case $D = -2$ we find $L_{-2}(1) = \frac{1}{2}\omega\sqrt[4]{2}$.

4. Proof of integrity

It is well known [21] that the division values of $\wp$ and $\wp'$ are in general algebraic functions of g_2 and g_3, and so in our particular case algebraic numbers. It follows from the results of §3 that $\omega^{-1}L_D(1)$ is an algebraic number; in this section we consider how nearly it is a rational integer. We deal only with the case when D is rational: but similar arguments could be used in the Gaussian case. Our object is to prove

Theorem 1. *Let $D > 1$ be a positive integer, fourth power free and not divisible by 4. Then*

$$\frac{D^{\frac{1}{4}} L_D(1)}{\omega} \text{ and } \frac{(4D)^{\frac{1}{4}} L_{-D}(1)}{\omega}$$

are rational integers.

The theorem holds for $D = 2$, from the explicit values given at the end of §3. Setting aside this case, we may therefore make use of (3.9) and its consequences (3.13) to (3.17). The proof of the theorem now falls into two parts. The first part (Lemmas 1–5) is concerned with integrity and is entirely elementary; the second is concerned with rationality and makes use of results from classical class field theory.

For convenience, in dealing with algebraic integers, we shall use 'odd' to mean prime to 2, and 'even' to mean not odd.

Lemma 1. *Let α be an odd Gaussian integer. Then*

$$\wp(\alpha u) = P_\alpha(\wp(u)) / Q_a^2(\wp(u)),$$

where P_α, Q_α are polynomials of degrees $\alpha\bar{\alpha}$, $\frac{1}{2}(\alpha\bar{\alpha} - 1)$ respectively with Gaussian integer coefficients; P_α has leading coefficient 1 and Q_α has leading coefficient α and constant term ± 1 or $\pm i$.

We know that $\wp(\alpha u)$ is an even doubly-periodic function whose only singularities in the period parallelogram are $\alpha\bar{\alpha}$ double poles at the α-division points. These include the origin but none of the midpoints. It follows, by well-known results on doubly-periodic functions, that we have

$$\wp(\alpha u) = P_\alpha(\wp(u)) / Q_\alpha^2(\wp(u)),$$

where P_α, Q_α are polynomials of degrees $\alpha\bar{\alpha}$, $\frac{1}{2}(\alpha\bar{\alpha} - 1)$ respectively. Moreover, the zeros of Q_α are just the distinct finite values of $\wp(u)$ at the α-division points, while the zeros of P_α are the values of $\wp(u)$ at those $(1+i)\,\alpha$-division points that are not α-division points, taken with their correct multiplicity.

We now normalize so that Q_α has leading coefficient α. Since for small u we have

$$\wp(u) \sim u^{-2}, \; \wp(\alpha u) \sim \alpha^{-2} u^{-2},$$

it follows that P_α has leading coefficient 1. We shall prove the remaining statements in the lemma by induction on the value of $\alpha\bar{\alpha}$, since they obviously hold when α is a unit. In the identity

$$\wp(v+\omega)\,\wp(v-\omega) = \left\{\frac{\wp(v)\,\wp(\omega) + 1}{\wp(v) - \wp(\omega)}\right\}^2 \tag{4.1}$$

we write $v = \alpha u$, $\omega = (1+i)\,u$. Since

$$\wp(u(1+i)) = \frac{\wp^2(u) - 1}{2i\wp(u)} \tag{4.2}$$

we deduce

$$\frac{P_{\alpha+1+i}P_{\alpha-1-i}}{Q^2_{\alpha+1+i}Q^2_{\alpha-1-i}} = \left\{\frac{(\wp^2-1)\,P_\alpha + 2i\wp\,Q^2_\alpha}{(\wp^2-1)\,Q^2_\alpha - 2i\wp\,P_\alpha}\right\}^2.$$

Here the left hand side is in its lowest terms; for from the results above no P can have a zero in common with a Q. But the numerator and denominator on the right have (after squaring) the same degrees and leading coefficients as those on the left. Hence they are equal, giving

$$P_{\alpha+1+i}P_{\alpha-1-i} = \{(\wp^2-1)\,P_\alpha + 2i\wp\,Q^2_\alpha\}^2,$$
$$Q_{\alpha+1+i}Q_{\alpha-1-i} = (\wp^2-1)\,Q^2_\alpha - 2i\wp\,P_\alpha.$$

We have also the similar results obtained by writing $-i$ for i. Now therefore, if $\eta = 1, -1, i$ or $-i$ and if the assertions of the lemma hold for $\alpha - \eta\,(1+i)$ and for $\alpha - 2\eta\,(1+i)\,|$, then they hold for α. But it is geometrically obvious that if α is an odd integer not a unit then we can always choose η so that both $\alpha - \eta\,(1+i)$ and $\alpha - 2\eta\,(1+i)$ are strictly smaller than α. The truth of the lemma now follows by induction.

Lemma 2. *Let Δ be an odd square-free Gaussian integer, not a unit; and let β be a Gaussian integer prime to Δ. Let*

$$\varphi(\Delta) = \begin{cases} \Delta^{2/(\Delta\bar{\Delta}-1)} & \text{if } \Delta \text{ is prime,} \\ 1 & \text{otherwise.} \end{cases}$$

Then $\varphi(\Delta)\,\wp(\beta\omega/\Delta)$ is an algebraic unit.

Suppose first that Δ is prime. Then the $\wp(\beta\omega/\Delta)$ are just the roots of $Q_\Delta(x) = 0$. Moreover they all generate the same field K over $Q(i)$; for given any β_1, β_2 prime to Δ we can find an odd α such that $\beta_2 \equiv \alpha\beta_1 \bmod \Delta$, and so

$$\wp(\beta_2\omega/\Delta) = \wp(\alpha\beta_1\omega/\Delta)$$

is a rational function of $\wp(\beta_1\omega/\Delta)$ over $Q(i)$, by Lemma 1. Since they are the roots of $Q_\Delta(x) = 0$, the $\wp(\beta\omega/\Delta)$ each have numerator a unit and denominator a factor of Δ; and not all of them are integers. Let $\mathfrak{q}$ be a prime ideal in K which divides the denominator of $\wp(\beta_1\omega/\Delta)$ to exactly the rth power, where $r > 0$. For any β_2 prime to Δ we define α as before. Then α is prime to $\mathfrak{q}$ and hence, using Lemma 1 again, $\mathfrak{q}$ divides the denominator of

$$\wp(\beta_2\omega/\Delta) = P_\alpha(\wp(\beta_1\omega/\Delta))\,/Q^2_\alpha(\wp(\beta_1\omega/\Delta))$$

to exactly the rth power. It follows that all the $\wp(\beta\omega/\Delta)$ must have the same denominator, up to a unit; and hence this must be $\varphi(\Delta)$, for there are $\frac{1}{2}(\Delta\bar{\Delta}-1)$ such denominators and their product is Δ.

Now suppose that $\Delta = \Delta_1\Delta_2$, where Δ_1 and Δ_2 are coprime and not units. Then $\wp(\beta\omega/\Delta)$ is a zero of $Q_\Delta/Q_{\Delta_1}Q_{\Delta_2}$, which has integer coefficients the first and last of which are units. Hence $\wp(\beta\omega/\Delta)$ is itself a unit. This completes the proof of the lemma.

Lemma 3. *Let γ_1, γ_2 be odd Gaussian integers, γ_2 being square-free, and let $r > 1$ be a rational integer. Then $\wp(\gamma_1\omega/\gamma_2(1+i)^r)$ is an algebraic unit.*

Without loss of generality, we may assume γ_1, γ_2 coprime. We first prove that $\wp(\gamma_1\omega/\gamma_2(1+i))$ is an algebraic integer. If γ_2 is a unit, then

$$\wp(\gamma_1\omega/\gamma_2(1+i)) = \wp(\omega/(1+i)) = 0.$$

If not, we have

$$\wp(u)\,\wp(u+\omega/(1+i)) = -1 \tag{4.3}$$

since, after (4.2), the roots of

$$x^2 - 2ix\wp(u(1+i)) - 1 = 0 \tag{4.4}$$

are $\wp(u)$ and $\wp(u+\omega/(1+i))$. But if we write $u = \gamma_1\omega/\gamma_2(1+i)$ then

$$u + \frac{\omega}{1+i} = \frac{\omega(\gamma_1+\gamma_2)}{\gamma_2(1+i)}$$

is an argument to which we can apply Lemma 2, with γ_2 for Δ. This and (4.3) show that $\wp(\gamma_1\omega/\gamma_2(1+i))$ is $\varphi(\gamma_2)$ times an algebraic unit, and is therefore an algebraic integer.

The lemma now follows by induction on r; for since $\wp(u)$ is a root of (4.4) we see that $\wp(u)$ is an algebraic unit whenever $\wp(u(1+i))$ is an algebraic integer.

Lemma 4. *Let $r > 0$ be a rational integer; and let β, γ, Δ be Gaussian integers of which Δ is odd, square-free and not a unit, β is prime to Δ and γ is odd. Write $\lambda = 1 - 2^{1-r}$, and define $\varphi(\Delta)$ as in Lemma 2. Then*

$$2^{-\lambda}\varphi(\Delta)\{\wp(\beta\omega/\Delta) - \wp(\gamma\omega/(1+i)^r)\}$$

is an algebraic unit.

For $r = 1$ we have $\lambda = 0$, $\wp(\gamma\omega/(1+i)) = 0$, and the result is precisely that of Lemma 2. We now proceed by induction on r. From (4.2) we have

$$\wp(v(1+i)) - \wp(w(1+i)) = \frac{-i\{\wp(v) - \wp(w)\}\{\wp(v)\wp(w)+1\}}{2\wp(v)\wp(w)}. \tag{4.5}$$

In this we write $v = \beta\omega/\Delta$, $w = \gamma\omega/(1+i)^r$, and assume the Lemma true for $r-1$. Write for convenience ϑ for an arbitrary algebraic unit, not necessarily the same from one occurrence to the next. By the induction hypothesis we have

$$\wp(v(1+i)) - \wp(w(1+i)) = 2^{2\lambda-1}\vartheta/\varphi(\Delta).$$

By Lemma 3 with $\gamma_2 = 1$ and Lemma 2 we have

$$2\wp(v)\wp(w) = 2\vartheta/\varphi(\Delta).$$

By (4.1) and Lemma 3 with $\gamma_2 = \Delta$ we have

$$\wp(v)\wp(w) + 1 = \vartheta\{\wp(v) - \wp(w)\}.$$

Combining these three results with (4.5) we obtain

$$\{\wp(v) - \wp(w)\}^2 = 2^{2\lambda}\vartheta/\varphi^2(\Delta),$$

and this proves the lemma.

Lemma 5. *Let D be a rational integer, fourth power free and not divisible by 4. If $D \neq \pm 1$ then $2^{\frac{3}{4}} \Delta \omega^{-1} L_D(1)$ is an algebraic integer.*

Here Δ is defined as in §3. From the actual values of $L_D(1)$ we verify that the Lemma is true for $D = \pm 2$; thus we need only deal with the cases in which one of (3.13) to (3.17) hold. As in the proof of Lemma 4 we shall write ϑ for an algebraic unit, which need not be the same from one occurrence to the next. Now, using Lemma 4,

$$\wp^2 - 2\wp - 1 = \left\{\wp - \wp\left(\frac{1}{4}\omega\right)\right\}\left\{\wp - \wp\left(\frac{3}{4}\omega + \frac{1}{2}i\omega\right)\right\} = 2^{\frac{7}{4}}\vartheta/\varphi^2(\Delta)$$

and

$$\wp + 1 = \wp - \wp\left(\frac{1}{2}i\omega\right) = 2^{\frac{1}{2}}\vartheta/\varphi(\Delta).$$

Hence each term on the right of (3.13) or (3.15) is of the form $\omega\vartheta\varphi(\Delta)/2^{\frac{7}{4}}\Delta$ But in each of these sums the terms given by β and by $-\beta$ are equal. Hence the right hand sides are $\omega\varphi(\Delta)/2^{\frac{3}{4}}\Delta$ times algebraic integers, and since $\varphi(\Delta)$ is an algebraic integer this proves the lemma in these two cases.

In the cases given by (3.16) and (3.17) we argue similarly, using in addition the facts that

$$\left(\wp^2 - 2\wp - 1\right)^4 - 32\wp^2\left(\wp^2 - 1\right)^2 = 2^{\frac{31}{4}}\vartheta/\varphi^8(\Delta)$$

since the left hand side is the product of eight terms of the form

$$\left\{\wp - \wp\left(\frac{1}{8}\gamma\omega\right)\right\} \text{ with } \gamma \text{ odd;}$$

that

$$\wp^2 - 1 = 2\vartheta/\varphi^2(\Delta),$$
$$\wp^2 + 1 = 2^{\frac{3}{2}}\vartheta/\varphi^2(\Delta),$$
$$\wp - 1 = 2^{\frac{1}{2}}\vartheta/\varphi(\Delta)$$

and

$$\wp^2 + 2\wp - 1 = 2^{\frac{7}{4}}\vartheta/\varphi^2(\Delta),$$

all similarly derived from Lemma 4. The rest of the proof goes through exactly as before.

Lemma 5 is the strongest result which we can obtain from elementary divisibility arguments. We now turn to the question of rationality. Here we need to quote classical results of class field theory over $Q(i)$. The relevant part of the Kronecker Jugendtraum (see Fueter [11] or Hasse [13]) is normally stated as follows:

Lemma 6. *Let μ be a Gaussian integer, not a unit. Then the strahlklass field mod μ over $Q(i)$ is generated by $\wp^2(\alpha\omega/\mu)$, where α is any Gaussian integer prime to μ. If this field is K, and if p is any Gaussian prime not dividing μ, then the Artin symbol $\left(\frac{K/Q(i)}{p}\right)$ is the automorphism which takes $\wp^2(\alpha\omega/\mu)$ into $\wp^2(\alpha p\omega/\mu)$.*

The fact that p is only defined up to a power of i is irrelevant; for the value of $\wp^2$ is not changed by multiplying its argument by a power of i. Unfortunately Lemma 6 is not in the most convenient form for us, since it deals with $\wp^2$ rather than with $\wp$. We therefore rephrase the relevant special case of it as follows:

Lemma 7. *Let Δ be an odd square-free Gaussian integer, not a unit. Then the strahlklass field mod 2Δ over $Q(i)$ is generated by $\wp(\alpha\omega/\Delta)$ where α is any Gaussian integer prime to Δ. If this field is K, and if p is any Gaussian prime not dividing 2Δ, normalized so that $p \equiv 1$ mod 2, then the Artin symbol $\left(\frac{K/Q(i)}{p}\right)$ is the automorphism which takes $\wp(\alpha\omega/\Delta)$ into $\wp(\alpha p\omega/\Delta)$.*

Let K be the strahlklass field mod 2Δ over $Q(i)$. Since

$$\wp\left(\frac{\omega(2\alpha+\Delta)}{2\Delta}\right) = \frac{\wp(\alpha\omega/\Delta)+1}{\wp(\alpha\omega/\Delta)-1},$$

the field generated by $\wp(\alpha\omega/\Delta)$ over $Q(i)$ contains $\wp^2(\beta\omega/2\Delta)$ for some β prime to 2Δ, and so contains K.

To deduce Lemma 7 from Lemma 6, we have still to show that $\wp(\alpha\omega/\Delta)$ is in K and is correctly transformed by the Artin symbol. Since we can alter α by a multiple of Δ, we may assume that $\alpha \equiv \Delta$ mod 2. We shall now prove – which is clearly sufficient for our needs – that $\wp(\alpha\omega/\Delta)$ can be written as a rational function of $\wp^2(\alpha\omega/2\Delta)$, the function being defined over $Q(i)$ and being independent of α.

For any Gaussian integer ν, $\wp(\nu u)$ is a rational function of $\wp(u)$ defined over $Q(i)$; this follows from (4.2) and Lemma 1. Moreover, it is an odd function; for writing iu for u changes the signs of both $\wp(u)$ and $\wp(\nu u)$. Hence $\wp^2(\nu u)$ is a rational function of $\wp^2(u)$, defined over $Q(i)$. Let $\gamma = \frac{1}{2}(\alpha - \Delta)$ and choose ν so that $2 \nmid \nu$ and $\Delta | (\nu - 1)$. Then $\wp^2(\gamma\omega/\Delta) = \wp^2(\nu\alpha\omega/2\Delta)$ is a rational function of $\wp^2(\alpha\omega/2\Delta)$, defined over $Q(i)$ and independent of α. Moreover

$$\wp(\alpha\omega/\Delta) = \frac{2\{\wp^2(\alpha\omega/2\Delta)+1\}\{\wp^2(\gamma\omega/\Delta)+1\}}{\{\wp^2(\alpha\omega/2\Delta)-1\}\{\wp^2(\gamma\omega/\Delta)-1\}}.$$

Combining these results, we see that $\wp(\alpha\omega/\Delta)$ can be written as a rational function of $\wp^2(\alpha\omega/2\Delta)$ defined over $Q(i)$ and independent of α (since ν is so). This completes the deduction of Lemma 7.

Corollary. *Let $R(x)$ be a rational function of x defined over $Q(i)$, and let D be a rational integer. Then with the notation of* §3,

$$E^{\frac{1}{4}} \sum \left(\frac{\beta}{E}\right)_4 R(\wp(\beta\omega/\Delta)) \tag{4.6}$$

is in $Q(i)$.

For let K be defined as in the proof of the Lemma. Then $E^{\frac{1}{4}}$ is in K; for if p prime to $2E$ is a Gaussian prime it splits in $Q\left(i, E^{\frac{1}{4}}\right)/Q(i)$ if and only if $(E/p)_4 = 1$; and this depends only on the residue class of p modulo 2Δ, by the quartic reciprocity law and the rationality of E. Moreover, the general automorphism of $K/Q(i)$ is that which takes $\wp(\alpha\omega/\Delta)$ into $\wp(\lambda\alpha\omega/\Delta)$ for each α, where λ is prime to Δ and $\lambda \equiv 1$ mod 2. By the identification of the Artin symbol in Lemma 7, this also takes $E^{\frac{1}{4}}$ into $(\lambda/E)_4 E^{\frac{1}{4}}$, and hence merely permutes the terms of (4.6). Thus this expression, being unchanged by each automorphism of $K/Q(i)$, must be in $Q(i)$; and this proves the Corollary.

We can now return to the proof of Theorem 1. Let D be a rational integer, fourth power free and not divisible by 4, but not necessarily positive, and suppose $|D| > 2$. Then one of (3.13), (3.15), (3.16) or (3.17) is valid. It follows that $L_D(1)$ is real; for the set B over which the sum is taken may be made to consist of real numbers and complex conjugate pairs. Moreover $D^{\frac{1}{4}}\omega^{-1}L_D(1)$ is in $Q(i)$, by the Corollary to Lemma 7. Since $-4 = (1+i)^4$, it follows that the two expressions in the Theorem are rational numbers. But by Lemma 5 they are algebraic numbers whose denominator is at most $(2\Delta)^{\frac{3}{4}}$. Since the only rational integers which divide this are ± 1, it follows that the expressions in Theorem 1 are rational integers. This proves the Theorem.

5. Estimation of g

In order to produce Table 1 below, we need also a method of estimating g, the number of independent generators of infinite order of $\mathcal{A}$, the group of rational points on Γ_D. We cannot do this by the methods of our previous paper, since they would be intolerably slow when D is large. Instead, we use the fact that multiplication by 2 can be written as a product of isogenies. The arguments involved are well known, though we cannot find them explicitly in the literature (compare the theses of Billing [1] and Lind [17]): we have stated them in unusually pompous form to be able to use the analogy with the work of Cassels [5].

We extend the definition of ν-covering to the case where $\nu\colon \mathcal{C}_1 \to \mathcal{C}$ is an isogeny of elliptic curves instead of an endomorphism. We assume that ν is defined over the rationals and is an n to 1 mapping; and we write $\mathfrak{o}$, $\mathfrak{o}_1$ for the zeros of $\mathcal{C}$, $\mathcal{C}_1$ considered as Abelian groups. There is an isogeny $\nu_1\colon \mathcal{C} \to \mathcal{C}_1$ such that the composite maps $\nu\nu_1$ and $\nu_1\nu$ are multiplication by n on $\mathcal{C}$, $\mathcal{C}_1$ respectively; and ν_1 is also defined over the rationals. We say that there is a ν-covering of $\mathcal{C}$ if there is a curve $\mathcal{D}$ defined over the rationals and a commutative triangle

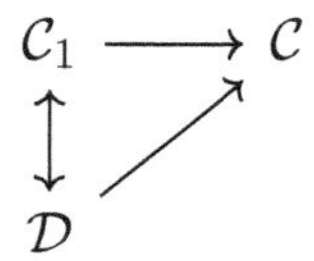

with associated generic points

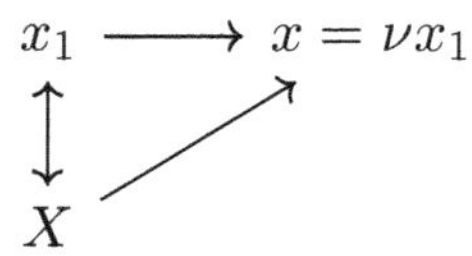

where $X \to x$ is over the rationals and $X \leftrightarrow x_1$ over the complex numbers. Another curve $\mathcal{D}'$ and its associated mappings give the same ν-covering if and only if there is a birational mapping $X \leftrightarrow X'$ over the rationals and a point $\mathfrak{d}_1$ on $\mathcal{C}_1$ with $\nu\mathfrak{d}_1 = \mathfrak{o}$ such that the diagram

(5.1)
$$\begin{array}{ccc} X & \longleftrightarrow & x_1 \\ \updownarrow & & \updownarrow \\ X' & \longleftrightarrow & x'_1 = x_1 + \mathfrak{d}_1 \end{array}$$

is commutative. There is a natural law of composition of ν-coverings inherited from the law of composition of homogeneous spaces over $\mathcal{C}_1$; and under it they form an Abelian group. We are interested in two subgroups: G_ν, the group of those coverings for which $\mathcal{D}$ has a point in every p-adic field including the reals, and G'_ν, the group of those coverings for which $\mathcal{D}$ actually has a rational point. If $\mathcal{A}$, $\mathcal{A}_1$ are the groups of rational points on $\mathcal{C}$, $\mathcal{C}_1$ respectively, then G'_ν is isomorphic to $\mathcal{A}/\nu\mathcal{A}_1$.

For convenience of notation, when we use n as an endomorphisrn it will always refer to the map $\mathcal{C} \to \mathcal{C}$. Now we have $n = \nu_1\nu$ and so, using square brackets for the order of groups,

$$(5.2) \qquad \begin{aligned}[\mathcal{A}/n\mathcal{A}] &= \frac{[\mathcal{A}/\nu\mathcal{A}_1]\,[\mathcal{A}_1/\nu_1\mathcal{A}]}{\text{Number of cosets of } \nu_1\mathcal{A} \text{ in } \mathcal{A}_1 \text{ which meet } \nu^{-1}\mathfrak{o}} \\ &= \frac{[G'_\nu]\,[G'_\nu]}{[\mathcal{A}\cap\nu_1^{-1}\mathfrak{o}_1]\,[\mathcal{A}_1\cap\nu^{-1}\mathfrak{o}]\,/\,[\mathcal{A}\cap n^{-1}\mathfrak{o}]}.\end{aligned}$$

Here the denominator only depends on the rational n-division points on $\mathcal{C}$, and so can easily be calculated.

We can obtain this result, and somewhat more, by a different argument. Suppose that $\mathcal{D}$ represents an n-covering of $\mathcal{C}$; then we can find $\mathcal{D}_1$ so that the extended diagram

(5.3)

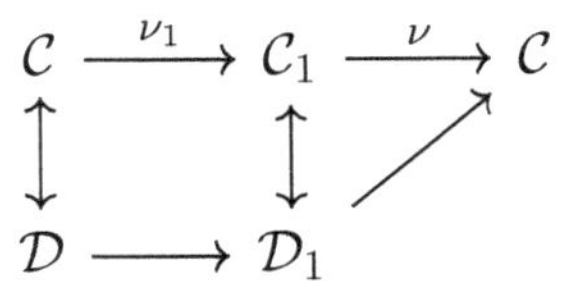

is commutative. Here $\mathcal{D} \to \mathcal{D}_1$ and $\mathcal{D}_1 \to \mathcal{C}$ are defined over the rationals, and $\mathcal{C}_1 \leftrightarrow \mathcal{D}_1$ is over the complex numbers. For let X be a generic point on $\mathcal{D}$ and let $\varphi\colon \mathcal{D} \to \mathcal{C}_1$ be the map (over the complex numbers) defined in the obvious way. The support of $\varphi^{-1}\{\varphi(X)\}$ is a divisor defined over $Q(X)$. Let its exact field of definition be $Q(X_1)$; then we may define $\mathcal{D}_1$ as the curve whose generic point is X_1, and now the assertions above are obvious.

The right-hand part of the last diagram provides a ν-covering of $\mathcal{C}$; and it is easy to show that this depends only on the original n-covering and not on its realisation. Moreover, this process is compatible with the law of composition of coverings. Thus we have homomorphisms $G_n \to G_\nu$ and $G'_n \to G'_\nu$; and the second of these is onto since any $\mathcal{D}_1$ which contains a rational point can be lifted back to a $\mathcal{D}$.

We now find the kernels of these mappings. If $\mathcal{D}$ is in the kernel of $G_n \to G_\nu$, then we may take $\mathcal{D}_1$ in (5.3) to be $\mathcal{C}_1$ and the map $\mathcal{D}_1 \leftrightarrow \mathcal{C}_1$ to be the identity. Thus any representation of an n-covering in Ker $(G_n \to G_\nu)$ is a representation of a ν_1-covering in G_{ν_1}, and vice versa. Since the condition for two representations to give the same covering is stricter for ν_1-coverings than for n-coverings, there is a homomorphism $G_{\nu_1} \to G_n$ such that $G_{\nu_1} \to G_n \to G_\nu$ is exact.

It remains to find the kernel of $G_{\nu_1} \to G_n$. If a representation of the trivial n-covering of $\mathcal{C}$ is given by

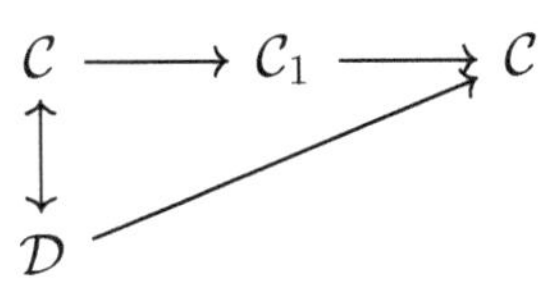

then $\mathcal{D}$ contains a rational point and may therefore be taken to be a copy of $\mathcal{C}$. Thus we get just n^2 effectively distinct representations, the birational maps $\mathcal{D} \leftrightarrow \mathcal{C}$ being $x \leftrightarrow x + \mathfrak{d}$ for the n^2 distinct n-division points on $\mathcal{C}$. A representation here gives rise to a representation of a ν_1-covering if and only if the map $\mathcal{D} \to \mathcal{C}_1$ is defined over the rationals; that is, if $\nu_1\mathfrak{d}$ is rational. Two of them give rise to the same ν_1-covering if and only if the difference of their $\mathfrak{d}$'s is the sum of a rational point and a ν_1-division point. Hence the order of the kernel is

$$\frac{[\mathcal{A}_1 \cap \nu^{-1}\mathfrak{o}]}{\text{number of cosets of } \nu_1^{-1}\mathfrak{o}_1 \text{ in } n^{-1}\mathfrak{o} \text{ which contain rational points}}.$$

This is clearly just the denominator of (5.2), and so we have

$$[G_n] \leqq \frac{[G_\nu]\,[G_{\nu_1}]}{[\mathcal{A} \cap \nu_1^{-1}\mathfrak{o}_1]\,[\mathcal{A}_1 \cap \nu^{-1}\mathfrak{o}]\,/\,[\mathcal{A} \cap n^{-1}\mathfrak{o}]}. \tag{5.4}$$

(The inequality sign here arises because the homomorphism $G_n \to G_\nu$ need not be onto.) Similar arguments work for the groups G'; and since $G'_n \to G'_\nu$ is onto, we have (5.2).

We now apply these arguments to the particular case that interests us, writing

$$\mathcal{C} = \Gamma_D\colon y^2 z = x^3 - Dxz^2,$$
$$\mathcal{C}_1 = \Gamma_{-4D}\colon y_1^2 z_1 = x_1^3 + 4Dx_1 z_1^2.$$

We take $n = 2$, and define ν as the map $\mathcal{C}_1 \to \mathcal{C}$ given by

$$(x_1,\, y_1,\, z_1) \to \left[2x_1\left(x_1^2 + 4Dz_1^2\right),\; y_1\left(x_1^2 - 4Dz_1^2\right),\; 8x_1^2 z_1\right]$$

and ν_1 as that given by $(x,\, y,\, z) \to \left[y^2 z,\; y\left(x^2 + Dz^2\right),\; x^2 z\right]$. Then we have

Lemma 8. *There is a natural isomorphism between the ν-coverings of Γ_D and the elements of Q^*/Q^{*2}. If (a) is any element of Q^*/Q^{*2}, the corresponding curve $\mathcal{D}$ may be taken to be*

$$\mathcal{D}\colon as^2 = a^2 t^4 - D. \tag{5.5}$$

We follow the proof of Lemma 4 of Cassels [5] II. Let $x_1 = (x_1,\, y_1,\, z_1)$ be generic on $\mathcal{C}_1$ and write

$$x = \nu x_1,\; t = y_1/2a^{\frac{1}{2}}x_1.$$

Then we have $y_1^2/x_1^2 = 4x/z = 4at^2$; and if $\mathcal{D}$ is the locus of (x, t) it follows that $\mathcal{D}$ is defined over the rationals and is in birational correspondence with $\mathcal{C}_1$ over the complex numbers, and that the induced map $\mathcal{D} \to \mathcal{C}$ maps (x, t) on x. Hence we have a ν-covering. If we write $y/z = ast$ then $\mathcal{D}$ takes the form (5.5).

Conversely, let $\mathfrak{d}$ be the point (0, 0, 1) on Γ_D, and let $\mathcal{D}$ be the curve of the ν-covering with generic point X corresponding to x_1. The inverse image of $\mathfrak{o}^{-1}\mathfrak{d}$ on $\mathcal{D}$ is rational and linearly equivalent to zero; hence there is a function $f(X)$ defined over the rationals with this divisor; and since x/z has divisor $\left(\mathfrak{o}^{-1}\mathfrak{d}\right)^2$ it follows that $x/z = af^2(X)$ for some rational a. The class of a in Q^*/Q^{*2} does not depend on the choice of f; and it is easy to show that the two correspondences we have set up are inverses. This proves the Lemma.

We note that if $\mathcal{D}$ is everywhere locally soluble then we may take a to be a factor of D; for if p is a simple prime factor of a not dividing D then the three terms in (5.5) are all divisible by different powers of p and so the equation is impossible. If $a|D$ then we need only examine the solubility of (5.5) in the reals and in those p-adic fields with $p|2D$; for the other fields it is trivially soluble. Hence it is easy to find G_ν, and similarly G_{ν_1}.

If, as in our previous paper, we write 2^t for the number of elements of order 2 in the Tate-Šafarevič group of Γ_D, so that 2^t is the order of G_2/G'_2, then we can express (5.2) and (5.4) as follows:

Lemma 9. *Write 2^λ for the order of G_ν, and define λ', λ_1, λ_1', similarly. Then we have*

$$g = \lambda' + \lambda_1' - 2, \; g + t \leqq \lambda + \lambda_1 - 2. \tag{5.6}$$

In proving this we have to split cases. If D is a rational square, the order of G_n is 2^{g+t+2} and the denominator of (5.2) is 1; otherwise the order of G_n is 2^{g+t+1} and the denominator of (5.2) is 2. Similarly for (5.4).

For historical reasons, $\lambda + \lambda_1 - 2$ is called the number of first descents; it is easy to show directly that it is non-negative. In general, there need not be equality in the second equation (5.6). We shall write G_ν^* for the group of those ν-coverings of Γ_D which can be lifted back to everywhere locally soluble 2-coverings, and define λ^* by analogy with λ; similarly for λ_1^*. Then

$$g + t = \lambda^* + \lambda_1 - 2,$$

and it may be shown by the methods of Cassels that $\lambda - \lambda^*$ is an even integer. Since Γ_D and Γ_{-4D} may have different Tate-Šafarevič groups, $\lambda - \lambda^*$ and $\lambda_1 - \lambda_1^*$ need not be equal; we have many examples of this phenomenon.

Most of the cases with $g = 0$ in Table 1 were obtained at once from Lemma 9; but when $t > 0$ for one of the curves Γ_D or Γ_{-4D} it was necessary to do a further descent. The next step is to find λ^*, and for this we use

Lemma 10. *Let (5.5) be a ν-covering of Γ_D which is everywhere locally soluble, and let σ, τ be rational numbers such that*

$$a\sigma^2 = a^2\tau^2 - D. \tag{5.7}$$

Then there is a one-one correspondence between the elements of Q^*/Q^{*2} and the ways of extending (5.5) to a 2-covering of Γ_D; if (b) is any element of Q^*/Q^{*2}, the corresponding curve $\mathcal{D}$ may be taken to be

$$as^2 = a^2t^4 - D, \; as\sigma - a^2t^2\tau + D = bu^2. \tag{5.8}$$

We know that (5.7) is soluble; for it is everywhere locally soluble by comparison with (5.5). Now let x_2 be generic on the left-hand copy of $\mathcal{C}$ in the diagram

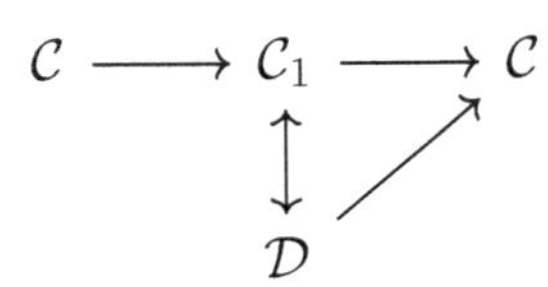

and write $x_1 = \nu_1 x_2$; retain otherwise the notation of Lemma 8. If we write

$$u = \left(\frac{a^{\frac{1}{2}}\sigma - a\tau}{b}\right)^{\frac{1}{2}} \left(\frac{x_2^2 - Dz_2^2 - 2x_2z_2\left(a^{\frac{1}{2}}\sigma + a\tau\right)}{2y_2z_2}\right)$$

we may after some trouble verify that

$$as\sigma - a^2t^2\tau + D = bu^2.$$

The rest of the proof is analogous to that of Lemma 8, and is therefore omitted. We note that the correspondence is not natural, that is, it depends on the choice of σ and τ.

By eliminating s, we may replace the equations (5.8) by the single equation

$$b^2u^4 + 2bu^2\left(a^2t^2\tau - D\right) + Da^2\left(t^2 - \tau\right)^2 = 0. \tag{5.9}$$

From this we may find necessary conditions on b for $\mathcal{D}$ to be everywhere locally soluble, analogous to those for a which follow the proof of Lemma 8. We may assume b to be a square free integer. Then if p is an odd prime which divides the denominators of σ, τ to an odd power and does not divide D, it must divide b; and the only other primes that may divide b are those that divide $2D$.

If we use Lemma 10 to find the 2-coverings of Γ_D explicitly, some care is needed to avoid duplication. In general, there are several essentially different ways of extending a ν_1-covering to the same n-covering; the number is equal to the order of the kernel of the map $G_\nu \to G_n$ described earlier. In our actual case, if D is not a square there are 2 distinct extensions giving rise to each 2-covering and they correspond to the values b_0 and Db_0 for b. If D is a square, there is only one extension for each 2-covering.

If $t > 0$ for both of the curves Γ_D and Γ_{-4D}, then in order to find g it is necessary to go on at least to the third descent. We have in fact worked out the formulae for this. Unfortunately, mainly owing to machine development, we have not been able to use these formulae, as we had intended, to fill in some of the gaps in Table 1. We therefore omit all the details of the third descent; we will be happy to supply the formulae to any intending computer. An alternative method is given in [10].

6. The Tamagawa number

In this section we define the Tamagawa number that we use, and show how to calculate it. Since we are dealing with an unusually simple particular case, we have tried to make our description at least formally self-contained; for a general account see [23].

Let V be a group-variety of dimension n, defined over the rationals. There exists on V an nth order differential ω which is defined over the rationals and invariant under left translation; and ω is unique up to multiplication by a non-zero rational. (This much freedom does not matter, for the rational factor turns out to be self-cancelling.) If x_1, ..., x_n are local coordinates at the origin on V, then we may write

$$\omega = g dx_1 \cdots dx_n.$$

For any prime p, including infinity, let V_p be the set of points of V defined over the p-adic numbers; then V_p is a topological group and ω induces on it a Haar measure

$$\omega_p = |g|_p (dx_1)_p \ldots (dx_n)_p$$

invariant under left translation. Here $|g|_p$ is the normal p-adic valuation, and for finite p we normalise $(dx)_p$ by the condition

$$\int_{\mathfrak{o}_p} (dx)_p = 1,$$

the integral being taken over the ring of p-adic integers; we note that this is equivalent to

$$\int_{\text{units}} (dx)_p = \frac{p-1}{p}.$$

Suppose now for simplicity that V is complete. We may define the 'adèle variety' V_A to be $\prod_p V_p$, the product being taken over all primes including infinity; then $\omega_A = \prod \omega_p$ is a measure on V_A. Since $\prod |a|_p = 1$ for non-zero rational a, ω_A remains invariant when co is multiplied by a non-zero rational. Write V_Q for the subgroup of points of V with rational coordinates; there is a natural imbedding of V_Q into V_A. The Tamagawa number may be defined by

$$t(V) = \int_{V_A/V_Q} \omega_A. \tag{6.1}$$

In the more usual case, V is not complete and one must proceed with more care; see [23]. Tamagawa and others have shown that for the classical groups the Tamagawa number is a small positive integer. For example, the orthogonal group of a quadratic form not representing zero has Tamagawa number 2; and this fact is equivalent to Siegel's theorem, see for instance [24].

In our particular case, V is the elliptic curve Γ_D, with differential

$$\omega = dx/y.$$

In the main case we consider, Γ_D has only finitely many rational points. We have found it more convenient not to factor them out, so we define formally

$$\tau(D) = \prod_p \int_{\Gamma_p} \omega_p, \tag{6.2}$$

where the product is over all primes including infinity and Γ_p is the p-adic completion of the projective curve Γ_D.

Various difficulties may arise in interpreting a product like (6.2). It may be necessary to insert convergence factors, one to each term of the product; we do not do this here, but expect that we shall need to do so when discussing curves Γ_D with $g > 0$ Whether or not this has been done, there are now three useful possibilities for the infinite product. It may converge absolutely, or it may converge conditionally provided the factors are taken in their natural order. Both these cases occur in the work of Tamagawa. Failing these, it is necessary to ascribe a value to the product (6.2) by a suitable summation method; and this we have to do here. Formally, the product (6.2) differs only in finitely many factors from that for $[L_D(1)]^{-1}$, by Lemma 12 below; and we can therefore deduce a value for $\tau(D)$ from that of $L_D(1)$. More rigorously, we regard the infinite product in (6.2) as the formal value of the Euler product form of a Dirichlet series at $s = 1$, and ascribe to it the value at $s = 1$ of the underlying function. This gives rise to a satisfactory summation method, closely allied to that of Abel.

There are no difficulties of principle in calculating the separate integrals in (6.2), though the details are tedious. For convenience we shall as usual assume that D is fourth power free; however we allow D to be divisible by 4 since $\tau(D)$ is not invariant under isogeny.

Lemma 11. *If p is odd and $p|D$ then $\int \omega_p = 2$ unless $p^2||D$ and D is a p-adic square, in which case $\int \omega_p = 4$.*

Suppose first that $p||D$; then we have

$$V_p = \bigcup_{n=0}^{\infty} F_n \cup \bigcup_{n=0}^{\infty} G_n,$$

where F_n is the set of points with $p^{2n+1}||x$, $p^{n+1}||y$ and G_n is the set of points with $p^{-2n}||x$, $p^{-3n}||y$. The point (x, y) can belong to F_n only if $-Dx$ is a p-adic square, and to each such x correspond two values of y; hence the contribution to the integral from F_n is $p^{n+1}(p-1)/p^{2n+2}$. Similarly, (x, y) can belong to G_n only if x is a p-adic square, and to each such x correspond two values of y; hence the contribution to the integral from G_n is $(p-1)/p^{n+1}$. Adding all these together, we find that $\int \omega_p = 2$.

The case $p^3||D$ is similar; we have

$$V_p = \bigcup_{n=1}^{\infty} F_n \cup \bigcup_{n=0}^{\infty} G_n$$

where G_n is as before but F_n is the set of points with $p^{2n+1}||x$, $p^{n+2}||y$. Now the contribution to the integral from F_n is $(p-1)/p^n$; and adding up as before we find that $\int \omega_p = 2$.

Finally, suppose that $p^2||D$. Then we have

$$V_p = \bigcup_{n=1}^{\infty} F_n \cup \bigcup_{n=0}^{\infty} G_n \cup \bigcup_{n=2}^{\infty} H_n,$$

where F_n is the set of points with $p^{2n}||x$, $p^{n+1}||y$, G_n is as before and H_n is the set of points with $p||x$, $p^n||y$. The contribution to the integral from F_n is $(p-1)/p^n$, and that from G_n is as usual $(p-1)/p^{n+1}$. H_n is null unless D is a p-adic square. If $D = c^2$ then the point (x, y) can belong to H_n only if $p^{2n-2}||(x \pm c)$ and $2(x \pm c)$ is a p-adic square; and to each such x correspond two values of y. Hence in this case the contribution to the integral from H_n is $2(p-1)/p^{n-1}$. Adding everything up, we obtain the result stated in the lemma.

Lemma 12. *If p is odd and $p \nmid D$ then $\int \omega_p = N_p/p$.*

This is a special case of Theorem 2.2.5 of [23]; alternatively it may be proved in the same way as Lemma 11, remembering that in finding N_p we include the point at infinity.

Lemma 13. *If D $\equiv$ 3 mod 4 then $\int \omega_2 = \frac{1}{2}$; if D $\equiv$ 4, 12, 36 or 60 mod 64 then $\int \omega_2 = 2$; in every other case $\int \omega_2 = 1$.*

We omit the proof of the lemma, which is similar to that of Lemma 11. It is necessary to consider a very large number of distinct cases, and though none of them individually presents much difficulty, the succession becomes wearisome.

Lemma 14.

$$\int \omega_\infty = \begin{cases} 4\omega D^{-\frac{1}{4}} & \text{for } D > 0, \\ 4\omega(-4D)^{-\frac{1}{4}} & \text{for } D < 0. \end{cases}$$

In the statement and proof of this lemma, ω denotes the real period of $\wp(u)$ as defined in §3. If $D > 0$ we have

$$\int \omega_\infty = 2\left\{\int_{-\sqrt{D}}^{0} + \int_{\sqrt{D}}^{\infty}\right\} \frac{dx}{\sqrt{(x^3 - Dx)}} = 4\omega D^{-\frac{1}{4}};$$

if $D < 0$ we have

$$\int \omega_\infty = 2\int_0^\infty \frac{dx}{\sqrt{(x^3 - Dx)}} = 4\omega(-4D)^{-\frac{1}{4}}.$$

Here the factor 2 arises because there are two values of y for each acceptable x; for the evaluation of the elliptic integrals see [4], formulae 233.00, 237.00 and 239.00.

We can now express the Tamagawa number $\tau(D)$ in terms of $\sigma(D)$ which we have already shown how to compute. By Lemma 12 and the results of §2, we have formally

$$\prod \int \omega_p = \prod (N_p/p) = [L_D(1)]^{-1}, \tag{6.3}$$

where the products are taken over all finite primes not dividing $2D$. Define $\sigma(D)$ by (1.6) even when $4|D$, so that we have

$$\begin{aligned} \sigma(-4D) &= 2\sigma(D) \quad \text{for } D > 0, \\ \sigma(-4D) &= \sigma(D) \quad \text{for } D < 0. \end{aligned}$$

Then by (1.6), (6.3) and Lemma 14

$$\tau(D)\,\sigma(D) = 4\prod \int \omega_p, \tag{6.4}$$

where the product is taken over the finitely many primes which divide $2D$. The right hand side of (6.4) can be evaluated in any particular case by Lemmas 11 and 13.

We can similarly ascribe a Tamagawa number to any elliptic curve with finitely many rational points; and it is in each case a computable multiple of the formal product (1.2). The analogues of Lemmas 11 and 13 are very messy, and we have in general no neat way of evaluating the formal product.

7. The numerical evidence

In this section we describe the calculations that we have actually carried out on EDSAC 2, and the extent to which they confirm our conjectures. Let D be a positive integer not divisible by a fourth power or by 4; define $\sigma(D)$ by (1.6) and Δ as in §3. After Theorem 1, we know that $\sigma(D)$ is a rational integer for $|D| > 1$; by means of the formulae (3.13) to (3.17) we have computed its value whenever $|\Delta| < 108$ or $|\Delta| = 113, 165, 195, 231$.

The details of the calculation are of little interest, except for one point of organisation. Each of the Δ^2 terms in the formula used involves a quartic residue and a value of $\wp(u)$; and nearly all the time taken by the computation is spent in evaluating these. We therefore gain efficiency by calculating $\sigma(D)$ simultaneously for all D corresponding to a given Δ. If $|\Delta|$ is a prime p then we obtain the 12 cases

$$D = \pm 2^a p^b \,(a = 0,\ 1;\ b = 1,\ 2,\ 3)\,;$$

if $|\Delta|$ is a product of two or three primes, we obtain respectively 36 or 108 cases. For a given value of Δ, the complete run took about $\Delta^2/20$ seconds. There is an automatic check against machine (or program) error, in that the final value of $\sigma(D)$ has to be an integer to within the accuracy of the calculation.

For comparison with our main conjecture (1.5), we have also found the value of g for as many of the curves as we conveniently could. For each curve we have carried out the first descent by the methods of §5, and where necessary the second descent also; we had hoped to carry out the third descent, but as explained in §5 we were unable to do so. The calculations are straightforward except for the recurrent need to solve equations of the form

$$ax^2 + by^2 + cz^2 = 0 \tag{7.1}$$

over the rationals. One of Lagrange's proofs of his theorem [16] gives a constructive method of solving (7.1) when it is possible. Alternatively we may search for a solution, since Holzer [14] has shown that we need only examine the region

$$x^2 \leqq |bc|\,,\ y^2 \leqq |ca|\,,\ z^2 \leqq |ab|\,.$$

The local solubility of the coverings was checked by methods similar to those described in our previous paper. In searching for rational points on the coverings we considered only relatively small values of the variables; for we have found that time spent on extending such a search is only meagrely rewarded.

For each of the values of D considered, we give in Table 1 the corresponding values of σ, g, λ and λ_1, the last two being the numbers of first descents defined in §5. These are also the values of σ, g, λ_1 and λ respectively corresponding to $-4D$. Let τ be the Tamagawa number of Γ_D as defined in §7; and let $\eta(D)$ be the number of rational points on Γ_D of finite order. (Thus within the limits of our table $\eta = 4$ when D is -4 or a square, and $\eta = 2$ otherwise.) Rather than take up space printing the values of $\tau(D)$ and $\tau\,(-4D)$, we have marked the value of σ with an asterisk whenever $\sigma \neq 0$ and $\tau\,(D) \neq \eta^2\,(D)$ or $\tau\,(-4D) \neq \eta^2\,(-4D)$. These cases are further described in Table 2; according to our conjectures they are just those in which $g = 0$ and the Tate-Šafarevič group is non-trivial; Table 2 is possibly less reliable than Table 1.

We have given the value of g whenever we know it. A gap instead of the value of g indicates that we do not know it, and indeed have no information beyond that given by the general inequality

$$0 \leqq g \leqq \lambda + \lambda_1 - 2.$$

An entry such as '1−' indicates that (because of the second descent) we know that $g \leqq 1$; in these cases there is a non-trivial Tate-Šafarevič group. An entry such as '1+' indicates that we know that $g \geqq 1$ since we have actually found a generator.

In Table 1 we have packed onto each line the entries corresponding to four linked values of D. In each line the first column gives an odd positive integer D_0. The remaining sixteen columns form four groups of four. The first group gives the values of σ, g, λ and λ_1 for $D = D_0$; and the other three groups correspond to the cases $D = -D_0$, $D = 2D_0$ and $D = -2D_0$ in that order.

Let us write $\gamma = \eta^2/\tau$; then the main conjectures of this paper are that $\gamma \neq 0$ if and only if $g = 0$, and that in this case γ is the order of the Tate-Šafarevič group. The first part of this agrees with our calculations whenever we know g. Moreover, the work of Cassels [5] strongly suggests that g should have the parity of $\lambda + \lambda_1$; and in our tables $\gamma = 0$ whenever $\lambda + \lambda_1$ is odd. For the second part of the conjecture, we note that in our tables γ is always a square; and Cassels has shown that when the order of the Tate-Šafarevič group is finite it is a square. We have made no attempt to calculate p-coverings for any prime $p > 2$; and the practical difficulties appear to us formidable. But we do known the 2-component of the Tate-Šafarevič group in every case for which we know that $g = 0$; and this yields always the correct power of 2 in γ. (The few cases where $\gamma \neq 0$ but we do not know that $g = 0$ are ones in which the second descent is indecisive; it is further corroboration of the conjecture that in these cases γ is divisible by a high power of 2.)

In the course of these calculations, we observed another identity which we conjectured to hold always; this is

$$\tau(-4D)/\tau(D) = 2^{\lambda_1 - \lambda}; \tag{7.2}$$

in fact, without some such identity our conjectures for $\tau(-4D)$ and $\tau(D)$ could hardly be consistent. This relation need not be confined to the case $\gamma \neq 0$, for the left hand side may be written as a finite product. In fact we deduce from the results of §6 that

$$\tau(-4D)/\tau(D) = \prod_{p|2D\infty} \varrho_p,$$

where, for each p, ϱ_p is short for $\int_{\Gamma_D} \omega_p / \int_{\Gamma_{-4D}} \omega_p$, and may be read off explicitly from Lemmas 11, 13 and 14.

We are glad to say that Cassels [7] has been able to prove a more general theorem including (7.2), which gives our conjectures considerable support. The identity has also been verified more directly by E. Forrest [10].

Table 1

D	D			$-D$			$2D$			$-2D$		
	σ	g	$\lambda.\lambda_1$	σ	g	$\lambda.\lambda_1$	σ	g	$\lambda.\lambda_1$	σ	g	$\lambda.\lambda_1$
1	$\frac{1}{4}$	0	1.1	$\frac{1}{2}$	0	0.2	0	1	2.1	1	0	1.1
3	1	0	1.1	0	1	1.2	0	1	2.1	2	0	1.1
5	0	1	2.1	0	1	1.2	0	1	2.1	2	0	1.1
7	0	1	1.2	2	0	1.1	0	1	2.1	0	2	2.2
9	1	0	2.0	0	1	0.3	2	0	1.1	0	1	2.1
11	1	0	1.1	2	0	1.1	0	1	2.1	2	0	1.1
13	2	0	1.1	0	1	1.2	0	1	2.1	2	0	1.1
15	0	1	1.2	0	1	1.2	0	1	2.1	4	0	1.1
17	0	2	2.2	4*	0	1.3	0	1	3.2	0	2	2.2
19	1	0	1.1	0	1	1.2	0	1	2.1	2	0	1.1

D	D			$-D$			$2D$			$-2D$		
	σ	g	$\lambda.\lambda_1$	σ	g	$\lambda.\lambda_1$	σ	g	$\lambda.\lambda_1$	σ	g	$\lambda.\lambda_1$
21	0	1	2.1	0	1	1.2	0	1	2.1	4	0	1.1
23	0	1	1.2	2	0	1.1	0	1	2.1	0	2	2.2
25	0	1	2.1	2	0	0.2	0	1	2.1	2	0	1.1
27	1	0	1.1	2	0	1.1	0	1	2.1	2	0	1.1
29	2	0	1.1	0	1	1.2	0	1	2.1	2	0	1.1
31	0	1	1.2	0	1	1.2	0	1	2.1	8*	0	2.2
33	4	0	1.1	0	2	1.3	0	1	2.1	0	2	2.2
35	2	0	1.1	0	1	1.2	0	1	2.1	4	0	1.1
37	0	1	2.1	0	1	1.2	0	1	2.1	2	0	1.1
39	0	1	1.2	0	2	2.2	0	1	2.1	4	0	1.1
41	0	1	2.1	4*	0	1.3	0	3	3.2	8*	0	2.2
43	1	0	1.1	2	0	1.1	0	1	2.1	2	0	1.1
45	0	1	1.2	4	0	1.1	0	2	3.1	0	1	1.2
47	0	1	1.2	0	1	1.2	0	1	2.1	0	2	2.2
49	0	1	2.1	0	1	0.3	4*	0	3.1	0	1	1.2
51	2	0	1.1	0	1	1.2	0	1	2.1	4	0	1.1
53	0	1	2.1	0	1	1.2	0	1	2.1	2	0	1.1
55	0	1	1.2	0	2	2.2	0	1	2.1	4	0	1.1
57	0	1	2.1	8*	0	1.3	0	1	2.1	0	2	2.2
59	1	0	1.1	2	0	1.1	0	1	2.1	2	0	1.1
61	2	0	1.1	0	1	1.2	0	1	2.1	2	0	1.1
63	4	0	1.1	0	2	1.3	4	0	1.1	0	1	2.1
65	0	2	2.2	0	2	1.3	0	1	2.1	4	0	1.1
67	1	0	1.1	0	1	1.2	0	1	2.1	2	0	1.1
69	0	1	2.1	0	1	1.2	0	1	2.1	0	2	2.2
71	0	1	1.2	2	0	1.1	0	1	2.1	8*		2.2
73	0	1	2.1	0	2	1.3	0	1	3.2	8*		2.2
75	2	0	1.1	4	0	1.1	0	1	2.1	0	2	2.2
77	0	2	2.2	0	1	1.2	0	1	2.1	0	2	2.2
79	0	1	1.2	0	1	1.2	0	1	2.1	0	2	2.2
83	1	0	1.1	0	1	1.2	0	1	2.1	2	0	1.1
85	0	1	2.1	0	1	1.2	0	1	2.1	4	0	1.1
87	0	1	1.2	4	0	1.1	0	1	2.1	4	0	1.1
89	0	1	2.1	0	2	1.3	0	1	3.2	0	2	2.2

D	D			$-D$			$2D$			$-2D$		
	σ	g	$\lambda.\lambda_1$	σ	g	$\lambda.\lambda_1$	σ	g	$\lambda.\lambda_1$	σ	g	$\lambda.\lambda_1$
91	2	0	1.1	4	0	1.1	0	1	2.1	4	0	1.1
93	4	0	1.1	0	1	1.2	0	1	2.1	4	0	1.1
95	0	1	1.2	0	1	1.2	0	1	2.1	4	0	1.1
97	0	2	2.2	4*	0	1.3	0	1	3.2	8*		2.2
99	0	1	1.2	0	2	2.2	8*	0	3.1	0	1	1.2
101	0	1	2.1	0	1	1.2	0	1	2.1	2	0	1.1
103	0	1	1.2	2	0	1.1	0	1	2.1	0	2	2.2
105	0	1	2.1	0	2	1.3	0	1	2.1	8	0	1.1
107	1	0	1.1	2	0	1.1	0	1	2.1	2	0	1.1
113	8*		2.2	0	2	1.3	0	3	3.2	0	2	2.2
117	0	2	3.1	8*	0	1.3	4	0	1.1	0	1	2.1
121	1	0	2.0	0	1	0.3	2	0	1.1	0	1	2.1
125	2	0	1.1	0	1	1.2	0	1	2.1	2	0	1.1
135	0	1	1.2	4	0	1.1	0	1	2.1	4	0	1.1
147	0	1	1.2	8	0	1.1	4	0	1.1	0	1	2.1
153	4	0	1.1	0	1	1.2	0	2	3.1	0	1	1.2
165	0	1	2.1	0	1	1.2	0	1	2.1	8	0	1.1
169	0	1	2.1	2	0	0.2	0	1	2.1	2	0	1.1
171	0	1	2.1	0	1	1.2	0	2	2.2	0	1	2.1
175	0	1	1.2	0	1	1.2	0	1	2.1	8	0	1.1
189	4	0	1.1	0	1	1.2	0	1	2.1	4	0	1.1
195	4	0	1.1	0	1	1.2	0	1	2.1	0	2	2.2
207	0	2	1.3	8	0	1.1	8*	0	3.1	0	1	1.2
225	0	1	2.1	0	1	0.3	4	0	1.1	0	1	2.1
231	0	1	1.2	8	0	1.1	0	1	2.1	8	0	1.1
245	4	0	1.1	4	0	1.1	4	0	1.1	0	1	2.1
261	4	0	1.1	4	0	1.1	0	2	3.1	0	1	1.2
275	0	2	2.2	0	1	1.2	0	1	2.1	4	0	1.1
279	4	0	1.1	0	1	1.2	4	0	1.1	0	1	2.1
289	4*	0	2.2	0	2	1.3	0	1	2.1	4	0	1.1
297	0	1	2.1	8*	0	1.3	0	1	2.1	0	2	2.2
315	0	1	1.2	0	1	2.1	0	2	3.1	0	1	1.2
325	0	1	2.1	0	1	1.2	0	1	2.1	0	2	2.2
343	0	1	1.2	2	0	1.1	0	1	2.1	8*		2.2

D	D			$-D$			$2D$			$-2D$		
	σ	g	$\lambda.\lambda_1$	σ	g	$\lambda.\lambda_1$	σ	g	$\lambda.\lambda_1$	σ	g	$\lambda.\lambda_1$
351	0	1	1.2	0	1	1.2	0	1	2.1	4	0	1.1
361	1	0	2.0	0	1	0.3	2	0	1.1	0	1	2.1
363	0	1	2.1	0	1	1.2	0	2	2.2	0	1	2.1
375	0	1	1.2	4	0	1.1	0	1	2.1	4	0	1.1
425	0	1	2.1	0	2	1.3	0	1	2.1	8	0	1.1
441	0	1	3.0	8*	0	0.4	0	1	2.1	8	0	1.1
459	2	0	1.1	4	0	1.1	0	1	2.1	4	0	1.1
475	4	0	1.1	0	2	2.2	0	1	2.1	4	0	1.1
495	8	0	1.1	0	2	1.3	8	0	1.1	0	1	2.1
507	0	2	2.2	8	0	1.1	0	1	2.1	4	0	1.1
513	4	0	1.1	0	2	1.3	0	1	2.1	16*		2.2
525	0	2	2.2	0	1	1.2	0	1	2.1	8	0	1.1
529	0	1	2.1	0	1	0.3	4*	0	3.1	0	1	1.2
539	0	1	2.1	0	1	1.2	8*	0	3.1	0	1	1.2
585	8	0	1.1	0	1	1.2	0	2	3.1	0	1	1.2
605	0	1	2.1	8*	0	1.3	4	0	1.1	0	1	2.1
621	0	2	2.2	0	1	1.2	0	1	2.1	16*		2.2
637	0	1	1.2	4	0	1.1	4	0	1.1	0	1	2.1
675	2	0	1.1	0	1	1.2	0	1	2.1	8	0	1.1
693	8	0	1.1	8	0	1.1	0	2	3.1	0	1	1.2
735	8	0	1.1	0	2	1.3	0	2	3.1	0	1	1.2
783	0	1	1.2	0	1	1.2	0	1	2.1	4	0	1.1
825	0	1	2.1	0	2	1.3	0	1	2.1	16	0	1.1
837	0	1	2.1	0	1	1.2	0	1	2.1	4	0	1.1
841	0	1	2.1	2	0	0.2	0	1	2.1	18*	0	1.1
845	4	0	1.1	0	1	1.2	0	1	3.2	8	0	1.1
847	0	2	1.3	8	0	1.1	8*	0	3.1	0	1	1.2
867	2	0	1.1	0	1	1.2	0	1	2.1	4	0	1.1
875	2	0	1.1	4	0	1.1	0	1	2.1	4	0	1.1
945	8	0	1.1	0	2	1.3	0	1	2.1	8	0	1.1
961	0	1	2.1	0		0.3	0	2	3.1	0		1.2
975	0	1	1.2	0	1	1.2	0	1	2.1	8	0	1.1
1029	0	1	2.1	0	1	1.2	0	1	2.1	4	0	1.1
1083	0	1	1.2	0	1	2.1	8*	0	3.1	0	1	1.2

D	D			$-D$			$2D$			$-2D$		
	σ	g	$\lambda.\lambda_1$	σ	g	$\lambda.\lambda_1$	σ	g	$\lambda.\lambda_1$	σ	g	$\lambda.\lambda_1$
1089	4	0	2.0	8*	0	0.4	0	1	1.2	0	2	3.1
1125	4	0	1.1	4	0	1.1	8*	0	3.1	0	1	1.2
1183	0	1	1.2	0	1	1.2	0	1	2.1	8	0	1.1
1225	4	0	2.0	0	1	0.3	0	2	3.1	0	1	1.2
1323	0	1	1.2	0	1	2.1	4	0	1.1	0	1	2.1
1331	1	0	1.1	0	1	1.2	0	1	2.1	18*	0	1.1
1369	0	1	2.1	2	0	0.2	0		2.1	2	0	1.1
1375	0	1	1.2	0	1	1.2	0	1	2.1	4	0	1.1
1445	0	1	2.1	0	1	1.2	0	1	2.1	4	0	1.1
1485	8	0	1.1	0	1	1.2	0	1	2.1	8	0	1.1
1521	0	1	2.1	0	1	0.3	4	0	1.1	0	1	2.1
1575	8	0	1.1	0	1	1.2	0	2	2.2	0	1	2.1
1587	0	1	2.1	16*	0	1.3	8*	0	3.1	0	1	1.2
1617	0	1	1.2	0	1	1.2	8	0	1.1	0	1	2.1
1625	0	1	2.1	8*	0	1.3	0	1	2.1	4	0	1.1
1681	0	2	2.2	8*		1.3	0	1	2.1	4	0	1.1
1715	2	0	1.1	0	1	1.2	0	1	2.1	4	0	1.1
1755	4	0	1.1	8	0	1.1	0	1	2.1	0	2	2.2
1805	0	1	2.1	0	2	1.3	4	0	1.1	0	1	2.1
1815	8	0	1.1	0	1	1.2	8	0	1.1	0	1	2.1
1849	9*	0	2.0	0	1	0.3	2	0	1.1	0	1	2.1
2079	0	1	1.2	0	1	1.2	0	1	2.1	8	0	1.1
2125	4	0	1.1	0	1	1.2	0	1	2.1	4	0	1.1
2197	0	1	2.1	0		1.2	0	1	2.1	2	0	1.1
2205	0	2	1.3	0	1	2.1	0	1	2.1	16	0	1.1
2209	0	1	2.1	0		0.3	16*		3.1	0	1	1.2
2375	0	1	1.2	0	2	2.2	0	1	2.1	4	0	1.1
2475	0	1	1.2	0	1	2.1	0	2	3.1	0	1	1.2
2523	18*	0	1.1	4	0	1.1	0	1	2.1	8	0	1.1
2535	0	1	1.2	8	0	1.1	0	1	2.1	0	2	2.2
2541	0	1	1.2	8	0	1.1	0	2	3.1	0	1	1.2
2601	4	0	2.0	0	1	0.3	0	2	2.2	0	1	2.1
2625	8	0	1.1	16*	0	1.3	0	1	2.1	8	0	1.1
2809	0		2.1	2	0	0.2	0		2.1	2	0	1.1

D	D			$-D$			$2D$			$-2D$		
	σ	g	$\lambda.\lambda_1$	σ	g	$\lambda.\lambda_1$	σ	g	$\lambda.\lambda_1$	σ	g	$\lambda.\lambda_1$
2883	0		1.2	8	0	1.1	4	0	1.1	0	1	2.1
2925	0	1	2.1	0	2	1.3	0	2	2.2	0	1	2.1
3025	0	1	2.1	0	1	0.3	4	0	1.1	0	1	2.1
3087	4	0	1.1	16*	0	1.3	4	0	1.1	0	1	2.1
3249	4	0	2.0	8*	0	0.4	0	1	1.2	0	2	3.1
3267	0	1	2.1	16*	0	1.3	16*		2.2	0	1	2.1
3375	0	1	1.2	0	1	1.2	0	1	2.1	4	0	1.1
3481	1	0	2.0	0		0.3	2	0	1.1	0		2.1
3675	0	1	1.2	0	1	2.1	16	0	1.1	0	1	2.1
3721	0	1	2.1	2	0	0.2	0	1	2.1	18*	0	1.1
3773	16*		2.2	0	1	1.2	0	1	2.1	16*		2.2
3993	0	1	2.1	8*	0	1.3	0	1	2.1	16*		2.2
4125	8	0	1.1	0	1	1.2	0	1	2.1	8	0	1.1
4225	0	2	2.2	8	0	0,2	0	1	2.1	4	0	1.1
4459	2	0	1.1	4	0	1.1	0		2.1	4	0	1.1
4489	1	0	2.0	0		0.3	18*	0	1.1	0	1	2.1
4563	16*		2.2	0	1	1.2	0	1	2.1	4	0	1.1
4725	0	1	2.1	0	1	1.2	0	1	2.1	8	0	1.1
4761	0	1	3.0	0	2	0.4	0	1	2.1	32*		2.2
4851	8	0	1.1	0	1	1.2	0	3	4.1	32*	0	1.3
4875	4	0	1.1	8	0	1.1	0	1	2.1	0	2	2.2
4913	8*		2.2	4*	0	1.3	0	1	3.2	8*		2.2
5041	0	1	2.1	0		0.3	4*	0	3.1	0		1.2
5145	0	1	2.1	16*	0	1.3	0	1	2.1	8	0	1.1
5329	4*	0	2.2	8*		1.3	0	1	3.2	0	2	2.2
5445	0	1	2.1	0	1	1.2	0	1	2.1	16	0	1.1
5929	0	1	3.0	0	2	0.4	0	1	2.1	0	2	2.2
6125	0	1	1.2	4	0	1.1	4	0	1.1	0	1	2.1
6241	0		2.1	0		0.3	0	2	3.1	0		1.2
6591	0	1	1.2	0	1	1.2	0	1	2.1	36*	0	1.1
6615	8	0	1.1	0	1	1.2	16*	0	3.1	0	1	1.2
6655	0	1	1.2	0	1	1.2	0	1	2.1	4	0	1.1
6859	9*	0	1.1	2	0	1.1	0		2.1	2	0	1.1
6889	1	0	2.0	0	1	0.3	18*	0	1.1	0	1	2.1

D	D			$-D$			$2D$			$-2D$		
	σ	g	$\lambda.\lambda_1$	σ	g	$\lambda.\lambda_1$	σ	g	$\lambda.\lambda_1$	σ	g	$\lambda.\lambda_1$
7225	0	1	2.1	8	0	0.2	0	1	3.2	0	2	2.2
7425	8	0	1.1	16*	0	1.3	0	1	2.1	16	0	1.1
7569	0	1	2.1	0	1	0.3	4	0	1.1	0	1	2.1
7605	8	0	1.1	8	0	1.1	0	2	3.1	0	1	1.2
7623	0	1	1.2	16	0	1.1	0	1	2.1	16	0	1.1
7803	18*	0	1.1	4	0	1.1	0	1	2.1	4	0	1.1
7875	0	1	1.2	16	0	1.1	0	2	3.1	0	1	1.2
7921	4*	0	2.2	8*		1.3	0	1	3.2	16*		2.2
8281	4	0	2.0	0	1	0.3	0	2	3.1	0		1.2
8575	0	1	1.2	0	1	1.2	0	1	2.1	8	0	1.1
8649	0	1	3.0	8*	0	0.4	0		2.1	8	0	1.1
8775	0	1	1.2	16	0	1.1	0	1	2.1	8	0	1.1
9025	0	1	2.1	0	1	0.3	4	0	1.1	0	1	2.1
9075	0	1	2.1	0	2	1.3	16	0	1.1	0	1	2.1
9261	4	0	1.1	0	1	1.2	0	1	2.1	36*	0	1.1
9317	0	1	2.1	0	1	1.2	0	1	2.1	16*		2.2
9409	4*	0	2.2	0	2	1.3	0		2.1	4	0	1.1
9747	0	1	1.2	32*		2.2	8*	0	3.1	0	1	1.2
10201	0		2.1	2	0	0.2	0		2.1	18*	0	1.1
10609	0		2.1	0		0.3	4*	0	3.1	0	1	1.2
10985	0	1	2.1	8*	0	1.3	0	1	2.1	4	0	1.1
11025	16	0	2.0	0	2	0.4	0	1	2.1	16	0	1.1
11319	0	1	1.2	8	0	1.1	0		2.1	8	0	1.1
11449	9*	0	2.0	0		0.3	18*	0	1.1	0	1	2.1
11979	0	1	1.2	0	1	2.1	8*	0	3.1	0	1	1.2
12167	0	1	1.2	18*	0	1.1	0		2.1	8*		2.2
12375	8	0	1.1	0	1	1.2	8	0	1.1	0	1	2.1
12675	8	0	1.1	0	1	1.2	0	1	2.1	16	0	1.1
12769	16*		2.2	0	2	1.3	0	1+	3.2	0	2	2.1
14283	0		2.1	0	1	1.2	8*	0	3.1	0		1.2
14553	8	0	1.1	0	1	1.2	8	0	1.1	0	1	2.1
14625	0	1	1.2	0	1	1.2	16*	0	3.1	0	1	1.2
14739	2	0	1.1	0	1	1.2	0		2.1	4	0	1.1
15125	0	2	3.1	8*	0	1.3	4	0	1.1	0	1	2.1

D	D			$-D$			$2D$			$-2D$		
	σ	g	$\lambda.\lambda_1$	σ	g	$\lambda.\lambda_1$	σ	g	$\lambda.\lambda_1$	σ	g	$\lambda.\lambda_1$
15379	18*	0	1.1	0	1	1.2	0		2.1	4	0	1.1
15435	0	1	1.2	0	1	2.1	16*	0	3.1	0	1	1.2
16335	8	0	1.1	0	2	1.3	8	0	1.1	0	1	2.1
17787	8	0	1.1	16	0	1.1	0	1	1.2	32*	0	3.1
18375	8	0	1.1	0	1	1.2	0	2	3.1	0	1	1.2
19773	0	1	2.1	8*	0	1.3	4	0	1.1	0	1	2.1
19965	8	0	1.1	0	1	1.2	0	1	2.1	8	0	1.1
20577	4	0	1.1	32*		1.3	0		2.1	0	2	2.2
21125	0	1	2.1	0		1.2	0	1	2.1	0	2	2.2
22707	2	0	1.1	0	1	1.2	0		2.1	0	2	2.2
22815	0	1	1.2	0	1	1.2	0	1	2.1	0	2	2.2
22869	8	0	1.1	8	0	1.1	16*	0	3.1	0	1	1.2
23625	0	1	2.1	16*	0	1.3	0	1	2.1	8	0	1.1
24389	0		2.1	0		1.2	0		2.1	2	0	1.1
24565	0	1	2.1	0	1	1.2	0		2.1	4	0	1.1
25725	16	0	1.1	0	1	1.2	0	1	2.1	8	0	1.1
25947	0	1	1.2	0		2.1	36*	0	1.1	0	1	2.1
27225	0	1	3.0	0	2	0.4	0	1	1.2	32*	0	3.1
27951	0	1	1.2	0	1	1.2	0	1	2.1	72*	0	1.1
29791	0		1.2	0		1.2	0		2.1	8*		2.2
32955	4	0	1.1	8	0	1.1	0	1	2.1	32*		2.2
33075	0	1	1.2	16	0	1.1	0	2	2.2	0	1	2.1
33275	16*		2.2	8	0	1.1	0		2.1	4	0	1.1
33957	8	0	1.1	8	0	1.1	0	2	3.1	0	1	1.2
34295	0		1.2	16*		2.2	0	1	2.1	4	0	1.1
35937	4	0	1.1	32*		1.3	0		2.1	16*		2.2
36125	4	0	1.1	0		1.2	0		2.1	4	0	1.1
36501	0	1	2.1	0		1.2	0		2.1	16*		2.2
37125	0	1	2.1	0	1	1.2	0		2.1	8	0	1.1
38025	16	0	2.0	0	1	0.3	8	0	1.1	0	1	2.1
41503	32*		1.3	8	0	1.1	8*	0	3.1	0	1	1.2
42875	18*	0	1.1	4	0	1.1	0	1	2.1	4	0	1.1
43875	4	0	1.1	0	1	1.2	0	1	2.1	32*		2.2
44217	4	0	1.1	0		1.2	32*		3.1	0	1	1.2

D	D			$-D$			$2D$			$-2D$		
	σ	g	$\lambda.\lambda_1$	σ	g	$\lambda.\lambda_1$	σ	g	$\lambda.\lambda_1$	σ	g	$\lambda.\lambda_1$
45125	8*	0	3.1	32*		1.3	4	0	1.1	0		2.1
45375	8	0	1.1	32*	0	1.3	8	0	1.1	0	1	2.1
46305	8	0	1.1	16*	0	1.3	0		2.1	8	0	1.1
50653	18*	0	1.1	0		1.2	0	1	2.1	2	0	1.1
53361	0	1	3.0	0	1	0.5	16	0	1.1	0	1	2.1
54925	4	0	1.1	0		1.2	0	1	3.2	8	0	1.1
55125	0	1	1.2	0	1	2.1	0	1	2.1	16	0	1.1
57967	0	1	1.2	0	1	1.2	0	1	2.1	8	0	1.1
59319	0		1.2	16*		2.2	0		2.1	4	0	1.1
59895	8	0	1.1	0	1	1.2	8	0	1.1	0	1	2.1
61731	0		2.1	16*	0	1.3	16*		2.2	0		2.1
63375	0	1	1.2	0	1	1.2	0	1	2.1	16	0	1.1
65219	0		2.1	16*	0	1.3	8*	0	3.1	0	1	1.2
68921	0		2.1	4*	0	1.3	0	1+	3.2	32*		2.2
73167	0		1.2	0		1.2	0		2.1	4	0	1.1
77175	8	0	1.1	0	1	1.2	16	0	1.1	0	1	2.1
79507	1	0	1.1	0		1.2	0		2.1	2	0	1.1
81675	0		2.1	0	1	1.2	16	0	1.1	0	1	2.1
83853	0	1	1.2	8	0	1.1	16*	0	3.1	0		1.2
89373	36*	0	1.1	0	1	1.2	0		2.1	4	0	1.1
98865	0	1	1.2	0	1	1.2	0	2	3.1	0	1	1.2
99825	8	0	1.1	16*	0	1.3	0	1	2.1	16	0	1.1
101871	0	1	1.2	0	1	1.2	0	1	2.1	8	0	1.1
103823	0		1.2	0		1.2	0		2.1	0	1+	- 2.2
107653	36*	0	1.1	4	0	1.1	4	0	1.1	0		2.1
109503	32*		1.3	8	0	1.1	8*	0	3.1	0	1	1.2
114075	8	0	1.1	0	2	2.2	0	1	2.1	64*		2.2
122825	0		2.1	0	2	1.3	0	1	2.1	8	0	1.1
124509	0	1	1.2	8	0	1.1	64*		3.1	0	1	1.2
128625	8	0	1.1	0	2	1.3	0	1	2.1	72*	0	1.1
132651	2	0	1.1	36*	0	1.1	0		2.1	36*	0	1.1
136125	16	0	1.1	0	1	1.2	0	1	2.1	16	0	1.1
148877	2	0	1.1	0		1.2	0		2.1	18*	0	1.1
160083	72*	0	1.1	0	1	1.2	0	1	1.2	0	2	3.1

D	D			$-D$			$2D$			$-2D$		
	σ	g	$\lambda.\lambda_1$	σ	g	$\lambda.\lambda_1$	σ	g	$\lambda.\lambda_1$	σ	g	$\lambda.\lambda_1$
164775	0	1	1.2	16	0	1.1	0	1	2.1	8	0	1.1
165375	8	0	1.1	32*	0	1.3	16*	0	3.1	0		1.2
166375	0		1.2	16*		2.2	0		2.1	4	0	1.1
171475	4	0	1.1	0	1	1.2	0	1	2.1	36*	0	1.1
179685	0	1	2.1	0		1.2	0		2.1	8	0	1.1
185193	0		2.1	8*	0	1.3	0		2.1	16*		2.2
190125	0	1	1.2	8	0	1.1	32*	0	3.1	0	1	1.2
195657	8	0	1.1	0		1.2	72*	0	1.1	0	1	2.1
205379	9*	0	1.1	0		1.2	0		2.1	2	0	1.1
219501	0	1	1.2	4	0	1.1	8*	0	3.1	0		1.2
226981	0		2.1	0		1.2	0		2.1	98*	0	1.1
231525	0	1	2.1	0	1	1.2	0		2.1	8	0	1.1
251559	0		1.2	8	0	1.1	0		2.1	8	0	1.1
268119	4	0	1.1	0	1	1.2	100*	0	1.1	0	1	2.1
274625	16*		2.2	32*		1.3	0		2.1	36*	0	1.1
296595	4	0	1.1	0	1	1.2	0	1	2.1	32*		2.2
299475	0	1	1.2	32	0	1.1	64*		3.1	0	1	1.2
300763	49*	0	1.1	2	0	1.1	0		2.1	18*	0	1.1
328509	16*		2.2	0		1.2	0		2.1	0	2	2.2
357911	0	1	1.2	18*	0	1.1	0		2.1	0	1+	2.2
373527	0	1	1.2	16	0	1.1	0		2.1	16	0	1.1
385875	0		1.2	16	0	1.1	16*	0	3.1	0		1.2
389017	0		2.1	64*		1.3	0	1–	3.2	8*		2.2
408375	8	0	1.1	0	1	1.2	8	0	1.1	0		2.1
456533	0		2.1	0		1.2	0		2.1	64*		2.2
493039	0		1.2	0	1	1.2	0		2.1	32*		2.2
494325	16*	0	3.1	64*		1.3	16	0	1.1	0	1	2.1
499125	0		2.1	0	1	1.2	0		2.1	72*	0	1.1
570375	0		1.2	8	0	1.1	0		2.1	16	0	1.1
571787	9*	0	1.1	50*	0	1.1	0		2.1	2	0	1.1
586971	8	0	1.1	16	0	1.1	0	1	4.1	128*		1.3
614125	4	0	1.1	0		1.2	0		2.1	36*	0	1.1
658503	0		1.2	4	0	1.1	0		2.1	100*	0	1.1
704969	0		2.1	0	2	1.3	0	1	3.2	32*		2.2

D	D			$-D$			$2D$			$-2D$		
	σ	g	$\lambda.\lambda_1$	σ	g	$\lambda.\lambda_1$	σ	g	$\lambda.\lambda_1$	σ	g	$\lambda.\lambda_1$
753571	2	0	1.1	0		1.2	0		2.1	100*	0	1.1
804357	0		2.1	0		1.2	0		2.1	4	0	1.1
823875	4	0	1.1	0	1	1.2	0		2.1	0	2	2.2
857375	0	1	1.2	0		1.2	0		2.1	36*	0	1.1
898425	0		2.1	0	2	1.3	0		2.1	16	0	1.1
912673	72*		2.2	4*	0	1.3	0	1−	3.2	0	1+	2.2
1030301	18*	0	1.1	0		1.2	0		2.1	50*	0	1.1
1092727	0	1	1.2	98*	0	1.1	0		2.1	8*		2.2
1120581	8	0	1.1	72*	0	1.1	16*	0	3.1	0		1.2
1157625	0		2.1	64*		1.3	0		2.1	8	0	1.1
1225043	81*	0	1.1	0		1.2	0		2.1	18*	0	1.1
1369599	0		1.2	0		1.2	0		2.1	8	0	1.1
1442897	8*		2.2	64*		1.3	0		3.2	0	1+	2.2
1482975	0	1	1.2	0	1	1.2	0		2.1	72*	0	1.1
1497375	8	0	1.1	32*	0	1.3	8	0	1.1	0		2.1
1760913	0		1.2	0	1	1.2	8	0	1.1	0		2.1
2471625	8	0	1.1	0	1	1.2	16*	0	3.1	0		1.2
4108797	0	1	1.2	200*	0	1.1	16*	0	3.1	0	1	1.2
4492125	72*	0	1.1	0		1.2	0		2.1	8	0	1.1
7414875	36*	0	1.1	72*	0	1.1	0		2.1	32*		2.2
12326391	0		1.2	72*	0	1.1	0		2.1	72*	0	1.1

Table 2

We write $\gamma(D) = \gamma^2(D)/\tau(D)$ as in §7. Then in the range of Table 1,

$$\gamma(D) = \gamma(-4D) = 0 \text{ or } 1,$$

except in the cases marked by asterisks. In the exceptional cases, γ takes the following values:

(i) $\gamma(D) = 1$, $\gamma(-4D) = 4$
for $D =$ 98, 198, 414, 1058, 1078, 1694, 2166, 2250, 3174, 10082, 13230, 19494, 21218, 23958, 28566, 29250, 30870, 45125, 45738, 83006, 130438, 167706, 219006, 330750, 380250, 439002, 494325, 771750, 2241162, 4943250, 8217594;
and for $-D =$ 35374, 54450.

(ii) $\gamma(D) = 4$, $\gamma(-4D) = 1$ for $D =$ 289, 5329, 7921, 9409;

and for $-D =$ 17, 41, 57, 97, 117, 297, 441, 605, 1089, 1587, 1625, 2625, 3087, 3249, 3267, 3993, 4913, 5145, 7425, 8649, 9702, 10985, 15125, 19773, 23625, 45375, 46305, 61731, 65219, 68921, 99825, 165375, 185193, 912673, 1497375.

(iii) $\gamma(D) = \gamma(-4D) = 4$
for $D =$ 113, 3773, 4563, 4913, 6534, 33275, 123462, 274625, 328509, 1442897;
and for $-D =$ 62, 82, 142, 146, 194, 686, 1026, 1342, 1681, 5329, 7546, 7921, 7986, 9522, 9747, 9826, 15842, 18634, 24334, 34295, 59319, 59582, 65910, 71874, 73002, 87750, 166375, 228150, 370386, 593190, 778034, 2185454, 14829750.

(iv) $\gamma(D) = 4$, $\gamma(-4D) = 16$ for $D =$ 4418, 88434, 249018, 598950.

(v) $\gamma(D) = 16$, $\gamma(-4D) = 4$ for $D =$ 12769, 41503, 109503;
and for $-D =$ 20577, 35937, 45125, 274625, 494325, 1157625, 1173942.

(vi) $\gamma(D) = \gamma(-4D) = 16$ for $-D =$ 137842, 913066, 986078.

(vii) $\gamma(D) = 64$, $\gamma(-4D) = 16$ for $-D =$ 389017, 1442897.

(viii) $\gamma(D) = \gamma(-4D) = 64$ for $-D =$ 1409938.

(ix) $\gamma(D) = \gamma(-4D) = 9$
for $D =$ 1849, 2523, 6859, 7803, 8978, 11449, 13778, 15379, 22898, 42875, 50653, 51894, 89373, 107653, 160083, 205379, 391314, 571787, 1030301, 4492125, 7414875;
and for $-D =$ 1682, 2662, 7442, 12167, 13182, 18522, 20402, 55902, 132651, 257250, 265302, 297754, 342950, 357941, 549250, 601526, 998250, 1120581, 1228250, 1714750, 2450086, 2965950, 7414875, 12326394, 24652782.

(ix) $\gamma(D) = \gamma(-4D) = 36$ for $D = 912673$.

(x) $\gamma(D) = \gamma(-4D) = 81$ for $D = 1225043$.

(xi) $\gamma(D) = \gamma(-4D) = 25$ for $D = 536238$;
and for $-D =$ 571787, 1317006, 1507142, 2060602, 4108797.

(xii) $\gamma(D) = \gamma(-4D) = 49$ for $D = 300763$;
and for $-D =$ 453962, 1092727.

References

[1] *G. Billing,* Beiträge zur arithmetischen Theorie ebener kubischer Kurven. Nova Acta Reg. Soc. Scient. Upsaliensis (4) **11** (1938), 1–165.

[2] *B. J. Birch* and *H. P. F. Swinnerton-Dyer,* Notes on elliptic curves I. J. reine angew. Math. **212** (1963), 7–25.

[3] *B. J. Birch,* Conjectures on elliptic curves, Amer. Math. Soc., Proceedings of Symposia in Pure Mathematics, **8**: Theory of Numbers, Pasadena 1963.

[4] *P. F. Byrd* and *M. D. Friedmann,* Handbook of elliptic integrals for engineers and physicists. Berlin-Göttingen-Heidelberg 1954.

[5] *J. W. S. Cassels,* Arithmetic on curves of genus 1. I. On a conjecture of Selmer. J. reine angew. Math. **202** (1959), 52–99. II. A general result. J. reine angew. Math. **203** (1960), 174–208. III. The Tate-Šafarevič and Selmer groups. Proc. London Math. Soc. (3) **12** (1962), 259–296. IV. Proof of the Hauptvermutung. J. reine angew. Math. **211** (1962), 95–112.
[6] *J. W. S. Cassels,* Arithmetic on an elliptic curve. Proc. Intern. Congress Math., Stockholm **1962**, 234–246.
[7] *J. W. S. Cassels,* Arithmetic on curves of genus 1. VIII. On conjectures of Birch and Swinnerton-Dyer. J. reine angew. Math. **217** (1965), 180–199.
[8] *H. Davenport* and *H. Hasse,* Die Nullstellen der Kongruenzzetafunktionen in gewissen zyklischen Fällen. J. reine angew. Math. **172** (1934), 151–182.
[9] *M. Deuring,* Die Zetafunktion einer algebraischen Kurve von Geschlechte Eins. I, II, III, IV. Nachr. Akad. Wiss. Göttingen (**1953**) 85–94; (**1955**) 13–42; (**1956**) 37–76; (**1957**) 55–80.
[10] *E. Forrest,* Number theory of elliptic curves. M. Sc. thesis, Cambridge 1964.
[11] *R. Fueter,* Vorlesungen über die singulären Moduln und die komplexe Multiplikation der elliptischen Funktionen. Leipzig 1924.
[12] *H. Hasse,* Bericht über neuere Untersuchungen und Probleme aus der Theorie der algebraischen Zahlkörper. II. Reziprozitätsgesetz. Jahresber. D. M. V. 1930 (Ergänzungsband 6), §19.
[13] *H. Hasse,* Neue Begründung der komplexen Multiplikation I, II. J. reine angew. Math. **157** (1927), 115–139; **165** (1931), 64–88.
[14] *L. Holzer,* Minimal solutions of Diophantine equations. Canadian J. Math. **2** (1950), 238–244.
[15] *E. Hecke,* Eine neue Art von Zetafunktionen und ihre Beziehungen zur Verteilung der Primzahlen. II. Math. Zeitschrift **6** (1920), 11–51.
[16] *J. L. Lagrange,* Oeuvres VII, 102–114.
[17] *Carl-Erik Lind,* Untersuchungen über die rationalen Punkte der ebenen kubischen Kurven vom Geschlecht Eins. Thesis, Univ. Uppsala 1940.
[18] *T. Ono,* On the Tamagawa number of algebraic tori. Annals of Math. **78** (1963) 47–73. On the relative theory of Tamagawa numbers. Bull. Amer. Math. Soc. **70** (1964), 325–326.
[19] *C. L. Siegel,* Über die analytische Theorie der quadratischen Formen. I, II, III. Ann. of Math. **36** (1935), 527–606; **37** (1936), 230–263; **38** (1937), 212–291.
[20] *J. Tate,* Duality theorems in Galois cohomology over number fields. Proc. Intern. Cong. Math., Stockholm **1962**, 288–295.
[21] *H. Weber,* Lehrbuch der Algebra, 2. Auflage, Braunschweig 1898.
[22] *A. Weil,* Jacobi sums as "Grössencharaktere". Trans. Amer. Math. Soc. **73** (1952), 487–495.
[23] *A. Weil,* Adèles and algebraic groups. Mimeo notes taken by M. Demazure and T. Ono, Princeton 1961.
[24] *A. Weil,* Sur la théorie des formes quadratiques. Colloque sur la Théorie des Groupes Algebriques, Brussels 1962, 9–22.
[25] *E. T. Whittaker* and *G. N. Watson,* A course of Modern Analysis. Cambridge 1902.
[26] *Ju. I. Manin,* Theory of commutative formal groups over fields of finite characteristic, Uspehi Math. Nauk. (Russian Math. Surveys) **18** (1963), 3–90.
[27] *J. Tate,* Algebric cohomology classes, Lecture delivered at the Summer Research Institute on Algebraic Geometry, Woods Hole 1964.

The conjecture of Birch and Swinnerton-Dyer

By *Yukako Kezuka* at Tokyo and *Don Zagier* at Bonn

Abstract. The two seminal papers by Birch and Swinnerton-Dyer played a key role in the development of the theory of elliptic curves, number theory in general, and the use of scientific computation in mathematics, demonstrating how large-scale numerical experiments can yield profound theoretical insights. The Birch–Swinnerton-Dyer conjecture remains one of the central unsolved problems in number theory, asserting a deep connection between the rank of an elliptic curve over $\mathbb{Q}$ and the behavior of its associated L-function at $s = 1$. This paper provides an overview of the conjecture, tracing its origins through computational experiments on the EDSAC 2 computer, examining its implications for Diophantine problems, and surveying modern computational and theoretical advances in assessing its validity.

1. Introduction

The Birch–Swinnerton-Dyer (BSD) conjecture is one of the most celebrated unsolved problems in number theory. In the year 2000, it was selected by the Clay Mathematics Institute as one of its seven million-dollar Millennium Prize Problems. The conjecture concerns mathematical objects called elliptic curves. Due to its depth and the wide range of connections to other areas, the theory of elliptic curves is considered one of the most beautiful in mathematics.

Elliptic curves are curves of genus one with a marked point (the point at infinity), and are the simplest examples of abelian varieties (varieties with an abelian group structure given by polynomial functions). Unless otherwise specified, an "elliptic curve" will refer to one defined over the rational numbers $\mathbb{Q}$. The addition law on the set of rational points on an elliptic curve E endows this set with the structure of an abelian group, called the *Mordell–Weil group*, denoted by $E(\mathbb{Q})$. It is a finitely generated abelian group, and its rank $r \geq 0$ is called the *rank* of E. The possible values and distribution of r remain largely mysterious. The key idea regarding the rank of an elliptic curve is contained in the conjecture made in the early 1960s by B. J. Birch and H. P. F. Swinnerton-Dyer. Motivated by the work of Siegel, which had shown that the density of rational points on a quadratic surface can be expressed in terms of the densities of p-adic points, Birch and Swinnerton-Dyer discovered by numerical calculations that there seems to be something similar happening for elliptic curves. As they put it, "one hopes that if for most p the curve has unusually many points in the finite field with p elements, then it will have a lot of rational points". The computations were carried out using the Cambridge computer EDSAC 2, which had a mere 2178 bytes of memory capacity, compared to the gigabytes available on modern laptops. Let N_p be the number of points on E modulo p. One can associate to E

a function $L(E, s)$ called the L-function of E, which is defined as an Euler product, and its factor at a prime p is defined in terms of N_p. The Birch–Swinnerton-Dyer conjecture states that the rank of an elliptic curve, which measures the "magnitude of infinity" of the number of rational solutions of E, is equal to the order of vanishing of the L-function $L(E, s)$ at $s = 1$, which is determined purely by information about E at each prime number p. Showing this part of the conjecture (weak BSD) is sufficient to win the Clay prize. In addition, the conjecture gives an exact formula for the leading Taylor coefficient about $s = 1$ in terms of arithmetic objects associated to the elliptic curve, including the order of the Tate–Shafarevich group. A more precise formulation is given in Section 4. In Section 2, we go through some basics of elliptic curves, and in Section 3, we discuss the original papers of Birch and Swinnerton-Dyer which gave rise to the conjecture. We will then list some evidence and developments on the conjecture in Section 5, and in Section 6 we end on a note on number theory and computations.

2. What are elliptic curves and what are their invariants?

Number theorists are interested in finding rational or integer solutions to Diophantine equations, that is, algebraic equations in variables $x, y, \ldots$. The name refers to Diophantus of Alexandria (ca. 200 A.D.), who first studied such problems. His analysis was so far ahead of his time that when 6 of the 13 books of his *Arithmetika* were rediscovered in 1570 (4 more have been found since) and published in Latin 50 years later, they were still considered revolutionary and inspirational; for example, Fermat studied his copy carefully and famously wrote his Last Theorem in the margin.

Diophantine problems played a central role in number theory, and they include many of the most classical problems in mathematics, such as Fermat's Last Theorem, the congruent number problem, and the problem of writing a number as a sum of two cubes. The BSD conjecture concerns the class of elliptic curves, which can be given by Diophantine equations of degree 3 in two variables. Equations of degrees 1 and 2 we understand completely and know how to solve. In fact, there are three classes of plane curves—roughly speaking, those of degree ≤ 2 (curves of genus 0), degree 3 (generically of genus 1, though not always with a rational point), and degree ≥ 4 (which generically define curves of genus > 1). In the first case, the equation either has no solution or the solutions are in 1-to-1 correspondence with $\mathbb{Q} \cup \{\infty\}$. Moreover, given one nontrivial rational solution, all other solutions can be systematically generated, and the Hasse–Minkowski theorem allows us to determine precisely which of the two situations we are in. In the second case, the equation can have no solution, finitely many, or infinitely many solutions, and there is a beautiful theory for the structure of the solutions. All of these situations can occur, but for a given elliptic curve, we currently do not know definitively which one applies. A way to predict precisely whether it has finitely or infinitely many solutions is given by the BSD conjecture. In the third case, as conjectured by Mordell in 1922 and proved by Faltings in 1983, equations of genus ≥ 2 have at most finitely many rational solutions, though not in an algorithmic or effective way. Elliptic curves thus occupy a unique position between equations that are easy to solve and those for which finding a general algorithm is currently out of reach, bridging the gap with a rich structure that makes them both accessible and profoundly challenging.

An elliptic curve can be given as a plane curve with an equation of the form

$$E: y^2 = x^3 - Ax - B, \tag{2.1}$$

where $A, B \in \mathbb{Q}$, and $4A^3 - 27B^2 \neq 0$ to ensure that the curve is non-singular. We are interested in rational solutions of this equation. This seemingly simple question leads to unexpectedly profound insights in number theory. We will give an overview of this aspect of the theory, describing the questions currently of greatest interest to us, the progress made in recent years, and the applications both within and outside number theory. To fully appreciate modern developments, one must understand some of the background of elliptic curves, which we will briefly describe now.

Diophantus developed a systematic method for solving quadratic equations in two variables and addressing cubic ones. For quadratic equations, his method is quite straightforward: one starts with a known solution, writes the equation of a rational line passing through this point, and then obtains a quadratic equation in one variable for which one rational solution is already known; the second solution is then automatically also rational. For example, to find the solutions of $x^2 + y^2 = 1$, one starts with the known solution $(x, y) = (1, 0)$, writes the equation of a line through this point as $x = 1 - \frac{n}{m}y$, and calculates as follows:

$$x = \frac{m^2 - n^2}{m^2 + n^2}, \qquad y = \frac{2mn}{m^2 + n^2}.$$

After rescaling, this gives the general solution (known long before Diophantus) of the Pythagorean equation $a^2+b^2 = c^2$. Diophantus also applied his method to some cubic equations (e.g. to the equation $y(6 - y) = x^3 - x$ in Problem 24 of Book IV): in modern terminology, he writes the equation for a line through an already known rational point and then adjusts the slope until that point becomes a double point (tangent), in which case the remaining solution of the cubic equation is a new rational point.

Fermat extended Diophantus's method into his famous *method of infinite descent.* The method is based on the principle that new rational solutions can be generated from existing ones, e.g., the tangent line at a known point P intersects the curve at a new rational point (and similarly, the secant line through two known points P and Q intersects the curve at a third point, as observed later by Newton). This process generally yields solutions with more complicated coordinates, with numerators and denominators much larger than those of P. Fermat's strategy of infinite descent consists in reversing this procedure. One shows that any solution with sufficiently large coordinates must arise from a "smaller" solution via such a construction, and by repeating this process, one obtains successively smaller solutions, leading to a contradiction if no small solutions exist, or to a description of all solutions if there are small solutions. For example, in 1643, Fermat challenged Mersenne to find a right-angled triangle whose hypotenuse c and sum of the legs $a + b$ were integer squares and used descent to find the smallest such solution:

$$\begin{gathered}(a, b, c) = (1061652293520, 4565486027761, 4687298610289)\\ a + b = 2372159^2, \qquad c = 2165017^2, \qquad a^2 + b^2 = c^2.\end{gathered}$$

Another classical problem is the congruent number problem, one of the oldest unsolved problems in mathematics. An integer is called a *congruent number* if it represents the area of a right triangle with three rational sides, i.e., if it can be written as $\frac{1}{2}ab$ for rational numbers a and b the sum of whose squares is a square. The problem is to determine whether a given number is congruent. For example, the second author showed in the 1970s, by a pure-thought descent argument which reduced the problem to a search over small integers, that the number 157 is congruent, the smallest solution being given by $a = \frac{6803298487826435051217540}{411340519227716149383203}$, $b = \frac{314}{a}$.[1)]

What Diophantus, Fermat and Newton had found was a first indication of the addition law on elliptic curves, although this was only realized much later, with the group law on the points of the curve (over either $\mathbb{R}$ or $\mathbb{Q}$) being defined by the requirement that the three points of intersection of a line with a plane cubic curve sum to zero (and with the identity element of the group being the point at infinity if the plane curve is given in Weierstrass form (2.1)). Around the turn of the twentieth century, Poincaré studied the rational points on an elliptic curve and saw them as a group (but with the group law being seen via the realization of $E(\mathbb{C})$ as the quotient of $\mathbb{C}$ by a lattice, so via elliptic functions, rather than in the algebraic model (2.1)). He treated the set of rational points on a cubic curve as if it were generated by finitely many points, a fact later proved by Mordell in 1922 and now known as Mordell's theorem. (For a detailed overview of the history of the group law on elliptic curves, we refer the reader to the beautiful article [35] by Norbert Schappacher.) It is the existence of this additional structure that is responsible for the richness of the theory of elliptic curves, which stems from the interplay between the relatively simple geometric structures and the deep arithmetic properties of the underlying fields.

The set of rational points on E (as a projective curve, so including the point at infinity) is denoted $E(\mathbb{Q})$. By Mordell's theorem, later generalized by Weil to abelian varieties over arbitrary number fields, it is a finitely generated abelian group, called the Mordell–Weil group of E. By the structure theorem for finitely generated abelian groups, it can be written as:

$$E(\mathbb{Q}) \simeq \mathbb{Z}^r \oplus T$$

where $r \geq 0$ is an integer, called the *rank* of E, and T is the finite abelian group $E(\mathbb{Q})_{\mathrm{tor}}$, called the *torsion group* of E. The Nagell–Lutz theorem from the 1930s provides a concrete way to find all the torsion points, i.e., to determine T for a given curve. Much later, in 1977, Mazur identified the possible structures for T: it is isomorphic to either a cyclic group $\mathbb{Z}/n\mathbb{Z}$ for $n = 1, 2, 3, \ldots, 10$ or 12, or a direct product of cyclic groups $\mathbb{Z}/2\mathbb{Z} \times \mathbb{Z}/2n\mathbb{Z}$ for $n = 1, 2, 3, 4$. These fifteen possibilities completely classify the torsion subgroups of elliptic curves, and each occurs infinitely often.

But how do we find the rank r or the generators of $E(\mathbb{Q})$? This is what the BSD conjecture tries to answer. To approach this problem, we study the number of solutions modulo p for prime numbers p. Set

$$N_p = \#(\widetilde{E}(\mathbb{F}_p)), \tag{2.2}$$

where $\widetilde{E}$ is the reduction of E modulo p over the finite field $\mathbb{F}_p$ with p elements. Then set

$$a_p = 1 + p - N_p.$$

The quantity a_p measures the deviation of the number of points from the value for a curve of genus 0. We now introduce the key players in the BSD conjecture, which are the L-function of E and the Tate–Shafarevich group of E.

The L-function in the complex variable s is defined for $\mathrm{Re}(s) > \frac{3}{2}$ by the Euler product

$$L(E, s) = \prod_{p \mid N} \frac{1}{1 - a_p p^{-s}} \prod_{(N,p)=1} \frac{1}{1 - a_p p^{-s} + p^{1-2s}},$$

1) This point can be found far more rapidly using mock Heegner points in a version discovered by Paul Monsky [29] and worked out more fully in recent work of Henri Cohen and Bill Allombert.

where N denotes the *conductor* of E, a product of powers of the primes of bad reduction for E. It is related to the Hasse–Weil zeta function of E by $\zeta_E(s) = \frac{\zeta(s)\zeta(s-1)}{L(E,s)}$. Thanks to the modularity theorem, we now know that $L(E,s)$ extends to the whole complex plane and satisfies a functional equation relating values at s and $2-s$. Roughly speaking, this theorem asserts that every elliptic curve over $\mathbb{Q}$ can be parametrized by modular functions. However, at the time Birch and Swinnerton-Dyer proposed their conjecture, $L(E,s)$ was not even known to be defined at $s=1$. For elliptic curves that admit *complex multiplication* (CM), that is, curves endowed with an additional endomorphism structure allowing "multiplication" of points by certain complex numbers, such as $i\cdot(x,y) = (-x,iy)$ on $y^2 = x^3 - x$, modularity had long been known by a theorem of Deuring, which expresses $L(E,s)$ in terms of Hecke L-functions. The proof of modularity for semistable elliptic curves came only much later through the work of Wiles and Taylor–Wiles [40, 43] in the context of Wiles's proof of Fermat's Last Theorem, and was later extended to general elliptic curves by Breuil et al. [9].

The Tate–Shafarevich group, denoted $\text{Ш}(E)$, can be defined as the group of isomorphism classes of torsors (principal homogeneous spaces) for E that are everywhere locally soluble. It measures the failure of the local-to-global principle for the existence of rational points, and plays a central role in the arithmetic study of elliptic curves and the BSD conjecture. The problem is that we do not know in general that $\text{Ш}(E)$ is finite. What we do know is that if it is finite, then a theorem of Cassels tells us that its order is a square, because $\text{Ш}(E)$ is self-dual under the Cassels–Tate pairing. The BSD conjecture includes the *Tate–Shararevich conjecture*, which states that $\text{Ш}(E)$ is always finite.[2)]

Tate wrote about the BSD conjecture [38]: "This remarkable conjecture relates the behavior of a function L at a point where it is not at present known to be defined to the order of a group $\text{Ш}(E)$ which is not known to be finite!"

3. The two Crelle papers

In this section, we discuss the two seminal papers by Birch and Swinnerton-Dyer, published in *Crelle*, titled *Notes on Elliptic Curves. I.* and *II.*

In the opening paragraphs of Paper I, the authors explain how they build upon Siegel's work, which established that the density of rational points on a quadric surface can be expressed in terms of p-adic densities, a principle later extended by Tamagawa and Kneser. They propose a similar relationship for elliptic curves: an abundance of points over the finite field with p elements for most primes p may indicate the presence of many rational points.

Although initial hand calculations provided early support for this conjecture, the authors used the Cambridge University EDSAC 2 computer to perform extensive computations, enabling them to refine their conjectures into an exact form and back them with substantial numerical evidence. Despite the robustness of their data, they conceded that no proof was in sight, stating, "it has become clear to us that we are unlikely to be able to prove these conjectures." Recognizing the importance of their findings, they present them in a "series of notes", emphasizing the interplay between theoretical insights and machine computation.

They focus on elliptic curves E written in the form (2.1) (they write Γ for E, $\mathcal{A}$ for $E(\mathbb{Q})$ and g for r, but we will not do this). Based on Mordell's theorem and the Nagell–Lutz theorem

---

2) We mention in passing that there is a possible weakening of the BSD conjecture in which both the finiteness of $\text{Ш}(E)$ and the equality of the Mordell–Weil and analytic ranks can be violated, but only simultaneously and in a very specific way [25].

(for which they quote only Lutz; Nagell's paper was independent), the authors turn their attention to the rank r of the Mordell–Weil group. Their main goals were:

(a) to find an efficient way to compute (or sometimes bound) r for a given curve, and

(b) to correlate r with $\#E(\mathbb{F}_p)$ as p varies.

Roughly, Paper I is on (a), and Paper II is on (b).

Though the BSD conjecture does not appear until Paper II, Paper I gives an insight into how the conjecture was developed. The primary objective of Paper I was to establish an effective algorithm for bounding the rank of an elliptic curve E. This was achieved by bounding the associated 2-Selmer group $\mathrm{Sel}_2(E/\mathbb{Q})$, which is a subgroup of the Galois cohomology group $H^1(\mathbb{Q}, E[2])$ classifying 2-coverings of E. These are pairs (C, θ), where C is a curve defined over $\mathbb{Q}$ and $\theta : C \to E$ is a morphism such that the pair (C, θ) becomes isomorphic to $(E, [2])$ over an algebraic closure of $\mathbb{Q}$. A 2-covering represents an element in $\mathrm{Sel}_2(E/\mathbb{Q})$ if and only if it is everywhere locally solvable, meaning it has a rational point over every completion of $\mathbb{Q}$. It can be expressed as a quartic curve $y^2 = g(x)$, where the coefficients of $g(x)$ are rational and share invariants with those of E. The set of 2-coverings forms a group, and the authors distinguish two important subgroups: G, consisting of coverings that have points in all local fields (i.e., the 2-Selmer group $\mathrm{Sel}_2(E/\mathbb{Q})$), and G', comprising coverings that have a rational point. In particular, $G' \simeq E(\mathbb{Q})/2E(\mathbb{Q})$ and has order 2^{r+r_2}, where r_2 is the number of independent generators of $E(\mathbb{Q})$ of finite even order. Now, $r_2 = 0$, 1 or 2 according as $x^3 - Ax - B = 0$ has 0, 1 or 3 rational roots. Thus, in order to find r it suffices to study $E(\mathbb{Q})/2E(\mathbb{Q})$. The order of G is 2^{r+r_2+t}, where $\#\left(\text{Ш}(E)[2]\right) = 2^t$ is the number of elements of order 2 in $\text{Ш}(E)$.

The authors describe algorithms that produce a list of all everywhere locally solvable 2-coverings, allowing them to compute $r + t$, and in some cases r, for specific families of elliptic curves. The results, presented in detailed tables, demonstrate the practical effectiveness of their methods while acknowledging areas where further verification is needed. For example, $y^2 = x^3 + 197x$ has a rational point of infinite order because a certain double cover has an explicit point, while $y^2 = x^3 - 309$ should "presumably" have a point of infinite order since it has a single locally solvable 2-covering and also a single locally solvable 3-covering, but they had not found one. (It was found later.)

Paper I contains four tables. In Table 1 (pp. 20–21), they list the values of r and t for elliptic curves of the form (2.1) with $|A| \le 20$, $|B| \le 30$. Paper I concludes with Tables 2–4 (pp. 23–25), covering two types of CM elliptic curves: (2.1) with $A = 0$ and $|B| \le 400$ or with $B = 0$ and $|A| \le 200$, together with a few additional examples for larger values. For $\#\left(\text{Ш}(E)[2]\right) = 1$, the table includes curves with rank 0 (394 cases), rank 1 (578 cases), rank 2 (166 cases), and rank 3 (9 cases). For $\#\left(\text{Ш}(E)[2]\right) = 4$, the table lists curves with rank 0 (42 cases) or rank 1 (7 cases); here no curves with $r > 1$ were found.

Paper II was published in 1965. Motivated by Siegel's work, they introduce

$$f(P) = \prod_{p \le P} N_p/p \tag{3.1}$$

where the product runs over all primes $\le P$ and N_p is the number of rational points on the curve defined by (2.1), considered as a projective curve over the finite field $\mathbf{F}_p$ with p elements, as seen in (2.2). Note that $\frac{N_p}{p}$ is the value at 1 of the reciprocal of the p-th Euler factor in $L(E, s)$ if $p \nmid N$; if $p \mid N$, the correct factor is $\frac{\#E_{\mathrm{ns}}(\mathbf{F}_p)}{p}$, where $E_{\mathrm{ns}}(\mathbf{F}_p)$ denotes the set of non-singular points over $\mathbf{F}_p$ on a minimal Weierstrass model for E at p.

Starting in the fall of 1958, Birch and Swinnerton-Dyer had conducted extensive computer calculations on EDSAC 2 and observed a striking correlation between:

- the rate of growth of $f(P)$ as $P \to \infty$,
- the rank r of E, and
- the asymptotic behavior of the L-series $L(E,s)$ (or zeta-function ζ_E; they use both),

leading to the following conjecture:

Conjecture (A). *Let E be an elliptic curve, and let f be as defined in* (3.1). *Then*

$$f(P) \sim C(\log P)^r \tag{3.2}$$

as $P \to \infty$, where r is the Mordell–Weil rank, and

$$\zeta_E(s) \sim C'(s-1)^{-r-1} \tag{3.3}$$

for some constants C, C' depending on E.

There is a typo in the original paper: they wrote $(s-1)^{g-1}$ instead of $(s-1)^{-g-1}$. In addition, they did not specify the constants C, C', or their product CC'. They write "The value of $f(P)$ oscillates vigorously as P increases, and there seems little hope of being able to find the constant C". They also acknowledge that the factors at the "primes p dividing $6(4A^3-27B^2)$ are almost certainly the wrong ones". (They were right.) Equation (3.3) implies that r is equal to the analytic rank $r^{\mathrm{an}} := \mathrm{ord}_{s=1} L(E,s)$, and is then equivalent (assuming that $L(E,s)$ is holomorphic at $s=1$) to $C' = -\frac{1}{2}r!/L^{(r)}(E,1)$.

In Paper II, they focus exclusively on the cases $E_D : y^2 = x^3 - Dx$ (where $j = 1728$), which correspond to elliptic curves with CM by $\mathbb{Q}(i)$. This restriction lets them:

(a) exploit known analytic behavior (without which $L(E,1)$ is not even defined), and

(b) use explicit formulas for a_p, essentially derived from the work of Gauss.

Based on calculations for E_D, they then formulate a second conjecture for the rank 0 case:

Conjecture (B). *If $r = 0$, then*

$$\#\left(\text{Ш}(E_D)\right) = \frac{\eta(D)^2}{\tau(D)}.$$

Here, $\eta(D)$ is the number of rational points on E_D, and $\tau(D)$ is the "Tamagawa number", defined formally as a product over *all* primes, including infinity. The factor at infinity, denoted Ω_∞ in (4.1) below, is given by the positive real period of E_D multiplied by the number of connected components of $E_D(\mathbb{R})$, which in this case is equal to 1 or 2 depending on the sign of D. At primes of good reduction, it coincides with the reciprocal of the Euler factor of $L(E_D,1)$.[3)] Conjecture (B) essentially presents the full modern formulation of the BSD conjecture in this special case, but it does not propose any relation involving heights of generators or a regulator, not even when $r = 1$.

[3)] At a prime of bad reduction, an extra correction factor, known as the Tamagawa factor, appears (see (4.1)). There are $\mathbb{Q}_p$-points on E, $\mathbb{Q}_p$-points on the connected component E^0 of the identity in the Néron model, and $\mathbb{Q}_p$-points on the quotient E^0/E^1, which corresponds to the $\mathbb{F}_p$-points of the Néron model. Here, $E^1(\mathbb{Q}_p)$ is the kernel of the reduction modulo p map $E^0(\mathbb{Q}_p) \to \widetilde{E}(\mathbb{F}_p)$. These are the ones that actually appear in the Euler factor of the L-function. The "fudge" factor (the product of Tamagawa factors at bad primes) essentially indicates that we should instead include the extra lifts coming from E/E^1 rather than just E^0/E^1 in the usual L-factor.

In Sections 2 and 3, they exploit the fact that E_D admits CM to write $L(E_D, s)$ in terms of Hecke L-series. This allows them to express the central L-values $L(E_D, 1)$ in a finite, computable form using quartic reciprocity, the Weierstrass $\wp$-function and its derivative. It then follows from the fact that $\wp$ and $\wp'$ are algebraic functions of g_2 and g_3 that $L(E_D, 1)/\omega$ is an algebraic number. They use this to show in Theorem 1 on p. 87 of Section 4 that the analytic order of the Tate–Shafarevich group (the conjectural order expressed in terms of $L(E, 1)$ in the case $r = 0$) is indeed a rational integer. Section 5 (pp. 92–96) is lengthy and focuses on the estimation of r. The authors state that the methods from Paper I would be "intolerably slow when D is large" and instead use the fact that the isogeny "multiplication by 2" factors non-trivially (as $(1 + i)(1 - i)$). This relies in part on earlier results by Billing, Lind, Cassels, and others. In Section 6, the Tamagawa number $\tau(D)$ is defined and computed in the context of the paper. They present four lemmas (Lemmas 11–14) covering different cases: when p is an odd prime dividing D, when p is an odd prime not dividing D, when $p = 2$, and when $p = \infty$.

In the final section, the authors describe the numerical computations they performed using EDSAC 2 to test their conjectures. Table 1 (pp. 101–107) presents values of r for certain values of D up to 12,326,391 (according to Birch, Swinnerton-Dyer wanted a specific number of examples). They marked some exceptional cases with an asterisk, which are further described in Table 2 (p. 108). According to their conjecture, these exceptional cases correspond to instances where $r = 0$ and the Tate–Shafarevich group is non-trivial.

4. The Birch–Swinnerton-Dyer conjecture

The conjecture of Birch and Swinnerton-Dyer, as stated in the Millennium Prize Problem, asserts that the rank of $E(\mathbb{Q})$ equals the order of vanishing of $L(E, s)$ at the central point $s = 1$ of the functional equation. The following modern version of BSD, which includes the regulator factor and is the form usually quoted as opposed to the original conjecture formulated in the Crelle papers, was introduced by Tate [38]. This version gives a beautiful exact formula for the order of the Tate–Shafarevich group, which is conjecturally finite, in terms of the leading coefficient of the Taylor expansion of $L(E, s)$ about $s = 1$.

BSD Conjecture. *Let E be an elliptic curve with rank r, and let $r^{\mathrm{an}} = \mathrm{ord}_{s=1} L(E, s)$ be the order of vanishing of the complex L-series of E at $s = 1$. Then $r = r^{\mathrm{an}}$, and*

$$(4.1)\qquad (\Omega_\infty R_E)^{-1} \lim_{s\to 1} \frac{L(E, s)}{(s-1)^r} = \prod_p c_p \cdot \frac{\#(Ш(E))}{\#(E(\mathbb{Q})_{\mathrm{tor}})^2}.$$

Here, $\Omega_\infty = \int_{E(\mathbb{R})} |\omega|$, where ω is a Néron differential of any global minimal Weierstrass equation for E, R_E is the regulator given by the absolute value of the determinant of the height pairing matrix on a basis of $E(\mathbb{Q})/E(\mathbb{Q})_{\mathrm{tor}}$, $c_p = [E(\mathbb{Q}_p) : E^0(\mathbb{Q}_p)]$ (equal to 1 at primes of good reduction), where $E^0(\mathbb{Q}_p)$ denotes the subgroup of $E(\mathbb{Q}_p)$ consisting of points of non-singular reduction modulo p, and $Ш(E)$ is the Tate–Shafarevich group.

We note that this full version of the BSD conjecture was *not* in their original paper, which was only stated for the case $r = 0$. Furthermore, the Euler factors in this modern formulation at $p|N$ were chosen differently by Birch and Swinnerton-Dyer, but this does not affect the conjecture, since the constants C and C' have not yet been specified in Conjecture (A).

Later, Goldfeld [17] noticed that if (3.2) in Conjecture (A) holds with $C > 0$, then the relation $C = -2\sqrt{2}e^{r\gamma}C'$ holds where γ is the Euler constant, and in addition, $r = r^{\mathrm{an}}$ and the Riemann Hypothesis holds for $L(E,s)$ (i.e., $L(E,s) \neq 0$ outside of $\mathrm{Re}(s) = 1$, which was then stated under the assumption of modularity). In particular, C is different from the reciprocal of the leading coefficient of $L(E,s)$: even when $r = 0$, it has an unexpected factor of $\sqrt{2}$.

We will present two alternative versions of this modern formulation of the BSD conjecture, or rather of the various factors appearing in (4.1), one by Bloch and one by the second author. The first is expressed in terms of adelic volumes and Tamagawa numbers, providing a new geometric and analytic perspective on this conjecture, and the second connects the profound arithmetic nature of the BSD conjecture with elementary Diophantine ideas, presenting it through the lens of point-counting functions and Dirichlet series.

In 1980, Bloch gave a volume-theoretic interpretation of the BSD conjecture using adelic methods, Tamagawa numbers, and height pairings [8]. He begins with an algebraic group G defined over a number field k and studies adelic points $G(\mathbb{A}_k)$ and their quotient by rational points $G(k)$. He interprets the BSD conjecture in terms of the volume of adelic groups and their Tamagawa measures.

Using adelic measures, Bloch connects the volume of $G(\mathbb{A}_k)/G(k)$ with the Tamagawa number $\tau(G)$, defined using the L-function $L(G,s)$ evaluated at $s = 1$. Bloch's interpretation builds on the adelic and Tamagawa-number framework for linear algebraic groups developed by André Weil [42], in which $G(k)$ is viewed as a lattice of finite covolume in $G(\mathbb{A}_k)$ with respect to a canonical Haar measure. For an abelian variety A, this adelic Tamagawa viewpoint provides a natural conceptual setting for BSD. In the special case $\mathrm{rank}(A) = 0$, it essentially yields the BSD formula directly: the covolume $\mathrm{vol}\, A(\mathbb{A}_k)$ is given in terms of $\mathrm{vol}\, A(k \otimes_{\mathbb{Q}} \mathbb{R})$ and $L(A/k,1)^{-1}$. Bloch's work extends this adelic framework to the case $\mathrm{rank}(A) > 0$, where the regulator and the height pairing enter, and where one must pass to the universal split torus extension X of A, for which $X(k)$ is again discrete in $X(\mathbb{A}_k)$.

Bloch constructs a pairing between the Mordell–Weil group $A(k)$ and the dual abelian variety $A'(k)$ using a volume-theoretic approach. He connects the global height pairing with local pairings, giving it a coherent adelic interpretation. He showed that we "should" in fact interpret the Tamagawa number as

$$\tau(X) = \prod_p c_p \cdot \lim_{s\to 1} \frac{(s-1)^r}{L(E,s)} \cdot \Omega_\infty \frac{R_E}{\#E(\mathbb{Q})_{\mathrm{tor}}},$$

where X is a group scheme that extends E by $\mathbb{G}_m^r$. Thus, the BSD formula becomes

$$\tau(X) = \frac{\#\mathrm{Pic}(X)_{\mathrm{tor}}}{\#\text{Ш}(X)}.$$

Bloch and Kato later generalized this to make sense of Tamagawa numbers in the setting of motives, extending the conjecture beyond abelian varieties.

In a completely different direction, the second author wrote a semi-expository paper [44] providing an alternative interpretation of the BSD conjecture, emphasizing a more elementary Diophantine perspective. The goal is to express the conjecture using invariants derived from "counting rational solutions" of elliptic curves, rather than relying heavily on abstract group-theoretic or cohomological frameworks. This draws an analogy between the BSD conjecture and the class number formula for quadratic fields, where key quantities emerge from counting solutions of quadratic equations. One replaces the L-function of E, whose p-th Euler factor involves the number of points of the reduction of E modulo p for all *primes* p, with a Dirichlet series $D(E, s)$ defined in terms of the number of solutions of (2.1) modulo n for all *integers* n.

$$D(E, s) = \sum_{n=1}^{\infty} \frac{\#E(\mathbb{Z}/n\mathbb{Z})}{n^s},$$

where $\#E(\mathbb{Z}/n\mathbb{Z})$ is the number of points on the elliptic curve E modulo n. One then shows that it has an Euler product whose relationship to that of $L(E, s)$ miraculously produces the "fudge" factor (the product of Tamagawa factors). More precisely, the function

$$D(E, s)L(E, s)\frac{\zeta(2s-1)}{\zeta(s)\zeta(s-1)}$$

has an analytic continuation to $\mathrm{Re}(s) > \frac{5}{6}$ and is equal to $\prod_p c_p$ at $s = 1$.

The BSD conjecture is then rephrased in terms of the asymptotic behavior of counting functions and Dirichlet series, rather than purely group-theoretic or height-pairing terms. It says that $D(E, s)$ has a meromorphic continuation to $s = 1$, and it satisfies

$$D(E, s) \sim -\frac{\kappa}{(s-1)^r \cdot \#\text{Ш}(E)}, \quad (s \to 1)$$

for some non-negative integer r and positive real number κ. If one assumes all known results, then this r *is* the analytic rank r^{an}, and κ can be calculated by $\Omega_E^{-1} Z(E, s)^2 \sim \kappa(\pi/s)^r$ as $s \to 0$, where the series $Z(E, s)$ is defined for $\mathrm{Re}(s)$ large as the sum of $(x^2 + y^2 + z^2)^{-s/2}$ over all primitive integral solutions of the homogeneous form of (2.1).

5. Numerical evidence, theoretical progress and later developments

Numerical evidence supporting the BSD conjecture has accumulated over decades, and all available data supports every aspect of the conjecture. The pioneering work of Birch and Swinnerton-Dyer in Paper II included two large tables containing the first systematic computations in its support. Earlier works by Selmer and Cassels also contributed crucial insights which helped shape its formulation. Later computations have further reinforced this evidence, the first numerical verification in a higher rank case having been in 1985 by Buhler, Gross, and the second author for a curve of level 5077 and rank 3 [10], and computations by the second author and Kramarz [45], which explored sums of two cubes and cases where the rank is greater than 1. Far more extensive computations were performed later by Cremona, Voight, Stein and many others. Cremona, in particular, produced tables of extensive data on elliptic curves [15] (now continued and expanded online through resources such as the LMFDB) that have spurred many of the later developments.

A major theoretical breakthrough in the study of the BSD conjecture came from the framework of Iwasawa theory. Introduced around 1960 in the context of class groups of cyclotomic and other $\mathbb{Z}_p$-extensions of number fields for a prime number p, Iwasawa theory was later extended to elliptic curves, beginning with Mazur's work [27] in the early 1970s, allowing the investigation of p-primary Selmer groups and their relation to p-adic L-functions (p-adic variation of complex L-functions). Coates and Wiles [14] recognized its powerful applications to the BSD conjecture. Their approach made use of elliptic units, introduced by Robert in 1973.

They showed that if E admits CM by an imaginary quadratic field K of class number 1 and is defined over a number field F (where F can be $\mathbb{Q}$ or K), then the non-vanishing of its L-function $L(E/F, s)$ at $s = 1$ implies that the group $E(F)$ of F-rational points is finite. This provided one of the first rigorous results suggesting a deep connection between the rank of an elliptic curve and the behavior of its L-function. Their work was later extended by Greenberg [18], who applied Iwasawa theory to study the behavior of various Selmer groups and Iwasawa invariants in $\mathbb{Z}_p$-extensions.

In 1986, Gross and the second author proved that the heights of so-called "Heegner points" on a modular curve (certain points constructed essentially by Birch from CM theory), are specific multiples of the derivatives of L-functions at $s = 1$, showing in particular that if $r^{\mathrm{an}} = 1$ then $r \geq 1$. The next major development, only three years later, was the introduction of Euler systems by Kolyvagin [26]. Using Heegner points, he constructed an intricate system of cohomology classes to prove that if $r^{\mathrm{an}} \leq 1$, then $r = r^{\mathrm{an}}$ and $\text{Ш}(E)$ is finite.

Iwasawa theory and Euler systems have continued to play an important role in the modern understanding of the BSD conjecture. Rubin [34] established the *p-part of the BSD conjecture*, which asserts that the p-adic valuation on both sides of the BSD formula (4.1) agree, for CM elliptic curves with $r^{\mathrm{an}} = 0$, for almost all p. This was achieved by constructing an Euler system of elliptic units, and proving a so-called "main conjecture" of Iwasawa theory. The prime $p = 2$ was eschewed in Rubin's work, as was $p = 3$ for elliptic curves with CM by $\mathbb{Q}(\sqrt{-3})$, as the classical Iwasawa theory fails at these primes. These cases were later studied by the first author, obtaining Tamagawa number divisibility of algebraic L-values, and the p-part of BSD in certain special cases with Y.-X. Li [22, 23].

Kato [20] extended Euler system techniques to modular forms. He constructed an Euler system from Siegel units and the K-theory of modular curves. His work established one divisibility of the main conjecture of Iwasawa theory for modular forms, providing a major partial result toward BSD for a large class of non-CM elliptic curves (for CM elliptic curves, Rubin's Euler system is used). The other divisibility was shown by Skinner and Urban [37], assuming the existence of Galois representations attached to automorphic forms on the group $U(2, 2)$, establishing the p-part of BSD for a large class of non-CM elliptic curves in the case $r^{\mathrm{an}} = 0$.

The weak BSD conjecture (the equality $r = r^{\mathrm{an}}$) can be further weakened to the *parity conjecture*, which says that r and r^{an} are congruent mod 2. Since the parity of r^{an} is determined by the root number $\epsilon(E/\mathbb{Q})$ of E (the sign of the functional equation of $L(E, s)$ at $s = 1$) an equivalent formulation is $\epsilon(E/\mathbb{Q}) = (-1)^r$. This would imply, in particular, that $E(\mathbb{Q})$ must be infinite whenever $\epsilon(E/\mathbb{Q}) = -1$. The root number can be expressed in terms of the eigenvalues of the Atkin–Lehner involutions which, in turn, are determined, in the case of squarefree level N, by the nature of the multiplicative reduction (split or non-split) for p dividing N. On the algebraic side, a 2-descent argument provides a matching expression for $(-1)^r$, assuming the finiteness of $\text{Ш}(E)$, giving a heuristic justification for the parity conjecture. The parity conjecture is still far from being proven, but thanks to the work of Nekovář, T. Dokchitser, V. Dokchitser [16] and others, we have the *p-parity conjecture* (agreement of the parities of the p^∞-Selmer rank and r^{an}) for all primes p.

The possible values and distribution of r remain largely mysterious. The current record for the highest known rank is at least 29, achieved by Elkies and Klagsbrun in 2024, surpassing the previous record of at least 28, set by Elkies in 2006. This rank is confirmed to be exactly 29 assuming the Generalized Riemann Hypothesis. It is conjectured that 50% of all elliptic curves (ordered by height) have rank 0, 50% have rank 1, and all remaining ranks consist of a total of 0% of all elliptic curves. Goldfeld's conjecture suggests that the average analytic rank of elliptic curves in a quadratic twist family tends to $1/2$, though it does not immediately imply a result for Mordell–Weil ranks. Bhargava and collaborators have made significant progress on understanding the distribution of ranks. They showed that a positive density of elliptic curves over $\mathbb{Q}$ have rank 0 or 1 [2, 3] and that a majority (more than 66%) have analytic rank 0 or 1 [5]. There is also a heuristic suggesting that there are only finitely many E of rank greater than 21 [30].

The BSD conjecture has inspired several far-reaching generalizations. In particular, there is a version for abelian varieties over global fields due to Tate [39]. Rubin's result has recently been significantly extended by Burungale and Flach [11], building on the work of Kings and Johnson-Leung. They showed that if an elliptic curve E defined over a number field F admits CM by the ring of integers of an imaginary quadratic field K and if the extension $F(E_{\mathrm{tor}})/K$ is abelian, then the non-vanishing of $L(E/F, 1)$ implies the full BSD conjecture for E/F. For a large family of quadratic twists of the "$\mathbb{Q}$-curves" (elliptic curves isogenous to all of their Galois conjugates over $\mathbb{Q}$) originally studied by Gross in his thesis, which have CM by an imaginary quadratic field K and are defined over the Hilbert class field H of K, Coates, Li and the first author [13, 24] have shown that the central L-value does not vanish and that the Tate–Shafarevich group is finite. Combined with the work of Burungale and Flach, this establishes the BSD conjecture for these curves over H. For an abelian variety A over a global function field K, we have the following surprising result: if the ℓ-primary part of the Tate–Shafarevich group of A is finite for some prime number ℓ, then it is finite for every prime number and in fact the full BSD conjecture holds for A over K. This built on foundational work by Tate and others, who had established the result for primes ℓ different from the characteristic of K; the proof was completed by Kato and Trihan [21], who handled the more difficult case where ℓ equals the characteristic.

In 1986, Mazur, Tate, and Teitelbaum published "On p-adic analogues of the conjectures of Birch and Swinnerton-Dyer" [28]. Unlike the archimedean case, it is unknown whether the p-adic height pairing is non-degenerate. A p-adic analogue of the Gross–Zagier theorem, due to Perrin-Riou [31], was developed in the framework of Iwasawa theory. More recently, Bertolini, Darmon, and Prasanna [1] have advanced our understanding of BSD in the context of p-adic Hodge theory and modular forms. The interplay between BSD, Iwasawa theory, and Euler systems continues to shape research, with many open questions. Another notable generalization is the Bloch–Kato Conjecture, which extends BSD to motives, predicting deep connections between special values of L-functions and arithmetic invariants. Furthermore, the Equivariant BSD Conjecture, which generalizes BSD in the context of a Galois extension K/F, was developed within the framework of non-abelian Iwasawa theory and equivariant Tamagawa number conjectures. It seeks to describe the behavior of the Tate–Shafarevich group and special values of L-functions in an equivariant setting, incorporating additional Galois symmetry.

In a different direction, the BSD conjecture, even in its weak form, has many important applications. In particular, it provides a practical method to determine the rank of an elliptic curve E by linking it to the vanishing order of the L-function $L(E, s)$ at $s = 1$. This gives us a terminating (though not in *a priori* bounded time) algorithm for finding the rank of a given elliptic curve. Computing the L-derivatives numerically can detect nonzero values, providing an upper bound on the analytic rank. At some point, the lower bound on r computed using point searches will equal the upper bound computed using the L-series.

Many Diophantine problems are closely related to the arithmetic of elliptic curves. A proof of the BSD conjecture would give a uniform way to determine when certain Diophantine equations, including several classical ones, have infinitely or only finitely many solutions. One notable example is the congruent number problem that was discussed in Section 2. It can be reformulated using elliptic curves: a positive integer n is congruent if and only if the congruent number curve $E^{(n)} : y^2 = x^3 - n^2x$ has a rational point of infinite order, i.e., a point (x, y) with $y \neq 0$. Tunnell [41] provided an explicit formula for determining when $L(E^{(n)}, 1)$ vanishes. Using Waldspurger's formula, he showed that an even integer n, assumed without loss of generality squarefree, satisfies $L(E^{(n)}, 1) = 0$ if and only if it has equally many representations as $a^2 + 2b^2 + 8c^2$ with c even or c odd. There is a similar condition for determining odd congruent numbers. This gives a complete solution to the problem if the BSD conjecture holds.

A second example is Sylvester's problem (1897) of determining which primes are sums of two rational cubes. Sylvester conjectured that every prime number p congruent to 4, 7, or 8 modulo 9 can be expressed as the sum of two rational cubes. The equation $x^3 + y^3 = p$ defines an elliptic curve E_p. We have $E_2(\mathbb{Q})_{\text{tor}} \simeq \mathbb{Z}/2\mathbb{Z}$ and $E_p(\mathbb{Q})_{\text{tor}} = \{0\}$ for $p \geq 3$. A 3-descent argument shows that $\text{rank}(E_p(\mathbb{Q})) = 0$ if $p \equiv 2, 5 \bmod 9$ (Pépin, Lucas, Sylvester), ≤ 1 if $p \equiv 4, 7, 8 \bmod 9$, and ≤ 2 if $p \equiv 1 \bmod 9$. At the same time, the root number of E_p is equal to -1 if $p \equiv 4, 7, 8 \bmod 9$, and $+1$ otherwise. Thus, the BSD conjecture predicts $\text{rank}(E_p(\mathbb{Q})) = 1$ for $p \equiv 4, 7, 8 \bmod 9$. Dasgupta and Voight, using mock Heegner points, showed that for primes $p \equiv 4, 7 \bmod 9$ for which 3 is not a cube modulo p, we indeed have $\text{rank}(E_p(\mathbb{Q})) = 1$. The most interesting case is $p \equiv 1 \bmod 9$, when the BSD conjecture predicts that the rank can be 0 or 2 (we know that the 3^∞-Selmer rank and the analytic rank have the same parity by the 3-parity conjecture proven by Monsky in this case or by the Dokchitsers more generally [16]) with both cases occurring very often. Here, Rodriguez Villegas and the second author [32] were able to give exact formulas determining exactly whether $L(E_p, 1)$ vanishes, and hence resolving the remaining open case if the BSD conjecture is assumed.

Beyond number theory, elliptic curves are indispensable in cryptography, particularly in elliptic curve cryptography (ECC). With rapid advances in quantum computing, the demand for a deeper understanding of their arithmetic is at the forefront of digital security.

6. Number theory, computation, and applications of the BSD conjecture

The BSD conjecture represents a remarkable shift in the philosophy of modern number theory, not only through its deep mathematical implications but also in its influence on the mindset of pure mathematicians. It exemplifies the interplay between theoretical insight and empirical evidence, reinforcing the idea that numerical experimentation can reveal profound underlying structures.

The reliance on computation as a research tool in mathematics is hardly new—already the Babylonians recorded examples of Pythagorean triples, and Gauss engaged in large-scale computations, e.g., of class numbers of negative discriminants, sometimes with the help of the human computer Zacharias Dase. Gauss was also an extraordinarily fast human computer himself. At the age of 16, he amused himself by often counting all primes within a "chiliad" (a block of 1000 consecutive integers), leading him to conjecture the Prime Number Theorem, perhaps the first deep theorem in number theory discovered by numerical computations. By the early 20th century, mechanical, then electric, and eventually electronic calculators became integral to mathematical research. A striking example is the work of D. H. Lehmer, who pioneered computational techniques in number theory, using devices such as the Lehmer sieve and early electronic computers like SWAC to study prime numbers and factor large integers.

Yet, the scale and sophistication of modern computations have fundamentally transformed mathematical methodology. The BSD conjecture has shaped the way mathematicians approach problems, reinforcing the idea that computation is not merely a tool for verification but an essential component of mathematical discovery itself. Over the past 50 years, numerical computation and the development of powerful mathematical software such as GP/PARI, SageMath and Magma have led to many significant discoveries in all parts of mathematics, and especially in number theory. Notable examples include refinements of the Cohen–Lenstra conjectures on the distribution of class groups, counterexamples to long-standing conjectures such as Mertens's conjecture (disproven using the LLL algorithm), and breakthroughs in prime number distribution, including Yitang Zhang's 2013 result on bounded gaps between primes, which led to further computational refinements by the Polymath project, reducing the bound from 70 million to 246 and later to 6 in specific settings.

Computational verification has also paved the way for a related but distinct development: machine-assisted proofs. One of the earliest and most famous examples is the proof of the four-color theorem by Appel and Haken (1976), which relied on exhaustive computer checks of reducible configurations. More recently, a collaborative effort led in 2007 to the successful "mapping" of the E_8 Lie group, that is, the explicit computation of its character table, which describes its fundamental structure and symmetries. This monumental task involved a 248-dimensional algebraic structure, resulting in a matrix with 453,060 rows and columns. If printed on paper, the calculations would cover an area comparable to Manhattan.

A related, but again different, development is the machine verification and formalization of mathematical proofs using "proof assistants" like Coq, Isabelle, and Lean. These tools, often enhanced by AI algorithms, help mathematicians ensure the correctness of intricate proofs by checking each logical step systematically. The proof of the four-color theorem was later simplified and made fully verifiable by computer, and it has been formalized in Coq. Additionally, a project to formalize the proof of Fermat's Last Theorem in Lean has been announced.

More generally, AI, particularly machine learning (ML), is increasingly being applied to mathematical research, including the study of invariants of elliptic curves. ML techniques have been applied to predict ranks, identify statistical behaviors, and even uncover new patterns by analyzing vast numerical datasets. A striking example of an ML-driven discovery related to elliptic curves and the BSD conjecture emerged in 2021, when He, Lee, Oliver, and Pozdnyakov unexpectedly found a curious fluctuation in average Frobenius traces a_p of elliptic curves within a given conductor interval depending on the rank, a phenomenon now known as murmuration. (The term is inspired by the visual similarity of the plotted a_p values to the coordinated yet unpredictable movements of large flocks of birds.) While this behavior remains incompletely understood, further insights have been provided by the works of Sutherland and Zubrilina. In particular, using a trace formula for Hecke operators, Zubrilina demonstrated that if we replace the average over elliptic curves of conductor N with an average over all weight 2 cuspidal Hecke eigenforms of level N, the resulting values yield a provable curve matching well with the experimental data for elliptic curves. Stated differently, the striking behavior observed for elliptic curves is nothing other than a provable property of all cusp forms of a given type: the Fourier coefficients of the L-series of an elliptic curve, appropriately averaged, resemble those of a randomly chosen Hecke eigenform of the same weight and level. What initially appeared to be a deep structural property of elliptic curves turns out to be, in fact, a manifestation of randomness!

As AI continues to reshape the mathematical landscape, the boundary between human and machine-generated mathematics may become increasingly subtle. AI is evolving beyond a mere computational assistant, emerging as an active collaborator in mathematical research. One might ask whether, with more advanced ML techniques and vast datasets of elliptic curves, AI might not independently detect the correlations between ranks and L-values and thus rediscover some form of the BSD conjecture, and maybe even find ideas towards a proof.

Let us close on a human note. Paper II, in which Birch and Swinnerton-Dyer formulated the BSD conjecture, was dedicated to *Gina*—who was a colleague of Swinnerton-Dyer, had assisted in an essential way with their computations, and had married Birch. We have seen a glimpse of the many people involved in the BSD conjecture. The list goes on, and it is impossible to name them all. Their collective efforts continue to shape our understanding of the mysteries surrounding the BSD conjecture and inspire young people to pursue the field. In the end, it is human ideas, stories, and lives which have driven and continue to advance the progress of mathematics.

Acknowledgement. The authors would like to thank Manjul Bhargava, Mike Daas, and Gerard van der Geer for their kind and encouraging comments on this article, as well as for their many helpful suggestions that have improved both its content and presentation. The authors are also sincerely grateful to the editorial team for their attentive oversight and helpful guidance throughout the refereeing process, and to the three exceptionally careful referees for their detailed reports and valuable remarks. The first author was supported by JSPS KAKENHI Grant JP25K17227.

References

[1] *M. Bertolini, H. Darmon and K. Prasanna,* Generalized Heegner cycles and p-adic Rankin L-series, Duke Math. J. **162** no. 6 (2013), 1033–1148.

[2] *M. Bhargava and C. Skinner,* A positive proportion of elliptic curves over $\mathbb{Q}$ have rank one, J. Ramanujan Math. Soc. **29** (2014), 221–242.

[3] *M. Bhargava and A. Shankar,* Ternary cubic forms having bounded invariants, and the existence of a positive proportion of elliptic curves having rank 0, Ann. Math. **181** no. 2 (2015), 587–621.

[4] *M. Bhargava and A. Shankar,* The average size of the 5-Selmer group of elliptic curves is 6, and the average rank is less than 1, Preprint.

[5] *M. Bhargava, C. Skinner and W. Zhang,* A majority of elliptic curves over $\mathbb{Q}$ satisfy the Birch and Swinnerton-Dyer conjecture, Preprint.

[6] *B. J. Birch and H. P. F. Swinnerton-Dyer,* Notes on elliptic curves. I, J. reine angew. Math. **212** (1963), 7–25.

[7] *B. J. Birch and H. P. F. Swinnerton-Dyer,* Notes on elliptic curves. II, J. reine angew. Math. **218** (1965), 79–108.

[8] *S. Bloch,* A note on height pairings, Tamagawa numbers, and the Birch and Swinnerton-Dyer conjecture, Invent. Math. **58** (1980), 65–76.

[9] *C. Breuil, B. Conrad, F. Diamond and R. Taylor,* On the modularity of elliptic curves over $\mathbb{Q}$: wild 3-adic exercises, J. Amer. Math. Soc. **14** no. 4 (2001), 843–939.

[10] *J. Buhler, B. Gross and D. Zagier,* On the conjecture of Birch and Swinnerton-Dyer for an elliptic curve of rank 3, Math. Comp. **44** no. 170 (1985), 473–481.

[11] *A. Burungale and M. Flach,* The conjecture of Birch and Swinnerton-Dyer for certain elliptic curves with complex multiplication, Cambridge J. Math. **12** (2024), 357–415.

[12] *J. Cassels,* Arithmetic on curves of genus 1. IV. Proof of the Hauptvermutung, J. reine angew. Math. **211** (1962), 95–112.

[13] *J. Coates and Y.-X. Li,* Non-vanishing theorems for central L-values of some elliptic curves with complex multiplication, Proc. Lond. Math. Soc. (3) **121** no. 6 (2020), 1531–1578.

[14] *J. Coates and A. Wiles,* On the conjecture of Birch and Swinnerton-Dyer, Invent. Math. **39** (1977), 223–251.

[15] *J. E. Cremona,* Algorithms for modular elliptic curves, 2nd ed., Cambridge Univ. Press, Cambridge, 1997.

[16] *T. Dokchitser and V. Dokchitser,* On the Birch–Swinnerton-Dyer quotients modulo squares, Ann. Math. **172** (2010), 567–596.

[17] *D. Goldfeld,* Sur les produits eulériens attachés aux courbes elliptiques, C. R. Acad. Sci. Paris Sér. I Math. **294** (1982), 471–474.

[18] *R. Greenberg*, Iwasawa theory for elliptic curves, in Arithmetic Theory of Elliptic Curves (Cetraro, 1997), Lecture Notes in Math. **1716**, Springer, 1999, 51–144.
[19] *B. Gross and D. Zagier*, Heegner points and derivatives of L-series, Invent. Math. **84** no. 2 (1986), 225–320.
[20] *K. Kato*, p-adic Hodge theory and values of zeta functions of modular forms, Astérisque **295** (2004), 117–290.
[21] *K. Kato and F. Trihan*, On the conjectures of Birch and Swinnerton-Dyer in characteristic $p > 0$, Invent. Math. **153** (2003), 537–592.
[22] *Y. Kezuka*, Tamagawa number divisibility of central L-values of twists of the Fermat elliptic curve, J. Théor. Nombres Bordeaux **33** no. 3.2 (2021), 945–970.
[23] *Y. Kezuka and Y.-X. Li*, A classical family of elliptic curves having rank one and the 2-primary part of their Tate–Shafarevich group non-trivial, Doc. Math. **25** (2020), 2115–2147.
[24] *Y. Kezuka and Y.-X. Li*, Non-vanishing of central L-values of the Gross family of elliptic curves, J. Lond. Math. Soc. **111** no. 6 (2025).
[25] *Y. Kezuka and D. Zagier*, A Birch–Swinnerton-Dyer dichotomy, in Moduli spaces and modular forms (G. Chenevier, G. Farkas and G. van der Geer, eds.), EMS Press, to appear.
[26] *V. A. Kolyvagin*, Finiteness of $E(\mathbb{Q})$ and $\text{Ш}(E, \mathbb{Q})$ for a subclass of Weil curves, Math. USSR Izvestiya **32** no. 3 (1989).
[27] *B. Mazur*, Rational points of abelian varieties with values in towers of number fields, Invent. Math. **18** (1972), 183–266.
[28] *B. Mazur, J. Tate and J. Teitelbaum*, On p-adic analogues of the conjecture of Birch and Swinnerton-Dyer, Invent. Math. **84** (1986), 1–49.
[29] *P. Monsky*, Mock Heegner points and congruent numbers, Math. Z. **204** (1990), 45–68.
[30] *J. Park, B. Poonen, J. Voight and M. M. Wood*, A heuristic for boundedness of ranks of elliptic curves, J. Eur. Math. Soc. **21** no. 9 (2019), 2859–2903.
[31] *B. Perrin-Riou*, Points de Heegner et dérivées de fonctions L p-adiques, Invent. Math. **89** (1987), 455–510.
[32] *F. Rodriguez Villegas and D. Zagier*, Which primes are sums of two cubes?, in Number Theory (Halifax, NS, 1994), CMS Conf. Proc. **15**, Amer. Math. Soc., Providence, RI (1995), 295–306.
[33] *K. Rubin*, Tate–Shafarevich groups and L-functions of elliptic curves with complex multiplication, Invent. Math. **89** no. 3 (1987), 527–559.
[34] *K. Rubin*, The "main conjectures" of Iwasawa theory for imaginary quadratic fields, Invent. Math. **103** no. 1 (1991), 25–68.
[35] *N. Schappacher*, Développement de la loi de groupe sur une cubique, Séminaire de Théorie des Nombres, Paris 1988/89, Progress in Math. **91**, Birkhäuser, 1991, 159–184.
[36] *E. S. Selmer*, The diophantine equation $X^3 + Y^3 = DZ^3$, Acta Math. **85** (1951), 203–362.
[37] *C. Skinner and E. Urban*, The Iwasawa Main Conjecture for GL_2, Invent. Math. **195** no. 1 (2014), 1–277.
[38] *J. Tate*, The arithmetic of elliptic curves, Invent. Math. **23** (1974), 179–206.
[39] *J. Tate*, On the conjectures of Birch and Swinnerton-Dyer and a geometric analog, Séminaire Bourbaki **1965/66**, Exp. 306.
[40] *R. Taylor and A. Wiles*, Ring-theoretic properties of certain Hecke algebras, Ann. Math. **141** no. 3 (1995), 553–572.
[41] *J. B. Tunnell*, A classical Diophantine problem and modular forms of weight 3/2, Invent. Math. **72** no. 2 (1983), 323–334.
[42] *A. Weil*, Adèles and Algebraic Groups, Progress in Math. **23**, Birkhäuser, 1982.
[43] *A. Wiles*, Modular elliptic curves and Fermat's last theorem, Ann. Math. **141** no. 3 (1995), 443–551.
[44] *D. Zagier*, The Birch–Swinnerton-Dyer conjecture from a naive point of view, in Arithmetic Algebraic Geometry, Progress in Math. **89** (1990), 377–389.
[45] *D. Zagier and G. Kramarz*, Numerical investigations related to the L-series of certain elliptic curves, J. Indian Math. Soc. (N.S.) **52** (1987), 51–69.

Yukako Kezuka, Graduate School of Mathematical Sciences, The University of Tokyo
3-8-1 Komaba, Meguro-ku, Tokyo 153-8914, Japan
e-mail: kezuka@ms.u-tokyo.ac.jp

Don Zagier, Max Planck Institute for Mathematics, Vivatsgasse 7, 53111, Bonn, Germany
and the International Centre for Theoretical Physics, Strada Costiera, 11, 34151, Trieste, Italy
e-mail: don.zagier@mpim-bonn.mpg.de

1976–2000

J. reine angew. Math. **398** (1989), 169–179
DOI 10.1515/crll.1989.398.169

Journal für die reine und angewandte Mathematik

Spectral curves and the generalised theta divisor

By *Arnaud Beauville* at Paris, *M.S. Narasimhan* at Bombay and *S. Ramanan* at Bombay

1. Statement of results

Let X be a smooth, irreducible, projective curve over an algebraically closed field of characteristic 0 and let $g = g_X \geq 2$ be its genus. We show in this paper that a generic vector bundle on X of rank n and any fixed degree can be obtained as the direct image of a line bundle on an n-sheeted (ramified) covering of X. From this we deduce results concerning linear systems on the moduli space of vector bundles of rank n and degree 0.

Let $\mathcal{U}_X(n,d)$, or simply $\mathcal{U}(n,d)$ (resp. $\mathcal{SU}(n,\zeta)$) denote the moduli space of semistable vector bundles of rank n and degree d (resp. with determinant isomorphic to a fixed line bundle ζ) on X. When ζ is the trivial line bundle, we simply write $\mathcal{SU}(n)$ for $\mathcal{SU}(n,\zeta)$. When we deal with $\mathcal{SU}(n,\zeta)$, but the particular ζ that we fix is not of importance, we denote it by $\mathcal{SU}(n,d)$. Let Y be any smooth curve and $\pi\colon Y \to X$ a finite morphism. We can then associate to any line bundle ξ on Y a vector bundle, namely the direct image $\pi_*(\xi)$, on X. This yields a morphism of the open set $\mathcal{T}_Y^\delta$ of J_Y^δ into $\mathcal{U}(n,d)$, where (a) J_Y^δ is the space of line bundles of degree δ on Y, (b) $\mathcal{T}_Y^\delta$ is the set of those ξ for which $\pi_*(\xi)$ is semistable, and (c) d and δ are connected by the relation $d = \delta + \deg(\pi_*(\mathcal{O}_Y))$.

Theorem 1. *For any pair of integers n and d, there exists an n-sheeted covering $\pi\colon Y \to X$ with Y smooth and irreducible, such that the morphism $\pi_*\colon \mathcal{T}_Y^\delta \to \mathcal{U}_X(n,d)$ is dominant with $\delta = d - \deg \pi_*(\mathcal{O}_Y)$.*

Before stating the next result, we need to introduce the notion of the *generalised theta divisor.* It is easy to define a divisor Θ in $\mathcal{U}(n,d)$ with $d = n(g-1)$ supported on the set of bundles E with $\Gamma(E) \neq 0$. This defines an invertible sheaf [2] on $\mathcal{U}(n, n(g-1))$, which we may denote by $\mathcal{O}(\Theta)$ In the case $n = 1$, this divisor is the wellknown Riemann theta divisor on the variety J^{g-1}.

Theorem 2. *We have dim* $\Gamma(\mathcal{U}(n,n(g-1)), \mathcal{O}(\Theta)) = 1$.

This theorem is proved by considering the dominant morphism $\pi_*\colon \mathcal{T}_Y^\delta \to \mathcal{U}(n,d)$ whose existence is asserted in Theorem 1, with $d = n(g-1)$. One can determine the pullback of Θ

Second author was supported by DFG.

by the morphism π_*. In fact this turns out to be the Riemann theta divisor. Now one shows that the complement of $\mathcal{T}_Y^\delta$ in J_Y^δ is of codimension at least 2 so that we get an injection of $\Gamma(\mathcal{U}(n,d),\Theta)$ into $\Gamma(T_Y^\delta,\Theta)=\Gamma(J_y,\Theta)$, proving our assertion.

There is a close relationship [9] between the moduli space $\mathcal{SU}(n)$ and the linear System of the invertible sheaf $\mathcal{O}(n\Theta)$ on J^{g-1}. Indeed, associate to a general vector bundle E, the divisor D_E given by $\{\xi\in J^{g-1}:\Gamma(E\otimes\xi)\neq 0\}$. This gives a rational map of $\mathcal{SU}(n)$ into the linear system of $n\Theta$ on J^{g-1}. On the other hand, let ξ be an element of J^{g-1} and let $\Delta_\xi=\{E\in\mathcal{SU}(n):\Gamma(E\otimes\xi)\neq 0\}$. One can show [2] that Δ_ξ defines a Cartier divisor and the corresponding line bundle h generates $\mathrm{Pic}(\mathcal{SU}(n))$

Theorem 3. *The (rational) map, which associates to a general point e of $\mathcal{SU}(n)$ the divisor D_E in J^{g-1} where E is a semistable bundle in the class of e, induces an isomorphism of $\Gamma(J^{g-1},\mathcal{O}(n\Theta))$ with $\Gamma(\mathcal{SU}(n),h)^*$, and the rational map is defined by this linear system. In particular, $\Gamma(\mathcal{SU}(n),h)$ has dimension n^g.*

The proof of this theorem is similar to that of Theorem 2, except that the Jacobian of Y is in this case replaced by the Prym variety associated to the covering.

In the case of vector bundles of rank 2, these results have been proved by A. Beauville [1]. In the rank n case, vector bundles arising from line bundles on an étale covering have been studied by Narasimhan and Ramanan [8]. Ideas of Hitchin [4], and the existence of 'very stable' vector bundles proved by Drinfeld and Laumon [5] have been useful to us.

2. Preliminaries

2.1. Twisted endomorphisms. Let E be a vector bundle of rank n, L a line bundle on X and $\varphi\colon E\to L\otimes E$ a homomorphism. Then one can define its trace as an element of $\Gamma(L)$ by interpreting it to be a map $E\otimes E^*\to L$ and taking the image of the identity section of $E\otimes E^*$. More generally, its characteristic coefficients $a_i\in\Gamma(X,L^i)$, for $0\leq i\leq n$, may be defined by setting $a_i=(-1)^i\mathrm{Tr}\,\Lambda^i\varphi$. The Cayley-Hamilton theorem then asserts that φ satisfies its characteristic equation. This means that $\sum a_i\varphi^{n-i}$, interpreted as a homomorphism $E\to L^n\otimes E$, is zero. In particular, if all a_i are zero for $1\leq i\leq n$, then $\varphi^n=0$, and we may say that φ is *nilpotent.*

2.2. Generalised theta divisor. Let $\mathcal{U}(n,n(g-1))$ be the moduli space of vector bundles of rank n and degree $n(g-1))$. Then by the Riemann-Roch theorem, we see that $\chi(E)=0$ for all E in $\mathcal{U}(n,n(g-1))$. Let $\{E_t\}_{t\in T}$ be a family of vector bundles, given by a vector bundle E on $T\times X$, with $E_t=E|_{t\times X}$. semistable for all $t\in T$. Let $p\colon T\times X\to T$ denote the first projection. The first direct image R^1p_*E is wellbehaved under base change and is supported on the set $\{t\in T:\Gamma(E_t)\neq 0\}$. Then $(\det R^1p_*E)\otimes(\det p_*E)^{-1}$ is defined to be the invertible sheaf $\mathcal{O}(\Theta_T)$. It can be shown [2] that there exists an invertible sheaf $\mathcal{O}(\Theta)$ on $\mathcal{U}(n,d)$ such that $\mathcal{O}(\Theta_T)$ is the pullback of $\mathcal{O}(\Theta)$ by the canonical classifying morphism $T\to\mathcal{U}(n,d)$.

2.3. Remarks on principally polarised Abelian varieties. Suppose (A,Θ) is a principally polarised Abelian variety and N an Abelian subvariety of A. Consider the subgroup P of A consisting of all $a\in A$ such that $T_a^*\mathcal{O}(\Theta)$ and $\mathcal{O}(\Theta)$ restrict to isomorphic line bundles on N. We call P the *orthogonal* of N. It is an Abelian subvariety of A such that the maps $P \to A/N$ and $N \to A/P$ are isogenies. Let $\mathcal{O}(\Theta_N)$ and $\mathcal{O}(\Theta_P)$ be the restrictions of $\mathcal{O}(\Theta)$ to N and P respectively. One can then define a rational map ψ of P into $|\Theta_N|$ by setting $\psi(p) = T_p^*(\Theta)_N$. On the other hand, we also have a rational map $\varphi\colon P \to |\Theta_P|^*$, defined by the linear system $|\Theta_P|$.

2.4. Proposition. *There is a canonical isomorphism* $\iota\colon |\Theta_P|^* \to |\Theta_N|$ *such that* $\iota \circ \varphi = \psi$.

Proof. Consider the isogeny $\pi\colon N \times P \to A$ sending (n, p) to $n + p$. The pullback $\pi^*\mathcal{O}(\Theta)$ is isomorphic to $p_N^*\mathcal{O}(\Theta_N) \otimes p_P^*\mathcal{O}(\Theta_P)$, by the theorem of the square. The canonical section of $\mathcal{O}(\Theta)$ pulls back to an element of $\Gamma(N, \mathcal{O}(\Theta_N)) \otimes \Gamma(P, \mathcal{O}(\Theta_P))$. One can now check that the induced map $\iota\colon \Gamma(P, \mathcal{O}(\Theta_P))^* \to \Gamma(N, \mathcal{O}(\Theta_N))$ is an isomorphism. This follows from the fact that the theta group of $\pi^*\mathcal{O}(\Theta)$ is the natural pushout of the direct product of the theta groups of $\mathcal{O}(\Theta_N)$ and $\mathcal{O}(\Theta_P)$ with respect to the imbedding of the multiplicative group $\mathbf{G}_m$ by $\lambda \mapsto (\lambda, \lambda^{-1})$ and that the irreducible theta action is given by that on $\Gamma(N, \mathcal{O}(\Theta_N)) \otimes \Gamma(P, \mathcal{O}(\Theta_P))$ It is easy to identify the unique 1-dimensional subspace fixed by the maximal level subgroup $\mathrm{Ker}(\pi)$, and see that it gives a nondegenerate element as claimed. The rational map φ of the variety P into $|\Theta_P|^*$ is characterised by its equivariance under the action of the theta group. Now one can check that $\iota^{-1} \circ \psi$ does have this property, thus proving the equality $\iota \circ \varphi = \psi$.

2.5. Polarisation on the Prym variety of a covering.. Let $\pi\colon Y \to X$ be a finite morphism of smooth, irreducible curves. Let us denote by b the line bundle $(\pi_*(\mathcal{O}_y))^{-1}$. The relative duality theorem gives an isomorphism of

$$\pi_*(\mathcal{O}_Y)^* \text{ with } \pi_*(K_Y \otimes \pi^* K_X^{-1}).$$

The ramification divisor Δ is given by a section of $\pi^* K_X^{-1} \otimes K_{Y'}$ so that we have $\mathfrak{d} = \det \pi_*(\mathcal{O}(\Delta))$. For any line bundle L on Y, we may associate the sheaf $\pi_*(L)$ on X. It is actually a vector bundle of rank n if π is n-sheeted. Then one can show easily that $\det \pi_*(L) = \mathrm{Nm}(L) \otimes \mathfrak{d}^{-1}$, where Nm is the norm map associated to the covering π. In particular, $\det \pi_*(\mathcal{O}(\Delta)) = \mathrm{Nm}(\mathcal{O}(\Delta)) \otimes \mathfrak{d}^{-1}$ and hence $\mathcal{O}(\Delta) = \mathfrak{d}^2$.

Suppose that the induced homomorphism π^* of J_X into J_Y is *injective*. Then we may set $N = \pi^* J_X$ and $A = J_Y$ and apply the considerations of 2.3 and 2.4. The norm homomorphism $\mathrm{Nm}\colon J_Y \to J_X$ can be identified with the transpose of π^* and consequently has connected kernel [7]. This is the Prym variety of the map π and we would like to take for P this kernel. However, in order to avoid making too many choices, we will consider the varieties $N' = \pi^* J_X^{g-1}$, $A' = J_Y^{h-1}$ and $P' = \mathrm{Nm}^{-1}(\mathfrak{d})$, where $g = g_X$, $h = g_Y$ and Nm is the norm map $J_Y^{\deg(\mathfrak{d})} \to J_X^{\deg(\mathfrak{d})}$. Then we have a natural addition map $N' \times P' \to A'$ (namely tensor product). Up to translation, this is the same as the map defined in 2.3. Hence the pullback of the line bundle $\mathcal{O}(\Theta_Y)$ by this map is the tensor product of a line bundle on N' and one on P'. In fact, under our assumption, N' is isomorphic to J_X^{g-1} on which is defined Riemann's theta divisor Θ_X. Now it is easy to deduce that the restriction of $\mathcal{O}(\Theta_Y)$ to $N' \times \{\alpha\}$ is isomorphic to $\mathcal{O}(n\Theta)$ for any $\alpha \in P'$. Hence there is a natural line bundle τ on P' such that the pullback mentioned above is $p_1^*\mathcal{O}(n\Theta_X) \otimes p_2^*\tau$. Now the pullback of the natural section of $\mathcal{O}(\Theta_Y)$ gives rise, in view of our remarks above, to a nondegenerate element of $\Gamma(P', \tau) \otimes \Gamma(J_X^{g-1}, \mathcal{O}(n\Theta))$, that is to say there is a natural isomorphism of $\Gamma(P', \tau)^*$ with $\Gamma(J_X^{g-1}, \mathcal{O}(n\Theta))$. Thus we have

2.6. Proposition. *Let $\pi\colon Y \to X$ be a finite morphism of projective nonsingular curves such that the induced morphism $\pi^*\colon J_X \to J_Y$ is injective. Then with the notation above, there is a natural isomorphism of $\Gamma(J_X^{g-1}, \mathcal{O}(n\Theta))$ with $\Gamma(P', \tau)^*$ such that the following diagram is commutative.*

$$\begin{array}{ccc} P' & \longrightarrow & \mathbb{P}\Gamma(P',\tau)^* \\ & \searrow & \downarrow \cong \\ & & \mathbb{P}\Gamma(\mathcal{J}_X^{g-1}, \mathcal{O}(n\Theta)) \end{array}$$

Here the rational map $P' \to \mathbb{P}\Gamma(P',\tau)^*$ is given by the linear system of τ, while the rational map $P' \to \mathbb{P}\Gamma(J_X^{g-1}, \mathcal{O}(n\Theta))$ is given by $p \mapsto T_p^*\Theta_Y \cap \pi^* J_X^{g-1}$.

2.7. Remark. One can show in fact that under the assumption that $\pi^*\colon J_X \to J_Y$ is injective, the Polarisation on P (and on P' as well) is of type $(1, 1, \ldots, n, n, \ldots)$ where the n's are repeated g times. In particular its Pfaffian is n^g which is of course the dimension of $\Gamma(J_X^{g-1}, \mathcal{O}(n\Theta))$. It is easy to show that τ is a primitive element in the group of algebraic equivalence classes.

3. Spectral curves

Spectral curves have been treated in recent years by several authors, in connection with some special kinds of nonlinear differential equations. In our context it was used by Hitchin [4]. Let L be a line bundle on X and $s = (s_k)$ be sections of L^k for $k = 1, 2, \ldots, n$. Then we will construct a scheme X_s and a finite morphism $\pi\colon X_s \to X$. Let $p\colon P = \mathbb{P}(\mathcal{O} \oplus L^*) \to X$ be the natural projection and $\mathcal{O}(1)$ the relatively ample bundle, or what may be called "the hyperplane bundle along the fibres". Then $p_*(\mathcal{O}(1))$ is naturally isomorphic to $\mathcal{O} \oplus L^*$ which has a canonical section, namely, the constant section 1 of $\mathcal{O}$. This gives a section of $\mathcal{O}(1)$ over P which we denote by y. On the other hand, $p_*(p^*L \otimes \mathcal{O}(1))$ is isomorphic, by the projection formula, to $L \otimes (\mathcal{O} \oplus L^*) = L \oplus \mathcal{O}$. Hence it also has a canonical section and we denote the corresponding section of $p^*L \otimes \mathcal{O}(1)$ by x. Note that the two sections x and y have as their zero schemes the natural subvarieties $\mathbb{P}(\mathcal{O})$ and $\mathbb{P}(L^*)$ of P. Now consider the section

$$x^n + (p^*s_1) \cdot y \cdot x^{n-1} + \cdots + (p^*s_n) \cdot y^n$$

of $p^*L^n \otimes \mathcal{O}(n)$. Its zero scheme is the scheme X_s we have in mind. It is clear that the restriction π of p to X_s is finite and that at any point v of X the fibre of π is the subscheme of $\mathbb{P}^1$ given by

$$x^n + a_1 y x^{n-1} + \cdots + a_n y^n = 0,$$

where (x, y) is a homogeneous coordinate system, and a_i is the value of s_i at v on identifying the fibre of L at P with the residue field at v.

3.1. Remark. It is easily seen that the set of all $(s) \in \Gamma(L) \oplus \Gamma(L^2) \oplus \cdots \oplus \Gamma(L^n)$ for which the scheme X_s is integral (i.e. reduced and irreducible), is open. Also, it is easy to see that it is nonempty, whenever L^n admits a section whose divisor is not of the form mD for some integer m dividing n. For, in that case we can take all s_i for $i \leq n-1$ to be 0 and s_n to be a section as postulated.

Hereafter we will assume that (s) has been so chosen that X_s is integral. Since the zeros of x and y are disjoint in P, it follows that the restriction of y to X_s is everywhere nonzero and hence the restriction of $\mathcal{O}(1)$ to X_s is trivial. The restriction of x to X_s can therefore be considered as a section of $\pi^*(L)$. An alternative description of X_s is given as follows. The direct image $\pi_*(\mathcal{O})$ is naturally isomorphic to $\mathcal{O} \oplus L^{-1} \oplus \cdots \oplus L^{-(n-1)}$ Its algebra structure is better understood by interpreting this to be Sym $(L^{-1})/\mathcal{J}$ where $\mathcal{J}$ is the ideal sheaf generated by the image of the homomorphism $u \colon L^n \to \mathrm{Sym}(L^{-1})$ given as the sum of the imbeddings $L^{-n} \to L^{-(n-i)}$ defined by s_i. In terms of this sheaf of algebras, X_s is simply Spec $(\mathrm{Sym}(L^{-1}/\mathcal{J})$.

3.2. Remark. The genus g' of X_s may be computed from the above Interpretation. In fact, $1-g'$ is given by

$$\begin{aligned}\chi(X_s, \mathcal{O}) = \chi(X, \pi_*(\mathcal{O})) &= \textstyle\sum \chi(X, L^i) \\ &= (\deg L) \cdot (0 - 1 - \cdots - (n-1)) + n(1-g) \\ &= -(\deg L) \cdot n(n-1)/2 + n(1-g)\end{aligned}$$

and hence $g' = (\deg L) \cdot n(n-1)/2 + n(g-1) + 1$.

3.3. Remark. Assume that X_s is integral. Then consider the resultant of the polynomial

$$x^n + s_1 y x^{n-1} + s_2 y^2 x^{n-2} + \cdots + s_n y^n,$$

and its derivative

$$n x^{n-1} + (n-1) s_1 y x^{n-2} + \cdots + s_{n-1} y^{n-1}.$$

This can be interpreted to be a section of $(p^* L \otimes \mathcal{O}(1))^{n(n-1)/2}$. One easily verifies that the map $\pi \colon X_s \to X$ is étale at a point if and only if this resultant does not vanish at that point. In other words, the resultant gives the ramification divisor of the map π.

3.4. Example. Let L be a line bundle with L^n isomorphic to $\mathcal{O}$. We could then take s_i to be 0 for $i \leq n-1$, and s_n to be any nonzero section of L^n. In this case one can conclude from 3.3 that the map π is actually étale. However, the scheme X_s is connected (and therefore irreducible) if and only if the section s_n is not a power of a section of L^r for some r which divides n. This would happen if and only if L^r is trivial. In other words, X_s is irreducible if and only if L is of order n in the Jacobian of X.

3.5. Remark. Let us assume that X_s is integral. Then using the Jacobian criterion it is not difficult to prove that X_s is smooth if and only if at every point which is a multiple zero of the section s_n, the section s_{n-1} is nonzero. In particular, the set of sections (s) such that the scheme X_s is smooth, is open on the one hand, and is nonempty if L^n admits a section without multiple zeros, on the other.

3.6. Proposition. *Let X be a curve and L any line bundle on it. Let $s = (s_i)$ be a set of sections of L^i for $1 \leq i \leq n$. Assume that the corresponding scheme X_s is integral. Then there is a bijective correspondence between isomorphism classes of torsion free sheaves of rank 1 on X_s and isomorphism classes of pairs (E, φ) where E is a vector bundle of rank n and $\varphi\colon E \to L \otimes E$ a homomorphism with characteristic coefficients s_i. The correspondence is given by associating to any line bundle M on X_s the sheaf $\pi_*(M)$ on X and the natural isomorphism*

$$\pi_*(M) \to L \otimes \pi_*(M) \approx \pi_*(\pi^* L \otimes M)$$

given by the section x of $\tau^*(L)$.

Proof. If M is a torsion free sheaf of rank 1 on X_s, then the sheaf $\pi_* M$ is a vector bundle of rank n on X, endowed with a $\pi_*(\mathcal{O}_{X_s})$-module structure. Conversely, since π is affine, a vector bundle E of rank n with a $\pi_*\mathcal{O}_{X_s}$-module structure defines a sheaf M with $\pi_*(M)$ isomorphic to E. Now M is torsion free of rank 1 because X_s is integral.

The $\mathcal{O}_X$-algebra $\pi_*\mathcal{O}_{X_s}$ is isomorphic to $\mathrm{Sym}(L^{-1})/\mathcal{J}$, so that the data of a $\pi_*(\mathcal{O}_{X_s})$-module structure on E is equivalent to an algebra homomorphism

$$\mathrm{Sym}(L^{-1})/\mathcal{J} \to \mathcal{E}nd(E),$$

that is to say an $\mathcal{O}_X$-homomorphism $\varphi\colon E \to L \otimes E$ such that $P_s(\varphi) = 0$, where P_s is the polynomial determined by (s). Since X_s is integral, P_s is irreducible over the function field of X, and is hence the characteristic polynomial of φ. Conversely, if E is a vector bundle of rank n on X, and $\varphi\colon E \to L \otimes E$ a linear homomorphism with characteristic coefficients s_i, then $P_s(\varphi) = 0$ by the Cayley-Hamilton theorem. This proves the proposition.

3.7. Remark. The sequence

$$0 \longrightarrow M(-\Delta) \longrightarrow \pi^* E \xrightarrow{\pi^*\varphi - x} \pi^*(L \otimes E) \longrightarrow \pi^* L \otimes M \longrightarrow 0$$

is exact. In fact one could actually define the line bundle M corresponding to (E, φ) by this exact sequence. This is essentially the point of view in Hitchin [4].

3.8. Remark. When X_s is nonsingular, then 'torsion free sheaves of rank 1' may be replaced by 'line bundles' in Proposition 3.6.

3.9. Example. Let us get back to Example 3.5 If L is a line bundle of order n in the Jacobian of X, then according to Proposition 3.6, there is a natural bijection between line bundles on X_s and pairs (E, φ). Now there exists a map $\varphi\colon E \to L \otimes E$ with the given characteristic polynomial if and only if $L \otimes E$ is isomorphic to E. In this particular case this map was studied and indeed many of the computations in this paper were carried out in [8],

3.10. Remark. If M is a nontrivial line bundle on X, and $\pi^* M$ is trivial on X_s, then it follows that $\pi_*(\pi^* M) \cong M \otimes \pi_*(\mathcal{O}) \cong \sum M \otimes L^{-i}$ admits a nontrivial section. In particular this implies that the degree of L is 0. But since $s_i \in \Gamma(L^i)$, we must have for each i, either $s_i = 0$, or L^i is trivial Thus we see that except in this case (which includes Example 3.9), the induced map $\pi^*\colon J_X \to J_{X_s}$ is injective.

4. Applications to the moduli space of vector bundles

In what follows, we will take for the line bundle L the canonical line bundle K of a smooth, projective curve X of genus $g \geq 2$. Denoting by W the direct sum $\bigoplus_{k=1}^{k=n} \Gamma(X, K^k)$ we see immediately from the Riemann-Roch theorem that

$$\dim W = n^2(g-1) + 1,$$

which is also the dimension of $\mathcal{SU}(n,d)$ *Let* $s = (s_i)$ be an element of W such that the corresponding curve X_s is smooth and irreducible. By Remark 3.2, the genus of X_s is $n^2(g-1)+1$. Moreover, for any line bundle ξ on X_s we have

$$\det \pi_*(\xi) = \mathrm{Nm}(\xi) \otimes \det \pi_*(\mathcal{O}) = \mathrm{Nm}(\xi) \otimes K^{-n(n-1)/2}$$

where $\mathrm{Nm}\colon \mathrm{Pic}(X_s) \to \mathrm{Pic}(X)$ is the norm map associated to π. In particular, the degree of $\pi_*(\xi)$ is $\deg(\xi) - n(n-1)(g-1)$.

Let $\mathcal{U}$ be the moduli space of stable vector bundles of rank n and degree $d = \deg(\xi) - n(n-1)(g-1)$. Let $\Omega = \Omega^1$ denote its cotangent bundle. If E is a stable bundle of rank n and degree d, then the tangent space to $\mathcal{U}$ at the corresponding point can be canonically identified with $H^1(X, \mathcal{E}nd\ E)$ By duality, the cotangent space may be identified with $\Gamma(X, K \otimes \mathcal{E}nd\ E)$. Therefore a point of Ω can be seen as a pair (E, φ), where E is a vector bundle in $\mathcal{U}$ and $\varphi\colon E \to K \otimes E$ is a homomorphism. By associating to (E, φ) the characteristic polynomial of φ, we define a morphism $H\colon \Omega \to W$. We will call this map the Hitchin map [4].

4.1. *Proof of the Theorem* 1. Consider the map $\lambda\colon \Omega \to \mathcal{U} \times W$ where the first factor of λ is the projection onto $\mathcal{U}$ and the second factor is the Hitchin map H. By a theorem of Drinfeld and Laumon [5], there exists a stable bundle E_0 of rank n and degree d such that every nilpotent homomorphism $E_0 \to K \otimes E_0$ is zero. This implies that the inverse image by λ of $(E_0, 0)$ in $\mathcal{U} \times W$ consists of a single point, namely the zero cotangent vector at $E_0 \in \mathcal{U}$. Now dim $(\Omega) = \dim(\mathcal{U} \times W) = 2(n^2(g-1)+1)$. By the dimension theorem, the morphism λ is dominant and moreover its differential is an isomorphism at the generic point of Ω. It follows that the projection $H^{-1}(s) \to \mathcal{U}$ is dominant for all point s in an open set in W. Moreover we may assume that for s in this open set the spectral curve X_s is smooth and irreducible in view of Remark 3.5. Now by Proposition 3.6, $H^{-1}(s)$ consists of pairs (E, φ) where E is a stable bundle obtained as the direct image of a line bundle ξ on X_s and $\varphi\colon E \to K \otimes E$ is the homomorphism induced by the canonical section of $\pi^*(K)$ on X_s. This proves Theorem 1.

5. The basic linear system on $\mathcal{SU}(n)$

We apply the considerations of the previous section to the case when $d = n(g-1)$. Let us choose an n-sheeted covering Y of X such that most vector bundles of rank n and degree d are direct images of line bundles of degree δ with $\delta = g_Y - 1 = n^2(g-1)$. We will denote the Riemann theta divisor in J_X^{g-1} (whose support consists of effective divisor classes of degree $(g-1)$) by Θ_X. Let $\mathcal{T}_Y^\delta$ denote the open subset of the component of the Jacobian consisting of ξ such that $\pi_*(\xi)$ is semistable. Thus there is a natural morphism $\pi_*\colon \mathcal{T}_Y^\delta \to \mathcal{U}(n, n(g-1))$.

5.1. Proposition. (a) *For every point ξ belonging to $J_Y^\delta - \Theta_Y$, the bundle $\pi_*(\xi)$ is semi-stable.*

(b) *There is a point ξ on the theta divisor Θ_Y such that $\pi_*(\xi)$ is semistable.*

(c) *The complement of $\mathcal{T}_Y^\delta$ in J_Y^δ is of codimension greater than* 1.

(d) *There is a point ξ in Θ_Y such that $\pi_*(\xi)$ is stable and the map π_* has maximal rank at ξ.*

Proof. (a) By our assumption, $\Gamma(Y, \xi) = 0$. Hence $\Gamma(X, \pi_*\xi) = 0$. If $\pi_*(\xi)$ were not semistable, it would admit a subbundle F of degree greater than $\mathrm{rk}(F) \cdot (g-1)$. Now the Riemann-Roch theorem implies that $\Gamma(X, F) \neq 0$, which is a contradiction.

(b) If $\alpha \in J_X$, then the projection formula implies that $\pi^*(\alpha) \otimes \xi$ has semistable direct image for every ξ as in a). Since Θ is ample and $\pi^* \colon J_X \to J_Y$ is a finite map, it follows that it intersects the positive dimensional variety $\pi^*(J_X)\xi$.

(c) Since the complement of $\mathcal{T}_Y^\delta$ in J_Y^δ is contained properly in the irreducible divisor Θ, it follows that it is of codimension at least 2.

(d) We have assumed that the map π_* is dominant so that there exists some $\eta \in J_Y^\delta$ such that $\pi_*(\eta)$ is stable and π_* is of maximal rank at η. Now as in b) above, we can translate η by $\pi^*(\alpha)$ for an $\alpha \in J_X$ so that $\xi = \eta \otimes \pi^*\alpha$ belongs to Θ_Y. It is clear from the commutativity of the diagram

$$\begin{array}{ccc} \mathcal{T}_Y^\delta & \longrightarrow & \mathcal{U}(n,d) \\ {\scriptstyle \otimes\pi^*\alpha}\big\downarrow & & \big\downarrow{\scriptstyle \otimes\alpha} \\ \mathcal{T}_y^\delta & \longrightarrow & \mathcal{U}(n,d) \end{array}$$

that $\pi_*(\xi)$ is stable and that the map π_* is of maximal rank at ξ also.

5.2. Remark. The complement of the set S of points ξ of J_Y^δ such that $\pi_*\xi$ is stable can be shown to have codimension at least $2g-2$, if $n \geq 3$. In fact suppose that Z is the set of points ξ in J_Y^δ such that $\pi_*\xi$ has a quotient line bundle of degree $\leq d/n$. We will estimate the codimension of Z in J_Y^δ. By the duality theorem applied to the map π, the existence of a nonzero homomorphism of $\pi_*\xi$ into M, where M is a line bundle on X, is equivalent to the existence of a nonzero homomorphism of ξ into $\pi^*M \otimes \omega_\pi$ where ω_π is the relative dualising sheaf, namely $K_Y \otimes \pi^*K_X^{-1}$. Thus every element of Z is of the form $\pi^*M \otimes \omega_\pi(-D)$ where M is a line bundle on X with $\deg(M) \leq d/n$, and D is an effective divisor on Y. We have

$$\deg(D) = n \cdot \deg M + \deg \omega_n - \delta \leq d - \delta + 2\{g_Y - 1 - n(g_X - 1)\}.$$

It follows that codim $(Z) \geq (n-1)(g-1) \geq 2(g-1)$ if $n \geq 3$. If E is a general stable bundle on X, then by [10], Lemma 2.1, for every subbundle F of E we have

$$\deg(F)/\mathrm{rk}(F) \leq \{\deg(E) - (\mathrm{rk}(E) - rk(F))(g-1)\}/\mathrm{rk}(E).$$

Let now L be a general element of $J_Y^{\delta+(2g-3)}$ such that $\pi_*L = E$ satisfies the condition given above. If D is an effective divisor of degree $2g-3$ on Y, the bundle $\pi_*(L(-D))$ is of rank n and degree d and is a subsheaf of π_*L. If F is a subbundle of $\pi_*(L(-D))$ with $\mathrm{rk}(F) \leq n-2$, it follows that $\deg(F)/\mathrm{rk}(F) < d/n$.

On the other hand, as D varies, the line bundles $L(-D)$ form a subvariety W_L of J_Y^δ of dimension $(2g-3)$. Changing L amounts to applying a translation on W_L. Since the variety Z defined earlier is of codimension at least $2g-2$, we can choose L such that W_L does not intersect Z. For such a choice of L, W_L is contained in S. Now W_L is cohomologous to $\Theta^k/k!$ with $k = \operatorname{codim}(W_L)$ in J_Y^δ. Hence W_L intersects every subvariety of J_Y^δ of codimension $2g-3$. This proves our assertion.

5.3. Remark. For $n=2$, the codimension of $(J_Y^\delta - S)$ is at least $g-1$. Coupled with Remark 5.2., this implies that with the exception of the case $g=r=2$, this codimension is at least 2. Consider the Hitchin map $H\colon \Omega \to W$. The restriction of the natural symplectic form ω to the fibre $H^{-1}(a)$ over a general point $a \in W$ is the boundary of a holomorphic 1-form. On the other hand, $H^{-1}(a)$ can be identified with the complement in J_Y^δ of a subvariety of codimension at least 2 and hence this 1-form can be extended to the whole of J_Y^δ. It is therefore closed and so ω restricts to 0 on $H^{-1}(a)$. In other words, the map H defines a completely integrable Hamiltonian system. Moreover the Hamiltonian vector fields given by components (with respect to a basis of W) of H also extend to the Jacobian and thus the system is linearised on the variety J_Y^δ. This gives another proof of the results of Hitchin [4].

5.4. Remark. *The map $\pi_*\colon \mathcal{T}_Y^\delta \to \mathcal{U}(n,d)$ is of degree $2^{3g-3}\cdot 3^{5g-5}\dots n^{(2n-1)(g-1)}$.*

Proof. Indeed, the required degree is equal to the degree of the map $H_E\colon \Omega_E^1 \to W$, namely the restriction of H to the cotangent space to the moduli space at a sufficiently general point E. Since this is clearly defined by $3g-3$ polynomials of degree two, $5g-5$ polynomials of degree three, ... the assertion follows from Bezout's theorem.

5.5. *Proof of Theorem* 2. Let $d = n(g-1)$ and $\delta = n^2(g-1)$. We will compute on $\mathcal{T}_Y^\delta$ the pullback of the theta divisor on $\mathcal{U}(n,d)$. Indeed, the vector bundle on $\mathcal{T}_Y^\delta \times X$ which induces the map $\pi_*\colon \mathcal{T}_Y^\delta \to \mathcal{U}(n,d)$ is easy to describe. Consider a Poincaré bundle P on $\mathcal{T}_Y^\delta \times Y$. Then the required bundle is $U = (1\times\pi)_*(P')$. In order to compute the pullback of $\mathcal{O}(\Theta)$ we need to compute only $(p_1)_*(U)$ and $R^1(p_1)_*(U)$. Since $(1\times\pi)$ is an affine morphism these are simply the sheaves $(p_1)_*(P)$ and $R^1(p_1)_*(P)$. But the sheaf $(p_1)_*(P)$ is 0 and $\det(R^1(p_1)_*(P))$ is the Riemann theta divisor on J_Y^δ. Hence the pullback of $\mathcal{O}(\Theta)$ by π_* is isomorphic to the restriction of $\mathcal{O}(\Theta)$ to $\mathcal{T}_Y^\delta$. Since the complement of $\mathcal{T}_Y^\delta$ in J_Y^δ is of codimension ≥ 2 it follows that π induces an injective map $\Gamma(\mathcal{U}(n,d),\mathcal{O}(\Theta)) \to \Gamma(J_Y^\delta,\mathcal{O}(\Theta))$. This proves that the dimension of $\Gamma(\mathcal{U}(n,d),\mathcal{O}(\Theta))$ is at most 1. But it is clearly given by an effective divisor so that Theorem 2 is proved.

5.6. *Proof of Theorem* 3. Take $d=0$ and $\delta = (n^2-n)(g-1)$ and choose a covering $\pi\colon Y \to X$ as in Theorem 1. We will use here the notation of 2.3.–2.6. Firstly the rational map $J_Y^\delta \to \mathcal{U}(n,0)$ is dominant.

5.7. Proposition. *The direct image map induces a dominant rational map $P' \to \mathcal{SU}(n)$. Moreover, the complement of the open set $\mathcal{T}_Y^\delta \cap P'$ is of codimension at least 2.*

Proof. If $\pi_*(\alpha)$ is semistable, so is $\pi_*(\pi^*L\otimes\alpha)$ for any line bundle L on X and we can therefore conclude that there is an induced dominant rational map $P' \to \mathcal{SU}(n)$. The rest of the assertions is also proved in the same way, using Proposition 5.1.

Now, from the commutative diagram

$$\begin{array}{ccc} P' \times \mathcal{J}_X^{gx-1} & \longrightarrow & \mathcal{J}_Y^{gx-1} \\ \downarrow & & \downarrow \\ \mathcal{SU}(n) \times \mathcal{J}_X^{gx-1} & \longrightarrow & \mathcal{U}(n, n(g-1)) \end{array}$$

we conclude that the pullback of the line bundle $\mathcal{O}(\Theta)$ to $\mathcal{SU}(n) \times J_X^{g-1}$ is of the form $p_1^*M \otimes p_2^*\mathcal{O}(n\Theta)$ for some line bundle M. This implies that the pullback of M by the map $P' \to \mathcal{SU}(n)$ is τ. But τ is a primitive element in the group of algebraic equivalence classes of divisors in P'. Hence the same is true of M. In other words, the line bundle is the same as the ample generator h of the Picard group, which is isomorphic to $\mathbb{Z}[2]$. From Remark 5.4, Proposition 2.6 and the codimension computation in Proposition 5.1, we deduce, as in the proof of Theorem 2, that the map $\Gamma(\mathcal{SU}(n), h) \to \Gamma(P', \tau)$ is injective. Now the group of n division points of the Jacobian of X acts on $\mathcal{SU}(n)$ as well as on $\mathbb{P}(\Gamma(P', \tau))$ and it is easy to see that the above map is compatible with these actions. Now it is easily seen that $\Gamma(\mathcal{SU}(n), h) \neq 0$ so that we conclude that this map is bijective and in particular that $\dim \Gamma(\mathcal{SU}(n), h) = n^g$, since the action of the theta group on $\Gamma(P', \tau)$ is irreducible [6]. Finally, it remains to use the isomorphism of $\Gamma(P', \tau)^*$ with $\Gamma(J_X^{g-1}, \mathcal{O}(n\Theta))$, and identify the induced rational map with $E \mapsto D_E$. In fact, this only amounts to computing the rational map of P' into $\mathbb{P}\Gamma(J_X^{g-1}, \mathcal{O}(n\Theta))$. But, in view of Proposition 2.6, this means that we have to verify that for all p belonging to an open set in P', the divisor $T_p^*\Theta \cap J^{g-1}$ is the same as $D_{\pi_* p}$. The latter is given by $\{\xi \in J^{g-1} \colon \Gamma(\xi \otimes \pi_* p) \neq 0\}$. The projection formula implies that this consists of those ξ with $\Gamma(\pi^*\xi \otimes p) \neq 0$. This is then the divisor $\{\xi \in J^{g-1} : T_p\pi^*\xi \in \Theta_Y\} = T_p^*\Theta \cap J^{g-1}$, as was to be proved.

5.8. Remark. On the one hand, we have a natural rational map $\Delta \colon J^{g-1} \to \mathbb{P}\Gamma(\mathcal{SU}(n), h)$ given by mapping any $\xi \in J^{g-1}$ to the pullback of the theta divisor on $\mathcal{U}(n, n(g-1))$ by the morphism $E \mapsto E \otimes \xi$ of $\mathcal{SU}(n)$ into $\mathcal{U}(n, n(g-1))$. On the other, the linear system of $n\Theta$ gives a rational map of J^{g-1} into $\mathbb{P}\Gamma(J^{g-1}, \mathcal{O}(n\Theta))^*$. These maps fit into a commutative diagram

$$\begin{array}{ccc} \mathcal{J}^{g-1} & \xrightarrow{\Delta} & \mathbb{P}\Gamma(\mathcal{SU}(n), h) \\ & \searrow & \downarrow \\ & & \mathbb{P}\Gamma(\mathcal{J}^{g-1}, \mathcal{O}(n\Theta))^* \end{array}$$

in which the vertical map is given by the duality in Theorem 3.

5.9. Remark. The linear system of h in $\mathcal{SU}(n)$ is base point free for $n = 2$. On the other hand, there are examples when it does have base points. Indeed, by 5.8., it is spanned by the image of J^{g-1} under Δ. Therefore a point $E \in \mathcal{SU}(n)$ is a base point if and only if all the divisors in the image of Δ pass through E. This condition means that for all ξ in J^{g-1}, one has $\Gamma(X, E \otimes \xi) \neq 0$. There are examples of bundles with this property ([11], §3).

References

[1] *A. Beauville,* Fibrés de rang 2 sur une courbe, fibré déterminant et fonctions thêta, Bull. Soc. Math. France **116** (1988).

[2] *J. M. Drezet* and *M. S. Narasimhan,* Groupes de Picard des variétés de modules de fibrés semistables sur les courbes algébriques, to appear.

[3] *R. Hartshorne,* Algebraic Geometry, Berlin-Heidelberg-New York 1977.

[4] *N. Hitchin,* Stable bundles and integrable Systems, Duke Math. J. **54** (1987), 91–114.

[5] *G. Laumon,* Un analogue global du cône nilpotent, Duke Math. J. **57** (1988), 667–671.

[6] *D. Mumford,* Equations defining abelian varieties, Invent. Math, **1** (1966), 287–354.

[7] *D. Mumford,* Prym Varieties I, In Contributions to Analysis, London-New York-San Francisco 1974, 325–350.

[8] *M. S. Narasimhan* and *S. Ramanan,* Generalised Prym Varieties as fixed points, J. Ind. Math. Soc. **39** (1975), 1–19.

[9] *M. S. Narasimhan* and *S. Ramanan,* 2Θ-linear systems, Proc. of the Int. Colloq. on Vector Bundles, Bombay 1984, Oxford (1988).

[10] *H. Lange,* Zur Klassifikation von Regelmannigfaltigkeiten, Math. Ann. **262** (1983), 447–459.

[11] *M. Raynaud,* Sections des fibrés vectoriels sur une courbe, Bull. Soc. Math. France **110** (1982), 103–125.

Arnaud Beauville, Mathématiques - Bât. 425, Université Paris-Sud, 91405 Orsay Cedex, France
e-mail:

M.S. Narasimhan, Tata Institute of Fundamental Research, Homi Bhabha Road, Bombay 400005, India
e-mail:

Spectral curves and their impact

By *Nigel Hitchin* at Oxford

Abstract. The article is a review of the paper "Spectral curves and the generalised theta divisor" by A. Beauville, M. S. Narasimhan and S. Ramanan from a modern perspective, giving a brief account of its significance in a variety of contexts.

1. Introduction

There is a whole community of mathematicians nowadays for whom the letters "BNR" mean just one thing – the paper being reviewed here. Reading the first paragraph of the article one might think that it simply describes a way of producing vector bundles on an algebraic curve X by taking the direct image of a line bundle on a finite covering $\pi : Y \to X$. However, its impact is much more than that, a fact which becomes more apparent given the confluence of ideas at the time it was written. The relevance arises not just from considering one such covering Y but a whole family arising from a moduli space, and the construction is then central to the study of Higgs bundles [22], or Ngô's work on the Fundamental Lemma [39], or recent work on the geometric Langlands correspondence [2].

There are essentially two aspects to the paper: one is a rigorous account of the construction of a vector bundle E by a direct image, and the associated endomorphism Φ of E, twisted by a line bundle, the other is the introduction on the moduli space of stable bundles E over X of an analogue of the theta divisor on the Jacobian of a curve, and its relation to the theta divisor of the covering Y.

Before discussing its ramifications, it is useful to consider its origins in the mid 1980s.

2. Background

2.1. Spectral curves. It seems appropriate to begin with this author's own activity at that time. After some years working with various forms of the self-dual Yang-Mills equations, I found myself writing a paper on a two-dimensional version which are now called Higgs bundles (following C. T. Simpson [40]). A Higgs bundle over a smooth projective curve X of genus g is a pair consisting of a holomorphic vector bundle E and a "Higgs field" Φ, a holomorphic section of $\operatorname{End} E \otimes K$. The paper concerned hermitian metrics on E which yield connections whose curvature F satisfies the gauge-theoretic equation $F + [\Phi, \Phi^*] = 0$. The existence of such a metric is guaranteed by a stability condition on the pair. The stability condition for E

HTTPS://DOI.ORG/10.1515/9783111260440-016

alone (i.e. when Φ vanishes) had long been studied since the work of M. S. Narasimhan and C. S. Seshadri in the 1960s [36] and in the following 20 years properties of the moduli space $\mathcal{N}$ of stable bundles, a projective variety, were quite well known. This new moduli space $\mathcal{M}$ of stable *Higgs* bundles was quite different, non-compact and moreover had a natural hyperkähler metric which was my main focus, as an infinite-dimensional application of the hyperkähler quotient construction.

The author's paper [22] deals with the case where E has rank 2 and the determinant line bundle $\det E = \Lambda^2 E$ is fixed, so that locally Φ is a 2×2 matrix of holomorphic 1-forms. Fixing $\Lambda^2 E$, it is natural also to take Φ to have trace zero and hence take values in the Lie algebra $\mathfrak{sl}(2)$. Its determinant $\det \Phi$ is a well-defined quadratic differential q and defines a projection of the moduli space $\mathcal{M}$ onto the $(3g-3)$-dimensional vector space $H^0(X, K^2)$. The estimates I used to prove existence of the hermitian metric also showed that this projection was proper and, out of curiosity since it had no obvious relation to the hyperkähler metric, I observed that the generic fibre was an abelian variety, isomorphic to the $(3g-3)$-dimensional Prym variety of the curve of eigenvalues of Φ defined by $x^2 + \det \Phi = 0$, the *spectral curve.* In a rather laborious argument I showed that conversely, given $q \in H^0(X, K^2)$ with simple zeros, the direct image of a line bundle on a double covering branched over those zeros gave a Higgs bundle. Narasimhan took an interest in this work and sent his postdoc N. Nitsure to Oxford in 1987 to learn more. While there Nitsure developed a more algebraic approach.

This article was followed up by the paper [23] observing a similar feature for principal G-bundles over X together with explicit descriptions of the abelian variety when G is a classical group. The main thrust of that paper was the structure of a completely integrable Hamiltonian system for each curve X and group G, defined on the cotangent bundle of the moduli space of stable G-bundles. The determinant function for $\mathfrak{sl}(2)$ is now replaced by generators of the ring of invariant polynomials on the Lie algebra $\mathfrak{g}$, so for $G = GL(n)$ instead of a quadratic differential these are the coefficients of the characteristic polynomial $\det(x - \Phi) = 0$, sections of K^i for $1 \leq i \leq n$.

In the context of integrable systems, the direct image approach was already implicit in classical cases where the curve X is just the projective line $\mathbb{P}^1$, the vector bundle is trivial and the Higgs field is twisted by a line bundle $\mathcal{O}(n)$ instead of the canonical bundle. The standard example is Euler's equations for a spinning top where Φ is a 2×2 matrix of quadratic polynomials in an affine parameter z and the spectral curve is the elliptic curve $x^2 + \det \Phi = 0$. The spinning top equations are solved by elliptic functions, or more geometrically by a linear flow on the Jacobian of the spectral curve.

2.2. The generalised theta divisor. For a curve X of genus g the Jacobian variety J_X is a complex torus of dimension g. Making a choice of origin this can be regarded as the moduli space of line bundles L on X of degree $g-1$. A generic member of this family has no non-zero holomorphic sections, but the dimension is positive when (the equivalence class of) L lies on a hypersurface called the theta divisor Θ. If we replace the line bundle by a stable vector bundle E of rank r and trivial determinant bundle $\Lambda^r E$ then the locus in the moduli space $\mathcal{N}$ of stable bundles of those E for which $E \otimes L$ has a non-zero section is also in general a hypersurface, and this is a *generalised theta divisor* Θ_L. The line bundle $\mathcal{O}_{\mathcal{N}}(\Theta_L)$ on $\mathcal{N}$ is independent of L, and each L defines a holomorphic section. The space of all sections, and those of higher powers of this line bundle, were beginning to be of increasing interest to algebraic geometers at the time, especially due to the conjectured relationship with the formulas produced by the physicist E. Verlinde [43].

One may also fix E and dually consider the locus D_E of L in the Jacobian for which the bundle $E \otimes L$ has a non-zero section, and this, for rank 2 bundles, had been studied by A. Beauville [4], giving a rational map from $\mathcal{N}$ to the projective space $|2\Theta|$ of divisors on J_X, following a paper of Narasimhan and Ramanan [38] which in particular identified the genus 3 moduli space $\mathcal{N}$ with a quartic hypersurface in $\mathbb{P}^7 = |2\Theta|$.

In [5] Beauville recalls the origin of his association with the BNR paper: "For vector bundles of rank $r \geq 3$, the analogous statements make sense (with a rational map $\mathcal{N} \dashrightarrow |r\Theta|$), but the proof in rank 2 is not directly adaptable. After a few months I found a way to do it. Shortly after I met Narasimhan and Ramanan at the AMS Summer conference on theta functions (in 1987), and we realized we had had the same ideas – using the Hitchin fibration and the notion of very stable vector bundle introduced by Drinfeld. So we decided to write the joint paper. Narasimhan invited me to the Tata Institute; I spent one month there in 1988. By that time, we had essentially finished the paper..."

3. The BNR paper

For a finite covering $\pi : Y \to X$ of nonsingular projective curves X, Y the direct image $\pi_*(M)$ of a line bundle M on Y is defined by the property that for each open set $U \subseteq X$, $H^0(U, \pi_*(M)) = H^0(\pi^{-1}(U), M)$, giving a rank n vector bundle E over X if the covering has degree n. It follows that $\pi_*(\pi^*(L) \otimes M) \cong L \otimes \pi_*(M)$ for each line bundle L on X, so clearly if we wish to fix the determinant $\det E$, then this requires a restriction on M to some subvariety of the Jacobian J_Y.

Fixing $\det E$ was always part of the story of studying moduli spaces of stable bundles, removing the abelian dependence. Following earlier work the authors write $\mathcal{U}(n, d)$ for the moduli space of semistable bundles of rank n and degree d and $\mathcal{SU}(n, \zeta)$ for bundles with fixed $\det E \cong \zeta$. The notation reflects the celebrated theorem of Narasimhan and Seshadri [36] identifying these moduli spaces with representations of $\pi_1(X)$, or a central extension, in $U(n)$ or $SU(n)$. The paper begins with a discussion of the Prym variety as the kernel of the norm map $\mathrm{Nm} : J_Y \to J_X$ and the translate which is required to fix $\det \pi_*(M)$.

A section x of $\pi^* L$ on Y defines by multiplication a homomorphism from $H^0(\pi^{-1}(U), M)$ to $H^0(\pi^{-1}(U), M \otimes \pi^* L)$ or equivalently an element in $H^0(U, \mathrm{End}(\pi_* M) \otimes L)$ and this is how a Higgs field, twisted by L, is produced on the bundle $E = \pi_*(M)$. By the Cayley-Hamilton theorem it satisfies its characteristic polynomial $x^n + s_1 x^{n-1} + \cdots + s_n = 0$ where s_i is a section of L^i and this yields an equation for Y. In the paper, Y is the scheme X_s (where $s = (s_1, \ldots, s_n)$) on the projective bundle $p : \mathbb{P}(\mathcal{O} \oplus L^*) \to X$ given by such an equation and x is the restriction of the tautological section of $p^* L$. Furthermore, X_s is allowed to be only integral (i.e. reduced and irreducible) and M a rank one torsion-free sheaf.

This is all for a general line bundle L on X but taking $L = K$ the determinant fixing is shown to be $\det \pi_*(M) = \mathrm{Nm}(M) \otimes K^{-n(n-1)/2}$, moreover, a computation of the genus of Y shows that the Prym variety in this case has the same dimension as the moduli space of stable bundles. The rest of the paper is concerned with the rational map defined by the direct image construction which relates these two varieties.

The paper first considers the case of $\mathcal{U} = \mathcal{U}(n, d)$. Since the tangent space at E is $H^1(X, \mathrm{End}\, E)$ by Serre duality a Higgs field $\Phi \in H^0(X, \mathrm{End}\, E \otimes K)$ may be considered as a cotangent vector. Then there is an obvious map from the cotangent bundle $\lambda : T^*\mathcal{U} \to \mathcal{U} \times W$ where the second factor $W = H^0(X, K) \oplus H^0(X, K^2) \oplus \cdots \oplus H^0(X, K^n)$ gives the coefficients

s_i of the characteristic polynomial of Φ (which the authors christen the *Hitchin map*). The notion of *very stable bundles* due to V. G. Drinfeld and G. Laumon [30] implies that for a generic point $E \in \mathcal{U}$ there are no nilpotent Higgs fields. Then the inverse image of $(E,0) \in \mathcal{U} \times W$ is a single point, which shows that λ is dominant (the image is dense).

Taking a generic s so that X_s is nonsingular, and restricting to the open set of J_{X_s} for which $\pi_* M$ is semistable gives a morphism to $\mathcal{U}$. Its degree is the degree of the map of vector spaces $H^0(X, \operatorname{End} E \otimes K) \to W$ defined by $3g-3 = \dim H^0(X,K^2)$ polynomials of degree 2, $5g-5 = \dim H^0(X,K^3)$ of degree 3 etc. Hence there is a rational map $\lambda : J_{X_s} \dashrightarrow \mathcal{U}$ with

$$\deg \lambda = 2^{3g-3}3^{5g-5} \cdots n^{(2n-1)(g-1)}. \tag{3.1}$$

The line bundles M on X_s which produce semistable bundles are shown to be given by the complement of a subvariety of codimension ≥ 2 (and in fact $\geq 2g-2$ if $n \geq 3$) and this enables the authors to compare divisors on J_{X_s} and $\mathcal{U}$. In particular, the generalised theta divisor on $\mathcal{U}(n, n(g-1))$ corresponds to the theta divisor for X_s which leads to the theorem that $\dim H^0(\mathcal{U}(n,n(g-1)), \mathcal{O}(\Theta)) = 1$, and a similar result for bundles of fixed determinant and the Prym variety.

In the context of Beauville's motivation the duality relation results in the theorem:

Theorem 3.1. *The rational map which associates to a general point $E \in \mathcal{SU}(n,\zeta)$ the divisor D_E in J_X^{g-1} induces an isomorphism*

$$H^0(J_X^{g-1}, \mathcal{O}(n\Theta)) \cong H^0(\mathcal{SU}(n,\zeta), H)^*$$

where H is the generator of the Picard group. In particular, $\dim H^0(\mathcal{SU}(n,\zeta), H) = n^g$.

4. Impact

4.1. Strange duality. Most of the citations for the BNR paper refer to the spectral curve work, but the theorem just stated led to the investigation of a more general phenomenon, variously known as "strange duality" or "rank-level duality". The level here refers to the integer k in consideration of sections of the line bundle H^k on the moduli space $\mathcal{SU}(n,\mathcal{O})$. Thus the theorem above identifies the dual of the level one space of sections for rank n with the level n space in rank 1 – namely line bundles. Rank-level duality lifts this to higher rank bundles. More specifically the tensor product of bundles of rank r and k defines a natural map

$$\tau : \mathcal{SU}(r,\mathcal{O}) \times \mathcal{U}(k, k(g-1)) \to \mathcal{U}(kr, kr(g-1))$$

and the generalised theta divisor Θ_{kr} restricts to a section of $k(\Theta_r) + r(\Theta_k)$ on the product, or an element in $H^0(\mathcal{SU}(r,\mathcal{O}), H^k) \otimes H^0(\mathcal{U}(k,k(g-1)), H^r)$ which was conjectured to give an isomorphism between one factor and the dual of the other. This was proved by P. Belkale [7] for generic curves and A. Marian and D. Oprea [33] in general.

Replacing sections of line bundles by "conformal blocks", this feature was known earlier to physicists, but in that context it was supposed to be independent of the complex structure on the curve X. By this one means that the vector spaces $H^0(\mathcal{U}, H^r)$ are fibres of a vector bundle over the moduli space of curves of genus g and there is a (projectively) flat connection on this vector bundle, parallel translation identifying fibres. A proof that the duality is constant with respect to this connection was achieved in a further paper of Belkale [8]. One approach to this connection [25] in fact uses the integrable system related to spectral curves, with the quadratic functions arising from $\mathrm{tr}\,\Phi^2$ defining holomorphic heat operators and generalizing the heat equation satisfied by classical theta functions, though Belkale's proof does not use this directly.

4.2. The Fundamental Lemma. Perhaps the most surprising reference to the paper occurs in the proof of the Fundamental Lemma, part of the Langlands programme, which led to a Fields Medal for Ngô Bau Châu in 2010. He recalls [1] "I came across the Hitchin system in a rather casual way. I was reading an article by my thesis advisor, Gérard Laumon, in a journal and, right after it, there was *Stable bundles and integrable systems*, an article by Nigel Hitchin.... This paper was very popular and widely cited, so I was very curious about what it was about. It was my first encounter with this paper and I must say that on the first reading I understood it, but I, somehow, passed on why it was so important. However, after my PhD thesis, I got some ideas on how one should prove the fundamental lemma (proposed by Robert Langlands and Diana Shelstad), but there was a missing step that ruined my reasoning. Only four years after my first reading of Nigel's paper, I found that this paper had exactly what I needed. But it was not in the first reading, it was a long process to figure that out: our ideas came from very different mathematical disciplines." The two papers he refers to are [31] by Laumon and [23] by myself which appeared by chance together in a volume dedicated to Yuri Manin.

Ngô's application involves the field $k(X)$ of rational functions on a curve X over a finite field of large characteristic. There is no stability condition here and the Higgs field is twisted by a divisor D of large degree. The setting is quite different but it combines global geometry and arithmetic in a way that the trace formula which is central in the Langlands programme makes sense. The section in [32] describing the spectral curves clearly follows the account in BNR, but one of the new ingredients Ngô brought into play, a support theorem concerning direct images of the map $\mathcal{M} \to W$, became very important in later developments in the theory of Higgs bundles over the complex numbers.

4.3. The nonabelian Hodge correspondence. As noted in Section 2.1 the hyperkähler property of the moduli space of Higgs bundles was the initial motivation for [22] and pairs (E, Φ) only realized one structure I out of the three quaternionic complex structures I, J, K. Neverthless, it was observed that in complex structure J the connection $\nabla + \Phi + \Phi^*$ is an irreducible flat connection with holonomy $SL(2, \mathbf{C})$ (if $\det E$ is trivial). S. K. Donaldson [13], using the theory of harmonic maps, showed there was a one-to-one correspondence and so the moduli space $\mathcal{M}$ could also be interpreted as a moduli space of flat connections or equivalently as the character variety of reductive representations $\pi_1(X) \to SL(2, \mathbf{C})$ modulo conjugation. This is in some sense a complexification of the Narasimhan-Seshadri theorem.

The full picture, which is now known as the *nonabelian Hodge correspondence*, emerged in the work of Simpson [40] and K. Corlette [11], who extended the theory to a higher dimensional Kähler manifold (with the added condition $[\Phi, \Phi] = 0$, which holds automatically for curves). In this more general situation, for the Higgs bundle moduli space $\mathcal{M}$ the map to W is again proper and the generic fibre is an abelian variety. In the author's paper [23] and in BNR this work was not available at the time for bundles of rank greater than two over curves and the results are phrased in terms of the cotangent bundle of the stable bundle moduli space, which is an open dense subspace of $\mathcal{M}$. Simpson extended the theory in many directions and in particular [41] reinterpreted a Higgs bundle in terms of a generalization of the spectral curve, namely a sheaf on $\mathbb{P}(\mathcal{O} \oplus T^*)$ whose support is disjoint from the divisor at infinity.

One question then presents itself: what relationship is there between the spectral curve approach and the character variety? The underlying complex structures on the same moduli space are very different, in the Higgs bundle structure I there are compact positive-dimensional subvarieties, but the character variety, with complex structure J, is affine. One link is provided by considering the Higgs bundle interpretation of homomorphisms from $\pi_1(M)$ into a real form of the complex group G. The conjugation action taking J to $-J$ on the character variety becomes a holomorphic involution with respect to I and so the fixed point set is a holomorphic subvariety of the moduli space of Higgs bundles.

An example of this is derived from taking the direct image of the trivial bundle. This, as shown in the BNR paper, for $X_s \subset \mathbb{P}(\mathcal{O} \oplus L^*)$ is the bundle $E = \mathcal{O} \oplus L^{-1} \oplus \cdots \oplus L^{-(n-1)}$ and x, the tautological section of L, acts in the obvious way transforming one line bundle to another, giving a Higgs field on E in the form of a companion matrix. One can write down the companion matrix for any characteristic polynomial $x^n + s_1 x^{n-1} + \cdots + s_n$ and this gives a section of the projection $\mathcal{M} \to W$, regardless of whether the spectral curve is integral or not.

In the case where $L = K$, then tensoring with $K^{(n-1)/2}$ to make $\det E$ trivial, the corresponding flat connection has holonomy in $SL(n, \mathbf{R})$ and the image of this section is in fact a connected component of the character variety for the real group. The involution defining the real structure is not so obvious when the Higgs field is written as a companion matrix, but it is clearer when seen via the work of B. Kostant on the principal 3-dimensional subgroup, and this leads to a similar story for the split real form of any simple Lie group [24]. These examples have given rise to the currently active topic of *higher Teichmüller theory* [44], which now uses different methods to investigate properties of these representations.

The above example is a case where the fixed point set of the involution on the Prym variety is discrete, in fact isomorphic to the 2-torsion points, although this particular component of the $SL(n, \mathbf{R})$ character variety intersects a fibre in a single point. For other real forms the intersection may consist of translates of an abelian subvariety. Note that for real groups associated to the quaternions the eigenspaces of the Higgs field are 2-dimensional and one needs to take the direct image of a rank 2 bundle on the spectral curve [27].

A more serious link between the two complex structures I and J (now conventionally called $\mathcal{M}_{Dol}$ and $\mathcal{M}_B$ (*Dol* for Dolbeault and B for Betti), concerns a shadow of the fibration by abelian varieties appearing in the cohomology of the character variety as evidenced by the so-called $P = W$ conjecture formulated in 2012 by M. de Cataldo, T. Hausel and L. Migliorini [10] and now proved in a variety of ways [28], [14]. The initial versions require the moduli space to be nonsingular, achieved when (n, d) are coprime.

The W in the conjecture refers to the mixed Hodge structure on the cohomology of $\mathcal{M}_B$: a weight filtration

$$0 \subset W_0 \subset \cdots \subset W_{2i} = H^i(\mathcal{M}, \mathbf{Q})$$

where the quotients W_j/W_{j-1} have the usual (p,q) decomposition that holds for a compact Kähler manifold. The P in the conjecture refers to a perverse filtration of the cohomology using the complex structure $\mathcal{M}_{Dol}$, but rather than give its definition here we may note that, as shown by de Cataldo and Migliorini [9], the perverse filtration is

$$P^i H^j(\mathcal{M}, \mathbf{Q}) = \ker(H^i(X, \mathbf{Q}) \to H^i(h^{-1}(A^{j-i-1}), \mathbf{Q}))$$

where A^{j-i-1} is a general affine subspace of W of dimension $j - i - 1$ and $h : \mathcal{M}_{Dol} \to W$ is the Hitchin map. The theorem is that the two filtrations coincide, so that the structure of a fibration by abelian varieties makes its presence felt in the algebraic structure of the character variety.

The origin of the conjecture lies in the paper by Hausel and F. Rodriguez-Villegas [20] concerning the mixed Hodge polynomial $H_c(x, y, t)$ (the three-variable generating function for the dimension of the (p, q, i) part of the compactly supported cohomology). They calculate the associated E-polynomial $H_c(x, y, -1)$ by counting points in the character variety over a finite field, a technique used by G. Harder and Narasimhan [18] for the moduli space of stable bundles in 1975. The authors observed a curious type of Poincaré duality which ultimately was explained by the relation $P = W$.

4.4. The asymptotic metric. The initial appearance of the fibration stucture of $\mathcal{M}_{Dol}$ seemed to have no relation with the natural hyperkähler metric, but in 2016 R. Mazzeo, J. Swoboda, H. Weiss, and F. Witt [34] began a serious study of the asymptotic structure of the metric, following on from the physicist's more conceptual picture [17], which revealed its presence there. In the author's original paper [22] all that was shown was that the metric was complete and invariant under the circle action $\Phi \mapsto e^{i\theta}\Phi$, which clearly preserves the equation $F + [\Phi, \Phi^*] = 0$. This action is part of a $\mathbf{C}^*$-action and, considering the real part, if Φ is not nilpotent then, since $h : \mathcal{M}_{Dol} \to W$ is proper, $\Phi \mapsto \lambda\Phi$ goes to infinity as $\lambda \to \infty$. The spectral curve just changes by a rescaling of x, and so its Jacobian remains the same, and has a role to play in the asymptotic Riemannian geometry.

In the intervening years, the work of several authors has shown that, in the asymptotic region where the spectral curve is nonsingular, the fibration by abelian varieties defines directly a hyperkähler metric which is a close approximation to the genuine one. This metric is known as the semi-flat metric, the abelian varieties being metrically flat tori, and is defined by a *special Kähler structure* on the open set of W which defines the equation of a nonsingular spectral curve.

A special Kähler manifold [16] is a Kähler manifold together with a flat symplectic connection and with a compatibility condition for the complex structure I and the flat connection. In the situation relevant here I, considered as a section of $\operatorname{End} T \cong T \otimes T^*$, is the covariant derivative of a Hamiltonian vector field X.

In our case, as an integrable system, the fibres of $h : \mathcal{M} \to W$ are Lagrangian with respect to a holomorphic symplectic form and therefore Lagrangian as real submanifolds with respect to the real part. Then the tangent space at $s \in W$ on the base is isomorphic to the cohomology group $H^1(\mathcal{M}_s, \mathbf{R})$ of the fibre $\mathcal{M}_s$ and the Gauss-Manin connection defines the flat connection for the special Kähler structure. The Kähler form on $\mathcal{M}_{Dol}$ defines a polarization on $\mathcal{M}_s$ and hence a symplectic structure on this cohomology group. The vector field X on W is given by the circle action induced from $\Phi \mapsto e^{i\theta}\Phi$. Then, as in [16], one may use this data – a differential-geometric version of the variation of periods of a family of abelian varieties – to define a hyperkähler metric on the total space of tori over W.

This semi-flat metric is shown to be exponentially close to the actual metric in an asymptotic region [15], [3].

4.5. Mirror symmetry. In the paper [23] this author mistakenly claimed that the abelian variety fibres for the groups $SO(2m + 1)$ and $Sp(m)$ were the same, but somewhat later M. Thaddeus pointed out that they are in fact dual to each other (see [26] for the author's corrected version). This was not just an observation but a consequence of his viewing the fibration $h : \mathcal{M} \to W$ in the light of the Strominger-Yau-Zaslow approach to mirror symmetry of Calabi-Yau manifolds [42].

Originally envisaged as a geometric way to relate the symplectic geometry of one Calabi-Yau 3-manifold M with the holomorphic geometry of another $M^\vee$, SYZ proposed finding a special Lagrangian fibration of M by 3-dimensional real tori and replacing each fibre M_x by the moduli space of flat $U(1)$-bundles over it, another 3-dimensional torus $H^1(M_x, \mathbf{R}/\mathbf{Z})$. "Special" in this case means that the real part of the holomorphic 3-form vanishes as well as the Kähler 2-form. Note that since this torus has a distinguished origin, and M_x does not in general, applying the process a second time requires twisting with a gerbe, an important feature.

A hyperkähler manifold is a particular example of a Calabi-Yau manifold with any of its complex structures. Using J, the fibration by abelian varieties is a special Lagrangian fibration, and so one may try to apply this relationship. The underlying conjecture is that the SYZ mirror of the moduli space for group G should be the space for the Langlands dual group ${}^LG$, and since $SO(2m+1)$ and $Sp(m)$ are indeed dual, this was what provoked Thaddeus's question. For nonsingular fibres, R. Donagi and T. Pantev showed in fact [12] that the two groups gave dual abelian varieties, in this case using the *cameral cover* of the curve X: a Galois cover acted on by the Weyl group. The abelian variety fibre is then a subvariety of the Jacobian identified via the root system. For the classical groups (and G_2) the paper [26] describes the duality in terms of spectral curves. Evidence for the conjecture using the full moduli space, and not just the restriction to nonsingular fibres, came in the work of Hausel and Thaddeus [19], showing the equality of the appropriate Hodge numbers for the Langlands dual groups $SL(n)$ and $PSL(n)$.

Another aspect of the original SYZ was that, by a real version of the Fourier-Mukai transform, another special Lagrangian submanifold of M, together with a flat $U(1)$- bundle defined on it, should correspond to a vector bundle on $M^\vee$ with Hermitian-Einstein connection. The hyperkähler version of this is to pass from a holomorphic Lagrangian submanifold in $\mathcal{M}_{Dol}$ to a hyperholomorphic vector bundle on $\mathcal{M}^\vee$. Such a bundle has a connection whose curvature is a 2-form of type $(1,1)$ with respect to all complex structures. Finding evidence for this formed one of the motivations for the paper [21].

The obvious Lagrangian to take is the cotangent space T^*_E of a point $E \in \mathcal{N}$ in the open inclusion $T^*\mathcal{N} \subset \mathcal{M}_{Dol}$. For E to be very stable (i.e. no nilpotent Higgs fields) is equivalent to T^*_E being closed in $\mathcal{M}_{Dol}$ and hence $h : T^*_E \to W$ is a proper map of vector spaces of the same dimension. Its degree is given by counting the intersection points of a generic fibre of h which is the same as the degree $2^{3g-3}3^{5g-5}\dots$ of the rational map (3.1) in BNR. Over the locus in W of nonsingular spectral curves we have a relative Fourier-Mukai transform and this will define a vector bundle of this high rank on the corresponding open set in the mirror. The question, unanswered at the present time, is whether it extends as a vector bundle and if so, whether it has a hyperholomorphic connection. Instead one may try and look further for similar natural Lagrangians, in particular where the vector bundle has lower rank.

The Higgs bundle $(E, 0)$ for E stable is clearly a fixed point of the $\mathbf{C}^*$-action on $\mathcal{M}_{Dol}$ but there are others, where $(E, \Phi) \cong (E, \lambda\Phi)$ is achieved by an automorphism of E. By an adaptation of Bialynicki-Birula theory, the upward flow from such a point (the pairs (E', Φ') such that the limit $(E', \lambda\Phi)$ as $\lambda \to 0$ is the given fixed point) is a Lagrangian submanifold, and, like the cotangent space, it is isomorphic to an affine space of dimension $\dim W$. These examples are considered in [21] and, extending the notion of "very stable" to other fixed points, one finds cases where there is a genuine hyperholomorphic bundle on the mirror.

The first, rather trivial, example is the image of the section of $\mathcal{M} \to W$ defined by the trivial bundle on each spectral curve, the real form $GL(n, \mathbf{R})$ in the character variety. Here the degree of h is one and the mirror of the trivial flat line bundle on the Lagrangian is the trivial rank one hyperholomorphic bundle. More interesting cases are provided by Hecke transforms of these Higgs bundles performed at points $a_1, \dots, a_m \in X$. The relative Fourier-Mukai transform defines a bundle on the mirror which is a tensor product of exterior powers of the universal bundles at the points. These exist on the whole space and there is indeed a hyperholomorphic connection for rather general reasons.

The notion of Hecke modifications of vector bundles goes back to Narasimhan and Ramanan in 1978 [37]. In its simplest form, the modification E' of a vector bundle E of rank n at a point $a \in X$ is defined as the kernel sheaf

$$0 \to \mathcal{O}(E') \to \mathcal{O}(E) \to \mathcal{O}_a^r \to 0$$

for a given homomorphism of the fibre $f : E_a \to \mathbf{C}^r$. The determinant bundle changes and the modification of a stable bundle may not be stable.

We can transform a Higgs bundle (E, Φ) to (E', Φ') so long as Φ preserves the kernel of the homomorphism f. Since the two Higgs bundles are identical outside a the spectral curve is the same and so it is just the line bundle L with $\pi_* L = E$ which changes. If Φ preserves $\ker f$ then generically there are $n - r$ eigenspaces with eigenvalues $x_1, \dots, x_{n-r}$ and then the line bundle for (E', Φ') is $L(-(x_1 + \dots + x_{n-r}))$.

The paper of A. Kapustin and E. Witten [29] in 2007 was highly influential in placing the moduli spaces $\mathcal{M}$, Hecke transforms and mirror symmetry in a physicist's interpretation of the geometric Langlands programme.

4.6. The geometric Langlands programme. There is a vast literature on the geometric Langlands programme, many papers citing the paper under review, but we restrict ourselves here to one by Donagi, Pantev and Simpson [2] which is perhaps the closest since it looks in more detail at the rational map from the Prym variety of the spectral curve to the moduli space of stable bundles, the central feature of BNR. The task is to pass from a local system on the curve X to a local system on the open set $\mathcal{N}^{vs}$ of very stable bundles.

Donagi recalls in the paper the motivation, which concerned "... a small detail immediately noticed by the first author, in the Manin volume of the Duke Mathematical Journal. Hitchin's article on the moduli space of Higgs bundles [23] and Laumon's article on the geometric Langlands correspondence [31] were both in this volume. In Remarque 5.5.2, Laumon states that Deligne, in unpublished communication, calculated the multiplicity of the zero-section in the characteristic cycle of the $\mathcal{D}$-module: this is the same as the rank of the local system over $\mathcal{N}^{vs}$. The answer (for $SL(2)$-bundles on a curve of genus g) was 2^{3g-3}. This number was the same as the degree of the map from a general fiber of the Hitchin system to the moduli space of bundles."

The idea is that a flat connection with respect to the Langlands dual group ${}^L G$ defines a point in $\mathcal{M}^\vee$ and by projection a point $s \in W$ in the base. As the abelian variety fibres are dual, this point gives a line bundle L on the fibre over s in the moduli space $\mathcal{M}$ for G, an abelian variety A_s. The rational map $A_s \dashrightarrow \mathcal{N}$ to the moduli space of stable bundles is well-defined on an open set of A_s and the direct image of L is a vector bundle defined on the very stable locus. This has rank given by the degree of the map and is predicted to have a flat connection with singularities on $\mathcal{N} - \mathcal{N}^{vs}$.

The authors restrict their attention to bundles E of rank 2 over a curve X of genus 2, where the moduli spaces for fixed determinant are known by early work of Narasimhan and Ramanan to be $\mathbb{P}^3$ for fixed determinant of even degree and the intersection of two quadrics in $\mathbb{P}^5$ in odd degree. They then need to describe in detail the points in the Prym variety which give unstable bundles to resolve the rational map into a morphism, and then investigate the so-called wobbly locus – the complement of the very stable points. The modified abelian variety becomes a spectral variety for a Higgs bundle on the 3-dimensional moduli space. To pass to a flat connection the authors appeal to the work of T. Mochizuki [35] for the nonabelian Hodge correspondence applied to higher dimensional open varieties. The role of Hecke correspondence is a central aspect of this work. Working in these low dimensions might seem very restrictive, but it is worthwhile pointing out that an explicit form of the integrable system on the cotangent bundle of the intersection of two quadrics has only recently been found [6].

References

Archival sources

[1] Interview with Hitchin and Ngô `https://www.icmat.es/news/interviews/19-12-24-en/`

[2] *R. Donagi, T. Pantev* and *C. T. Simpson*, Twistor Hecke eigensheaves in genus 2, preprint `http://arXiv:2403.17045`.

[3] *T. Mochizuki*, Semi-flat metrics of the moduli spaces of Higgs bundles in the non-zero degree case, preprint 2025 `http://arXiv:2501.12741`.

Published material

[4] *A. Beauville*, Fibrés de rang 2 sur une courbe, fibré detérminant et fonctions thêta, Bull. Soc. Math. France **116** (1988), 431–448.

[5] *A. Beauville*, in: Remembering M. S. Narasimhan (1932–2021), Notices of the AMS **69** (2022), 1198–119.

[6] *A. Beauville, A. Höring, J. Liu* and *C. Voisin*, Symmetric tensors on the intersection of two quadrics and Lagrangian fibration, Moduli (2024);1:e4. doi:10.1112/mod.2024.3

[7] *P. Belkale*, The strange duality conjecture for generic curves, J. Amer. Math. Soc. **21** (2008), 235–258.

[8] *P. Belkale*, Strange duality and the Hitchin/WZW connection, J. Differential Geometry, **82** (2009), 445–465.

[9] *M. A. de Cataldo* and *L. Migliorini*, The perverse filtration and the Lefschetz hyperplane theorem, Ann. of Math. **171** (2010), 2089–2113.

[10] *M. de Cataldo, T. Hausel* and *L. Migliorini*, Topology of Hitchin systems and Hodge theory of character varieties: the case A_1, Ann. of Math. **175** (2012), 1329–1407.

[11] *K. Corlette*, Flat G-bundles with canonical metrics, J. Differential Geometry **28** (1988), 361–382.

[12] *R. Donagi* and *T. Pantev*, Langlands duality for Hitchin systems, Invent. Math. **189** (2012), 653–735.

[13] *S. K. Donaldson*, Twisted harmonic maps and the self-duality equations, Proc. London Math. Soc. **55** (1987), 127–131.

[14] *C. Felisetti*, $P = W$ phenomena in algebraic and enumerative geometry, Bollettino dell'Unione Matematica Italiana **18** (2025), 457–487.

[15] *L. Fredrickson*, Exponential decay for the asymptotic geometry of the Hitchin metric, Comm. Math. Phys. **375** (2020), 1393–1426.

[16] *D. S. Freed*, Special Kähler manifolds, Comm. Math. Phys. **203** (1999), 31–52.

[17] *D. Gaiotto, G. Moore* and *A. Neitzke*, Four-dimensional wall-crossing via three-dimensional field theory, Comm. Math. Phys. **299** (2010) 163–224.

[18] *G. Harder* and *M. S. Narasimhan*, On the cohomology groups of moduli spaces of vector bundles on curves, Math. Annalen **212** (1975) 215–248.

[19] *T. Hausel* and *M. Thaddeus*, Mirror symmetry, Langlands duality and Hitchin systems, Invent. Math. **153** (2003), 197–229.

[20] *T. Hausel* and *F. Rodriguez-Villegas*, Mixed Hodge polynomials of character varieties, Invent. Math. **174** With an appendix by N. M. Katz, (2008), 555–624.

[21] *T. Hausel* and *N. J. Hitchin*, Very stable Higgs bundles, equivariant multiplicity and mirror symmetry, Invent. Math. **228** (2022), 893–989.
[22] *N. J. Hitchin*, The self-duality equations on a Riemann surface, Proc. London Math. Soc. **55** (1987), 59–126.
[23] *N. J. Hitchin*, Stable bundles and integrable systems, Duke Math. J. **54** (1987), 91–114.
[24] *N. J. Hitchin*, Lie groups and Teichmüller space, Topology **31** (1992), 449 – 473.
[25] *N. J. Hitchin*, Flat connections and geometric quantization, Comm. Math. Phys. **131** (1994), 347–380.
[26] *N. J. Hitchin*, Langlands duality and G_2 spectral curves, Quart. J. Math. **58** (2007), 319–344.
[27] *N. J. Hitchin* and *L. P. Schaposnik*, Nonabelianization of Higgs bundles, J. Differential Geometry **97** (2014), 79–89.
[28] *V. Hoskins*, Two proofs of the $P = W$ conjecture [after Maulik-Shen and Hausel-Mellit-Minet-Schiffmann], Séminaire Bourbaki, **1213** (2023–2024).
[29] *A. Kapustin* and *E. Witten*, Electric-magnetic duality and the geometric Langlands program, Commun. Number Theory Phys. **1** (2007), 1– 236.
[30] *G. Laumon*, Un analogue global du ĉone nilpotent, Duke Math. J. **57** (1987), 647– 671.
[31] *G. Laumon*, Correspondance de Langlands géométrique pour les corps de fonctions, Duke Math. J. **54** (1987), 309–359.
[32] *G. Laumon* and *B.C. Ngô*, Le lemme fondamental pour les groupes unitaires, Ann. of Math. **168** (2008), 477–573.
[33] *A. Marian* and *D. Oprea*, The level rank duality for non-abelian theta functions, Invent. Math. **168** (2007), 225–247.
[34] *R. Mazzeo, J. Swoboda, H. Weiss* and *F. Witt*, Ends of the moduli space of Higgs bundles, Duke Math. J. **165** (2016), 2227–2271.
[35] *T. Mochizuki*, Asymptotic behaviour of tame harmonic bundles and an application to pure twistor $\mathcal{D}$-modules. I, II, Mem. Amer. Math. Soc., **185** (869, 870) (2007)
[36] *M. S. Narasimhan* and *C. S. Seshadri* , Stable and unitary vector bundles on a compact Riemann surface, Ann. of Math. **82** (1965), 540–567.
[37] *M. S. Narasimhan* and *S. Ramanan*, Geometry of Hecke cycles I, in: C.P. Ramanujam. - A tribute, Tata Inst. Fundam. Res., Stud. Math. **8** (1978), 291–345.
[38] *M. S. Narasimhan* and *S. Ramanan*, 2θ-linear systems on abelian varieties, in: Vector bundles on algebraic varieties (Bombay, 1984). Tata Inst. Fund. Res. Stud. Math., **11**, Tata Inst. Fund. Res., Bombay, (1987), 415–427.
[39] *B. C. Ngô*, Le lemme fondamental pour les algèbres de Lie, Inst. Hautes Études Sci. Publ. Math **111** (2010), 1–169.
[40] *C. T. Simpson*, Higgs bundles and local systems, Inst. Hautes Études Sci. Publ. Math. **75** (1992), 5–95.
[41] *C. T. Simpson*, Moduli of representations of the fundamental group of a smooth projective variety. II, Inst. Hautes Études Sci. Publ. Math **80** (1994), 5–79.
[42] *A. Strominger, S-T. Yau* and *E. Zaslow*, Mirror symmetry is T-duality, Nucl. Phys. **B 479** (1996), 243–259.
[43] *E. Verlinde*, Fusion rules and modular transformations in 2D conformal field theory, Nucl. Phys. **B 300** (1988), 360–376.
[44] *A. Wienhard*, An invitation to higher Teichmüller theory, Proceedings of the International Congress of Mathematicians (ICM 2018), World Scientific (2019), 1013–1039.

Nigel Hitchin, Mathematical Institute, University of Oxford, Woodstock Road, OX2 6GG, Oxford, UK
e-mail: Nigel.Hitchin@maths.ox.ac.uk

2001–2025

J. reine angew. Math. **545** (2002), 201–233
DOI 10.1515/crll.2002.036

Journal für die reine und angewandte Mathematik

Dynamics on K3 surfaces: Salem numbers and Siegel disks

By *Curtis T. McMullen* at Cambridge

Abstract. This paper presents the first examples of K3 surface automorphisms $f\colon X \to X$ with Siegel disks (domains on which f acts by an irrational rotation). The set of such examples is countable, and the surface X must be non-projective to carry a Siegel disk.

These automorphisms are synthesized from Salem numbers of degree 22 and trace -1, which play the role of the leading eigenvalue for $f^* \mid H^2(X)$. The construction uses the Torelli theorem, the Atiyah-Bott fixed-point theorem and results from transcendence theory.

1. Introduction

The first dynamically interesting automorphisms of compact complex manifolds arise on K3 surfaces.

Indeed, automorphisms of curves are linear (genus 0 or 1) or of finite order (genus 2 or more). Similarly, automorphisms of most surfaces (including $\mathbb{P}^2$, surfaces of general type and ruled surfaces) are either linear, finite order or skew-products over automorphisms of curves. Only K3 surfaces, Enriques surfaces, complex tori and certain non-minimal rational surfaces admit automorphisms of positive topological entropy [Ca2]. The automorphisms of tori are linear, and the Enriques examples are double-covered by K3 examples.

Over $\mathbb{R}$. To give an idea of the richness of dynamics on K3 surfaces, consider a smooth hypersurface

$$X \subset \mathbb{P}^1 \times \mathbb{P}^1 \times \mathbb{P}^1$$

of degree $(2, 2, 2)$, defined by the affine equation

$$(1.1) \qquad (1 + x^2)(1 + y^2)(1 + z^2) + Axyz = 2, \ A \in \mathbb{R}.$$

By the adjunction formula, the canonical bundle of X is trivial, and by the Lefschetz hyperplane theorem $b_1(X) = 0$, so X is a K3 surface. Its real points $X(\mathbb{R})$ for $A = 8$ are shown in Figure 1.

Research supported in part by the NSF.

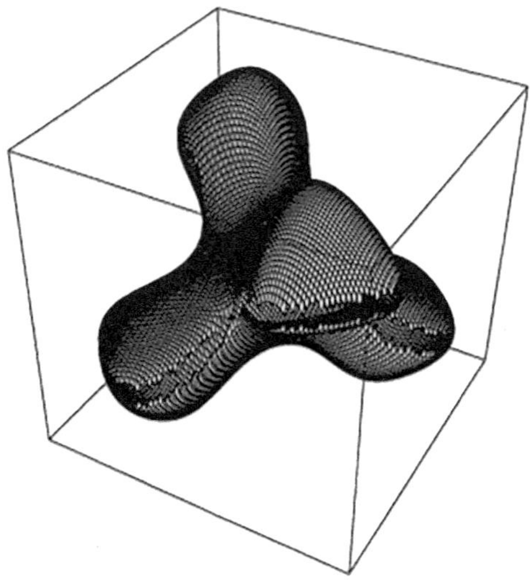

Figure 1. A K3 surface in $\mathbb{P}^1 \times \mathbb{P}^1 \times \mathbb{P}^1$.

A line through $p \in X$ parallel to the x-axis passes through a unique second point $\iota_x(p) \in X$, and similarly for y and z. For our dynamical system on X, we take the composition of these 3 involutions,

$$f = \iota_x \circ \iota_y \circ \iota_Z.$$

The automorphism $f|X(\mathbb{R})$ is area-preserving; indeed, any K3 surface carries a nowherezero holomorphic $(2,0)$-form η, which in turn determines an invariant measure $|\eta|$ on $X(\mathbb{R})$.

Orbits of $f\colon X(\mathbb{R}) \to X(\mathbb{R})$, as viewed from above, are shown in Figure 2 for $A = 2$, 2.5 and 8. When $A = 2$ (top) the dynamics is dominated by the elliptic islands predicted by KAM theory; there are many invariant disks and no dense orbits. For $A = 8$ the dynamics on $X(\mathbb{R})$ seems to be ergodic. The intermediate parameter $A = 2.5$ exhibits a mixture of behaviors: elliptic islands seem to coexist with an ergodic component of positive measure.

The dynamics in these real examples is typical for area-preserving maps on surfaces. Aside from the existence of elliptic islands, little is rigorously known about the ergodic theory of such mappings; cf. [Mak1], [Mak2].

Over $\mathbb{C}$. In this paper we will be concerned with the dynamics of automorphisms of complex K3 surfaces, $f\colon X \to X$.

Every automorphism preserves the measure $\eta \wedge \bar{\eta}$ on X determined by a nowhere zero holomorphic $(2,0)$-form η. Using Kummer surfaces, it is easy to construct ergodic K3 surface automorphisms from complex tori (§4). It is natural to ask if all automorphisms of positive topological entropy are ergodic, or at least have a dense orbit.

Siegel disks. Let us say a linear map $F(z_1, z_2) = (\lambda_1 z_1, \lambda_2 z_2)$ is an *irrational rotation* if $|\lambda_1| + |\lambda_2| = 1$ and F has dense orbits on $S^1 \times S^1$. A domain $U \subset X$ is a *Siegel disk* for f if $f(U) = U$ and $f|U$ is analytically conjugate to $F|\Delta^2$ for some irrational rotation F. (Here $\Delta = \{z\colon |z| < 1\}$.)

Like an elliptic island on $X(\mathbb{R})$, a Siegel disk on X is an obstruction to ergodicity and to the existence of dense orbits. The main result of this paper, achieved in §10, is:

Figure 2. Dynamics on real K3 surfaces.

Theorem 1.1. *There exist* K3 *surface automorphisms with Siegel disks.*

Every such automorphism has positive topological entropy.

Unfortunately, these Siegel disks are invisible to us: they live on non-projective K3 surfaces, and we can only detect them implicitly, through Hodge theory and dynamics on the cohomology. Indeed, in §7 we show:

Theorem 1.2. *There are at most countably many* K3 *surface automorphisms with Siegel disks, up to isomorphism; and there are no Siegel disks on projective* K3 *surfaces.*

Application: Kähler-Einstein metrics. Recall that any class in the Kähler cone $C_X \subset H^{1,1}(X)_{\mathbb{R}}$ of a K3 surface is represented by the symplectic form ω of a unique Kähler-Einstein metric. It would be interesting to understand the behavior of these metrics as $[\omega]$ tends to ∂C_X. In §11 we use dynamics to show that $[\omega]$ can collapse on an open set.

Theorem 1.3. *Let $f\colon X \to X$ be a K3 surface automorphism with a Siegel disk U. Then there exists a sequence of Kähler-Einstein symplectic forms such that*

$$[\omega_n] \to \xi \in \partial C_X, \quad \xi \neq 0,$$

while $\omega_n | U \to 0$ uniformly on compact sets.

Action on cohomology. We now turn to a more detailed account of the dynamics of automorphisms.

For any K3 surface, the space $H^2(X, \mathbb{Z})$ is an even, unimodular lattice of signature $(3, 19)$ with respect to the intersection form. By the Torelli theorem for K3 surfaces, $\operatorname{Aut}(X)$ acts faithfully on $H^2(X)$. Our approach to dynamics on X is via the action on $H^2(X)$.

Two useful invariants, discussed in §3, are:

$\lambda(f)$—the spectral radius of $f^* \mid H^2(X)$; and

$\delta(f)$—the eigenvalue of f^* on the line $H^{2,0}(X) = \mathbb{C} \cdot \eta$.

The topological entropy of f is given by $h(f) = \log\lambda(f) \geqq 0$. When f has positive entropy, $\lambda(f) > 1$ is the unique eigenvalue of $f^* \mid H^2(X)$ outside the unit circle. Thus $\lambda(f)$ is a *Salem number*: an algebraic unit whose conjugates other than $\lambda(f)^{\pm 1}$ lie on the unit circle.

The map f preserves $\int_X \eta \wedge \bar{\eta}$, and so $|\delta(f)| = 1$. We refer to $\delta(f)$ as the *determinant* of f, since

$$\det Df_p = \delta(f)$$

at any fixed-point p of X. When X is projective, $\delta(f)$ is a root of unity, and thus p cannot be the center of a Siegel disk. This is why we must look to non-projective surfaces for Siegel disk examples.[1)]

Construction of automorphisms. Here is a sketch of the construction of automorphisms with Siegel disks.

1. Let $\lambda > 1$ be a degree 22 Salem number with minimal polynomial $S(x) \in \mathbb{Z}[x]$. Our first goal is to construct a K3 surface automorphism $f\colon X \to X$ such that $\lambda(f) = \lambda$ and $S(x)$ is the characteristic polynomial of $f^* \mid H^2(X)$.

2. Let $B = \mathbb{Z}[y]/(S(y))$, and let K be the field of fractions of B. Let $U(x)$ be a unit in the subring of B generated by $x = y + y^{-1}$. We make B into a lattice by defining the inner product

$$\langle g_1, g_2 \rangle_{B(U)} = \operatorname{Tr}_{\mathbb{Q}}^{K}\left(\frac{U(x) g_1(y) g_2(y^{-1})}{R'(x)}\right),$$

where $R(x)$ is the minimal polynomial of $\lambda + \lambda^{-1}$. Assuming $|S(\pm 1)| = 1$ (that is, λ is *unramified*) and U is suitably chosen, the above inner product makes $B(U)$ into an even, unimodular lattice of signature (3,19). See §8.

1) 'Resonant' rotation domains, where $\det Df_p$ is a root of unity, seem unlikely to exist (§7), but we do not know how to rule them out.

3. Let $F\colon B \to B$ be multiplication by y. Then F is an isometry of the lattice $B(U)$. There is a unique eigenspace $E \subset B(U) \otimes \mathbb{C}$ of $F + F^{-1}$ with signature (2, 0). Writing

$$E = H^{2,0} \oplus H^{0,2}$$

as a sum of eigenspaces of F with eigenvalues δ and δ^{-1} respectively, we obtain a Hodge structure on $B(U)$ invariant under F.

4. By surjectivity of the period mapping, there is a K3 surface X and an isomorphism of lattices $B(U) \cong H^2(X, \mathbb{Z})$ respecting the Hodge structure. The subspace $H^{1,1}(X)$ is sufficiently irrational that $\mathrm{Pic}(X) = 0$; in particular, X is nonprojective.

5. By the Torelli theorem, there is an automorphism $f\colon X \to X$ making the diagram

$$\begin{array}{ccc} B(U) & \xrightarrow{F} & B(U) \\ \downarrow & & \downarrow \\ H^2(X) & \xrightarrow{f^*} & H^2(X) \end{array}$$

commutative (§9). Thus $\lambda(f) = \lambda$ and $\delta(f) = \delta$. The fixed-points of f are isolated because the only subvarieties of X are points.

6. Now suppose the trace of λ is -1. Then f has a unique fixed-point $p \in X$, since its Lefschetz number is given by

$$L(f) = \mathrm{Tr} f^*|(H^0 \oplus H^2 \oplus H^4) = 1 - 1 + 1 = 1.$$

The Atiyah-Bott formula relates $\mathrm{Tr}\, Df_p$ to $\mathrm{Tr} f^* \mid H^{2,0}(X) = \delta$, and we already know that $\det Df_p = \delta$. Thus the eigenvalues α, β of Df_p are determined by δ. See §6.

7. For suitable values of δ, Df_p is an irrational rotation. That is, its eigenvalues α, β lie on S^1 and are *multiplicatively independent*, meaning

$$\alpha^i = \beta^j \Leftrightarrow (i, j) = (0, 0).$$

The eigenvalues lie on S^1 if $\tau = \delta + \delta^{-1} > 1 - 2\sqrt{2}$, and they are multiplicatively independent if τ has a conjugate $\tau' < 1 - 2\sqrt{2}$. See §7.

8. Assume now that the algebraic numbers α and β are multiplicatively independent. Then they are jointly Diophantine, by a result of Fel'dman. That is, there exist $C, M > 0$ such that

$$|\alpha^i \beta^j - 1| > C(|i| + |j|)^{-M}$$

for all $(i, j) \neq (0, 0)$. The proof uses transcendence theory and the Gel'fond-Baker method.

9. By a result of Siegel and Sternberg, once the eigenvalues of Df_p are jointly Diophantine, f is locally linearizable. We conclude that f has a Siegel disk centered at p. See §5.

10. To complete the construction, we must exhibit unramified degree 22 Salem polynomials $S(x)$ of trace -1, and units $U(x)$, such that the root δ of $S(x)$ satisfies the bounds required in step 7. We note that Salem numbers with trace -1 are rather rare; there are only finitely many such numbers of degree 22, and there are no known Salem numbers of trace < -1.

The desired polynomials, located by an intensive computer search, are presented in §10.

Questions. In conclusion we remark that many basic questions concerning dynamics on *projective* K3 surfaces remain open. For example, let $f\colon X \to X$ be an automorphism of a projective K3 surface with positive entropy. Can f have a rotation domain? Does f have a dense orbit? Is f ergodic?

Notes and references. For more about dynamics on complex surfaces, see the work of Cantat [Ca2], [Ca1]. Dynamical systems on certain cubic surfaces are considered by Fried in [Fr]; see also [Br].

For an arithmetic perspective on K3 surface dynamics, see work of Silverman, Mazur and Wang: [Sil], [Maz], [Wa]. Silverman studies the intersection of $(1, 1)$ and $(2, 2)$ hypersurfaces in $\mathbb{P}^2 \times \mathbb{P}^2$, while Wang studies $(2, 2, 2)$ surfaces in $\mathbb{P}^1 \times \mathbb{P}^1 \times \mathbb{P}^1$ as we do above.

Acknowledgements. I would like to thank B. Mazur and B. Gross for many stimulating discussions. Mazur introduced me to dynamics on K3 surfaces in 1992, including the examples of $(2, 2, 2)$ hypersurfaces presented above; this paper grew out of our dialog. Gross helped engineer the number-theoretic construction of unimodular lattices; these matters are developed in more detail in [GM]. I would also like to thank D. Allcock, W. Stein and the referees.

2. K3 surfaces

This section provides a resumé of basic facts about K3 surfaces and maps between them; for details see [BPV], [Bv2]. We start with some terminology concerning lattices, to streamline the discussion of the intersection form.

Lattices. A *lattice* is a finitely-generated free abelian group $L \cong \mathbb{Z}^n$ equipped with a non-degenerate symmetric bilinear form (or inner product) $\langle x, y\rangle_L \in \mathbb{Z}$. The lattice L is *even* if $\langle x, x\rangle_L \in 2\mathbb{Z}$ for all $x \in L$; otherwise L is *odd*. If the inner product gives an isomorphism between L and $L^* = \mathrm{Hom}(L, \mathbb{Z})$, then L is *unimodular*. We say L has *signature* (p, q) if the quadratic form $\langle x, x\rangle_L$ on $L \otimes \mathbb{R} \cong \mathbb{R}^n$ is equivalent to

$$x_1^2 + \cdots + x_p^2 - x_{p+1}^2 - \cdots - x_{p+q}^2.$$

If $(p, q) = (n, 0)$ or $(0, n)$, then L is *definite*; otherwise L is an *indefinite* lattice.

Any two even, indefinite, unimodular lattices with the same signature are isomorphic. There exists an even unimodular lattice with signature (p, q) iff $p \equiv q \bmod 8$. See [Ser1], §5, [MH].

K3 surfaces. A *complex surface* X is a connected compact complex manifold with $\dim_{\mathbb{C}} X = 2$. If X is simply-connected and the canonical bundle of X is trivial, then X is a K3 *surface*. All K3 surfaces are diffeomorphic.

Intersection form. Let X be a K3 surface, and let

$$(C, D) \mapsto C \cdot D \in \mathbb{Z}$$

denote the intersection form on the middle-dimensional cohomology group $H^2(X, \mathbb{Z})$. This form makes $H^2(X, \mathbb{Z})$ into an even unimodular lattice of signature (3, 19). As remarked above, all such lattices are isomorphic. Thus in an appropriate basis, the intersection form is given by $C \cdot D = CQD^t$ where $Q = 3H \oplus 2(-E_8)$, $H = \begin{pmatrix} 0 & 1 \\ 1 & 0 \end{pmatrix}$ and

$$E_8 = \begin{pmatrix} 2 & 0 & 0 & -1 & 0 & 0 & 0 & 0 \\ 0 & 2 & -1 & 0 & 0 & 0 & 0 & 0 \\ 0 & -1 & 2 & -1 & 0 & 0 & 0 & 0 \\ -1 & 0 & -1 & 2 & -1 & 0 & 0 & 0 \\ 0 & 0 & 0 & -1 & 2 & -1 & 0 & 0 \\ 0 & 0 & 0 & 0 & -1 & 2 & -1 & 0 \\ 0 & 0 & 0 & 0 & 0 & -1 & 2 & -1 \\ 0 & 0 & 0 & 0 & 0 & 0 & -1 & 2 \end{pmatrix}$$

defines the positive-definite even unimodular form of rank 8. For brevity we write $C^2 = C \cdot C$.

Since X is simply-connected, its other integral cohomology groups are given by $H^0(X, \mathbb{Z}) = H^4(X, \mathbb{Z}) = \mathbb{Z}$ and $H^1(X, \mathbb{Z}) = H^3(X, \mathbb{Z}) = 0$.

Signatures. The intersection form determines a Hermitian inner product on $H^2(X, \mathbb{C}) = H^2(X, \mathbb{Z}) \otimes \mathbb{C}$. We say a subspace $S \subset H^2(X, \mathbb{C})$ has *signature* (p, q) if $S = \bar{S}$ and the intersection pairing on S is non-degenerate, with signature (p, q). In this case we have $H^2(X, \mathbb{C}) = S \oplus S^\perp$, and $S^\perp$ has signature $(3 - p, 19 - q)$.

Hodge structure. The complex structure of X is reflected in the Hodge decomposition

$$H^2(X, \mathbb{C}) = H^{2,0}(X) \oplus H^{1,1}(X) \oplus H^{0,2}(X).$$

Here $H^{1,1} = \overline{H^{1,1}}$ and $H^{2,0} = \overline{H^{0,2}}$. Since the canonical bundle of X is trivial, we have

$$H^{2,0}(X) = \mathbb{C} \cdot \eta$$

where η is a nowhere vanishing holomorphic $(2, 0)$-form, locally given by

$$\eta = \eta(z_1, z_2) dz_1 \wedge dz_2.$$

Noting that

$$\eta \cdot \bar{\eta} = \int_X \eta \wedge \bar{\eta} = 4 \int_X |\eta|^2 > 0,$$

we find the orthogonal subspaces $H^{1,1}$ and $H^{2,0} \oplus H^{0,2}$ have signatures $(1, 19)$ and $(2, 0)$ respectively.

Projective surfaces. The *Picard group* of X is given by

$$\operatorname{Pic}(X) = H^{1,1}(X) \cap H^2(X, \mathbb{Z});$$

it classifies holomorphic line bundles on X. A K3 surface is projective iff there is a line bundle $L \in \operatorname{Pic}(X)$ with $L^2 > 0$ ([BPV], IV.5).

The Kähler cone. Every K3 surface is Kähler. The *Kähler cone*

$$C_X \subset H^{1,1}(X)_{\mathbb{R}}$$

is the set of all classes represented by the symplectic forms of Kähler metrics on X. (Here $V_{\mathbb{R}} = \{v \in V : v = \overline{v}\}$.)

The shape of the Kähler cone can be made more explicit as follows. Let

$$\Delta(X) = \{D \in \operatorname{Pic}(X) : D^2 = -2\},$$

and

$$W(X) = \{\omega \in H^{1,1}(X)_{\mathbb{R}} : \omega^2 > 0 \text{ and } \omega \cdot \mathrm{D} \neq 0 \text{ for all } D \in \Delta(X)\}.$$

Since the intersection form on $H^{1,1}(X)_{\mathbb{R}}$ has signature $(1, 19)$, $W(X)$ is the cone over two copies of the hyperbolic space $\mathbb{H}^{19}$, with a configuration of hyperplanes corresponding to $\Delta(X)$ deleted.

It is known that the Kähler cone C_X coincides with a component or *chamber* of $W(X)$. The automorphisms of $H^2(X, \mathbb{Z})$ preserving the intersection form and Hodge structure act transitively on the set of chambers. (Observe that such automorphisms include the reflections $C \mapsto C + (C \cdot D)D$ through the hyperplanes defined by $D \in \Delta(X)$.)

Torelli theorem. The Torelli theorem asserts that a K3 surface is determined up to isomorphism by its Hodge structure. More precisely we have:

Theorem 2.1. *Let X and Y be* K3 *surfaces, and let*

$$F : H^2(X, \mathbb{Z}) \to H^2(Y, \mathbb{Z})$$

be an isomorphism preserving the intersection pairing. Extend F to $H^2(X, \mathbb{C})$ by tensoring with $\mathbb{C}$; then:

1. *if F sends $H^{2,0}(X)$ to $H^{2,0}(Y)$, then X and Y are isomorphic;*
2. *if F also sends C_X to C_Y, then $F = f^*$ for a unique isomorphism $f : Y \to X$.*

Marked K3 surfaces. Next we discuss the space of all possible Hodge structures on a K3 surface.

Let L be a fixed even, unimodular lattice of signature $(3, 19)$. A *Hodge structure* on L is a splitting

$$L \otimes \mathbb{C} = H^{2,0} \oplus H^{1,1} \oplus H^{0,2}$$

such that $\dim H^{2,0} = 1$, $H^{2,0} = \overline{H^{0,2}}$, $H^{2,0} \oplus H^{0,2}$ has signature (2, 0) and

$$H^{1,1} = (H^{2,0} \oplus H^{0,2})^{\perp}.$$

The space of Hodge structures on L is parameterized by the *period domain*

$$\Omega(L) = \{[\eta] \in \mathbb{P}(L \otimes \mathbb{C}) : \eta \cdot \eta = 0 \text{ and } \eta \cdot \bar{\eta} > 0\},$$

via the correspondence

$$H^{2,0} \oplus H^{1,1} \oplus H^{0,2} = \mathbb{C} \cdot \eta \oplus \{\eta, \bar{\eta}\}^{\perp} \oplus \mathbb{C} \cdot \bar{\eta}.$$

The period domain is an open subset of a smooth, 20-dimensional quadric hypersurface in $\mathbb{P}^{21}$.

Now let X be a K3 surface. A *marking* for X is an isomorphism

$$\iota\colon H^2(X,\mathbb{Z}) \cong L$$

preserving the intersection pairing. Every K3 surface admits a marking. Two marked surfaces (X_1,ι_1), (X_2,ι_2) are *equivalent* if there is an isomorphism $f\colon X_1 \to X_2$ such that $\iota_2 = \iota_1 \circ f^*$.

Let $\mathcal{M}(L)$ be the *moduli space* of equivalence classes of K3 surfaces marked by L. The *period mapping*

$$\pi\colon \mathcal{M}(L) \to \Omega(L) \subset \mathbb{P}(L \otimes \mathbb{C})$$

is defined by $\pi(X,\iota) = [\iota(\eta)]$, where $\eta \neq 0$ is a holomorphic $(2,0)$-form on X. The image of π lies in the period domain because $\eta \wedge \eta = 0$ and $\eta \cdot \bar{\eta} > 0$.

The next result complements the Torelli theorem by showing all possible Hodge structures on K3 surfaces actually arise:

Theorem 2.2. *The period mapping $\pi\colon \mathcal{M}(L) \to \Omega(L)$ is surjective.*

Note: the period mapping is not injective, because the marked Hodge structure on $H^2(X)$ does not uniquely determine the Kähler cone C_X.

3. Automorphisms of K3 surfaces

Given a K3 surface X, let $\mathrm{Aut}(X)$ denote the group of biholomorphic maps $f\colon X \to X$. By the Torelli theorem, the map $f \mapsto f^* \mid H^2(X)$ gives an isomorphism

$$\mathrm{Aut}(X) \cong \mathrm{Aut}(H^2(X,\mathbb{Z}), H^{2,0}(X), C_X).$$

That is, $\mathrm{Aut}(X)$ acts faithfully on $H^2(X,\mathbb{Z})$, preserving the intersection form, the Hodge decomposition and the Kähler cone C_X; and conversely, any automorphism of $H^2(X,\mathbb{Z})$ preserving these three structures is induced by an automorphism of X.

In this section we study the eigenvalues of $f^* \mid H^2(X)$, and the synthesis of automorphisms with a given action on $H^2(X)$.

Algebraic numbers. A complex number λ is *algebraic* if $P(\lambda) = 0$ for some irreducible, monic polynomial $P(t) \in \mathbb{Q}[t]$. The *degree* of λ is the degree of $P(t)$, and the roots of $P(t)$ in $\mathbb{C}$ are the *conjugates* of λ. If we have $P(t) \in \mathbb{Z}[t]$ then λ is an *algebraic integer*; and if λ^{-1} is also an algebraic integer, then λ is a *unit*. If λ is a root of unity, then $P(t)$ is a *cyclotomic polynomial*. An algebraic integer is a root of unity iff all its conjugates lie on the unit circle $S^1 \subset \mathbb{C}$.

Salem numbers. A *Salem number* is a unit $\lambda > 1$ whose conjugates other than $\lambda^{\pm 1}$ lie on the unit circle. The irreducible polynomial $S(t)$ of λ is a *Salem polynomial*; it has even degree, with roots of the form

$$\{\lambda, \lambda^{-1}, \alpha_1, \bar{\alpha}_1, \ldots, \alpha_d, \bar{\alpha}_d\},$$

where $|\alpha_i| = 1$. (For our purposes it is natural to permit quadratic Salem numbers; these were excluded in Salem's original definition [Sa], III.3.)

A *Salem trace* is an algebraic integer $\tau > 2$ whose other conjugates lie in $[-2, 2]$; its irreducible polynomial is a *Salem trace polynomial*. Salem traces and Salem numbers correspond bijectively, via the relation $\tau = \lambda + \lambda^{-1}$.

Orthogonal groups. Let $\mathrm{O}(p,q) \subset \mathrm{GL}_n(\mathbb{R})$ denote the orthogonal group of the real quadratic form

$$x_1^2 + \cdots + x_p^2 - x_{p+1}^2 - \cdots - x_{p+q}^2$$

of signature (p,q). Given $T \in \mathrm{O}(p,q)$ we obtain a T-invariant splitting

$$\mathbb{C}^{p+q} = \oplus E(\lambda),$$

where $E(\lambda) = \bigcup \mathrm{Ker}(\lambda I - T)^n$ is the generalized λ-eigenspace of T. The intersection pairing is trivial between $E(\lambda)$ and $E(\mu)$ unless $\lambda\mu = 1$; thus the set of eigenvalues of T is invariant under $\lambda \mapsto \lambda^{-1}$.

Lemma 3.1. *A transformation $T \in \mathrm{O}(p,q)$ has at most $\min(p,q)$ eigenvalues outside the unit circle, counted with their multiplicities.*

Proof. The subspace $S = \bigoplus_{|\lambda|>1} E(\lambda)$ is isotropic and defined over $\mathbb{R}$, so

$$\dim S \leqq \min(p,q).$$

But $\dim S$ is at least as large as the number of eigenvalues outside S^1. □

Automorphisms. Now let $f\colon X \to X$ be an automorphism of a K3 surface, let $f^* = f^* \mid H^2(X, \mathbb{Z})$, and let

$$\lambda(f) = \sup\{|\lambda|\colon \det(\lambda I - f^*) = 0\}$$

be the spectral radius of f^*. It is known that the topological entropy of f is given by $h(f) = \log\lambda(f)$; thus f has positive entropy iff $\lambda(f) > 1$ ([Ca1], Theorem 2.1.5, see also [Gr], [Ym], [Frl]).

Since f^* is invertible and preserves the integral structure on H^2, its eigenvalues are algebraic integers, in fact units.

Theorem 3.2. *Either all eigenvalues of $f^* \mid H^2(X)$ are roots of unity, or there is a unique, simple eigenvalue λ with $|\lambda| > 1$, and λ is a Salem number.*

Proof. Since the conjugates of eigenvalues are again eigenvalues, if they all lie on S^1 then they are all roots of unity.

Now suppose f^* has an eigenvalue with $|\lambda| > 1$. Since f^* stabilizes the subspace $H^{1,1}(X) \subset H^2(X)$ of signature (1, 19), it is conjugate to a transformation

$$T \in \mathrm{O}(2,0) \times \mathrm{O}(1,19).$$

By Lemma 3.1, f^* has at most one eigenvalue outside the unit circle, so λ is unique, and therefore real.

As remarked above, λ is a unit. Since f^* preserves the Kähler cone, its $\mathrm{O}(1,19)$ part does not interchange the sheets of the light-cone in $H^{1,1}_{\mathbb{R}}$, and therefore $\lambda > 1$. Applying the same reasoning to f^{-1}, we find that $\lambda^{-1} > 0$ is the unique eigenvalue of f^* inside the unit circle. Since all other eigenvalues of f^* lie on the unit circle, λ is a Salem number. □

Corollary 3.3. *The irreducible factors of the characteristic polynomial of* $f^* \mid H^2(X)$ *include at most one Salem polynomial; and the remaining factors are cyclotomic.*

Synthesis of automorphisms. The next theorem provides the key to building examples of K3 surface automorphisms. It reduces the construction of automorphisms to a problem in integral quadratic forms. It also represents the first step towards determining which Salem numbers can arise as $\lambda(f)$.

Theorem 3.4 (Synthesis). *Let* $F\colon L \to L$ *be an automorphism of an even, unimodular lattice of signature* $(3,19)$. *Suppose* $S(t) = \det(tI - F)$ *is a Salem polynomial. Then there is:*

- *a* K3 *surface automorphism* $f\colon X \to X$; *and*
- *a marking* $\iota\colon H^2(X,\mathbb{Z}) \to L$, *such that* $F = \iota \circ f^* \circ \iota^{-1}$.

Proof. Since F has only two eigenvalues off the unit circle, while the signature of L is $(3,19)$, there exists an eigenvector $\eta \in L \otimes \mathbb{C}$ such that $T(\eta) = \delta\eta$, $|\delta| = 1$ and $\eta \cdot \bar{\eta} > 0$. By surjectivity of the period mapping (Theorem 2.2), there exists a K3 surface X and a marking $\iota\colon H^2(X,\mathbb{Z}) \to L$ such that $\iota\big(H^{2,0}(X)\big) = \mathbb{C} \cdot \eta$.

Let $T\colon H^2(X,\mathbb{Z}) \to H^2(X,\mathbb{Z})$ be the automorphism given by $T = \iota^{-1} \circ F \circ \iota$. Then T respects the intersection pairing and the Hodge structure on $H^2(X)$, and its characteristic polynomial is also $S(t)$.

We claim that $\mathrm{Pic}(X) = 0$. Indeed, since $S(t)$ is irreducible, T has no proper rational invariant subspace, and thus $H^{1,1}(X) \cap H^2(X,\mathbb{Z}) = (0)$. In particular, $\Delta(X) = \emptyset$, so the Kähler cone $C_X \subset H^{1,1}(X)_{\mathbb{R}}$ is simply one of the two components of the space $W(X) = \{\omega\colon \omega^2 > 0\}$.

Since the leading eigenvalue of T is a Salem number $\lambda > 1$, T does not interchange the components of $W(X)$, and therefore $T(C_X) = C_X$. By the Torelli theorem (Theorem 2.1), there is a unique automorphism $f\colon X \to X$ such that $f^* \mid H^2(X,\mathbb{Z}) = T$. □

Remarks. The marked K3 surface X constructed above is unique up to complex conjugation. Note that X is non-projective, since $\mathrm{Pic}(X) = (0)$.

The determinant $\delta(f)$. Let $\eta \neq 0$ be a holomorphic $(2,0)$-form on X. Since $\dim H^{2,0}(X) = 1$, η is always an eigenvector for f^*. Let

$$\delta(f) = \mathrm{Tr} f^* \mid H^{2,0}(X) = f^*\eta/\eta$$

denote the corresponding eigenvalue. The norm $\eta \cdot \bar{\eta} > 0$ is preserved by f, so we have $|\delta(f)| = 1$.

Now suppose $p \in X$ is a fixed-point of f, and let $Df_p\colon T_pX \to T_pX$ denote the complex derivative. Since $\eta(p)$ lies in $\wedge^2 T_p^*(X)$, we have

$$\delta(f) = \det Df_p.$$

In particular, $\det Df_p$ is the same at all fixed-points. For this reason, we call $\delta(f)$ the *determinant* of f.

We conclude this section with two results about the eigenvalue $\delta(f)$.

Theorem 3.5. *If f is an automorphism of a projective* K3 *surface X, then $\delta(f)$ is a root of unity.*

Proof. Since X is projective, there is a $D \in \operatorname{Pic}(X) \subset S$ with $D^2 > 0$. The subspace $H^{1,1}(X) \cap D^\perp$ is negative-definite, with signature $(0, 19)$, and contains $\operatorname{Pic}(X) \cap D^\perp$. Thus the intersection form on $\operatorname{Pic}(X) \otimes \mathbb{R}$ has signature $(1, n)$ for some n, $0 \leqq n \leqq 19$.

Consequently, the rational f^*-invariant subspace

$$S = \operatorname{Pic}(X)^\perp \supset H^{2,0}(X) \oplus H^{0,2}(X)$$

has signature $(2, 19 - n)$. Now $f^*|S$ preserves the signature (2,0)-subspace on the right, so it is conjugate to an element of $\mathrm{O}(2) \times \mathrm{O}(19 - n)$. Thus all eigenvalues of $f^*|S$, including $\delta(f)$, lie on the unit circle. But $f^*|S$ also preserves the lattice $S \cap H^2(X, \mathbb{Z})$, so its characteristic polynomial lies in $\mathbb{Z}[t]$, and therefore its eigenvalues are roots of unity. □

Theorem 3.6. *Up to isomorphism, there are only countably many pairs (X, f) where $\delta(f)$ is not a root of unity.*

Proof. Assume $\delta = \delta(f)$ is not a root of unity. By Corollary 3.3, δ has multiplicity one as a root of the characteristic polynomial for $f^* \mid H^2(X, \mathbb{Z})$. Thus $H^{2,0}(X)$ is an eigenspace for f^*, and therefore f^* determines the Hodge structure on $H^2(X)$ up to finitely many choices. By the Torelli theorem, the Hodge structure on $H^2(X)$ together with $f^* \mid H^2(X)$ determines (X, f) up to isomorphism. Since there are only countably many possibilities for $f^* \mid H^2(X, \mathbb{Z})$, there are only countably many possibilities for (X, f). □

In §4 we will construct countably many examples where $\delta(f)$ is not a root of unity. The invariant $\delta(f)$ is studied on more general manifolds in [Bv1], which includes a version of Theorem 3.5.

4. Ergodic dynamics on Kummer surfaces

Every K3 surface X carries a natural probability measure μ, defined by

$$\mu(A) = \int_A \eta \wedge \bar{\eta},$$

where η is a holomorphic $(2, 0)$-form on X, scaled so $\mu(X) = 1$. All automorphisms of X preserve the measure μ.

An automorphism $f\colon X \to X$ is *ergodic* if every f-invariant Borel set satisfies $\mu(A) = 0$ or 1. It is *mixing* if any pair of Borel sets satisfy

$$\mu(f^n(A) \cap B) \to \mu(A)\mu(B)$$

as $|n| \to \infty$. Mixing implies ergodicity.

In this section we discuss an elementary construction of mixing automorphism of K3 surfaces, using complex tori. The same construction yields automorphisms where $\delta(f)$ is not a root of unity. For example we will show:

Theorem 4.1. *For every $a = 0, 1, 2 \ldots$, there exists a mixing automorphism of a* K3 *surface $f\colon X \to X$ such that $\lambda(f)$ and $\delta(f)$ are roots of the Salem polynomial*

$$S(t) = t^6 - at^5 - t^4 + (2a-1)t^3 - t^2 - at + 1.$$

Complex tori. An n-dimensional *complex torus* is a compact complex Lie group of the form $Y = \mathbb{C}^n/\Lambda$, where Λ is a lattice in $\mathbb{C}^n$. As in the case of K3 surfaces, the canonical bundle of Y is trivial; $H^{n,0}(Y)$ is 1-dimensional, spanned by the holomorphic form $\xi = dz_1 \wedge \cdots \wedge dz_n$.

Let $\mathrm{Aut}(Y)$ denote the automorphisms of Y as a complex Lie group. Elements $F \in \mathrm{Aut}(Y)$ correspond bijectively to linear maps

$$\tilde{F}\colon \mathbb{C}^n \to \mathbb{C}^n$$

satisfying $\tilde{F}(\Lambda) = \Lambda$. The eigenvalues $(\lambda_1, \ldots, \lambda_n)$ of $\tilde{F}$ coincide with the eigenvalues of $F^* \mid H^1(Y)$. Defining $\lambda(F)$ as the spectral radius of $F^* \mid H^*(Y)$ and $\delta(F)$ as $F^*\xi/\xi$, one can easily check that:

$$\lambda(F) = \prod \{|\lambda_i|^2 \colon |\lambda_i| \geqq 1\}$$

and

$$\delta(F) = \lambda_1 \lambda_2 \cdots \lambda_n.$$

We have $|\delta(F)| = |\det(F)| = 1$.

The measure $\xi \wedge \bar{\xi}$ on Y is $\mathrm{Aut}(Y)$-invariant. Using Fourier analysis on the space of square-integrable functions $L^2(Y)$, one finds (cf. [Man], Chapter II, 3.1, 8.5 and 8.6):

Theorem 4.2. *An automorphism $F \in \mathrm{Aut}(Y)$ is ergodic iff F is mixing iff no eigenvalue of $F^* \mid H^1(Y)$ is a root of unity.*

Kummer surfaces. Now let $Y = \mathbb{C}^2/\Lambda$ be a 2-dimensional complex torus, and let $\iota\colon Y \to Y$ be the involution $\iota(y) = -y$. Blowing up the 16 double points of the quotient space Y/ι, we obtain the *Kummer surface* $X = \kappa(Y)$; it comes equipped with a degree 2 rational map $Y \dashrightarrow X$. The complex manifold X is simply-connected, and the form $\xi = dz_1 \wedge dz_2$ on Y descends to a nowhere-vanishing holomorphic $(2,0)$-form η on X; thus X is a K3 surface.

Since every $F \in \mathrm{Aut}(Y)$ commutes with the involution ι, it gives rise to an automorphism $f = \kappa(F) \in \mathrm{Aut}(X)$.

Theorem 4.3. *Let F be an automorphism of a complex 2-torus Y, and f the corresponding automorphism of the Kummer surface $X = \kappa(Y)$. Then*

$$\delta(F) = \delta(f), \quad \lambda(F) = \lambda(f),$$

and F is mixing $\Leftrightarrow$ f is mixing.

Proof. The rational map $Y \dashrightarrow X$ determines a natural isomorphism

$$H^2(X) = H^2(Y) \oplus \mathbb{C}^{Y[2]},$$

where $Y[2] \subset Y$ denotes the 16 points of order 2 [BPV], VIII.5. Under this isomorphism, the action of $f^* \mid H^2(Y)$ corresponds to the action of $F^* \mid H^2(Y)$ together with the action of F by permutations on $Y[2]$; thus $\lambda(f) = \lambda(F)$. Moreover $H^{2,0}(X)$ corresponds to $H^{2,0}(Y)$, so $\delta(f) = \delta(F)$. For the assertions on mixing, use the fact that $L^2(X)$ is canonically identified with the set of even functions in $L^2(Y)$. □

Example: $\mathbf{SL_2(\mathbb{Z})}$. As a first example, let $Y = E \times E$ where $E = \mathbb{C}/\Lambda$ is a 1-dimensional complex torus. Then each matrix $A = \begin{pmatrix} a & b \\ c & d \end{pmatrix} \in \mathrm{SL}_2(\mathbb{Z})$ acts on Y by

$$F_A(e_1, e_2) = (ae_1 + be_2, ce_1 + de_2),$$

and F_A is mixing iff A has no eigenvalues on S^1. Thus each matrix $A \in SL_2(\mathbb{Z})$ with $|\mathrm{Tr}\, A| > 2$ gives rise to a mixing automorphism $f = \kappa(F_A)$ on the Kummer surface $X = \kappa(E \times E)$.

Synthesis of dynamics. These $\mathrm{SL}_2(\mathbb{Z})$ examples satisfy $\delta(f) = 1$. To obtain a greater variety of values for $\delta(f)$, we will *synthesize* an automorphism from the desired characteristic polynomial for $F^* \mid H^1(Y)$.

Theorem 4.4. *Let $P(t) = t^{2n} + a_1 t^{2n-1} + \cdots + a_{2n-1} t + 1 \in \mathbb{Z}[t]$ be a polynomial whose roots occur in conjugate pairs*

$$\{\alpha_1, \bar{\alpha}_1, \ldots, \alpha_n, \bar{\alpha}_n\}.$$

Then there is a complex torus Y and an $F \in \mathrm{Aut}(Y)$ such that

$$P(t) = \det\left(tI - F^*|H^1(Y)\right)$$

and

$$\delta(F) = \alpha_1 \alpha_2 \cdots \alpha_n.$$

Proof. Choose a semisimple transformation $T \in \mathrm{SL}_{2n}\mathbb{Z}$ whose characteristic polynomial is $P(t)$; this is possible because $P(0) = 1$. Since the roots of $P(t)$ occur in conjugate pairs, we can find a basis $(v_i, \overline{v}_i)_1^n$ for $\mathbb{C}^{2n}$ such that $Tv_i = \alpha_i v_i$. The pairs of conjugate eigenvectors determine a splitting $\mathbb{R}^{2n} = \bigoplus_1^n E_i$ where $E_i \otimes \mathbb{C} = \mathbb{C}v_i \oplus \mathbb{C}\overline{v}_i$. We can then choose an $\mathbb{R}$-linear isomorphism $\phi \colon \mathbb{R}^{2n} \to \mathbb{C}^n$ sending each E_i to a coordinate axis, such that $\phi(Tv) = \tilde{F}(\phi(v))$, where $\tilde{F}(z_1, \ldots, z_n) = (\alpha_1 z_1, \ldots, \alpha_n z_n)$. Setting $\Lambda = \phi(\mathbb{Z}^{2n})$ and $Y = \mathbb{C}^n/\Lambda$, and letting $\tilde{F}$ descend to $F \in \mathrm{Aut}(Y)$, we obtain the required automorphism. □

Examples: degree 6 Salem numbers. For any integer $a \geqq 0$, it is easy to check that

$$P(t) = t^4 + at^2 + t + 1$$

has only complex roots, say $\{\alpha, \beta, \bar{\alpha}, \bar{\beta}\}$ with $|\alpha| > |\beta|$. By the preceding result, there is a complex 2-torus Y and an automorphism $F \in \mathrm{Aut}(Y)$ such that

$$\delta(F) = \alpha\beta, \quad \lambda(F) = |\alpha|^2,$$

and $P(t)$ is the characteristic polynomial of $F^* \mid H^1(Y)$.

The products of pairs of distinct roots of $P(t)$ give the roots of the characteristic polynomial

$$S(t) = t^6 - at^5 - t^4 + (2a - 1)t^3 - t^2 - at + 1$$

of $F^* \mid H^2(Y) = \wedge^2 H^1(Y)$. Thus $\delta(F)$ and $\lambda(F)$ are roots of $S(t)$. Similarly,

$$\tau = \lambda(F) + \lambda(F)^{-1} > 2$$

is a root of the cubic polynomial

$$R(t) = (t - a)(t^2 - 4) - 1.$$

See Figure 3. (The formulas for $S(t)$ and $R(t)$ come from a straightforward calculation with determinants and the companion matrix of $P(t)$.)

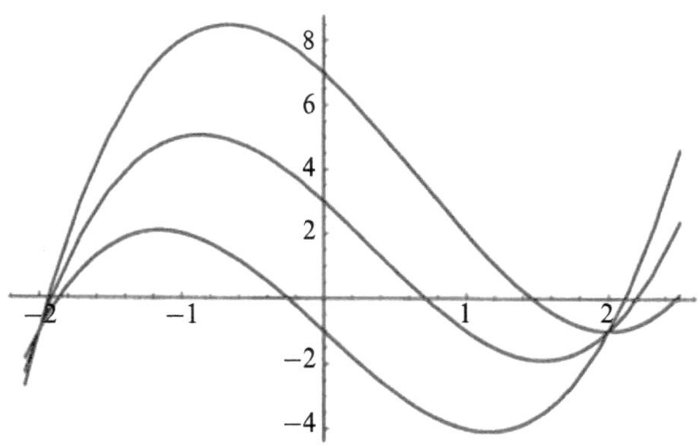

Figure 3. Salem trace polynomials $R(t) = (t - a)(t^2 - 4) - 1$ for $a = 0, 1, 2$.

Since $R(-2) = R(2) = -1$ while $R(-1) > 0$, the roots of $R(t)$ other than τ lie in the interval $[-2, 2]$, and since $R(n) \neq 0$ for $n \in \mathbb{Z}$, $R(t)$ is irreducible. Thus $R(t)$ is a Salem trace polynomial. Therefore $\lambda(F)$ is a sextic Salem number, and $S(t)$ is a Salem polynomial.

Proof of Theorem 4.1. In the examples just discussed, pass to the Kummer surface $X = \kappa(Y)$. □

Note that in these examples, $\delta(F) = \alpha\beta$ is not a root of unity, since it is conjugate to $\lambda(F) = |\alpha|^2 > 1$.

Lattès examples. We remark that the automorphisms of Kummer surfaces discussed above are higher-dimensional relatives of the *Lattès examples* of rational maps $f: \mathbb{P}^1 \to \mathbb{P}^1$. To construct f, start with an endomorphism $F(y) = ny$ of a complex 1-torus $Y = \mathbb{C}/\Lambda$. Observing that the quotient $X = Y/(y \sim -y)$ is isomorphic to $\mathbb{P}^1$, one finds that F descends a rational map $f: \mathbb{P}^1 \to \mathbb{P}^1$ of degree n^2, and f is ergodic if $n > 1$. These are among the simplest rational maps with Julia set the whole Riemann sphere [Mil], §7, [Mc], §3.5.

Towards Siegel disks. As discussed above, the ergodicity of an automorphism of a complex torus can be detected by its action on cohomology: F is ergodic iff the eigenvalues of $F^* \mid H^1(Y)$ include no roots of unity.

Similarly, the next three sections develop a cohomological criterion for the existence of a *Siegel disk* on a K3 surface, based on the action of f^* on $H^2(X)$.

5. Siegel disks and transcendence theory

This section applies transcendence theory to establish the existence of Siegel disks in an arithmetic setting. The main result is:

Theorem 5.1. *Let X be a complex n-manifold, and let $f: X \to X$ be a holomorphic map fixing $p \in X$. If the eigenvalues (λ_i) of Df_p are algebraic, multiplicatively independent and satisfy $|\lambda_i| = 1$, then f has a Siegel disk at p.*

The proof uses results originally developed to address Hilbert's seventh problem, on the transcendence of numbers like $2^{\sqrt{2}}$.

Definitions. Let $(\lambda_1, \ldots, \lambda_n)$ be nonzero complex numbers. We say (λ_i) are *multiplicatively independent* if the only solution to

$$\lambda_1^{k_1} \cdots \lambda_n^{k_n} = 1$$

with $k = (k_1, \ldots, k_n) \in \mathbb{Z}^n$ is $k = 0$. The numbers (λ_i) are *jointly Diophantine* if there exist $C, M > 0$ such that for all integral exponents $(k_1, \ldots, k_n) \in \mathbb{Z}^n$, not all zero, we have

$$|\lambda_1^{k_1} \cdots \lambda_n^{k_n} - 1| > C(\max|k_i|)^{-M} > 0. \tag{5.1}$$

Theorem 5.2. *Let $(\lambda_1, \ldots, \lambda_n)$ be multiplicatively independent algebraic numbers on S^1. Then $(\lambda_1, \ldots, \lambda_n)$ are jointly Diophantine.*

Proof. By multiplicative independence, the numbers $(2\pi i, \log \lambda_1, \ldots, \log \lambda_n)$ are linearly independent over $\mathbb{Q}$. Using the Gel'fond-Baker method, Fel'dman shows that linear independence over $\mathbb{Q}$ implies linear independence over any number field, in the following effective sense [Fe]: for any algebraic numbers (k_i) not all zero, we have

$$|k_0 2\pi i + k_1 \log \lambda_1 + \cdots + k_n \log \lambda_n| > \exp(-M(d + \log H)), \tag{5.2}$$

where d is the degree of the field $\mathbb{Q}[k_0, \ldots, k_n, \lambda_1, \ldots, \lambda_n]$, $M = M(\lambda_i, d)$ is a constant depending only on the numbers (λ_i) and d, $H = \max H(k_i)$, and the *height* $H(k) = \sum |a_j|$ if

$p(k) = 0$ where $p(x) = \sum_0^s a_j x^j$ is an irreducible polynomial with relatively prime coefficients $a_j \in \mathbb{Z}$.

For $k_i \in \mathbb{Z}$ we have $H = \max|k_i|$, and M depends only on (λ_i); therefore

$$\exp(-M(d + \log H)) = C(\max|k_i|)^{-M}$$

for some $C > 0$. Thus (5.2) implies the Diophantine condition (5.1). □

Remark. Taking $\lambda_1^{k_1} = \alpha^\beta$, Fel'dman's bound (5.2) implies the Gel'fond-Schneider theorem: α^β is transcendental for all algebraic numbers $\alpha \neq 0, 1$ and $\beta \notin \mathbb{Q}$.

Siegel disks. Let $F\colon \mathbb{C}^n \to \mathbb{C}^n$ be a linear map of the form

$$F(z_1, \ldots, z_n) = (\lambda_1 z_1, \ldots, \lambda_n z_n)$$

with $|\lambda_i| = 1$. We say F is an *irrational rotation* if the numbers (λ_i) are multiplicatively independent; equivalently, if the orbits of F on $(S^1)^n$ are dense.

Now let $f\colon X \to X$ be a holomorphic endomorphism of a complex n-manifold, and suppose $f(p) = p$. We say f has a *Siegel disk* U, centered at p, if there is a neighborhood U of p and an analytic isomorphism to a polydisk,

$$\phi\colon (U, p) \to (\Delta^n, 0),$$

sending the action of $f|U$ to the action of an irrational rotation $F|\Delta^n$.

Here is a local criterion for the existence of a Siegel disk.

Theorem 5.3. *If the derivative $Df_p\colon T_pX \to T_pX$ has jointly Diophantine eigenvalues $(\lambda_1, \ldots, \lambda_n) \in (S^1)^n$, then f has a Siegel disk at p.*

The proof is due to Siegel for $n = 1$ and to Sternberg for $n > 1$ ([St], p. 465).

Proof of Theorem 5.1. Combine the results of Fel'dman and Sternberg. □

Notes. See [Ti] and [Her] for more on the Gel'fond-Baker method and analytic linearization respectively. In Theorem 5.3, (5.1) can be replaced by the weaker condition $\inf_j |\lambda_1^{k_1} \cdots \lambda_n^{k_n} - \lambda_j| > C|k_1 + \cdots + k_n|^{-M}$ for all $(k_1, \ldots, k_n) \in \mathbb{Z}^n$ with $k_i \geqq 0$ and $\sum k_i \geqq 2$.

6. Holomorphic Lefschetz numbers

Let $f\colon X \to X$ be an automorphism of a complex manifold with a unique fixed-point p. In this section we show the derivative Df_p is determined by the action of f^*on $H^{r,s}(X)$. In the special case of K3 surfaces, we obtain:

Theorem 6.1. *Let $f\colon X \to X$ be an automorphism of a* K3 *surface with a unique transverse fixed-point p. Then we have*

$$\operatorname{Tr} Df_p = \frac{1 + \delta + \delta^2}{1 + \delta} \tag{6.1}$$

where $\delta = \delta(f)$.

The Atiyah-Bott fixed-point theorem. Let $f: X \to X$ be a holomorphic automorphism of a compact complex manifold X with real dimension n. Then f acts on the singular cohomology groups $H^q(X), 0 \leqq q \leqq n$, as well as the Dolbeault cohomology groups $H^{r,s}(X)$. We define the *Lefschetz number* of f by

$$L(f) = \sum_{q=0}^{n} (-1)^q \mathrm{Tr} f^* \mid H^q(X),$$

and the *r*th *holomorphic Lefschetz number* of f by

$$L^r(f) = \sum_{s=0}^{n} (-1)^s \mathrm{Tr} f^* \mid H^{r,s}(X).$$

It is known that $L(f) = \sum (-1)^r L^r(f)$, and if f has a finite set of fixed-points then their number, counted with multiplicity, is $L(f)$.

We say f has *transverse fixed-points* if the graph of f is transverse to the diagonal in $X \times X$. In this case, the set of fixed-points $p \in X$ is finite, every fixed-point has multiplicity one, and by transversality the complex derivative

$$Df_p : T_p(X) \to T_p(X)$$

satisfies $\det(I - Df_p) \neq 0$. The Atiyah-Bott fixed-point theorem states ([AB], (4.9–4.10)):

Theorem 6.2. *If f has transverse fixed-points, then*

$$L^r(f) = \sum_{f(p)=p} \frac{\mathrm{Tr} \wedge^r Df_p}{\det(I - Df_p)}.$$

Corollary 6.3. *If f has a unique, transverse fixed-point p, then we have*

$$\det(tI - Df_p) = L^0(f)^{-1} \sum_{0}^{n} (-1)^r L^r(f) t^{n-r}.$$

Proof. Apply the formula $\det(I - A) = \sum (-1)^r \mathrm{Tr} \wedge^r A$. □

Proof of Theorem 6.1. From the Atiyah-Bott formula with $r = 2$ and the fact that $\delta = \mathrm{Tr} f^* \mid H^{2,0}(X) = \det Df_p$, we obtain

$$L^2(f) = 1 + \delta = \frac{\mathrm{Tr} \wedge^2 Df_p}{\det(I - Df_p)} = \frac{\delta}{1 - \mathrm{Tr} Df_p + \delta},$$

which gives (6.1). □

7. Siegel disks on K3 surfaces

With the results of the preceding sections in hand, we can now establish a cohomological criterion for the existence of a Siegel disk on a K3 surface.

Theorem 7.1. *Let $f: X \to X$ be an automorphism of a* K3 *surface. Suppose that:*

1. *the characteristic polynomial of* $f^* \mid H^2(X)$ *is a Salem polynomial*;
2. $\mathrm{Tr} f^* \mid H^2(X) = -1$;
3. *the determinant* $\delta = \delta(f)$ *satisfies* $\tau = \delta + \delta^{-1} > 1 - 2\sqrt{2}$; *and*
4. *the algebraic integer* τ *has a conjugate* $\tau' < 1 - 2\sqrt{2}$.

Then f *has a unique fixed-point* p, *and* p *is the center of a Siegel disk.*

We begin by remarking that Siegel disks are rare.

Theorem 7.2. *An automorphism of a projective* K3 *surface never has a Siegel disk. Moreover, there are at most countably many* K3 *surface automorphisms* (X, f) *with Siegel disks, up to isomorphism.*

Proof. Suppose f has a Siegel disk centered at p, and let α, β be the eigenvalues of Df_p. Since α and β are multiplicatively independent, $\delta(f) = \alpha\beta$ is not a root of unity. By Theorems 3.5 and 3.6, X is non-projective, and its Hodge structure is rigid, limiting (X, f) to a countable set. □

Resonances. Theorem 7.2 does not rule out the possibility of a 'resonant' rotation domain on a projective K3 surface. That is, $f\colon X \to X$ might conceivably have a linearizable fixed-point p where the eigenvalues λ_1, λ_2 of Df_p lie on S^1 but are *not* multiplicatively independent. Nevertheless, such a fixed-point seems unlikely to exist. Generically, a resonant fixed-point is not even formally linearizable; and if $\lambda_1^{k_1}\lambda_2^{k_2} = 1$, then $z_1^{k_1} z_2^{k_2}$ is invariant under the linear map $(z_1, z_2) \mapsto (\lambda_1 z_1, \lambda_2 z_2)$, and thus f must preserve a pencil of analytic curves near p.

We remark that Siegel disks can exist only for automorphisms of positive entropy.

Theorem 7.3. *If* $f\colon X \to X$ *is a* K3 *surface automorphism with a Siegel disk, then* $\lambda(f) > 1$.

Proof. Since $\delta(f)$ is an eigenvalue of $f^* \mid H^2(X)$ and not a root of unity, there must also be an eigenvalue off the unit circle. □

To obtain a Siegel disk we need conditions for the existence of a fixed-point p and for multiplicative independence of the eigenvalues of Df_p.

Theorem 7.4. *Suppose* $\mathrm{Tr} f^* \mid H^2(X) = -1$ *and* $(I - f^*)$ *is invertible on* $H^{1,1}(X)$. *Then* f *has a unique, transverse fixed-point* $p \in X$.

Proof. First note that fixed-points of f are isolated; otherwise the locus $f(p) = p$ would contain a 1-dimensional analytic cycle, representing an f^*-invariant class

$$C \in H^{(1,1)}(X).$$

Since every K3 surface is Kähler, we would have $C \neq 0$, contrary to our hypothesis on $I - f^*$.

Next, recall that every isolated fixed-point of a holomorphic map has positive multiplicity. Since $L(f) = 2 + \mathrm{Tr} f^* \mid H^2(X) = 1$, f has a unique fixed-point p, of multiplicity one and hence transverse. □

Lemma 7.5. *Let δ be a root of a Salem polynomial with $|\delta| = 1$, let $\tau = \delta + 1/\delta$ and let τ' be a conjugate of τ. Suppose we have*

$$\tau' < 1 - 2\sqrt{2} < \tau.$$

Then the roots (α, β) of the simultaneous equations

$$\alpha\beta = \delta \text{ and } \alpha + \beta = \frac{1 + \delta + \delta^2}{1 + \delta} \tag{7.1}$$

are multiplicatively independent, with $|\alpha| = |\beta| = 1$.

Proof. Since $\alpha\beta/\delta = 1$ and $|\delta| = 1$, we have $\alpha, \beta \in S^1 \Leftrightarrow (\alpha + \beta)^2/\delta \in [0, 4]$. By (7.1), this condition is the same as

$$\frac{(1 + \delta + \delta^2)^2}{\delta(1 + \delta)^2} = \frac{(\delta^{-1} + 1 + \delta)^2}{(\delta^{-1} + 2 + \delta)} = \frac{(\tau + 1)^2}{\tau + 2} \in [0, 4],$$

which holds iff $\tau \in (1 - 2\sqrt{2}, 1 + 2\sqrt{2})$. By assumption $\tau > 1 - 2\sqrt{2}$, and we have $\tau < 2 < 1 + 2\sqrt{2}$ because $|\delta| = 1$. Thus $|\alpha| = |\beta| = 1$.

By Galois theory, there exist conjugates $(\alpha', \beta', \delta')$ of (α, β, δ) also satisfying (7.1), with $\delta' + 1/\delta' = \tau'$. Since $\tau' < 2$ and δ' is a root of a Salem polynomial, we have $|\delta'| = 1$. But now $\tau' \notin (1 - 2\sqrt{2}, 1 + 2\sqrt{2})$, so we have $|\alpha'| = |\beta'|^{-1} \neq 1$.

Now suppose $\alpha^i\beta^j = 1$. Then $(\alpha')^i(\beta')^j = 1$ as well, and thus $i = j$. Therefore $(\alpha\beta)^i = \delta^i = 1$. Since δ satisfies a Salem polynomial, it is not a root of unity, and thus $i = 0$. Therefore α and β are multiplicatively independent. □

Proof of Theorem 7.1. By Theorem 7.4, f has a unique fixed-point $p \in X$. Via Theorem 6.1, the Atiyah-Bott formula implies the eigenvalues (α, β) of Df_p satisfy (7.1) with $\delta = \delta(f) = \det Df_p$. Since $\delta(f)$ is an algebraic number, so are α and β. The preceding lemma shows α and β lie on the unit circle and are multiplicatively independent. Theorem 5.1 states that such algebraic numbers satisfy the Diophantine condition for linearization, so f has a Siegel disk at p. □

8. Lattices in number fields

In this section we use number theory to construct lattice automorphisms with given characteristic polynomials. Via the Torelli theorem, these lattice automorphisms will give rise to K3 surface automorphisms.

To motivate the construction, suppose $F\colon B \to B$ is an automorphism of a lattice, with irreducible characteristic polynomial $s(y) = \det(yI - F) \in \mathbb{Z}[y]$. Since the pairing between the λ and μ eigenspaces of F is nontrivial iff $\lambda\mu = 1$, the eigenvalues of F occur in reciprocal pairs. Thus $s(y)$ has degree $2d$, and there is a degree d polynomial $r(x) \in \mathbb{Z}[x]$ whose roots have the form $\tau = \lambda + \lambda^{-1}$ as λ ranges over the roots of $s(y)$.

In this section we reverse the discussion. Starting from a degree d polynomial $r(x)$ such that the associated degree $2d$ polynomial $s(y)$ is irreducible, we construct a lattice B and an automorphism $F\colon B \to B$ with characteristic polynomial $s(y)$. The construction can be twisted by a unit $u \in \mathbb{Z}[x]/(r(x))$, yielding lattices $B(u)$ with a range of different of signatures.

Discriminants. Let $L \cong \mathbb{Z}^n$ be a lattice with inner product $\langle x, y\rangle_L \in \mathbb{Z}$. Given a basis $(e_i)_1^n$ for L, the *discriminant* of L is given by

$$\operatorname{disc}(L) = \det(\langle e_i, e_j\rangle);$$

it is independent of the choice of basis. The inner product allows us to regard L as a subgroup of $L^* = \operatorname{Hom}(L, \mathbb{Z})$, via $x(y) = \langle x, y\rangle_L$, and we have $|\operatorname{disc}(L)| = |L^*/L|$. Thus L is unimodular iff $|\operatorname{disc}(L)| = 1$.

A symmetric linear embedding $Q\colon L \to L$ determines a new lattice $L(Q)$ with the same underlying free abelian group as L, but with the new inner product

$$\langle x, y\rangle_{L(Q)} = \langle Qx, y\rangle_L.$$

We have $\operatorname{disc}\big(L(Q)\big) = \det(Q) \cdot \operatorname{disc}(L)$.

Trace forms. Lattices occur naturally in number rings. Let $r(x) \in \mathbb{Z}[x]$ be a degree d irreducible monic polynomial with roots $(x_i)_1^d$ in $\mathbb{C}$. Let A be the integral domain $\mathbb{Z}[x]/r(x)$ and let k be its field of fractions. Define an inner product on A by

$$\langle f_1, f_2\rangle_A = \operatorname{Tr}_{\mathbb{Q}}^k\left(\frac{f_1(x)f_2(x)}{r'(x)}\right) = \sum_1^d \left(\frac{f_1(x_i)f_2(x_i)}{r'(x_i)}\right)$$

(where $r'(x) = dr/dx$).

As was known to Euler, this inner product takes values in $\mathbb{Z}$ and makes A into a unimodular lattice. To prove this, one can use the residue theorem to compute:

$$\begin{aligned}\langle 1, x^n\rangle_A &= \sum \operatorname{Res}(x^n dx/r(x), x_i)\\ &= -\operatorname{Res}(x^n dx/r(x), \infty) = \begin{cases} 0, & 0 \leqq n < \deg(r) - 1,\\ 1, & n = \deg(r) - 1;\end{cases}\end{aligned}$$

compare [Ser2], §III.6.

Invariant forms. Now suppose $x^2 - 4$ is not a square in k. Let $K = k(y)$ be the quadratic extension of k obtained by adjoining a root of the equation

$$y + \frac{1}{y} = x.$$

Let $s(y) \in \mathbb{Z}[y]$ be the degree $2d$ minimal polynomial for y over $\mathbb{Q}$. Regarding $K \cong \mathbb{Q}[y]/\big(s(y)\big)$ as a space of polynomials in y, let $F\colon K \to K$ be the multiplication map

$$F\big(g(y)\big) = y \cdot g(y).$$

Then $s(y)$ is the characteristic polynomial for F as a linear endomorphism of $K/\mathbb{Q}$.

We will construct a lattice $B \subset K$ such that F is an *isometry* of B. For the underlying group, we take

$$B = \mathbb{Z}[y]/\big(s(y)\big) = A \oplus Ay \subset K.$$

Then $F(B) = B$.

The Galois group of K/k is generated by $\sigma(y) = 1/y$. Clearly $B^\sigma = B$, and the trace map

$$\mathrm{Tr}_k^K(g) = g + g^\sigma = g(y) + g(y^{-1})$$

sends B into A. We make B into a lattice by defining the inner product:

$$\langle g_1, g_2\rangle_B = \langle 1, \mathrm{Tr}_k^K(g_1 g_2^\sigma)\rangle_A = \mathrm{Tr}_{\mathbb{Q}}^K\left(\frac{g_1 g_2^\sigma}{r'(x)}\right).$$

Then $F\colon B \to B$ is an isometry, because

$$F(g_1)F(g_2)^\sigma = \big(yg_1(y)\big)\big(yg_2(y)\big)^\sigma = yg_1(y)y^{-1}g_2(y^{-1}) = g_1 g_2^\sigma.$$

Unimodularity. Our main concern is with automorphisms of unimodular lattices. Thus it is of interest to compute the discriminant of B.

Theorem 8.1. *The lattice B is even, with discriminant satisfying*

$$|\mathrm{disc}(B)| = |N_{\mathbb{Q}}^k(x^2 - 4)| = |r(2)r(-2)|.$$

Here $N_{\mathbb{Q}}^k\colon A \to \mathbb{Z}$ is the norm map, defined by $N_{\mathbb{Q}}^k(f) = \prod_1^d f(x_i)$.

Proof. The inner product on A makes A^2 into a unimodular lattice as well. Define $Q\colon A^2 \to A^2$ by

$$Q(a, b) = \begin{pmatrix} 2 & x \\ x & 2 \end{pmatrix}\begin{pmatrix} a \\ b \end{pmatrix}.$$

Then for $a + by \in A \oplus Ay = B$ we have

$$(a + by)(a + by)^\sigma = a^2 + b^2 + abx,$$

and therefore

$$\langle a + by, a + by\rangle_B = 2\langle 1, a^2 + b^2 + abx\rangle_A = \langle Q(a, b), (a, b)\rangle_{A^2}.$$

Thus B is even, and we have

$$|\mathrm{disc}(B)| = |\det(Q)| = |N_{\mathbb{Q}}^k(4 - x^2)| = |N_{\mathbb{Q}}^k(2 - x)N_{\mathbb{Q}}^k(2 + x)| = |r(2)r(-2)|,$$

since $N_{\mathbb{Q}}^k(n - x) = r(n)$. □

Corollary 8.2. *The following conditions are equivalent*: (a) B *is an even, unimodular lattice*; (b) $y + 1$ *and* $y - 1$ *are units in* B; (c) $x + 2$ *and* $x - 2$ *are units in* A; (d) $|r(\pm 2)| = 1$; (e) $|s(\pm 1)| = 1$.

Proof. Use the fact that $N_{\mathbb{Q}}^K(y \pm 1) = N_{\mathbb{Q}}^k(x \pm 2)$. □

Twisting by a unit. Let $u \in A^\times$ be a *unit* in A. Then multiplication by u is a symmetric automorphism of A with determinant ± 1, so the lattice $A(u)$ with inner product

$$\langle f_1, f_2 \rangle_{A(u)} = \langle u f_1, f_2 \rangle_A$$

is still unimodular. Similarly, the lattice $B(u)$ with inner product

$$\langle g_1, g_2 \rangle_{B(u)} = \langle u g_1, g_2 \rangle_B$$

is still even, with $|\mathrm{disc}(B(u))| = |\mathrm{disc}(B)|$; and $F\colon B(u) \to B(u)$ is still an isometry, since $u^\sigma = u$.

Signature. In general the signature of $B(u)$ varies with the unit u. To calculate the signature, first observe that the lattice $B(u)$ determines a Hermitian inner product on $B(u) \otimes \mathbb{C}$. Using the fact that $F\colon B(u) \to B(u)$ is an isometry, we obtain an orthogonal, F-invariant decomposition

$$B(u) \otimes \mathbb{C} = \bigoplus_{r(\tau)=0} E(\tau),$$

where $E(\tau) = \mathrm{Ker}(F + F^{-1} - \tau I)$ is 2-dimensional, and the eigenvalues λ, λ^{-1} of $F|E(\tau)$ satisfy $\lambda + \lambda^{-1} = \tau$.

Theorem 8.3. *Let τ be a zero of $r(x)$. For $\tau \in \mathbb{R}$, the subspace $E(\tau) \subset B(u) \otimes \mathbb{C}$ has signature*

$$\begin{aligned} &(2,0) \quad \text{if } |\tau| < 2 \text{ and } u(\tau) r'(\tau) > 0; \\ &(0,2) \quad \text{if } |\tau| < 2 \text{ and } u(\tau) r'(\tau) < 0; \text{ and} \\ &(1,1) \quad \text{otherwise.} \end{aligned}$$

For $\tau \notin \mathbb{R}$ the subspace $E(\tau) \oplus E(\bar\tau)$ has signature $(2,2)$.

Proof. First suppose $\tau \in \mathbb{R}$. Then using the isomorphism $B = A + Ay \cong A^2$ as in Theorem 8.1, we find the Hermitian inner product on $E(\tau) \cong \mathbb{C}^2$ comes from the complexification of the quadratic form

$$q(a,b) = 2u(\tau)(a^2 + b^2 + ab\tau)/r'(\tau)$$

on $\mathbb{R}^2$. The signature of the form $(a^2 + b^2 + ab\tau)$ is $(1,1)$ if $|\tau| > 2$ and $(2,0)$ if $|\tau| < 2$. The signature of $q(a,b)$ is the same, unless $u(\tau)/r'(\tau) < 0$, in which case it is reversed.

Now suppose $\tau \notin \mathbb{R}$, and let $S \subset E(\tau) \oplus E(\bar\tau)$ be the direct sum of the λ and $\bar\lambda$ eigenspaces for F, where $\lambda + \lambda^{-1} = \tau$. Since $\tau \notin \mathbb{R}$, $\bar\lambda$ is distinct from both λ and λ^{-1}; therefore S is a 2-dimensional isotropic subspace, and thus $E(\tau) \oplus E(\bar\tau)$ has signature $(2,2)$. □

Corollary 8.4. *The lattice $B(u)$ has signature $(d,d) + (p,-p) + (-q,q)$, where p is the number of roots of $r(x)$ in $[-2,2]$ satisfying $u(\tau)r'(\tau) > 0$, and q is the number satisfying $u(\tau)r'(\tau) < 0$.*

Unramified Salem polynomials. As an application, we show certain Salem numbers can be realized as eigenvalues of automorphisms of unimodular lattices. Following Corollary 8.2, we say a Salem polynomial $S(y)$ (or the corresponding Salem number) is *unramified* if $|S(\pm 1)| = 1$; and a Salem trace polynomial $R(x)$ is *unramified* if $|R(\pm 2)| = 1$.

Theorem 8.5. *Let $S(y)$ be an unramified Salem polynomial of degree $2d$. Then d is odd, and $S(y)$ is the characteristic polynomial of an automorphism of the even, unimodular lattice L of signature (d, d).*

Proof. Let $R(x)$ be the associated Salem trace polynomial of degree d, with roots $\tau_1, \dots, \tau_{d-1}$ in $(-2, 2)$ and $\tau_d > 2$. Since $R(x)$ has no roots in the interval $(-\infty, -2]$ and exactly one root in $(2, \infty)$, we have $R(2) < 0$ and $R(-2)(-1)^d > 0$. Since $S(y)$ is unramified, $|R(\pm 2)| = 1$, and thus $R(-2) = (-1)^d$, $R(2) = -1$. But $R(2) - R(-2) = 0 \bmod 4$ and therefore d is odd.

Now let $B = \mathbb{Z}[y]/S(y)$ and $F(g) = yg$ as above. Then F is an isometry of the even, unimodular lattice $L = B(1)$. Since $R'(x)$ alternates sign at the zeros of $R(x)$, the signatures of the summands $E(\tau_i)$ of $B(1) \otimes \mathbb{C}$ alternate between $(2, 0)$ and $(0, 2)$ for $i = 1, \dots, d-1$. The last summand $E(\tau_d)$ has signature $(1, 1)$, so altogether $B(1)$ has signature (d, d). □

Ramification and parity. Here is another explanation for the parity constraint on unramified Salem polynomials. If $S(y)$ has degree $2d$ and d is even, then $R(x)$ has an odd number of zeros in the range $[-2, 2]$, and therefore the lattice $B(1)$ has signature $(d-1, d+1)$. But an even unimodular lattice of signature (p, q) satisfies $p + q = 0 \bmod 8$, so $S(y)$ must be ramified.

Sextic Salem numbers, *reprise*. The simplest unramified Salem polynomials are the sextics of the form

$$S(y) = y^6 - ay^5 - y^4 + (2a-1)y^3 - y^2 - ay + 1, \quad a \geqq 0.$$

There are no unramified Salem polynomials of lower degree, and these are the only unramified Salem polynomials of degree 6.

In §4 we showed $S(y)$ arises as the characteristic polynomial of $F^* \mid H^2(Y, \mathbb{Z})$, where $F: Y \to Y$ is an automorphism of a complex torus. Noting that the intersection pairing makes $H^2(Y, \mathbb{Z})$ into an even, unimodular lattice of signature $(3, 3)$, we obtain a geometric instance of the result above.

9. From Salem numbers to automorphisms

Consolidating the preceding results, we can now give

– a general construction of K3 surface automorphisms from unramified Salem numbers, and

– a criterion for the resulting automorphism to have a Siegel disk.

Theorem 9.1. *Let $R(x)$, $U(x) \in \mathbb{Z}[x]$ be a pair such that:*

- *$R(x)$ is an unramified degree 11 Salem trace polynomial;*
- *$U(x)$ represents a unit in $\mathbb{Z}[x]/(R(x))$; and*

- *there is a unique root τ of $R(x)$ in $[-2, 2]$ such that $U(\tau)R'(\tau) > 0$.*

Then there exists a K3 *surface automorphism $f: X \to X$ such that*
- $\delta(f) + \delta(f)^{-1} = \tau$; *and*
- $S(x) = \det\big(xI - f^* \mid H^2(X)\big)$,

where $S(x)$ is the degree 22 Salem polynomial associated to $R(x)$.

Note that $S(x) = x^{11}R(x + x^{-1})$.

Proof. The trace-form construction of §8 yields a lattice automorphism

$$F: B(U) \to B(U)$$

with characteristic polynomial $S(x)$. Since $R(x)$ is unramified, $B(U)$ is an even, unimodular lattice, with signature $(3, 19)$ by Corollary 8.4.

Using the Torelli theorem, surjectivity of the period mapping, and uniqueness of the even unimodular $(3, 19)$ lattice, Theorem 3.4 (Synthesis) yields a K3 surface automorphism $f: X \to X$ and a marking $\iota: H^2(X, \mathbb{Z}) \to B(U)$ such that $F = \iota \circ f^* \circ \iota^{-1}$. Thus $S(x)$ is also the characteristic polynomial of $f^* \mid H^2(X)$.

By Theorem 8.3, the eigenspaces of $F + F^{-1}$ are 2-dimensional, and

$$E(\tau) = \mathrm{Ker}(F + F^{-1} - \tau I) \subset B(U) \otimes \mathbb{C}$$

is the unique eigenspace of $F + F^{-1}$ with signature $(2, 0)$. Similarly,

$$H^{2,0}(X) \oplus H^{0,2}(X) \subset H^2(X)$$

is the unique eigenspace of $f^* + (f^*)^{-1}$ with signature (2, 0). Thus

$$\iota^{-1}\big(E(\tau)\big) = H^{2,0}(X) \oplus H^{0,2}(X)$$

and therefore $\delta(f) + \delta(f)^{-1} = \tau$. □

Traces. The *trace* of a monic polynomial $P(x) = x^d + a_1x^{d-1} + \cdots + a_d$ is $-a_1$, the sum of its roots. The traces of $R(x)$ and $S(x)$ agree and coincide with $\mathrm{Tr}\big(f^* \mid H^2(X)\big)$. Continuing from the previous result, the cohomological criterion for a Siegel disk given by Theorem 7.1 implies:

Theorem 9.2. *Suppose in addition that $R(x)$ and τ satisfy:*
- *the trace of $R(x)$ is -1;*
- $\tau > 1 - 2\sqrt{2} = -1.8284271\ldots$; *and*
- *$R(x)$ has a root $\tau' < 1 - 2\sqrt{2}$.*

Then f has a unique fixed-point $p \in X$, and p is the center of a Siegel disk.

10. Examples of Siegel disks

In this section we exhibit polynomials $R(x)$ and $U(x)$ satisfying Theorem 9.2 and thereby prove:

Theorem 10.1. *There exist* K3 *surface automorphisms with Siegel disks.*

Table of Salem polynomials and units. Table 5 presents 10 examples of K3 surface automorphisms with Siegel disks. Each example occupies three rows.

The first two rows give the coefficients of a degree 22 Salem polynomial $S(x)$ and the associated degree 11 Salem trace polynomial $R(x)$. Since the coefficients of $S(x)$ are symmetric—that is, the coefficients of x^i and x^{22-i} agree—we only give them up to x^{11}.

The third row exhibits a unit $U(x)$ *compatible* with $R(x)$, meaning there is a unique root τ of $R(x)$ in $[-2, 2]$ such that $R'(\tau)U(\tau) > 0$. (In general there are many such units.)

For each example, the pair $(R(x), U(x))$ satisfies the hypotheses of Theorem 9.2, and so it determines a K3 surface automorphism $f\colon X \to X$ with a Siegel disk. The Salem polynomial $S(x)$ is the characteristic polynomial of $f|H^2(X)$; its largest root $\lambda(f)$ is a Salem number, given in the first column of the table. The topological entropy $\log\lambda(f)$ is a natural measure of the complexity of f, so we have ordered the table with $\lambda(f)$ increasing.

Detailed example. To explain the table more fully, we present some details regarding the degree 22 Salem number $\lambda \approx 1.37289$, the first example listed.
This Salem number is a root of the polynomial

$$\begin{aligned} S(x) =& 1 + x - x^3 - 2x^4 - 3x^5 - 3x^6 - 2x^7 + 2x^9 + 4x^{10} + 5x^{11} \\ & +4x^{12} + 2x^{13} - 2x^{15} - 3x^{16} - 3x^{17} - 2x^{18} - x^{19} + x^{21} + x^{22}. \end{aligned}$$

The corresponding Salem trace polynomial, satisfied by $\lambda + \lambda^{-1}$, is

$$\begin{aligned} R(x) =& -1 - 8x + 24x^2 + 42x^3 - 54x^4 - 66x^5 + 40x^6 + 42x^7 - 11x^8 \\ & -11x^9 + x^{10} + x^{11}. \end{aligned}$$

Note that both polynomials have trace -1 and are unramified—for example, $R(\pm 2) = -1$. The graph of $R(x)$, displaying its 11 real roots, is shown in Figure 4. All roots except $\lambda + \lambda^{-1} \approx 2.10128$ lie in $[-2, 2]$, as required.

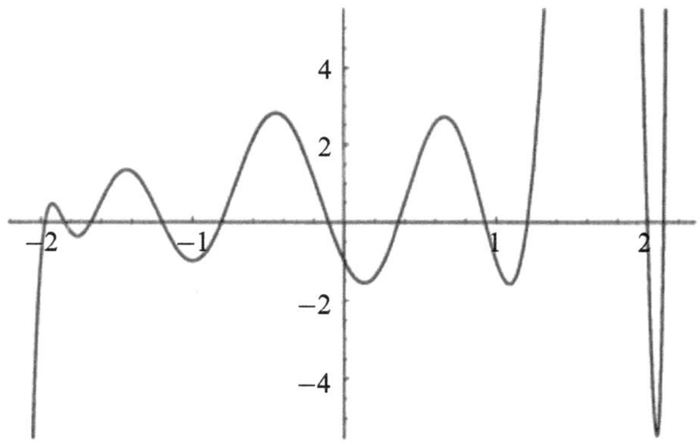

Figure 4. Graph of the degree 11 Salem trace polynomial $R(x)$.

A unit compatible with $R(x)$ is given by

$$U(x) = -2x + 6x^3 - 5x^5 + x^7.$$

To verify that $U(x) \in \mathbb{Z}[x]/(R(x))$ is a unit, one can check that $|\det U(M_R)| = 1$, where M_R is the 11×11 companion matrix of R.

Table 5. **Salem polynomials and units: (S(x), R(x), U(x)).**

λ	x^0	x^1	x^2	x^3	x^4	x^5	x^6	x^7	x^8	x^9	x^{10}	x^{11}
1.37289	1	1	0	-1	-2	-3	-3	-2	0	2	4	5
	-1	-8	24	42	-54	-66	40	42	-11	-11	1	1
	0	-2	0	6	0	-5	0	1				
1.45099	1	1	-1	-3	-3	-1	2	4	4	1	-3	-5
	-5	-4	49	37	-100	-81	58	50	-13	-12	1	1
	2	0	-9	14	23	-23	-25	9	9	-1	-1	
1.48115	1	1	0	-2	-3	-3	-2	-1	0	2	4	5
	-1	4	36	23	-73	-58	48	41	-12	-11	1	1
	0	-1	3	6	-4	-5	1	1				
1.52612	1	1	0	-2	-4	-4	-2	1	3	3	2	1
	-1	0	20	12	-65	-51	47	40	-12	-11	1	1
	0	-1	-5	17	37	-12	-33	2	10	0	-1	
1.55377	1	1	-1	-3	-3	-1	1	1	1	1	1	1
	-5	4	61	39	-103	-82	58	50	-13	-12	1	1
	3	3	-6	16	37	-12	-33	2	10	0	-1	
1.60709	1	1	0	-2	-4	-5	-4	-1	2	4	5	5
	-1	-4	20	21	-61	-53	46	40	-12	-11	1	1
	0	-7	8	34	-6	-32	1	10	0	-1		
1.6298	1	1	-1	-3	-3	-2	-1	0	2	3	2	1
	-9	-8	58	50	-98	-84	57	50	-13	-12	1	1
	-4	-6	16	14	-20	-7	8	1	-1			
1.6458	1	1	-1	-4	-5	-2	4	8	7	1	-6	-9
	-1	8	40	2	-110	-65	65	48	-14	-12	1	1
	-2	2	21	-4	-25	1	9	0	-1			
1.66566	1	1	-2	-5	-3	3	7	4	-2	-4	-1	1
	-1	32	112	36	-164	-103	78	59	-15	-13	1	1
	-8	-33	-23	49	62	-22	-42	3	11	0	-1	
1.69496	1	1	-1	-4	-5	-2	3	6	5	1	-3	-5
	-1	12	48	5	-112	-66	65	48	-14	-12	1	1
	0	1	-18	-7	48	14	-35	-7	10	1	-1	

There is a unique root $\tau \approx -1.66716$ of $R(x)$ in $[-2, 2]$ with $R'(\tau)U(\tau) > 0$. On the other hand, there are two roots of $R(x)$ smaller than $1 - 2\sqrt{2} \approx -1.82843$, the smallest being $\tau' \approx -1.97098$.

Thus the hypotheses of Theorem 9.2 are satisfied, and we obtain an example of a K3 surface automorphism $f\colon X \to X$ with a Siegel disk. In fact we obtain a pair of complex-conjugate examples, one for each root of the equation $\delta(f) + \delta(f)^{-1} = \tau$. For the root $\delta(f) \approx \exp(2.55635i)$, we obtain a Siegel disk centered at $p \in X$ with the eigenvalues of Df_p given by

$$\{\alpha, \beta\} \approx \{\exp(-2.81769i), \exp(-0.909141i)\},$$

according to equation (7.1).

Finiteness. Note that the set of Salem numbers of degree 22 and trace -1 is *finite*. More generally, the set of Salem polynomials $S(x)$ of degree d and trace t is finite, since the trace controls the size of the unique root outside the unit circle, and the size of the roots controls the size of the (integral) coefficients of $S(x)$.

One can also show that $S(x) = \det\big(xI - f^*|H^2(X)\big)$ determines the automorphism $f\colon X \to X$ up to a finite number of choices. Thus Theorem 9.2 yields only a finite number of Siegel disk examples up to isomorphism.

Searching for Salem numbers. Table 5 is the result of an intensive computer search for unramified degree 22 Salem numbers of trace -1. Altogether we located 51 such numbers, ranging from 1.37289 to 2.51503; the first 10 appear in the table. Sturm's method, based on alternations of signs [La], XI.2, was used to count the number of roots of a candidate Salem trace polynomial $R(x)$ in the interval $[-2, 2]$.

All 51 Salem numbers we found arise from K3 surface automorphisms with Siegel disks. That is, for each of the corresponding Salem trace polynomials $R(x)$, there was a compatible unit $U(x)$ satisfying Theorem 9.2.

Generally many units are compatible with a given $R(x)$, allowing one to vary $\tau = \delta(f) + \delta(f)^{-1}$ through different roots of $R(x)$ and to obtain several distinct K3 surface automorphisms for the same Salem polynomial. On the other hand, units which differ by a square give isomorphic lattice automorphisms $F\colon B(U) \to B(U)$, so only the class of U in the finite group $A^\times/(A^\times)^2$ is relevant.

To find $U(x)$, we used an algebra package to obtain a basis for the unit group of the order $A = \mathbb{Z}[x]/\big(R(x)\big)$ in the totally real field $k = \mathbb{Q}[x]/\big(R(x)\big)$. It was then straightforward to search the cosets of $A^\times/(A^\times)^2$ for a compatible unit.

Negative trace. We remark that Salem numbers with negative trace seem to be rather rare. Among the 630 degree 22 Salem polynomials with coefficients satisfying $|a_i| \leqq 1$, 596 have trace 1, 33 have trace 0 and only one has trace -1, namely

$$\begin{aligned} S(x) = &1 + x - x^3 - x^4 - x^5 - x^6 - x^7 - x^8 - x^9 - x^{10} - x^{11} \\ &- x^{12} - x^{13} - x^{14} - x^{15} - x^{16} - x^{17} - x^{18} - x^{19} + x^{21} + x^{22}. \end{aligned}$$

On the other hand, Smyth has shown there exist infinitely many Salem numbers with trace -1 [Sm].

Question. Are there any Salem numbers with trace less than -1?

11. Limits of Kähler-Einstein metrics

In this section we present an application of dynamics to the differential geometry of K3 surfaces.

Kähler-Einstein metrics. Let g be a Kähler metric on a compact complex manifold X, with associated symplectic $(1,1)$-form ω_g. Then g is a *Kähler-Einstein metric* if there is a constant $\kappa \in \mathbb{R}$ such that

$$\mathrm{Ric}(g) = \kappa \cdot \omega_g,$$

where $\mathrm{Ric}(g) = -\partial\bar{\partial} \log \det g$ is the Ricci form of g.

Now suppose X is a K3 surface. Then by triviality of the canonical bundle, g is a Kähler-Einstein metric $\Leftrightarrow$ $\mathrm{Ric}(g) = 0 \Leftrightarrow$ the volume form ω_g^2 of g is proportional to $\eta \wedge \bar{\eta}$, where $\eta \neq 0$ is a holomorphic $(2,0)$ form on X. Such metrics exist in abundance, according to:

Theorem 11.1 (Yau). *Every cohomology class in the Kähler cone $C_X \subset H^{1,1}(X)$ is represented by the $(1,1)$-form ω of a Kähler-Einstein metric.*

See [Yau], [Bo].

It would be interesting to understand the behavior of a Kähler-Einstein metric as its cohomology class $[\omega]$ tends to the boundary of Kähler cone. Here we will show that the metric can collapse to zero on an open set, if X comes from one of the Siegel disk examples of §10.

Theorem 11.2. *Let $f\colon X \to X$ be a K3 surface automorphism with a Siegel disk U. Then there exists a sequence of Kähler-Einstein metrics on X, with symplectic forms satisfying*

$$[\omega_n] \to \xi \in \partial C_X \subset H^{1,1}(X), \quad \xi \neq 0,$$

such that $\omega_n | U \to 0$ uniformly on compact sets.

Proof. Since f has a Siegel disk, we have $\lambda = \lambda(f) > 1$ (Theorem 7.3). Consider any Kähler-Einstein metric g_0 on X with $(1,1)$-form ω_0, and let $\omega_n = \lambda^{-n}(f^n)^*(\omega_0)$. Then ω_n is also the $(1,1)$-form of a Kähler-Einstein metric g_n, since the condition $\mathrm{Ric}(g_0) = 0$ is preserved under pulling back.

We claim the cohomology classes $[\omega_n]$ converge to an eigenvector $\xi \neq 0$ in ∂C_X satisfying $f^*(\xi) = \lambda\xi$. Indeed, f^* preserves the intersection form of signature $(1,19)$ on $H^{1,1}(X)$, and therefore it determines an isometry F of the hyperbolic space

$$\mathbb{H}^{19} \subset \mathbb{P}H^{1,1}(X)_{\mathbb{R}}.$$

Since we have $\lambda > 1$, F translates along a geodesic $\gamma \subset \mathbb{H}^{19}$ joining points $p_\pm \in \partial\mathbb{H}^{19}$, corresponding to the $\lambda^{\pm 1}$-eigenspaces of f. For any point $q \in \mathbb{H}^{19}$, including that represented by $[\omega_0]$, we have $F^n(q) \to p_+$. Since $[\omega_n]$ is $[(f^n)^*(\omega_0)]$ renormalized by the leading eigenvalue of f^*, it converges to a λ-eigenvector ξ.

Since $f|U$ is conjugate to a rotation, any compact set $K' \subset U$ can be enlarged to an f-invariant compact set $K \subset U$. Moreover there is a smooth f-invariant metric on U. Using this invariant metric to measure the size of ω_n, we have:

$$\|\omega_n | K\|_\infty = \lambda^{-n} \|\omega_0 | K\|_\infty \to 0$$

as claimed. (In fact $\omega_n | K \to 0$ in the C^∞ topology.) □

Eigencurrents. By a result of Cantat, the sequence $\omega_n = \lambda^{-n}(f^n)^*(\omega_0)$ just constructed actually converges to a natural current $\omega_\infty = C_+\alpha_+$ representing the cohomology class ξ.

Theorem 11.3 (Cantat). *Let $f\colon X \to X$ be a K3 surface automorphism with $\lambda = \lambda(f) > 1$. Then:*

- *there exists a pair of closed, positive $(1,1)$-currents α_-, α_+, unique up to scale, such that $f^*(\alpha_\pm) = \lambda^{\pm 1}\alpha_\pm$;*
- *for any other closed, positive $(1,1)$ current ω, there are constants $C_\pm$ such that $\lambda^{-|n|}(f^{\pm n})^*(\omega) \to C_\pm\alpha_\pm$ as $n \to \infty$.*

(See [Ca1], §4.2–4.4, or observe that [Ca2], Theorem 2, while stated for projective surfaces, also applies to Kähler surfaces.)

So we also have:

Theorem 11.4. *Let $f\colon X \to X$ be a K3 surface automorphism with a Siegel disk U. Then the eigencurrents α_+ and α_- vanish on U.*

References

[AB] *M. F. Atiyah* and *R. Bott*, A Lefschetz fixed point formula for elliptic complexes II, Ann. Math. **88** (1968), 451–491.
[BPV] *W. Barth, C. Peters*, and *A. van de Ven*, Compact Complex Surfaces, Springer-Verlag, 1984.
[Bv1] *Beauville*, Some remarks on Kähler manifolds with $c_1 = 0$, in: Classification of Algebraic and Analytic Manifolds (Katata 1982), Birkhäuser (1983), 1–26.
[Bv2] *A. Beauville* et al., Géométrie des surfaces K3: modules et périodes, Astérisque **126** (1985).
[Bo] *J. Bourguingon* et al., Première classe de Chern et courbure de Ricci: preuve de la conjecture de Calabi, Astérisque **58** (1978).
[Br] *R. J. Brown*, Anosov mapping class actions on the SU(2)-representation variety of a punctured torus, Ergod. Th. Dynam. Sys. **18** (1998), 539–554.
[Ca1] *S. Cantat*, Dynamique des automorphismes des surfaces complexes compactes, Thesis, ENS Lyon, 1999.
[Ca2] *S. Cantat*, Dynamique des automorphismes des surfaces projectives complexes, CRAS Paris Sér. I Math. **328** (1999), 901–906.
[Fe] *H. I. Fel'dman*, An improvement of the estimate of a linear form in the logarithms of algebraic numbers, Math. USSR Sb. **6** (1968), 393–406.
[Fr] *D. Fried*, Word maps, isotopy and entropy, Trans. Amer. Math. Soc. **296** (1986), 851–859.
[Frl] *S. Friedland*, Entropy of algebraic maps, in: Proceedings of the Conference in Honor of Jean-Pierre Kahane (Orsay 1993), J. Fourier Anal. Appl. (1995), 215–228.
[Gr] *M. Gromov*, On the entropy of holomorphic map, manuscript, 1980.
[GM] *B. Gross* and *C. McMullen*, Automorphisms of even unimodular lattices and unramified Salem numbers, J. Alg., to appear.
[Her] *M. Herman*, Recent results and some open questions on Siegel's linearization theorem of germs of complex analytic diffeomorphisms of $\mathbb{C}^n$ near a fixed point, in: VIIIth International Congress on Mathematical Physics (Marseille 1986), World Sci. Publishing (1987), 138–184.
[La] *S. Lang*, Algebra, Addison-Wesley, 1984.
[Mak1] *R. S. MacKay*, Hyperbolic structure in classical chaos, in: Quantum Chaos (Varenna 1991), North-Holland (1993), 1–50.
[Mak2] *R. S. MacKay*, Renormalisation in Area-Preserving Maps, World Scientific, 1993.
[Man] *R. Mañé*, Ergodic Theory and Differentiable Dynamics, Springer-Verlag, 1987.
[Maz] *B. Mazur*, The topology of rational points, J. Exp. Math. **1** (1992), 35–46.
[Mc] *C. McMullen*, Complex Dynamics and Renormalization, Ann. Math. Stud. **135** (1994).
[Mil] *J. Milnor*, Dynamics in One Complex Variable: Introductory Lectures, Vieweg, 1999.

[MH] *J. Milnor* and *D. Husemoller*, Symmetric Bilinear Forms, Springer, 1973.
[Sa] *R. Salem*, Algebraic Numbers and Fourier Analysis, Wadsworth, 1983.
[Ser1] *J. P. Serre*, A Course in Arithmetic, Springer-Verlag, 1973.
[Ser2] *J. P. Serre*, Local Fields, Springer-Verlag, 1979.
[Sil] *J. Silverman*, Rational points on K3 surfaces: A new canonical height, Invent. math. **105** (1991), 347–373.
[Sm] *C. J. Smyth*, Salem numbers of negative trace, Math. Comp. **69** (2000), 827–838.
[St] *S. Sternberg*, Infinite Lie groups and the formal aspects of dynamical systems, J. Math. Mech. **10** (1961), 451–474.
[Ti] *R. Tijdeman*, On the Gel'fond-Baker method and its applications, in: F. Browder, ed., Mathematical Developments Arising from Hilbert Problems, Proc. Symp. Pure Math. **28** (1976).
[Wa] *L. Wang*, Rational points and canonical heights on K3-surfaces in $\mathbb{P}^1 \times \mathbb{P}^1 \times \mathbb{P}^1$, in: Recent Developments in the Inverse Galois Problem, Amer. Math. Soc. (1995), 273–289.
[Yau] *S.-T. Yau*, On the Ricci-curvature of a complex Kähler manifold and the complex Monge-Ampère equation, Comment. Pure Appl. Math. **31** (1978), 339–411.
[Ym] *Y. Yomdin*, Volume growth and entropy, Israel J. Math. **57** (1987), 285–300.

Curtis T. McMullen, Mathematics Department, Harvard University, 1 Oxford St, Cambridge, MA 02138-2901
e-mail:

Dynamical degrees, entropy, and Fatou sets

By *Serge Cantat* at Rennes

Abstract. In 2002, the Journal für die reine und angewandte Mathematik published an article of C. T. McMullen on the dynamics of holomorphic diffeomorphisms of K3 surfaces. Nine years later, a second one followed. Both of them construct automorphisms of K3 surfaces with interesting dynamical features. The following pages present some of the main developments around the themes of research initiated in these two publications.

Introduction

Our main objects of study will be holomorphic diffeomorphisms $f\colon X \to X$ of compact, complex manifolds, in particular of complex projective surfaces. When the manifold X is projective, Chow's theorem implies that f is an algebraic transformation of X. Thus, for simplicity, holomorphic diffeomorphisms will be called *automorphisms* and holomorphic transformations will be called *endomorphisms*.

Let $X \subset \mathbb{P}^3_{\mathbf{C}}$ be a smooth projective surface defined by a homogeneous equation of degree d. When $d = 1$, X is a projective plane $\mathbb{P}^2_{\mathbf{C}}$ and its automorphism group $\mathsf{Aut}(X)$ is the group of linear projective transformations $\mathsf{PGL}_3(\mathbf{C})$. When $d = 2$, X is isomorphic to $\mathbb{P}^1_{\mathbf{C}} \times \mathbb{P}^1_{\mathbf{C}}$ and $\mathsf{Aut}(X)$ is the semi-direct product of $\mathsf{PGL}_2(\mathbf{C}) \times \mathsf{PGL}_2(\mathbf{C})$ and the involution $(x, y) \mapsto (y, x)$ permuting the two factors of $\mathbb{P}^1_{\mathbf{C}} \times \mathbb{P}^1_{\mathbf{C}}$. When $d = 3$ or $d \geq 5$, $\mathsf{Aut}(X)$ is finite. Thus, if $d \neq 4$, (a) $\mathsf{Aut}(X)$ is a complex algebraic group (with only finitely many components) and (b) the dynamics of any automorphism $f\colon X \to X$ is well understood. The situation changes drastically for some specific quartic surfaces, for instance for the Fermat quartic defined in homogeneous coordinates by the equation $x_1^4 + x_2^4 + x_3^4 + x_4^4 = 0$. Another example is mentioned by A. Weil in [60]: "*Early on,* he wrote, *I was intrigued by Severi's example, by means of a quartic in $\mathbb{P}^3$, of a surface with an infinite group of automorphisms related to the group of units of a real quadratic field; for a while, I even hoped to find a way there of generating abelian extensions of this field.*" As we shall see, the automorphisms of Segre's quartic surface have a rich and chaotic dynamics.

This quote from Weil comes from his commentary on his 1958 presentation of K3 surfaces. A *K3 surface* is a compact, complex surface X with first Betti number $b_1(X) = 0$ and with a non-vanishing holomorphic 2-form Ω_X (see [3]). Such a 2-form can be chosen to satisfy

This work is supported by the European Research Council (ERC GOAT 101053021).

$\int_X \Omega_X \wedge \overline{\Omega}_X = 1$ and is then unique up to multiplication by a complex number of modulus 1. K3 surfaces are simply connected. They form a continuous family of surfaces and, in particular, are all C^∞-diffeomorphic to the same manifold. The first examples are smooth quartic surfaces in $\mathbb{P}^3$, double covers of $\mathbb{P}^2$ ramified along a smooth sextic curve, and smooth surfaces of degree $(2, 2, 2)$ in $\mathbb{P}^1 \times \mathbb{P}^1 \times \mathbb{P}^1$. Another example is given by Kummer's construction: one starts with a torus $\mathbf{C}^2/\Lambda$, defined by a lattice $\Lambda \subset \mathbf{C}^2$, and takes the quotient by the involution $\eta(x, y) = (-x, -y)$; the quotient is a singular complex analytic manifold, but its minimal desingularization is a K3 surface (it is projective if and only if the torus is projective, a property which depends on the choice of the lattice Λ). As this example shows, K3 surfaces are not all projective. But all are Kähler.

There is a classification of compact complex surfaces by Enriques and Kodaira, and of projective surfaces over algebraically closed field of arbitrary characteristic by Bombieri and Mumford: K3 surfaces occupy a central position in both classifications.

Understanding the dynamical features of automorphisms of K3 surfaces is at the heart of the article of McMullen reproduced in this volume and its sibling published in the same journal in 2011 (see [46, 48]). Instead of focusing on their precise content, I will survey two of the main questions addressed by McMullen: which values can be achieved by the topological entropy of automorphisms of complex surfaces, or by dynamical degrees of birational maps? Can one find examples of intermingled dynamical features, that is, surfaces in which regions with chaotic dynamics and regions with tame dynamics coexist?

1. From topological entropy to dynamical degrees

1.1. Topological entropy. Consider a continuous transformation $f \colon X \to X$ of a compact metric space X. The orbit of a point $x \in X$ under the dynamics generated by f is the sequence defined recursively by $x_0 = x$ and $x_{n+1} = f(x_n)$; thus, $x_n = f^n(x)$ where f^n denotes the n-th iterate of f. In other words, X is the phase space, time takes discrete values $n \in \mathbf{N}$, and f determines the evolution of the system over time; one speaks of the *discrete dynamical system* generated by f.

As an example, take f to be the transformation of the unit circle $\mathbb{S}^1 = \{z \in \mathbf{C} \,;\, |z| = 1\}$ defined by $f(z) = z^d$, where d is some integer ≥ 2. Writing $z = \exp(2i\pi\theta)$ with $\theta \in \mathbf{R}/\mathbf{Z}$, this transformation reads $\theta \mapsto d\theta \mod (1)$. The orbit of z is finite if and only if its angle θ is rational; it is dense if and only if the d-adic expansion $\theta = \sum_{i\geq 1} c_i d^{-i}$ corresponds to a sequence $(c_i) \in \{0, 1, \ldots, d-1\}^{\mathbf{N}}$ in which every finite sequence of numbers between 0 and $d-1$ appears as a subsequence of consecutive numbers $(c_i, c_{i+1}, \ldots, c_{i+l})$; similarly, if the sequence (c_i) does not include the sequence $(0, 0, 1, 1)$ but includes any finite sequence satisfying this constraint, then the orbit of $\theta = \sum_{i\geq 1} c_i d^{-i}$ is dense in a Cantor subset of the circle. Thus, finite orbits, dense orbits, and orbits with a closure homeomorphic to a Cantor set coexist; in any non-empty open subset of $\mathbb{S}^1$, one finds starting points whose orbits realize any of these three possibilities. As a consequence, the behavior of an orbit is highly sensitive to the precise value of its initial condition, i.e. to the starting point x.

Topological entropy is a way to measure the complexity of such a system. To define it, we fix a distance dist on X which is compatible with its topology. Then, for any $\epsilon > 0$ and $t \geq 1$, we count the number of orbits that can be distinguished at scale ϵ during a period of observation t. To make this precise, we say that two points x and y are (t, ϵ)-distinguishable if

$$\mathrm{dist}(f^n(x), f^n(y)) > \epsilon$$

for some $0 \leq n < t$ and we define the *number of orbits* $N(t, \epsilon)$ as the maximal number of elements in a subset of X made of points which are pairwise distinguishable in time $< t$ (this number is finite because X is compact). The quantity

$$h(f; \epsilon) = \limsup_{t \to +\infty} \frac{1}{t} \log(N(t, \epsilon))$$

gives the exponential growth rate of $N(t, \epsilon)$ as the period of observation goes to $+\infty$. And the *topological entropy* is obtained by making finer and finer observations:

$$h_{\mathrm{top}}(f) = \lim_{\epsilon \to 0} h(f; \epsilon).$$

This quantity is non-negative and does not depend on the initial choice of distance dist. When X is a manifold and f is Lipschitz, $h_{\mathrm{top}}(f) \leq \dim(X) \log(\mathrm{Lip}(f))$. In the case of $f(z) = z^d$ on the unit circle, one easily gets $h_{\mathrm{top}}(f) = \log(d)$.

1.2. Theorems of Yomdin and Gromov. Assume now that X is a compact and connected manifold, and denote its integral cohomology groups by

$$H^*(X; \mathbf{Z}) = \bigoplus_{k=0}^{\dim(X)} H^k(X; \mathbf{Z}).$$

Then, f induces an endomorphism f^* of $H^*(X; \mathbf{Z})$ preserving each summand $H^k(X; \mathbf{Z})$. Taking real coefficients, $H^*(X; \mathbf{R})$ becomes a finite-dimensional real vector space. Given any norm $\| \cdot \|$ on $\mathsf{End}(H^*(X; \mathbf{R}))$, the *spectral radius* of f^* is defined by

$$\rho(f^*) = \lim_{n \to +\infty} \| (f^*)^n \|^{1/n} \tag{1.1}$$

$$= \max \{ |\lambda| \; ; \; \lambda \in \mathbf{C} \text{ is an eigenvalue of } f^* \text{ on } H^*(X; \mathbf{C}) \}. \tag{1.2}$$

Y. Yomdin proved in 1986 (see [38, 61]) that a C^∞-transformation f of a compact manifold X satisfies $h_{\mathrm{top}}(f) \geq \log(\rho(f^*))$, a result that had been conjectured by M. Shub around 1974. One can get a similar lower bound by looking at the action of f on the fundamental group of X, even when f is just continuous (see [16, 44]). On the other hand, in 1976, M. Gromov had obtained an upper bound for $h_{\mathrm{top}}(f)$ when

- X is a compact Kähler manifold. (That is, X is a compact complex manifold which admits at least one Kähler form κ, see [53].)
- f is a holomorphic transformation of X.

Combining Yomdin's and Gromov's theorems, we obtain the following statement.

Theorem 1.1 (Gromov-Yomdin). *The topological entropy of a holomorphic transformation f of a compact Kähler manifold X is equal to the logarithm of the spectral radius of the linear map $f^* \colon H^*(X; \mathbf{R}) \to H^*(X; \mathbf{R})$. That is, $h_{\mathrm{top}}(f) = \log(\rho(f^*))$.*

This is an amazing theorem: it provides a closed formula for a quantity which is usually extremely hard to estimate and it shows that the entropy is constant along holomorphic deformations (X_t, f_t).

1.3. Dynamical degrees. The proof of Gromov's upper bound gives more than just $h_{\rm top}(f) \leq \log \rho(f^*)$ (see [39]). To describe it, we endow X with the Riemannian metric induced by some fixed Kähler form κ. In a first step, Gromov introduces the iterated graph $\Gamma(f;n) = \{(x, f(x), \ldots, f^{n-1}(x))\ ;\ x \in X\} \subset X^n$ and proves that

$$h_{\rm top}(f) \leq \limsup_{n\to+\infty} \frac{1}{n} \log \left(\mathsf{vol}(\Gamma(f;n))\right)$$

where the volume is computed with respect to the product metric on X^n. This relies on the fact that complex analytic subsets of Kähler manifolds are minimal in the sense of Riemannian geometry[1]. In particular, we have the extraordinary fact that the iterated graphs $\Gamma(f;n)$ are always the "geometrically simplest representatives" of their homology classes in X^n.

The second step is a computation of $\mathsf{vol}(\Gamma(f;n))$. To describe it, recall that the cohomology group of a compact Kähler manifold comes with a Hodge decomposition: for each integer k between 0 and $2\dim_{\mathbf{C}}(X)$,

$$H^k(X;\mathbf{C}) = \bigoplus_{p+q=k} H^{p,q}(X;\mathbf{C}) \tag{1.3}$$

where $H^{p,q}(X;\mathbf{C})$ is the Dolbeault cohomology group of (closed forms of) type (p,q). The Kähler form κ is of type $(1,1)$ and its exterior powers κ^p are of type (p,p). The main input is Wirtinger's theorem saying that the volume of a complex analytic subset $Y \subset X$ of dimension ℓ is equal to the integral of κ^ℓ over Y; thus, it is also equal to the product $[Y]\cdot[\kappa]^\ell$ between the classes of Y in $H_{2\ell}(X;\mathbf{Z})$ and of κ^ℓ in $H^{\ell,\ell}(X;\mathbf{C})$. Gromov applies this theorem in X^n together with Künneth's formula, and computes $\mathsf{vol}(\Gamma(f;n))$ as an intersection product that involves only the action of f^* on $H^*(X;\mathbf{C})$ and exterior powers of κ. Since $[\kappa^p] \in H^{p,p}(X;\mathbf{C})$, one gets

$$\rho(f^*) = \max_p \lambda_p(f)$$

where $\lambda_p(f)$ is the spectral radius of f^* on $H^{p,p}(X;\mathbf{C})$.

Example 1.2. Consider an endomorphism g of $\mathbb{P}^m(\mathbf{C})$. It can be defined in homogeneous coordinates by $m+1$ homogeneous polynomial functions $g_i \in \mathbf{C}[\mathbf{z}_0, \ldots, \mathbf{z}_m]$ of the same degree d and with no non-trivial common zeros:

$$g[\mathbf{z}_0 : \ldots : \mathbf{z}_m] = [g_0(\mathbf{z}_0, \ldots, \mathbf{z}_m) : \cdots : g_m(\mathbf{z}_0, \ldots, \mathbf{z}_m)].$$

Each cohomology group $H^k(\mathbb{P}^m(\mathbf{C});\mathbf{Z})$ is either trivial (when k is odd) or cyclic (when $k = 2p$ is even), and the action of g^* on $H^{2p}(\mathbb{P}^m(\mathbf{C});\mathbf{Z})$ is given by multiplication by d^p. In particular, the topological degree of g is d^m, it is the largest eigenvalue of g^*, and

$$h_{\rm top}(g) = \log(\deg_{\rm top}(g)) = m\log(d).$$

For instance, if $g(\mathbf{z}) \in \mathbf{C}(\mathbf{z})$ is a rational function, viewed as an endomorphism of the Riemann sphere $\overline{\mathbf{C}} = \mathbb{P}^1(\mathbf{C})$, its topological entropy is the logarithm of its degree.

[1] This means that a small C^1-smooth perturbation V_ϵ of a complex submanifold V in a Kähler manifold will always have a volume larger than or equal to the volume of V.

1.4. Meromorphic transformations and dynamical degrees. Gromov's inequality $h_{\mathrm{top}}(f) \leq \max_p \log \lambda_p(f)$ can be extended to meromorphic transformations. The first issue is to define topological entropy when f has indeterminacy points, the second is to define the spectral radii $\lambda_p(f)$. To define topological entropy, the idea is simply to consider only the orbits that never visit the indeterminacy locus (for this we refer to [41]). Let us describe the definition of $\lambda_p(f)$ when $f\colon X \dashrightarrow X$ is a dominant meromorphic transformation of a compact Kähler manifold.

The graph of f provides a correspondence $\Gamma_f \subset X \times X$ such that the projections $\pi_1(x, y) = x$ and $\pi_2(x, y) = y$ are surjective; $\pi_1\colon \Gamma_f \to X$ is a bimeromorphic map, while the topological degree of π_2 coincides with the one of f. A differential form α on X can be pulled back to Γ_f by π_2 and then pushed forward as a current by π_1. This defines a linear operator $f^*\alpha = (\pi_1)_*\pi_2^*\alpha$, which induces a linear transformation of $H^*(X; \mathbf{Z})$ preserving the Hodge decomposition. Note, however, that if $g\colon X \dashrightarrow X$ is a meromorphic map, $(f \circ g)^*$ usually differs from $g^* \circ f^*$. For instance, the standard quadratic involution $s\colon \mathbb{P}^2(\mathbf{C}) \dashrightarrow \mathbb{P}^2(\mathbf{C})$, defined by $s[x_0 : x_1 : x_2] = [x_1x_2 : x_2x_0 : x_0x_1]$, acts on $H^2(\mathbb{P}^2(\mathbf{C}); \mathbf{Z})$ by multiplication by 2; but $s \circ s = \mathrm{id}$ acts by multiplication by 1 (not by 4).

With such a definition at hand, the p-th *dynamical degree* of f is defined by

$$\lambda_p(f) = \lim_{n\to+\infty} \| \, (f^n)^*_{H^{p,p}(X;\mathbf{C})} \, \|^{1/n}$$

where $\| \cdot \|$ is any norm on $\mathsf{End}(H^{p,p}(X; \mathbf{C}))$. The existence of this limit and the following properties are obtained in [30].

(1) $\lambda_p(f) = \lambda_p(\varphi^{-1} \circ f \circ \varphi)$ for any bimeromorphic map $\varphi\colon Y \dashrightarrow X$.

(2) For any Kähler form κ on X,

$$\lambda_p(f) = \lim_{n\to+\infty} \left(\int_X (f^n)^*\kappa^p \wedge \kappa^{\dim(X)-p} \right)^{1/n} \tag{1.4}$$

where $(f^n)^*\kappa^p$ and the integral are computed on the complement of the indeterminacy locus of f^n. In particular, this limit does not depend on κ.

(3) The sequence $p \mapsto \lambda_p(f)$ is log-concave, that is

$$\log(\lambda_p(f)) \geq \frac{1}{2}\log(\lambda_{p-1}(f)) + \frac{1}{2}\log(\lambda_{p+1}(f))$$

for any $1 \leq p \leq m - 1$.

T. C. Dinh and N. Sibony proved in [30] that the upper bound $h_{\mathrm{top}}(f) \leq \max_p \log(\lambda_p(f))$ still holds for dominant meromorphic transformations of compact Kähler manifolds. On the other hand, one cannot expect an equality as in Theorem 1.1 (see [41] for a simple example).

1.5. Rational transformations of projective varieties. When X is projective, defined over a field $\mathbf{k}$, one can replace the Kähler form κ in Property (2) above by a hyperplane section H of X and the integral of $(f^n)^*\kappa^p \wedge \kappa^{\dim(X)-p}$ in Equation (1.4) by the intersection product $((f^n)^*H^p \cdot H^{\dim(X)-p})$. Doing so, the definition of $\lambda_p(f)$ makes sense for any rational transformation f, defined over any field $\mathbf{k}$:

$$\lambda_p(f) = \lim_{n\to+\infty} \left((f^n)^*H^p \cdot H^{\dim(X)-p} \right)^{1/n}.$$

This is consistent with our initial definition when $\mathbf{k} = \mathbf{C}$, and Properties (1) to (3) are still satisfied in this context (see [28, 58]). Moreover, $\max_p \lambda_p(f)$ is equal to the limit of $\| [\Gamma_{f^n}] \|^{1/n}$ where Γ_{f^n} is the graph of f^n in $X \times X$, $[\Gamma_{f^n}]$ is its numerical class and $\| \cdot \|$ is any norm on the space of numerical classes. Thus, dynamical degrees are basic invariants describing the asymptotic complexity of the n-th iterate f^n. When $\mathbf{k}$ is a local field and f is regular, f defines a continuous transformation of the compact space $X(\mathbf{k})$. It turns out that the topological entropy of f on $X(\mathbf{k})$ is always bounded from above by $\max \log(\lambda_p(f))$ (see [35] for this and better statements), but this upper bound is usually sharp; for instance, there are many examples of automorphisms of real projective surfaces for which the entropy on $X(\mathbf{R})$ is strictly less than $\log(\lambda_1(f))$: Figure 2 below corresponds to such an example (see also [52]).

The first question we shall look at, in the next sections, is: what are the possible values of $\lambda_1(f)$ when f is an automorphism or a birational transformation of a surface?

2. From entropy to Salem and Pisot numbers

Let f be a dominant rational transformation of a closed Riemann surface X. Then f is an endomorphism and its dynamical degrees are $\lambda_0(f) = 1$ and $\lambda_1(f) = \deg_{\mathrm{top}}(f)$; in particular, they are positive integers and $h_{\mathrm{top}}(f) = \log(\deg_{\mathrm{top}}(f))$. When the genus of X is at least 2, f is an automorphism of finite order. Endomorphisms of elliptic curves $\mathbf{C}/\Lambda$ are induced by affine maps $z \mapsto az + b$, and their dynamical features are well understood. Endomorphisms of $\mathbb{P}^1$ of topological degree > 1 provide some of the most interesting examples of dynamical systems (see [1, 50]); subgroups of $\mathsf{Aut}(\mathbb{P}^1) = \mathsf{PGL}_2(\mathbf{C})$ are central to topology, geometry, and dynamics, but individually, each automorphism is just a homography $z \mapsto (az + b)/(cz + d)$. Thus, the study of invertible algebraic transformations and of dynamical degrees really starts in dimension 2. We shall see that, in the context of dimension 2, Salem and Pisot numbers play an important role.

2.1. Salem and Pisot numbers [12]. A *Pisot number* is a real algebraic integer $\lambda > 1$ such that all Galois conjugates of λ, except λ itself, are in the open unit disk. A *Salem number* is a real algebraic integer $\lambda > 1$ whose Galois conjugates are in the closed unit disk, with at least one of them on the unit circle. The minimal polynomial of a Salem number is reciprocal and its degree is even; in particular, $1/\lambda$ is one of the conjugates of λ. We include reciprocal quadratic integers in the set of Pisot numbers ([46] makes a different choice).

R. Salem proved that Pisot numbers form a closed subset of $\mathbf{R}$, and C. L. Siegel proved that the smallest Pisot number is the root $\lambda \simeq 1.324717$ of the cubic equation $t^3 = t + 1$. On the other hand, the set of Salem numbers is not closed (every Pisot number is a limit of Salem numbers on both sides) and a famous conjecture of D. Lehmer asks whether there is a Salem number below the Lehmer number $\lambda_{10} \simeq 1.17628$, defined as the unique root > 1 of

$$t^{10} + t^9 - t^7 - t^6 - t^5 - t^4 - t^3 + t + 1 = 0.$$

More generally, we shall denote by λ_d *the smallest Salem number of degree* d (d even). This number is well defined; indeed, it is easily seen that the Pisot and Salem numbers of degree at most d form a discrete subset of $\mathbf{R}$.

2.2. Intersection form. Let X be a compact Kähler surface. Let q_X denote the intersection form on $H^2(X;\mathbf{Z})$. On $H^{1,1}(X;\mathbf{R})$, by the Hodge index theorem, q_X is non-degenerate and of signature $(1,m)$ where $1+m = h^{1,1}(X)$; thus, its orthogonal group is isomorphic to $\mathsf{O}_{1,m}(\mathbf{R})$. The Kähler classes satisfy $q_X([\kappa],[\kappa']) > 0$ and the set $\mathrm{Kah}(X)$ of all Kähler classes is contained in exactly one component, denoted by $\mathrm{Pos}^+(X)$, of the set $\{u \in H^{1,1}(X;\mathbf{R})\,;\ q_X(u,u) > 0\}$; by definition, $\mathrm{Pos}^+(X)$ is the *positive cone.*

We denote by $\mathrm{NS}(X)$ the Néron-Severi group of X, defined as the $\mathbf{Z}$-module of all Chern classes of holomorphic line bundles on X. Its rank is the *Picard number* $\rho(X)$. For projective surfaces defined over some field $\mathbf{k}$, one can replace $\mathrm{NS}(X)$ by numerical classes of divisors, and $\mathrm{Pos}^+(X)$ by the component of the cone $q_X(u,u) > 0$ containing classes of hyperplane sections (i.e. very ample classes).

2.3. Dynamical degrees are algebraic numbers. The group $\mathsf{Aut}(X)$ acts linearly on $H^2(X;\mathbf{Z})$ and preserves q_X. In particular, it acts by isometries on $H^{1,1}(X;\mathbf{R})$, and it preserves the Kähler cone. If f is an automorphism with positive entropy, f^* has an eigenvalue $\lambda_1(f) > 1$ on $H^{1,1}(X;\mathbf{R})$, and the corresponding isometry is *loxodromic* (see [55], Chapter 5):

(a) f^* has an eigenvalue $\lambda_1(f) > 1$, another one $\lambda_1(f)^{-1} < 1$, and its remaining eigenvalues $\alpha \in \mathbf{C}$ have modulus 1;

(b) the multiplicity of $\lambda_1(f)$ is 1, and the corresponding eigenspace E_f^+ is a line contained in the isotropic cone of $(q_X)_{|H^{1,1}(X;\mathbf{R})}$; one can write $E_f^+ = \mathbf{R}\Theta_f^+$ for some Θ_f^+ in the boundary of the Kähler cone of X;

(c) if $u \in \mathrm{Pos}^+(X)$, the sequence
$$\frac{1}{\lambda_1(f)^n}(f^n)^*(u)$$
converges towards a non-zero multiple of Θ_f^+ as n goes to $+\infty$.

Analogous objects E_f^-, Θ_f^- are associated to $\lambda_1(f)^{-1}$. If κ is some fixed Kähler form on X, Θ_f^+ and Θ_f^- can be chosen to satisfy

$$q_X(\Theta_f^+,\Theta_f^-) = 1 \quad \text{and} \quad q_X(\Theta_f^\pm,[\kappa]) = 1. \tag{2.1}$$

With such a choice, $\lambda_1(f)^{-n}(f^n)^*u$ converges towards $q_X(u,\Theta_f^-)\Theta_f^+$ in Property (c).

If Ω is a non-zero holomorphic 2-form on X, $\int_X \Omega\wedge\overline{\Omega} > 0$; thus $q_X(v,\overline{v})$ is a positive definite Hermitian form on $H^{2,0}(X;\mathbf{C})$, and the image of $\mathsf{Aut}(X)$ in $\mathsf{GL}(H^{2,0}(X;\mathbf{C}))$ is contained in a compact group. The same holds for $H^{0,2}(X;\mathbf{C})$ and, using the Hodge decomposition (1.3), Property (a) extends to the action of f^* on $H^2(X;\mathbf{C})$.

Since f^* preserves $H^2(X;\mathbf{Z})$, we see that $\lambda_1(f)$ is an algebraic integer of degree $\leq b_2(X)$ with Galois conjugates equal to $1/\lambda_1(f)$ or of modulus 1. In other words, *$\lambda_1(f)$ is a Salem number of degree $\leq b_2(X)$ or a reciprocal quadratic integer.* The eigenvalues of f^* that are not Galois conjugate to $\lambda_1(f)$ are roots of unity; indeed, a lemma due to Kronecker says that an algebraic integer the conjugates of which are all in the closed unit disk must be a root of unity.

This analysis has been extended to birational transformations by J. Diller and C. Favre. If f is a birational transformation of a projective surface X, defined over an algebraically closed field $\mathbf{k}$, they find a birational model $\varphi\colon Y \dashrightarrow X$ of X such that $f_Y := \varphi^{-1}\circ f\circ\varphi$ satisfies

$(f_Y^*)^n = (f_Y^n)^*$ for all $n \geq 1$, where f_Y^* is defined as in Section 1.4. Then, $\lambda_1(f)$ coincides with the eigenvalue of the linear operator f_Y^* on the Néron-Severi group $\mathrm{NS}(X)$. The intersection form is not invariant, but the inequality $q_X(f_Y^* u, f_Y^* u) \geq q_X(u,u)$ is satisfied by all classes $u \in \mathrm{NS}(X)$ which are big and nef. This inequality turns out to be sufficient to show the following theorem (see [29] and [14]).

Theorem 2.1. *Let $f\colon X \dashrightarrow X$ be a birational transformation of a compact Kähler surface (resp. a projective surface defined over an algebraically closed field* $\mathbf{k}$*). Then,*

(1) *$\lambda_1(f)$ is equal to 1, to a Salem number or to a Pisot number;*

(2) *if $\lambda_1(f)$ is a Salem number, there is a birational model $\varphi\colon Y \dashrightarrow X$ such that the map $f_Y := \varphi^{-1} \circ f \circ \varphi$ is an automorphism of Y.*

This result is specific to invertible transformations of surfaces, since transcendental dynamical degrees appear for well chosen non-invertible rational maps in dimension 2 and birational maps in dimension ≥ 3 (see [8,9]).

Remark 2.2. For automorphisms of projective surfaces in arbitrary characteristic, $\lambda_1(f)$ was defined in Section 1.5 using hyperplane sections; equivalently, $\lambda_1(f)$ is the spectral radius of f^* on $\mathrm{NS}(X)$. Algebraic geometry provides other cohomology groups — for instance étale ℓ-adic, de Rham, or crystalline cohomologies — hence also new concepts of dynamical degrees. H. Esnault and V. Srinivas proved in [34] that they all coincide (the characteristic polynomial of f^* is always the product of the minimal polynomial of $\lambda_1(f)$ and cyclotomic polynomials).

3. Possible entropies, possible dynamical degrees

We now arrive at one of the main topics initiated in [46] and [48]: among Pisot and Salem numbers, which ones are realized as dynamical degrees of automorphisms of surfaces? For complex surfaces, the problem is to describe the possible values taken by the topological entropy of automorphisms. We shall enlarge the point of view by considering also birational transformations, defined over any algebraically closed field; in other words, we want to describe the *dynamical spectrum*

$$\Lambda = \{\lambda \in \mathbf{R}\ ;\ \lambda \text{ is the dynamical degree } \lambda_1(f) \text{ of a birational transformation of a surface}\}.$$

Elements of Λ larger than 1 are Pisot or Salem numbers. As we shall see below, many Pisot and Salem numbers are not in Λ, and the structure of this set is somewhat mysterious.

3.1. The classification of surfaces. The existence of an automorphism with positive entropy on a compact complex surface X imposes strong constraints on the geometry of X. This can be stated in two ways (see [22,24,26]).

(1) If X is a compact complex surface with an automorphism of positive entropy, then X is a blow up of $\mathbb{P}^2$, of a torus $\mathbf{C}^2/\Lambda$, of a K3 surface, or of an Enriques surface[2)]; moreover, in the

2) An Enriques surface X is, by definition, the quotient of some K3 surface X' by a fixed point free involution. The surface X' is the universal cover of X, and automorphisms of X lift to automorphisms of X'.

last three cases, the automorphism comes from an automorphism of the corresponding torus, K3 surface, or Enriques surface. Thus, one only needs to study automorphisms of rational surfaces, tori, and K3 surfaces to understand automorphisms with positive entropy.

(2) If X is a projective surface over an algebraically closed field $\mathbf{k}$ with a birational transformation f such that $\lambda_1(f) > 1$, then X is birationally equivalent to a rational surface, an abelian surface, a K3 surface or an Enriques surface. Moreover, any birational transformation of an abelian surface (resp. a torus), a K3 surface, or an Enriques surface is in fact an automorphism. Thus, to describe the possible values of $\lambda_1(f)$ one only needs to look at birational transformations of $\mathbb{P}^2$, automorphisms of tori, and automorphisms of Enriques and K3 surfaces.

To this, of course, one must add that there are infinitely many (pairwise non-conjugate) examples of such automorphisms or birational transformations in each of the above families, for instance on well chosen K3 surfaces.

3.2. The lowest dynamical degrees.

Theorem 3.1 (McMullen). *The Lehmer number λ_{10} is the smallest dynamical degree larger than 1 realized by birational transformations of compact Kähler surfaces. It is realized by an automorphism of a non-projective K3 surface, by an automorphism of a projective K3 surface, and by an automorphism of a rational (hence projective) surface.*

This statement combines Theorem 2.1(2) and the main results of [45–47, 49]. As a byproduct, $\log(\lambda_{10})$ is the minimal positive entropy for automorphisms of compact complex surfaces. For projective K3 surfaces, examples of automorphisms with $\lambda_1(f) = \lambda_{10}$ have been found over any algebraically closed field $\mathbf{k}$ of characteristic $\neq 2, 3, 7$ (see [19]).

The table below shows the smallest dynamical degree > 1 for the different types of compact Kähler surfaces in the classification of Enriques and Kodaira (see [48, 54]). For surfaces not birationally equivalent to one of this table, $\lambda_1(f) = 1$ for every birational transformation $f\colon X \dashrightarrow X$. A similar table has been obtained for automorphisms of projective surfaces in positive characteristic, except we don't know yet what is the smallest possible dynamical degree for automorphisms of (non-classical) Enriques surfaces in characteristic 2.

Type of surface	Minimal dynamical degree > 1
rational	Lehmer number $\lambda_{10} \simeq 1.17628$
Abelian	$\lambda_4 \simeq 1.72208$, a root of $t^4 - t^3 - t^2 - t + 1$
compact (non-projective) torus	$\lambda_6 \simeq 1.40126$, a root of $t^6 - t^4 - t^3 - t^2 + 1$
projective and non-projective K3	Lehmer number $\lambda_{10} \simeq 1.17628$
Enriques	$\lambda_E \simeq 1.59234$, a root of $t^6 - t^4 - 2t^3 - t^2 + 1$

This provides a complete description of the bottom of the dynamical spectrum Λ; its structure above λ_{10}, and in particular above its first accumulation points, is more mysterious.

3.3. The case of K3 surfaces. If X is a K3 surface, $\mathsf{Bir}(X)$ coincides with $\mathsf{Aut}(X)$ and, as shown above, the dynamical degree of any loxodromic automorphism $f\colon X \to X$ is a root of a quadratic or Salem polynomial of degree at most $22 = b_2(X)$, and at most the Picard number $\rho(X)$ when X is projective. More precisely,

- if X is projective over a field of characteristic 0, then $\rho(X) \leq 20$ hence $\deg \lambda_1(f) \leq 20$;
- McMullen constructs automorphisms of K3 surfaces for which $\deg \lambda_1(f) = 22$.
- in any characteristic $p > 0$, $\rho(X)$ reaches 22 for the so-called supersingular K3 surfaces. There are automorphisms of K3 surfaces in any characteristic $p > 0$ for which $\deg \lambda_1(f) = 22$; those do not lift to characteristic 0 (see [14, 17, 33, 57] for instance).

In particular, $\lambda_1(f) \geq \lambda_{10}$, because the Lehmer number λ_{10} is known to be the smallest Salem number of degree ≤ 22. One does not know yet which ones of the Salem numbers of degree ≤ 22 are realized by automorphisms of K3 surfaces, but a complete answer is provided by S. Brandhorst if we *stabilize* the question by allowing taking powers (see [18]):

Theorem 3.2 (Brandhorst). *Let λ be a reciprocal quadratic number or a Salem number of degree at most 20. Then, there is an integer $k \geq 1$, a complex projective $K3$ surface X, and an automorphism f of X such that $\lambda_1(f) = \lambda^k$.*

Tori, non-projective K3 surfaces, and Enriques surfaces are also dealt with in [18].

3.4. The case of rational surfaces. Let $\pi\colon X \to \mathbb{P}^2$ be any surface obtained by blowing up the projective plane m times. Pulling back a line from $\mathbb{P}^2$ to X and taking its class, one obtains an element $\mathbf{e}_0 \in \mathrm{NS}(X)$. Pulling back the exceptional divisors of the blow-ups, one obtains m additional classes $\mathbf{e}_i \in \mathrm{NS}(X)$.

Denote simply by $(\,\cdot\,)$, instead of $q_X(\;,\;)$, the intersection form on $\mathrm{NS}(X)$. The classes $\mathbf{e}_j$ form an orthogonal basis of $\mathrm{NS}(X)$ for the intersection form, with $(\mathbf{e}_0 \cdot \mathbf{e}_0) = +1$ and $(\mathbf{e}_i \cdot \mathbf{e}_i) = -1$ if $1 \leq i \leq m$. The vectors $r_0 = \mathbf{e}_0 - \mathbf{e}_1 - \mathbf{e}_2 - \mathbf{e}_3$ and $r_i = \mathbf{e}_i - \mathbf{e}_{i+1}$, for $1 \leq i \leq m-1$, satisfy $(r_j \cdot r_j) = -2$. The reflection with respect to the orthogonal complement $r_j^\perp$ is $s_j\colon u \mapsto u + (u \cdot r_j) r_j$, and these reflections generate a subgroup $W_m \subset O_{1,m}(\mathbf{Z})$ isomorphic to the Coxeter group attached to the diagram of type E_m drawn below.

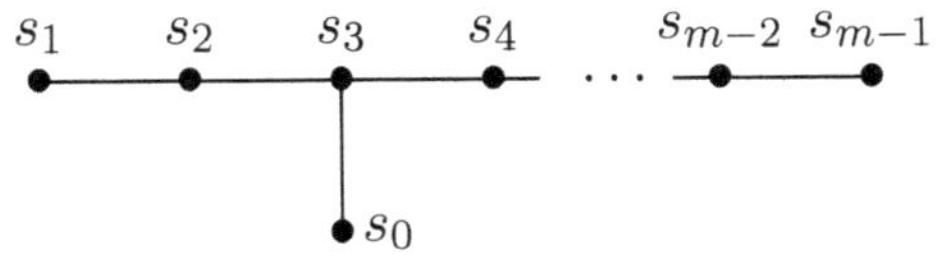

Figure 1. Coxeter-Dynkin diagram of type E_m.

When $m \leq 8$, W_m is finite; when $m = 9$, it contains an abelian subgroup of finite index isomorphic to $\mathbf{Z}^8$; and when $m \geq 10$, it contains a non-abelian free group $\mathbf{Z} \star \mathbf{Z}$. Moreover, W_m preserves the canonical class $k_m = -3\mathbf{e}_0 + \sum_{i \geq 1} \mathbf{e}_i$ and its action on $k_m^\perp$ is the Tits representation of W_m. By a theorem of M. Nagata, *the image of* $\mathsf{Aut}(X)$ *in* $\mathsf{GL}(\mathrm{NS}(X))$ *is contained in* W_m. In particular, dynamical degrees of automorphisms of rational surfaces are eigenvalues of elements of W_m (for some m).

Theorem 3.3 (T. Uehara [59]). *A real number $\lambda \geq 1$ is the dynamical degree of some automorphism of some complex rational surface if and only if there is an integer $m \geq 1$ and an element w in the Coxeter group W_m such that λ is the largest eigenvalue of w in its Tits representation.*

The integer m is arbitrarily large and the numbers λ obtained in this way do not form a discrete subset of $\mathbf{R}$. But some structure naturally shows up: *the dynamical spectrum* Λ *is a well ordered subset of* $\mathbf{R}$ (every decreasing sequence of dynamical degrees is finite), *its order type is* ω^ω, and *every number in this set is a limit of dynamical degrees of automorphisms* (see [14, 15]). For instance, the golden ratio $\gamma = (1+\sqrt{5})/2$ is the dynamical degree of the birational map $(x, y) \mapsto (y, xy)$; it is a limit from below of Pisot and Salem numbers contained in Λ; and there is an $\epsilon_0 > 0$ such that $]\gamma, \gamma + \epsilon_0[$ contains infinitely many Pisot and Salem numbers, but does not contain any dynamical degree.

4. Periodic points and the canonical volume form

Let X be a compact Kähler surface and f an automorphism of X with positive topological entropy. Let Θ_f^+ and Θ_f^- be the eigenvectors of f^* in $H^{1,1}(X;\mathbf{R})$ constructed in Section 2.3, with the normalization from Equation (2.1) for some Kähler form κ. Then, one easily shows that there is a unique closed positive current T_f^+ (resp. T_f^-) of type $(1,1)$ such that $[T_f^+] = \Theta_f^+$ (resp. $[T_f^-] = \Theta_f^-$). These currents satisfy $f^*T_f^\pm = \lambda_1(f)^{\pm 1}T_f^\pm$ and the value of $T_f^\pm$ on a smooth 2-form ω can be obtained dynamically as the limit

$$\langle T_f^\pm|\omega\rangle = \lim_{n\to+\infty} \frac{1}{\lambda_1(f)^n}\int_X \omega \wedge (f^{\pm n})^*\kappa.$$

Moreover, locally $T_f^\pm = dd^c g^\pm$ for some continuous functions $g^\pm$, which implies that the product $\mu_f = T_f^+ \wedge T_f^-$ is a well defined probability measure; this measure is invariant by f. Thus, Hodge theory and complex analysis provide a natural f-invariant probability measure. As we shall see now, μ_f describes the equidistribution of periodic points of f.

Denote by $\mathrm{Per}(f;n)$ the set of periodic points of f of period n: by definition x is an element of $\mathrm{Per}(f;n)$ if and only if $f^n(x) = x$ and $f^k(x) \neq x$ for $1 \leq k \leq n-1$. This algebraic set might contain a curve, and one denotes by $\mathrm{Per}^0(f;n)$ the finite subset of its isolated points. A point $q \in \mathrm{Per}(f;n)$ is a *saddle* if the eigenvalues α and β of the tangent map Df_q^n satisfy $|\alpha| > 1 > |\beta|$. Saddle periodic points are isolated (for, if 1 is not an eigenvalue of Df_q^n the graph of f^n in $X \times X$ is transverse to the diagonal). Averaging on the set of isolated periodic points one gets a probability measure

$$\mu_n = \frac{1}{|\mathrm{Per}^0(f;n)|}\sum_{q\in \mathrm{Per}^0(f;n)} \delta_q.$$

Averaging over the set $\mathrm{Per}^s(f;n)$ of saddle periodic points, one gets a second probability measure μ_n^s. The following equidistribution result is proven in [23, 24, 31].

Theorem 4.1. *Let X be a complex projective surface and f be an automorphism of X with positive entropy $h_{\mathrm{top}}(f) = \log(\lambda_1(f))$. Then, $|\mathrm{Per}^0(f;n)| \sim \lambda_1(f)^n$ and the sequence of probability measures (μ_n) converges towards μ_f as n goes to $+\infty$. Similarly $|\mathrm{Per}^s(f;n)| \sim \lambda_1(f)^n$ and (μ_n^s) converges towards μ_f.*

In particular, there are infinitely many isolated periodic points, a result that can also be derived from the Lefschetz fixed point formula (see [24]), and most of them are saddle periodic points. Moreover, saddle periodic points are dense in the support of μ_f, a compact set with positive Hausdorff dimension. On the support of μ_f, the dynamics is chaotic: the behavior of an orbit $(f^n(x))$ is sensitive to the initial condition x, as in the prototypical example $z \mapsto z^d$ from Section 1.1. To each saddle periodic point q are associated a stable manifold $W^{\mathrm{s}}(q)$ and an unstable manifold $W^{\mathrm{u}}(q)$; they are f^n-invariant, where n is the period of q; and they are parametrized by injective holomorphic entire curves $\xi^{\mathrm{s/u}}\colon \mathbf{C} \to X$ such that $f^n \circ \xi^{\mathrm{s}}(z) = \xi^{\mathrm{s}}(\beta z)$ (resp. $f^n \circ \xi^{\mathrm{u}}(z) = \xi^{\mathrm{u}}(\alpha z)$) where α and β are the eigenvalues of Df^n_q. Stable manifolds intersect unstable ones, creating horseshoe pictures; this, in turn, is well known to generate a chaotic behaviour in the dynamics of f (see [21,44]).

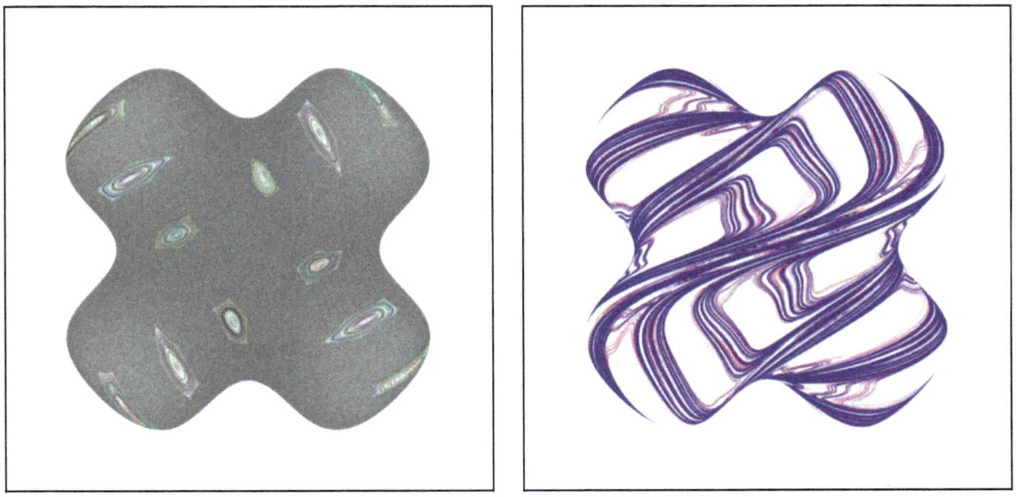

Figure 2. This is the real part of a K3 surface with an automorphism acting on it. On the left, distinct colors correspond to distinct orbits. Some are trapped in invariant (KAM) islands; some might a priori fill densely a subset of positive Lebesgue measure. On the right, the curve is (the real part of) a stable manifold $W^{\mathrm{s}}(q)$, for some saddle fixed point q. [COURTESY C. T. MCMULLEN AND V. PIT]

As explained in the introduction, a K3 surface admits a natural holomorphic 2-form Ω_X which is uniquely determined up to a scalar factor of modulus 1. Thus, the volume form $\mathsf{vol}_X = \Omega_X \wedge \overline{\Omega_X}$ is preserved by $\mathsf{Aut}(X)$. Kummer surfaces are the only ones for which $\mu_f = \mathsf{vol}_X$, and even the only ones for which μ_f is absolutely continuous with respect to the measure vol_X (see [27, 36]). Thus, natural questions arise. What are the possible stochastic properties of vol_X with respect to loxodromic automorphisms? What can be said on μ_f and its support? The study of Fatou sets is related to these questions.

5. Fatou sets and Siegel balls

If f is an automorphism of a complex projective surface X with positive entropy, its dynamics is chaotic. At least, this is true on the support of the invariant measure μ_f. But it might occur that this support is a "thin" subset of X, for instance with empty interior. Note, however, that Figure 1 is misleading: in this example $\mu_f(X(\mathbf{R})) = 0$, most periodic points are not real, and it might very well be the case that the support of μ_f is equal to $X(\mathbf{C})$. The notion of Fatou set, described below, is the right one to understand the complement of the support of μ_f or, more accurately, the complement of the supports of T^+_f and T^-_f.

5.1. The Fatou set. Let $f\colon X \to X$ be a dominant endomorphism of a compact complex manifold X. A point $x \in X$ is in the *Fatou set* $\mathrm{Fat}(f)$ if there is a compact neighborhood U of x on which the iterates $f^n\colon U \to X$ form a normal family in the sense of Montel. In other words, the family $((f^n)_{|U})$ is pre-compact: for any subsequence $((f^{n_i})_{|U})$, one can extract a further subsequence that converges uniformly to a holomorphic map $U \to X$; equivalently, on U the differential Df^n is bounded by a constant that does not depend on n. The Fatou set is open and $f^{-1}(\mathrm{Fat}(f)) = \mathrm{Fat}(f)$. The connected components of $\mathrm{Fat}(f)$ are called *Fatou components.*

The dynamics of f on $\mathrm{Fat}(f)$ is tame, in the sense that the trajectories of two nearby points in a Fatou component stay close forever. Also, $\mathrm{Fat}(f) = X$ if and only if f is invertible and contained in a compact subgroup of $\mathsf{Aut}(X)$; in that case, there is an f-invariant Riemannian metric on X, in particular the entropy of f vanishes. Thus, in the most interesting cases, $\mathrm{Fat}(f)$ is a proper subset of X, but it is usually hard, on a given example, to determine whether $\mathrm{Fat}(f)$ is empty or not.

When f is an endomorphism of $\mathbb{P}^1(\mathbf{C})$ of degree $d \geq 2$, a celebrated theorem of D. Sullivan says that

(1) the orbit $f^n(V)$ of every Fatou component V is ultimately periodic: there is a Fatou component $U = f^m(V)$ such that $f^\ell(U) = U$, for some $m, \ell \geq 1$;

(2) an f-invariant Fatou component is either a Siegel disk, a Herman ring, or an immediate attracting basin.

By a Siegel disk, we mean a domain $U \subset \mathbb{P}^1(\mathbf{C})$ for which there is a biholomorphic mapping $\varphi\colon \mathbb{D} \to U$ such that $\varphi^{-1} \circ f \circ \varphi$ is a rotation of the unit disk. Similarly, a Herman ring is a rotation domain which is biholomorphic to an annulus $\{z \in \mathbf{C}\,;\, a < |z| < b\}$. Siegel disks contain fixed points (periodic Siegel disks contain periodic points), but Herman rings do not. If q is a fixed point such that $|f'(q)| < 1$, its attracting basin is the open set of points $z \in \mathbb{P}^1(\mathbf{C})$ such that $f^n(z)$ converges towards q as n goes to $+\infty$; its immediate attracting basin is the component containing q.

Such a classification of Fatou components does not hold in higher dimension or for dynamics over complete non-archimedean fields. In particular, there are examples of Fatou components V such that the $f^n(V)$ are pairwise disjoint (see [2, 10, 11, 32]). (On a K3 surface, the volume form vol_X being invariant, Poincaré's recurrence theorem implies that every Fatou component is periodic.)

5.2. Constructing Siegel balls. Let f be a holomorphic transformation of a complex manifold X of dimension m. One way to construct a Fatou component is to conjugate locally the dynamics of f to the dynamics of a rotation around a fixed point of f. More precisely, let us denote by $\mathbb{B}^m$ the unit ball of $\mathbf{C}^m$ with respect to the standard Hermitian metric, and by U_m the unitary group. A *Siegel ball* for f is an open subset $U \subset X$ such that (i) $f(U) = U$ and (ii) there is a biholomorphism $\varphi\colon \mathbb{B}^m \to U$ that conjugates f to a rotation $R \in \mathsf{U}_m$. The center of the ball is fixed by the rotation, the point $x = \varphi(0)$ is fixed by f, and $D\varphi_0$ conjugates R to Df_x[3]. Conversely, assume f fixes a point $x \in X$ and Df_x is conjugate to a rotation $R \in \mathsf{U}_m$ by some linear change of coordinates. One can then try to conjugate f itself to R on some neighborhood of x. Let $\alpha_1, \ldots, \alpha_m$ be the eigenvalues of Df_x, repeated according to their

[3] In this situation, the open set U will be contained in a Fatou component V of f, but V is a priori not biholomorphic nor even homeomorphic to a round ball.

multiplicities (they all have modulus 1). By definition, the α_i are *multiplicatively independent* if the only solution to $\alpha_1^{k_1}\cdots\alpha_m^{k_m} = 1$ with $k = (k_1, \dots, k_m) \in \mathbf{Z}^m$ is the trivial solution $k = 0$; and they *satisfy a Diophantine condition* if

$$|\alpha_1^{k_1}\cdots\alpha_m^{k_m} - 1| \geq \frac{C}{(\max|k_i|)^M} \tag{5.1}$$

for some positive constants C and M and for all $k \in \mathbf{Z}^m \setminus \{0\}$. The following theorem is due to C. L. Siegel and S. Sternberg (see [42]).

Theorem 5.1. *Let g be a germ of holomorphic diffeomorphism of $\mathbf{C}^m$ fixing the origin 0. Suppose the eigenvalues $\{\alpha_i\}_{i=1}^m$ of Dg_0 have modulus 1 and satisfy a Diophantine condition. Then, there is a holomorphic diffeomorphism φ from $\mathbb{B}^m$ to an open and g-invariant neighborhood U of 0 in $\mathbf{C}^m$ that conjugates g to the rotation $R(z_1, \dots, z_m) = (\alpha_1 z_1, \dots, \alpha_m z_m)$.*

Coming back to holomorphic transformations of complex manifolds, we see that if f fixes x, Df_x is conjugate to a rotation R, and its eigenvalues satisfy a Diophantine condition, then f is locally conjugate to R on a neighborhood of x and $\mathrm{Fat}(f)$ contains x. When a Siegel ball U is obtained in this way, the images of the coordinate hyperplanes by the conjugacy φ are f-invariant; in their complement, each orbit is dense in an m-dimensional real torus.

6. Siegel balls in compact Kähler surfaces

Theorem 6.1. *There are examples of automorphisms $f\colon X \to X$ of compact complex surfaces with positive entropy and an invariant Siegel ball. Such examples can be constructed on rational surfaces and K3 surfaces.*

Examples on rational surfaces are constructed in [6,47], while [46] and [43] concern non-projective K3 surfaces. On the other hand, it is an open question to decide whether such an example can be constructed on a projective K3 or an Enriques surface (see Section 7 below). As explained in Sections 4 and 5, this theorem provides examples of automorphisms with positive entropy for which the support of μ_f is not equal to X, so that periodic points are not dense.

6.1. Proof strategy.

Here is how the proof goes for K3 surfaces.

The Torelli theorem. The Torelli theorem for K3 surfaces provides a way to construct a K3 surface X and an automorphism f of X from Hodge theoretic data.

The complex structure of X determines

- a Hodge decomposition $H^2(X;\mathbf{Z}) \otimes_{\mathbf{Z}} \mathbf{C} = H^{2,0}(X) \oplus H^{1,1}(X) \oplus H^{0,2}(X)$, and
- a Kähler cone $\mathrm{Kah}(X) \subset H^{1,1}(X;\mathbf{R})$ (made of all classes of Kähler forms).

Both are invariant under the action of $\mathsf{Aut}(X)$. Conversely, the Torelli theorem says that if an element F of $\mathsf{GL}(H^2(X;\mathbf{Z}))$ preserves q_X, the Hodge decomposition, and the Kähler cone, then $F = f^*$ for a unique $f \in \mathsf{Aut}(X)$.

To understand the third constraint, a description of $\mathrm{Kah}(X)$ is necessary. Firstly, $\mathrm{Kah}(X)$ is contained in $\mathrm{Pos}^+(X)$ (see Section 2.2). Secondly, if $C \subset X$ is a complex curve and κ is a Kähler form, then $\int_C \kappa > 0$; thus, identifying the homology class of C with its Poincaré dual $[C] \in H^{1,1}(X;\mathbf{R})$, we obtain $q_X([\kappa],[C]) > 0$. Conversely, these constraints determine $\mathrm{Kah}(X)$ completely: *a class in* $\mathrm{Pos}^+(X)$ *that intersects every curve positively is a Kähler class*. Moreover, by the Hodge index theorem, one only needs to test this last property on irreducible curves E with negative self-intersection and, on a K3 surface, such a curve is a smooth rational curve with $q_X(E,E) = -2$; conversely, if u is a class with $q_X(u,u) = -2$ then u or $-u$ is a sum of classes of smooth rational curves of self-intersection -2.

To sum up, each class $u \in \mathrm{NS}(X)$ with $q_X(u,u) = -2$ determines a hyperplane $u^\perp$. The arrangement of hyperplanes obtained in this way cuts $\mathrm{Pos}^+(X)$ into a certain (finite or infinite) number of chambers, and $\mathrm{Kah}(X)$ is one of these chambers.

Surjectivity of the period map. Since all K3 surfaces are diffeomorphic they all have the same cohomology group. With the intersection form q_X, it turns out that $H^2(X;\mathbf{Z})$ is isometric to the unique, even, unimodular lattice[4)] $\mathrm{II}_{3,19}$ of rank 22 and signature $(3,19)$. For simplicity, we denote this lattice by (L, q_L) and identify $(H^2(X;\mathbf{Z}), q_X)$ with it. Doing so, the Hodge decomposition gives a decomposition $L \otimes_{\mathbf{Z}} \mathbf{C} = L^{2,0} \oplus L^{1,1} \oplus L^{0,2}$ into three complex subspaces such that $\overline{L^{2,0}} = L^{0,2}$, $\dim(L^{1,1}) = 20$, and q_L is of signature $(1,19)$ on $L^{1,1}(\mathbf{R})$. In a nutshell, the *surjectivity of the period map* tells us that all such decompositions come from an identification of L with $H^2(X;\mathbf{Z})$ for some K3 surface.

The Lefschetz formulas. In the holomorphic setting, the Lefschetz fixed point formula becomes the conjunction of several equalities due to M. Atiyah and R. Bott, one for each integer $\ell \leq \dim(X)$ (see e.g. [37]).

Here is a consequence of those formulas when f is an automorphism of a K3 surface constructed from the Torelli theorem, that is, from $f^* \in \mathsf{GL}(H^2(X;\mathbf{Z}))$. Suppose the largest eigenvalue $\lambda_1(f)$ of f^* is a Salem number of degree 22. Then, 1 is not an eigenvalue of f^* on $H^2(X;\mathbf{Z})$, thus f does not fix any curve and its fixed points are isolated. If the trace of f^* is -1, the classical Lefschetz formula tells us that f has a unique fixed point $x_0 \in X$. The determinant of Df_{x_0} is the eigenvalue δ of f^* on $H^{2,0}(X)$ and the trace of Df_{x_0} can then be obtained from the holomorphic Lefschetz formulas. In particular, *the eigenvalues* α *and* β *of* Df_{x_0} *are algebraic numbers determined by* δ. For good choices of δ, they have modulus 1.

As a general fact (assuming $\lambda_1(f) > 1$), δ not being a root of unity is equivalent to δ being Galois conjugate to $\lambda_1(f)$ and to X not being a projective surface. Since we supposed $\deg(\lambda_1(f)) = 22 = b_2(X)$, we are in this situation. Moreover, the characteristic polynomial of f^* is the minimal polynomial of $\lambda_1(f)$, so the trace of f^* is the trace of $\lambda_1(f)$.

Gel'fond-Baker independence of logarithms. If the algebraic numbers α and β are multiplicatively independent[5)], a theorem of N. I. Feldman, based on the Gel'fond-Baker method, implies that they satisfy a Diophantine condition, as in Equation (5.1) (see [13] for a good introduction). From Theorem 5.1, we conclude that f has a Siegel ball centered at x_0.

---

4) This lattice $\mathrm{II}_{3,19}$ is isometric to $3U \oplus 2E_8(-1)$ where U is $\mathbf{Z}^2$ with the intersection form $2xy$ and $E_8(-1)$ is the lattice or rank 8 associated to the graph of type E_8, but with sign reversed to be negative definite. See [51, 56] for the theory of lattices.

5) McMullen shows in Theorem 6.1 and Lemma 7.5 of [48] that (1) the trace of Df_{x_0} is $(1+\delta+\delta^2)/(1+\delta)$, (2) α and β have modulus 1 if and only if $\delta + \delta^{-1} > 1 - 2\sqrt{2} \simeq -1.82842$, and (3) α and β are multiplicatively independent as soon as some Galois conjugate of $\delta + \delta^{-1}$ is $< 1 - 2\sqrt{2}$.

Lattices. So, doing some reverse engineering, one starts with a Salem number λ and one wants to construct a triple made of

(a) a linear map $F \in \mathsf{GL}(L; \mathbf{Z})$ preserving the intersection form q_L with spectral radius λ;

(b) an F-invariant decomposition $L \otimes_Z \mathbf{C} = L^{2,0} \oplus L^{1,1} \oplus L^{0,2}$ into complex subspaces such that $\overline{L^{2,0}} = L^{0,2}$, $\dim(L^{1,1}) = 20$, and the signature of q_L on $L^{1,1}(\mathbf{R})$ is $(1, 19)$;

(c) in the cone $\{v \in L^{1,1}(\mathbf{R})\,;\ q_L(v, v) > 0\}$, an F-invariant component K of the complement of the hyperplanes $\{u^\perp\,;\ u \in L^{1,1} \cap L,\ q_L(u, u) = -2\}$.

Then, the surjectivity of the period map and the Torelli theorem will provide a K3 surface X and an automorphism $f\colon X \to X$ with $f^* \simeq F$ and $H^{1,1}(X) \simeq L^{1,1}$.

On top of that, one wants λ to have degree 22 and trace -1 together with some additional conditions on the eigenvalue δ of F on $L^{2,0}$ to ensure that Df_{x_0} has multiplicatively independent eigenvalues of modulus 1. This is achieved in [46] with a beautiful blend of arithmetic and lattice theoretic arguments. In the construction of this example, Part (c) follows directly from $\deg(\lambda) = 22$. Indeed, since 22 is the rank of L, every proper F-invariant subspace of $L \otimes_\mathbf{Z} \mathbf{Q}$ is trivial. Thus, $L^{1,1} \cap L$ is empty and (c) is void.

So, the main issues are to construct adequate Salem numbers[6)] and then an isometry F of L with characteristic polynomial equal to the minimal polynomial $S(t)$ of λ. Once λ is given, the lattice is chosen to be the ring of integers $\mathbf{Z}[t]/\langle S(t)\rangle$, F is the multiplication by t, and the quadratic form q_L is $q_L(x, y) = \mathrm{Tr}(axy)$ where Tr is the trace function and $a \in \mathbf{Q}[t]/\langle S(t)\rangle$ is chosen to obtain an even, unimodular form of signature $(3, 19)$. (See [4,5,40] for a complete description of characteristic polynomials of isometries of even unimodular lattices.)

6.2. Other examples. To get Theorem 3.1 for K3 surfaces, one starts with the Lehmer number λ_{10}. The natural lattices associated to it have rank $10 = \deg(\lambda_{10})$ (for instance, λ_{10} is an eigenvalue of the Coxeter element in W_{10}, the group introduced in Section 3.4). Thus, to construct the pair (L, F) one needs to glue such a lattice M of rank 10 to an additional lattice M' of rank 12 (on which F will act as a finite order isometry), so as to get a lattice isometric to L. In general, $M \oplus M'$ is not unimodular and a finite index extension $M \oplus M' \subset L$ is needed in the gluing construction: this is a classical but tricky part, perfectly described in §2-4 of [48]. What makes the construction significantly more delicate is Part (c) related to the invariance of a candidate K for the Kähler cone. Typically, the hyperplanes $u^\perp$, with $u \in L^{1,1} \cap L$ satisfying $q_L(u, u) = -2$, cut the positive cone into infinitely many chambers and one must tune the construction of L, F, and the decomposition $L^{2,0} \oplus L^{1,1} \oplus L^{0,2}$ in such a way that at least one chamber is F-invariant. This problem is at the heart of [48,49,54].

This does not end the story. As the reader certainly noticed, even if the end product is an automorphism $f\colon X \to X$ of a projective surface, the surjectivity of the period map is an implicit statement that does not provide explicit models of X. Finding an embedding $X \subset \mathbb{P}^N_\mathbf{C}$, equations defining X, and formulas for f is an additional, nontrivial task. It has been achieved in [19] for one of the most interesting examples with $\lambda_1(f) = \lambda_{10}$ (see also [7,20,49]).

6) It is known now, by a theorem of J. McKee and C. Smyth, that every integer is the trace of some Salem numbers. When McMullen's paper appeared, the existence of infinitely many Salem numbers with trace -1 was a recent result of C. Smyth, published in 2002.

7. An open question

Does there exist an automorphism f of a complex projective K3 surface X such that $\lambda_1(f) > 1$ and the set of its periodic points is not dense in $X(\mathbf{C})$ for the euclidean topology?

A stronger formulation would be to require $\mathrm{Fat}(f) \neq \emptyset$. An example with a Siegel ball would provide a positive answer, but Siegel balls are much harder to construct because on a projective K3 surface the Jacobian determinant δ of any automorphism is a root of unity, which implies multiplicative dependence of the eigenvalues at periodic points.

One-dimensional Herman rings suggest another strategy, which leads to the following question. Does there exist a loxodromic automorphism $f\colon X_{\mathbf{R}} \to X_{\mathbf{R}}$ of a real projective K3 surface such that $X(\mathbf{R})$ is non-empty and contained in $\mathrm{Fat}(f)$? (See [25, 52].)

Acknowledgement. My warmest thanks to Simon Brandhorst, Romain Dujardin, Simion Filip, Daniel Huybrechts, Curtis McMullen, Giacomo Mezzedimi, and Joachim Schwermer for reading preliminary versions of this text and providing interesting feedback. A few sentences in this text are directly extracted from my email exchanges with Curtis McMullen.

References

[1] *D. S. Alexander*, A history of complex dynamics. From Schröder to Fatou and Julia, *Aspects of Mathematics*, volume E24, Friedr. Vieweg & Sohn, Braunschweig, 1994, URL `https://doi.org/10.1007/978-3-663-09197-4`.

[2] *M. Astorg, X. Buff, R. Dujardin, H. Peters* and *J. Raissy*, A two-dimensional polynomial mapping with a wandering Fatou component, Ann. of Math. (2) **184** (2016), no. 1, 263–313, , URL `https://doi.org/10.4007/annals.2016.184.1.2`.

[3] *W. P. Barth, K. Hulek, C. A. M. Peters* and *A. Van de Ven*, Compact complex surfaces, *Ergebnisse der Mathematik und ihrer Grenzgebiete. 3. Folge.*, volume 4, Springer-Verlag, Berlin, 2004, 2nd edition, URL `https://doi.org/10.1007/978-3-642-57739-0`.

[4] *E. Bayer-Fluckiger*, Isometries of lattices and Hasse principles, J. Eur. Math. Soc. (JEMS) **26** (2024), no. 9, 3365–3428, URL `https://doi.org/10.4171/jems/1334`.

[5] *E. Bayer-Fluckiger* and *L. Taelman*, Automorphisms of even unimodular lattices and equivariant Witt groups, J. Eur. Math. Soc. (JEMS) **22** (2020), no. 11, 3467–3490, URL `https://doi.org/10.4171/jems/991`.

[6] *E. Bedford* and *K. Kim*, Dynamics of rational surface automorphisms: rotation domains, Amer. J. Math. **134** (2012), no. 2, 379–405, URL `https://doi.org/10.1353/ajm.2012.0015`.

[7] *E. Bedford* and *K. Kim*, Dynamics of (pseudo) automorphisms of 3-space: periodicity versus positive entropy, Publ. Mat. **58** (2014), no. 1, 65–119, URL `http://projecteuclid.org/euclid.pm/1387570391`.

[8] *J. P. Bell, J. Diller* and *M. Jonsson*, A transcendental dynamical degree, Acta Math. **225** (2020), no. 2, 193–225, URL `https://doi.org/10.4310/ACTA.2020.v225.n2.a1`.

[9] *J. P. Bell, J. Diller, M. Jonsson* and *H. Krieger*, Birational maps with transcendental dynamical degree, Proc. Lond. Math. Soc. (3) **128** (2024), no. 1, Paper No. e12573, 47, URL `https://doi.org/10.1112/plms.12573`.

[10] *R. L. Benedetto*, Dynamics in one non-archimedean variable, *Graduate Studies in Mathematics*, volume 198, American Mathematical Society, Providence, RI, 2019, URL `https://doi.org/10.1090/gsm/198`.

[11] *P. Berger* and *S. Biebler*, Emergence of wandering stable components, J. Amer. Math. Soc. **36** (2023), no. 2, 397–482, URL `https://doi.org/10.1090/jams/1005`.

[12] *M.-J. Bertin* and *M. Pathiaux-Delefosse*, Conjecture de Lehmer et petits nombres de Salem, *Queen's Papers in Pure and Applied Mathematics*, volume 81, Queen's University, Kingston, ON, 1989.

[13] *Y. Bilu* and *Y. Bugeaud*, Démonstration du théorème de Baker-Feldman via les formes linéaires en deux logarithmes, J. Théor. Nombres Bordeaux **12** (2000), no. 1, 13–23, URL `http://jtnb.cedram.org/item?id=JTNB_2000__12_1_13_0`.

[14] *J. Blanc* and *S. Cantat*, Dynamical degrees of birational transformations of projective surfaces, J. Amer. Math. Soc. **29** (2016), no. 2, 415–471, URL `https://doi.org/10.1090/jams831`.

[15] *A. Bot*, The ordinal of dynamical degrees of birational maps of the projective plane, C. R. Math. Acad. Sci. Paris **362** (2024), 117–134, URL `https://doi.org/10.5802/crmath.540`.

[16] *R. Bowen*, Entropy and the fundamental group, in: The structure of attractors in dynamical systems (Proc. Conf., North Dakota State Univ., Fargo, N.D., 1977), Springer, Berlin-New York, 1978, *Lecture Notes in Math.*, volume 668, 21–29.

[17] *S. Brandhorst*, Automorphisms of Salem degree 22 on supersingular K3 surfaces of higher Artin invariant, Math. Res. Lett. **25** (2018), no. 4, 1143–1150, URL `https://doi.org/10.4310/MRL.2018.v25.n4.a4`.

[18] *S. Brandhorst*, On the stable dynamical spectrum of complex surfaces, Math. Ann. **377** (2020), no. 1-2, 421–434, URL `https://doi.org/10.1007/s00208-020-01978-7`.

[19] *S. Brandhorst* and *N. D. Elkies*, Equations for a K3 Lehmer map, J. Algebraic Geom. **32** (2023), no. 4, 641–675.

[20] *S. Brandhorst* and *M. Zach*, An explicit enriques surface with an automorphism of minimum entropy, 2025, URL `https://arxiv.org/abs/2502.19018`.

[21] *M. Brin* and *G. Stuck*, Introduction to dynamical systems, Cambridge University Press, Cambridge, 2015.

[22] *S. Cantat*, Dynamique des automorphismes des surfaces projectives complexes, C. R. Acad. Sci. Paris Sér. I Math. **328** (1999), no. 10, 901–906, URL `https://doi.org/10.1016/S0764-4442(99)80294-8`.

[23] *S. Cantat*, Dynamique des automorphismes des surfaces $K3$, Acta Math. **187** (2001), no. 1, 1–57, URL `https://doi.org/10.1007/BF02392831`.

[24] *S. Cantat*, Dynamics of automorphisms of compact complex surfaces, in: Frontiers in complex dynamics, Princeton Univ. Press, Princeton, NJ, 2014, *Princeton Math. Ser.*, volume 51, 463–514.

[25] *S. Cantat*, Automorphisms and dynamics: a list of open problems, in: Proceedings of the International Congress of Mathematicians—Rio de Janeiro 2018. Vol. II. Invited lectures, World Sci. Publ., Hackensack, NJ, 2018, 619–634.

[26] *S. Cantat*, The Cremona group, in: Algebraic geometry: Salt Lake City 2015, Amer. Math. Soc., Providence, RI, 2018, *Proc. Sympos. Pure Math.*, volume 97, 101–142.

[27] *S. Cantat* and *C. Dupont*, Automorphisms of surfaces: Kummer rigidity and measure of maximal entropy, J. Eur. Math. Soc. (JEMS) **22** (2020), no. 4, 1289–1351, URL `https://doi.org/10.4171/JEMS/946`.

[28] *N.-B. Dang*, Degrees of iterates of rational maps on normal projective varieties, Proc. Lond. Math. Soc. (3) **121** (2020), no. 5, 1268–1310, URL `https://doi.org/10.1112/plms.12366`.

[29] *J. Diller* and *C. Favre*, Dynamics of bimeromorphic maps of surfaces, Amer. J. Math. **123** (2001), no. 6, 1135–1169, URL `http://muse.jhu.edu/journals/american_journal_of_mathematics/v123/123.6diller.pdf`.

[30] *T.-C. Dinh* and *N. Sibony*, Regularization of currents and entropy, Ann. Sci. École Norm. Sup. (4) **37** (2004), no. 6, 959–971, URL `https://doi.org/10.1016/j.ansens.2004.09.002`.

[31] *R. Dujardin*, Laminar currents and birational dynamics, Duke Math. J. **131** (2006), no. 2, 219–247, URL `https://doi.org/10.1215/S0012-7094-06-13122-8`.

[32] *R. Dujardin*, Geometric methods in holomorphic dynamics, in: ICM—International Congress of Mathematicians. Vol. 5. Sections 9–11, EMS Press, Berlin, [2023] ©2023, 3460–3482.

[33] *H. Esnault*, *K. Oguiso* and *X. Yu*, Automorphisms of elliptic K3 surfaces and Salem numbers of maximal degree, Algebr. Geom. **3** (2016), no. 4, 496–507, URL `https://doi.org/10.14231/AG-2016-023`.

[34] *H. Esnault* and *V. Srinivas*, Algebraic versus topological entropy for surfaces over finite fields, Osaka J. Math. **50** (2013), no. 3, 827–846, URL `http://projecteuclid.org/euclid.ojm/1380287436`.

[35] *C. Favre*, *T. T. Truong* and *J. Xie*, Topological entropy of a rational map over a complete metrized field, 2022, to appear in the Annales of the ENS. URL `https://arxiv.org/abs/2208.00668`.

[36] *S. Filip* and *V. Tosatti*, Kummer rigidity for K3 surface automorphisms via Ricci-flat metrics, Amer. J. Math. **143** (2021), no. 5, 1431–1462, URL `http://dx.doi.org/10.1353/ajm.2021.0036`.

[37] *P. Griffiths* and *J. Harris*, Principles of algebraic geometry, Wiley-Interscience [John Wiley & Sons], New York, 1978.

[38] *M. Gromov*, Entropy, homology and semialgebraic geometry, 1987, 145-146, 5, 225–240. Séminaire Bourbaki, Vol. 1985/86.

[39] *M. Gromov*, On the entropy of holomorphic maps, Enseign. Math. (2) **49** (2003), no. 3-4, 217–235.

[40] *B. H. Gross* and *C. T. McMullen*, Automorphisms of even unimodular lattices and unramified Salem numbers, J. Algebra **257** (2002), no. 2, 265–290, URL `https://doi.org/10.1016/S0021-8693(02)00552-5`.

[41] *V. Guedj*, Entropie topologique des applications méromorphes, Ergodic Theory Dynam. Systems **25** (2005), no. 6, 1847–1855, URL `https://doi.org/10.1017/S0143385705000192`.

[42] *M.-R. Herman*, Recent results and some open questions on Siegel's linearization theorem of germs of complex analytic diffeomorphisms of $\mathbf{C}^n$ near a fixed point, in: VIIIth international congress on mathematical physics (Marseille, 1986), World Sci. Publishing, Singapore, 1987, 138–184.

[43] *K. Iwasaki* and *Y. Takada*, K3 surfaces, Picard numbers and Siegel disks, J. Pure Appl. Algebra **227** (2023), no. 3, Paper No. 107215, 31, URL `https://doi.org/10.1016/j.jpaa.2022.107215`.

[44] *A. Katok* and *B. Hasselblatt*, Introduction to the modern theory of dynamical systems, *Encyclopedia of Mathematics and its Applications*, volume 54, Cambridge University Press, Cambridge, 1995, , URL `https://doi.org/10.1017/CBO9780511809187`. With a supplementary chapter by Katok and Leonardo Mendoza.

[45] *C. T. McMullen*, Coxeter groups, Salem numbers and the Hilbert metric, Publ. Math. Inst. Hautes Études Sci. (2002), no. 95, 151–183, URL `https://doi.org/10.1007/s102400200001`.

[46] *C. T. McMullen*, Dynamics on $K3$ surfaces: Salem numbers and Siegel disks, J. reine angew. Math. **545** (2002), 201–233, URL `https://doi.org/10.1515/crll.2002.036`.

[47] *C. T. McMullen*, Dynamics on blowups of the projective plane, Publ. Math. Inst. Hautes Études Sci. (2007), no. 105, 49–89, URL `https://doi.org/10.1007/s10240-007-0004-x`.

[48] *C. T. McMullen*, K3 surfaces, entropy and glue, J. reine angew. Math. **658** (2011), 1–25, URL `https://doi.org/10.1515/CRELLE.2011.048`.

[49] *C. T. McMullen*, Automorphisms of projective K3 surfaces with minimum entropy, Invent. Math. **203** (2016), no. 1, 179–215, URL `https://doi.org/10.1007/s00222-015-0590-z`.

[50] *J. Milnor*, Dynamics in one complex variable, *Annals of Mathematics Studies*, volume 160, Princeton University Press, Princeton, NJ, 2006, 3rd edition.

[51] *J. Milnor* and *D. Husemoller*, Symmetric bilinear forms, *Ergebnisse der Mathematik und ihrer Grenzgebiete [Results in Mathematics and Related Areas]*, volume Band 73, Springer-Verlag, New York-Heidelberg, 1973.

[52] *A. Moncet*, Real versus complex volumes on real algebraic surfaces, Int. Math. Res. Not. IMRN (2012), no. 16, 3723–3762.

[53] *D. Mumford*, Algebraic geometry. I Complex projective varieties, Classics in Mathematics, Springer-Verlag, Berlin, 1995.

[54] *K. Oguiso* and *X. Yu*, Minimum positive entropy of complex Enriques surface automorphisms, Duke Math. J. **169** (2020), no. 18, 3565–3606, , URL `https://doi.org/10.1215/00127094-2020-0033`.

[55] *J. G. Ratcliffe*, Foundations of hyperbolic manifolds, *Graduate Texts in Mathematics*, volume 149, Springer, New York, 2006, 2nd edition.

[56] *J.-P. Serre*, A course in arithmetic, *Graduate Texts in Mathematics*, volume No. 7, Springer-Verlag, New York-Heidelberg, 1973. Translated from the French.

[57] *I. Shimada*, Automorphisms of supersingular $K3$ surfaces and Salem polynomials, Exp. Math. **25** (2016), no. 4, 389–398, URL `https://doi.org/10.1080/10586458.2015.1073641`.

[58] *T. T. Truong*, Relative dynamical degrees of correspondences over a field of arbitrary characteristic, J. reine angew. Math. **758** (2020), 139–182, URL `https://doi.org/10.1515/crelle-2017-0052`.

[59] *T. Uehara*, Rational surface automorphisms with positive entropy, Ann. Inst. Fourier (Grenoble) **66** (2016), no. 1, 377–432, URL `https://doi.org/10.5802/aif.3014`.

[60] *A. Weil*, Oeuvres scientifiques/collected papers. II. 1951–1964, Springer Collected Works in Mathematics, Springer, Heidelberg, 2014. Reprint of the 2009 [MR2883739] and 1979 [MR0537935] editions.

[61] *Y. Yomdin*, Volume growth and entropy, Israel J. Math. **57** (1987), no. 3, 285–300, URL `https://doi.org/10.1007/BF02766215`.

Serge Cantat, IRMAR, Rennes University, CNRS, bât. 22-23 du Campus de Beaulieu, F-35200, France
e-mail: serge.cantat@univ-rennes.fr